METHODS IN CHEMOSENSORY RESEARCH

METHODS & NEW FRONTIERS IN NEUROSCIENCE

Series Editors
Sidney A. Simon, Ph.D.
Miguel A.L. Nicolelis, M.D., Ph.D.

Published Titles

Apoptosis in Neurobiology
Yusuf A. Hannun, M.D., Professor of Biomedical Research and Chairman/Department of Biochemistry and Molecular Biology, Medical University of South Carolina
Rose-Mary Boustany, M.D., tenured Associate Professor of Pediatrics and Neurobiology, Duke University Medical Center

Methods for Neural Ensemble Recordings
Miguel A.L. Nicolelis, M.D., Ph.D., Professor of Neurobiology and Biomedical Engineering, Duke University Medical Center

Methods of Behavioral Analysis in Neuroscience
Jerry J. Buccafusco, Ph.D., Alzheimer's Research Center, Professor of Pharmacology and Toxicology, Professor of Psychiatry and Health Behavior, Medical College of Georgia

Neural Prostheses for Restoration of Sensory and Motor Function
John K. Chapin, Ph.D., Professor of Physiology and Pharmacology, State University of New York Health Science Center
Karen A. Moxon, Ph.D., Assistant Professor/School of Biomedical Engineering, Science, and Health Systems, Drexel University

Computational Neuroscience: Realistic Modeling for Experimentalists
Eric DeSchutter, M.D., Ph.D., Professor/Department of Medicine, University of Antwerp

Methods in Pain Research
Lawrence Kruger, Ph.D., Professor of Neurobiology (Emeritus), UCLA School of Medicine and Brain Research Institute

Motor Neurobiology of the Spinal Cord
Timothy C. Cope, Ph.D., Professor of Physiology, Emory University School of Medicine

Nicotinic Receptors in the Nervous System
Edward D. Levin, Ph.D., Associate Professor/Department of Psychiatry and Pharmacology and Molecular Cancer Biology and Department of Psychiatry and Behavioral Sciences, Duke University School of Medicine

Methods in Genomic Neuroscience
Helmin R. Chin, Ph.D., Genetics Research Branch, NIMH, NIH
Steven O. Moldin, Ph.D, Genetics Research Branch, NIMH, NIH

Methods in Chemosensory Research
Sidney A. Simon, Ph.D., Professor of Neurobiology, Biomedical Engineering, and Anesthesiology, Duke University
Miguel A.L. Nicolelis, M.D., Ph.D., Professor of Neurobiology and Biomedical Engineering, Duke University

The Somatosensory System: Deciphering the Brain's Own Body Image
Randall J. Nelson, Ph.D., Professor of Anatomy and Neurobiology, University of Tennessee Health Sciences Center

The Superior Colliculus: New Approaches for Studying Sensorimotor Integration
William C. Hall, Ph.D., Department of Neuroscience, Duke University
Adonis Moschovakis, Ph.D., Institute of Applied and Computational Mathematics, Crete

New Concepts in Cerebral Ischemia
Rick C. S. Lin, Ph.D., Professor of Anatomy, University of Mississippi Medical Center

DNA Arrays: Technologies and Experimental Strategies
Elena Grigorenko, Ph.D., Technology Development Group, Millennium Pharmaceuticals

Methods for Alcohol-Related Neuroscience Research
Yuan Liu, Ph.D., National Institute of Neurological Disorders and Stroke, National Institutes of Health
David M. Lovinger, Ph.D., Laboratory of Integrative Neuroscience, NIAAA

In Vivo Optical Imaging of Brain Function
Ron Frostig, Ph.D., Associate Professor/Department of Psychobiology, University of California, Irvine

Primate Audition: Behavior and Neurobiology
Asif A. Ghazanfar, Ph.D., Primate Cognitive Neuroscience Lab, Harvard University

Methods in Drug Abuse Research: Cellular and Circuit Level Analyses
Dr. Barry D. Waterhouse, Ph.D., MCP-Hahnemann University

Functional and Neural Mechanisms of Interval Timing
Warren H. Meck, Ph.D., Professor of Psychology, Duke University

Biomedical Imaging in Experimental Neuroscience
Nick Van Bruggen, Ph.D., Department of Neuroscience Genentech, Inc., South San Francisco
Timothy P.L. Roberts, Ph.D., Associate Professor, University of Toronto

The Primate Visual System
John H. Kaas, Department of Psychology, Vanderbilt University
Christine Collins, Department of Psychology, Vanderbilt University

Neurosteroid Effects in the Central Nervous System
Sheryl S. Smith, Ph.D., Department of Physiology, SUNY Health Science Center

METHODS IN CHEMOSENSORY RESEARCH

Edited by

Sidney A. Simon
Miguel A.L. Nicolelis

CRC PRESS

Boca Raton London New York Washington, D.C.

Special thanks for the front cover photographs — A. Caicedo and Stephen D. Roper, Department of Physiology and Biophysics, University of Miami, Miami, Florida; Matt Wachowiak, Ying-Wan Lam, Lawrence B. Cohen, and Michal R. Zochowski from the Department of Cellular and Molecular Physiology, Yale University School of Medicine, New Haven, Connecticut; Thomas A. Christensen and John G. Hildebrand from ARL Division of Neurobiology, University of Arizona, Tucson, Arizona.

Library of Congress Cataloging-in-Publication Data

Methods in chemosensory research / edited by Sidney A. Simon and Miguel A.L. Nicolelis.
p. cm. -- (Methods & new frontiers in neuroscience)
Includes bibliographical references and index.
ISBN 0-8493-2329-0 (alk. paper)
1. Chemical senses--Research--Methodology. I. Simon, Sidney A. II. Nicolelis, Miguel A. L. III. Methods & new frontiers in neuroscience series.

QP455 .M48 2001
612.8′6′072--dc21

2001035759

International Standard Book Number 0-8493-2329-0
Library of Congress Card Number 2001035759
2 3 4 5 6 7 8 9 0

Methods & New Frontiers in Neuroscience

Series Editors
Sidney A. Simon, Ph.D.
Miguel A. L. Nicolelis, M.D., Ph.D.

Our goal in creating the Methods & New Frontiers in Neuroscience Series is to present the insights of experts on emerging experimental techniques and theoretical concepts that are, or will be, at the vanguard of neuroscience. Books in the series cover topics ranging from methods to investigate apoptosis to modern techniques for neural ensemble recordings in behaving animals. The series also covers new and exciting multidisciplinary areas of brain research, such as computational neuroscience and neuroengineering, and describes breakthroughs in classical fields such as behavioral neuroscience. We want these to be the books every neuroscientist will use in order to get acquainted with new methodologies in brain research. These books can be given to graduate students and postdoctoral fellows when they are looking for guidance to start a new line of research.

Each book is edited by an expert and consists of chapters written by the leaders in a particular field. Books are richly illustrated and contain comprehensive bibliographies. Chapters provide substantial background material relevant to the particular subject. Hence, they are not just "methods books." They contain detailed "tricks of the trade" and information as to where these methods can be safely applied. In addition, they include information about where to buy equipment and Web sites helpful in solving both practical and theoretical problems.

We hope that as the volumes become available, the effort put in by us, the publisher, the book editors, and the individual authors will contribute to the further development of brain research. The extent to which we achieve this goal will be determined by the utility of these books.

Preface

To edit a book in the Methods and New Frontiers in Neuroscience series in which we are the series editors was an endeavor that was taken only after much consideration. Now that this project is completed, we feel that all the hard work has paid off handsomely. The stated goal of the books in this series is to produce a book in which each chapter contains not only information about the emerging methods, but also provides what information can be obtained from these methods. The disciplines of olfaction, taste, and the common chemical sense have exploded. This is partially the result of the identification of many of the genes and their products involved in these fields and partially because methods and analyses developed in other fields of neuroscience are now being applied to the chemical senses. The multidisciplinary nature of this field demanded that this book cover a large range of topics. We therefore have structured this book into three sections: behavioral, molecular and cell biology, and higher order studies. The latter two sections cover methods ranging from molecular biology, to multielectrode recordings from moths and rats, to NMR studies of the olfactory bulb and the cortex.

To the extent that this book is found to be useful, the authors of the chapters deserve the lion's share of the credit. The most enjoyable aspect of our task was having the privilege of interacting with them. We also acknowledge three other people who contributed significantly to the success of this project. Don Katz not only wrote a chapter in this book, but also read and provided insightful comments on many others. Doug Buchacek kept us organized and was a great help in communicating with CRC Press. Finally, we truly appreciate Barbara Norwitz at CRC Press who not only had to deal with us and this book, but with all others in the series. Without her support and faith in us, this series would not exist.

Sid Simon and Miguel Nicolelis

Editors

Sidney A. Simon is a professor of neurobiology, biomedical engineering, and anesthesiology at Duke University. Born in New York City, he received his B.S. in physics at IIT and his Ph.D. in 1973 in materials science at Northwestern. He then did a short postdoctoral stint with Robert MacDonald at Northwestern where he was introduced to the physical chemistry of membranes, a field in which he has continued to study for the past 30 years and, following this, he went to Duke University where he worked as a postdoctoral fellow with Daniel Tosteson on anesthetic mechanisms.

Although he has remained at Duke since that time, his interests have shifted over the years to investigating gap junctions, mammalian lens, epithelial transport, and most recently the chemical senses. His recent work has involved investigating the effects of the spice capsaicin on the gustatory and trigeminal systems and joining with his co-editor, Miguel Nicolelis, to elucidate the encoding of gustatory stimuli.

Miguel A.L. Nicolelis was born in Sao Paulo, Brazil, in 1961. He received an M.D. degree from the University of Sao Paulo Medical School in 1984 and a Ph.D. in physiology in 1988 from the Department of Physiology in the Institute of Biomedical Science at the University of Sao Paulo. From 1989 to 1992, Dr. Nicolelis was a postdoctoral fellow in the Department of Physiology at Hahnemann University. In 1994, Dr. Nicolelis joined the faculty of the Department of Neurobiology at Duke University, where he is currently a professor.

Dr. Nicolelis' research focuses on the investigation of the basic physiological mechanism through which populations of cortical and subcortical neurons encode tactile and motor information. His published work includes studies on the development, plasticity, and normal function of the mammalian somatosensory and motor systems, and the design of neuroprosthetic devices to restore normal brain function.

For his research efforts, Dr. Nicolelis received the Oswaldo Cruz Award for excellence in biomedical research (1984), the Whitehead Scholar Award (1994), a fellowship from the McDonnell Pew Foundation (1994), the Whitehall Young Investigator Fellowship (1994), and the Klingenstein Fellowship Award (1996).

Contributors

Linda A. Barlow
Department of Biological Sciences
University of Denver
and
Department of Cellular and Structural Biology
Rocky Mountain Taste and Smell Center
Denver, Colorado

Joseph G. Brand
Monell Chemical Senses Center
Veterans Affairs Medical Center
and
University of Pennsylvania
Philadelphia, Pennsylvania

Earl E. Carstens
Section of Neurobiology, Physiology and Behavior
University of California — Davis
Davis, California

Mirela Iodi Carstens
Section of Neurobiology, Physiology and Behavior
University of California — Davis
Davis, Callifornia

M. A. Chaput
Laboratoire de Neurosciences et Systèmes Sensoriels
Université Claude Bernard Lyon
Villeurbanne, France

Thomas A. Christensen
ARL Division of Neurobiology
University of Arizona
Tucson, Arizona

Lawrence B. Cohen
Department of Cellular and Molecular Physiology
Yale University School of Medicine
New Haven, Connecticut
and
RedShirtImaging, LLC
Fairfield, Connecticut

Yuri Danilov
University of Wisconsin
and
Wisconsin Regional Primate Center
Madison, Wisconsin

Vicktoria Danilova
University of Wisconsin
and
Wisconsin Regional Primate Center
Madison, Wisconsin

Jean-Marc Dressier
Section of Neurobiology, Physiology, and Behavior
and
Department of Food Science and Technology
University of California — Davis
Davis, California

Patricia DiLorenzo
Department of Psychology
State University of New York
Binghamton, New York

A. DuChamp
Laboratoire de Neurosciences et Systèmes Sensoriels
Université Claude Bernard Lyon
Villeurbanne, France

P. DuChamp-Viret
Laboratoire de Neurosciences et Systèmes Sensoriels
Université Claude Bernard Lyon
Villeurbanne, France

Donald Elmer
University of Wisconsin
and
Wisconsin Regional Primate Center
Madison, Wisconsin

Thomas E. Finger
Department of Cellular and Structural Biology
Rocky Mountain Taste and Smell Center
Denver, Colorado

Barry G. Green
The John B. Pierce Laboratory and
Department of Surgery (Otolaryngology)
Yale University School of Medicine
New Haven, Connecticut

Charles A. Greer
Department of Neurosurgery
Yale University School of Medicine
New Haven, Connecticut

Gören Hellekant
University of Wisconsin
and
Wisconsin Regional Primate Center
Madison, Wisconsin

M. Scott Herness
Department of Oral Biology
College of Dentistry
and
Department of Neuroscience
College of Medicine
The Ohio State University
Columbus, Ohio

John G. Hildebrand
ARL Division of Neurobiology
University of Arizona
Tucson, Arizona

Mark A. Hoon
National Institute of Dental and Craniofacial Research
National Institutes of Health
Bethesda, Maryland

Thomas Hummel
Department of Otorhinolaryngology
University of Dresden Medical School
Dresden, Germany

Steven L. Jinks
Section of Neurobiology, Physiology, and Behavior
University of California — Davis
Davis, California

Donald B. Katz
Department of Neurobiology
Duke University Medical Center
Durham, North Carolina

Birgit Kettenmann
Department of Experimental and Clinical Pharmacology
University of Erlangen — Nuremberg
Erlangen, Germany

Gerd Kobal
Department of Experimental and Clinical Pharmacology
University of Erlangen — Nuremberg
Erlangen, Germany

Ying-Wan Lam
Department of Cellular and Molecular Physiology
Yale University School of Medicine
New Haven, Connecticut

Christian H. Lemon
Department of Psychology
State University of New York
Binghamton, New York

Minghong Ma
Department of Neurobiology
Yale University School of Medicine
New Haven, Connecticut

Miguel A.L. Nicolelis
Department of Neurobiology
Duke University Medical Center
Durham, North Carolina

Carlos R. Plata-Salamán
Central Nervous System Research
The R.W. Johnson Pharmaceutical
Research Institute
Spring House, Pennsylvania

Thomas Roberts
University of Wisconsin
and
Wisconsin Regional Primate Center
Madison, Wisconsin

Nicholas J.P. Ryba
National Institute of Dental and
Craniofacial Research
National Institutes of Health
Bethesda, Maryland

Heather Schellinck
Department of Psychology
Dalhousie University
Halifax, Nova Scotia

Geoffrey M. Schoenbaum
Department of Psychology
Johns Hopkins University
Baltimore, Maryland

Thomas R. Scott
College of Sciences
San Diego State University
San Diego, California

Gordon M. Shepherd
Department of Neurobiology
Yale University School of Medicine
New Haven, Connecticut

Sidney A. Simon
Department of Neurobiology
Duke University Medical Center
Durham, North Carolina

Christopher T. Simons
Section of Neurobiology, Physiology,
and Behavior
and
Department of Food Science and
Technology
University of California — Davis
Davis, California

Burton M. Slotnick
Department of Psychology
American University
Washington, D.C.

Andrew I. Spielman
College of Dentistry
New York University
New York, New York
and
Monell Chemical Senses Center

Makoto Sudo
Department of Anesthesiology and
Resuscitology
Ehime University School of Medicine
Ehime, Matsuyama, Japan

Satoko Sudo
Department of Anesthesiology and
Resuscitology
Ehime University School of Medicine
Ehime, Matsuyama, Japan

Gulshan Sunavala
College of Dentistry
New York University
New York, New York

Matt Wachowiak
Department of Cellular and Molecular Physiology
Yale University School of Medicine
New Haven, Connecticut

F. Xu
Department of Diagnostic Radiology
Yale University School of Medicine
New Haven, Connecticut

Wentao Yan
College of Dentistry
New York University
New York, New York

Michal R. Zochowski
Department of Cellular and Molecular Physiology
Yale University School of Medicine
New Haven, Connecticut
and
Center for Complex Systems
Warsaw School of Advanced Social Psychology
Warsaw, Poland

Table of Contents

SECTION 3 Higher-Order Studies in Taste and Olfaction

Section 1

Behavioral Studies

1 Psychophysical Measurement of Oral Chemesthesis

Barry G. Green

CONTENTS

0-8493-2329-0/02/$0.00+$1.50

1.1 INTRODUCTION

The sensory irritancy of foods and beverages — such as the burn of red and black pepper, the coolness of menthol, and the bites of acids and salts — are mediated by chemically sensitive receptors of the somatosensory system. Until quite recently this class of sensations was attributed to an hypothesized cutaneous chemosensory system called the common chemical sense.[1] The common chemical sense was thought to be a warning system for noxious chemicals that was capable of encoding the intensity, but not the quality, of stimulation.[2,3] Evidence against this idea began to accumulate decades ago from studies of the neural basis of inflammatory pain and itch,[4,5] and electrophysiological investigations later confirmed that some types of pain fibers are chemically sensitive.[6-8] Although references to a common chemical sense disappeared from the somatosensory literature, the idea persisted among chemosensory researchers until additional research on the sensory physiology and psychophysics of capsaicin (and to a lesser extent menthol) made it clear that sensory irritation arises primarily from chemically sensitive neurons that encode information that is interpreted as pain and temperature.[9-11] In 1990, the term chemesthesis was coined to describe this multimodal sensibility.[12]

Because the trigeminal nerve (V) is the principal somatosensory nerve innervating the nose and mouth, chemosensory irritation is often referred to as trigeminal sensitivity. This terminology oversimplifies the neurophysiology of oral chemesthesis. Although the trigeminal nerve innervates the anterior areas of the mouth, it is not the sole mediator of somatosensation in the oral cavity. The glossopharyngeal nerve (IX), which contains tactile fibers, thermal fibers, and nociceptors as well as taste fibers, innervates the back of the tongue and parts of the pharynx, and the vagus nerve (X) innervates the remainder of the pharynx and larynx. Both IX and X contribute to somatosensation during ingestion, and recent evidence indicates that both play a significant role in chemesthesis whenever stimuli are swallowed.[13]

The multimodal nature and neural complexity of chemesthesis have profound implications for how it should be studied psychophysically. Its dependence on neurons that are sensitive to temperature, mechanical stimulation, or both,[9,11,14,15] and the existence of potent inhibitory mechanisms within the somatosensory system,[16,17] pose unique problems for psychophysical measurement. Standard methods and procedures of taste research are often inappropriate for studying chemesthesis, and their use without modification risks the introduction of serious artifacts.

This chapter is therefore intended as a guide to help researchers choose and modify psychophysical methods for use in oral chemesthesis. The first section provides an overview of the sensory properties of chemesthesis, and subsequent sections describe, critique, and recommend psychophysical procedures that can be used to measure chemesthetic sensitivity, intensity, and perceptual quality.

1.2 PROPERTIES OF ORAL CHEMESTHESIS THAT AFFECT ITS MEASUREMENT

1.2.1 Slow Onset and Decay

In contrast to taste and smell, the perceptual response to most chemical irritants develops much more slowly and lasts much longer. Depending on the chemical, its concentration, and where in the oral cavity it is applied, sensations of irritation can take tens of seconds to be felt.[18] Response times are shortest when stimuli contact the anterior edge of the tongue, where an abundance of trigeminal nerve endings rise close to the epithelial surface in fungiform taste papillae.[19] Throughout the rest of the mouth, irritants must penetrate the epithelium to reach somatosensory receptors much deeper in the dermis. This is particularly true for the middle of the tongue, where receptors are both deeper lying and less densely distributed.[20]

The greater depth of somatosensory receptors compared to gustatory receptors, together with the tendency for lipophilic substances to remain in the epithelium, causes most sensations of irritation to persist longer than sensations of taste. Even moderate concentrations of chemicals such as menthol and capsaicin can produce sensations that remain perceptible for 10 to 15 min or longer.[21,22] The slow rise and decay of chemesthetic sensations places constraints on the rate at which stimuli can be presented and the point at which subjects should be asked to rate their sensations. Rating too soon or too late can lead to underestimations of irritancy. It is advisable to use pilot tests to determine approximately when sensations reach a peak and how long they take to decay. In addition, this information is important for setting the minimum inter-stimulus interval (ISI) when experiments call for multiple stimulus presentations. However, other temporal properties that affect sensitivity over time can be even more critical for setting the temporal parameters of stimulation (see below).

1.2.2 Sensitization and Desensitization

In most sensory systems, constant or repeated stimulation causes a progressive decline in sensitivity, or adaptation. This is not always the case for chemesthesis. Continuous or recurrent stimulation can lead either to a progressive growth in irritation (sensitization) or to a decrease in sensitivity (desensitization). The irritant best known for inducing both sensitization and desensitization is capsaicin, but piperine, zingerone, eugenol, and menthol also produce similar effects. In general, shorter ISIs are conducive to sensitization[23] and longer ISIs are conducive to desensitization.[24,25] This means that waiting several minutes between stimuli, as is often done to allow sensations from the preceding stimulus to diminish, does not avoid desensitization. Desensitization therefore imposes very severe limitations on experimental design: stimuli delivered to the same site within the same experimental session, even as long as 24 h later,[26] cannot be assumed to be perceptually or physiologically independent.

When an experiment with capsaicin calls for maintaining a constant level of sensory irritation over many minutes of testing, the best strategy is to apply stimuli with an ISI of 1.5 to 2.0 min.[23] This rate of stimulation appears to balance the effects of sensitization and desensitization and leaves the average magnitude of irritation approximately constant. However, because individuals vary with respect to rate and amount of sensitization and desensitization,[27,28] precise control of sensation intensity requires that ISIs be tailored for each individual.

It bears reminding that not all irritants exhibit these effects, and that the time-courses of sensitization and desensitization vary among the irritants that do.[29] This makes it essential to search the literature for physiological or psychophysical information about the temporal characteristics of stimuli before designing an experiment, and to conduct pilot tests to ensure that the chosen temporal parameters produce the desired effects.

1.2.3 Temperature Sensitivity

Temperature can drastically influence chemosensory irritation. Thermal modulation of irritation is attributable primarily to the temperature sensitivity of neurons that respond to chemical irritants. The most common effects are for cooling to suppress chemical irritation and for heating to enhance it,[10,30,31] although other changes can also occur. For example, menthol's chemical cooling effect is stronger at cooler than at warmer temperatures.[32] Cooling can also attenuate perception of nociceptive (pain) stimulation indirectly, via inhibitory mechanisms in the CNS.[33] The latter effects can occur even if cooling takes place away from the immediate area of stimulation. A cardinal rule of chemesthetic research is therefore to minimize temperature change in the oral cavity during stimulation and throughout the response period. A common mistake is to deliver test solutions and water rinses at ambient temperature (20 to 22°C). This practice reduces the sensitivity to most irritants by cooling the mouth from its resting temperature of 35 to 37°C. Over the course of a testing session, oral temperature falls at a rate that depends on the temperature of the solution, the frequency of sips and rinses, and how long each is held in the mouth. Thus, use of room-temperature solutions creates an unpredictable time-order effect. This problem should be avoided by keeping test solutions and rinse water in beakers or bottles that are partially submerged in a 37°C water bath. Solutions and rinses can then be poured into medicine cups immediately before being presented.

Another common oversight is to allow subjects to leave the mouth open (and/or the tongue extended) after localized stimulation of the tongue. Because the tongue is wet with saliva, and because the stimulus itself is moist, evaporation immediately begins to cool the lingual surface. Inhaling over the tongue augments evaporation and intensifies cooling (just as taking a breath with a menthol lozenge triggers a flash of coolness in the mouth and throat). Retracting the tongue into the warm, humid environment of the closed mouth curtails evaporation and thus helps keep the tongue near normal oral temperature.

1.2.4 Tactile Inhibition

Tactile stimulation must also be controlled because, like cooling, mechanical stimulation can cause inhibition in pain pathways.[16] Mechanical stimulation as subtle as that produced during sipping and expectoration appears to be sufficient to attenuate capsaicin burn,[10] and tasting motions such as smacking the tongue and lips should definitely be avoided. Instructions to subjects should routinely include keeping the tongue immobile during stimulation and while making perceptual judgments. This prohibition includes talking, which has the potential to reduce perceived irritation both via mechanical stimulation and evaporative cooling. In experiments in which stimulation is directed at specific oral sites, keeping the mouth closed and motionless also reduces the risk of spreading stimulation to unwanted areas.

BOX 1.1
Common Procedural Pitfalls to Avoid

- Presenting stimuli at room temperature rather than at oral temperature (35 to 37°C)
- Allowing evaporative cooling to occur during stimulation (e.g., leaving the mouth open after a stimulus has been applied)
- Failing to take the potential for sensitization and desensitization into account in experimental designs
- Allowing tongue and mouth movement during stimulation and/or in the response period

1.3 STIMULUS DELIVERY AND CONTROL

The preponderance of psychophysical measurements of oral chemesthesis has used one of three approaches: whole-mouth stimulation (sip-and-spit), localized stimulation with filter paper, or localized stimulation with cotton swabs. The choice of which of these simple procedures to use depends primarily on the objective of the study, but each has pros and cons that need to be considered.

1.3.1 Whole-Mouth Sip-and-Spit

The sip-and-spit procedure, borrowed from taste psychophysics, involves nothing more than sipping a measured volume of stimulus (usually 10 to 20 ml), holding it in the front of the mouth for a fixed period of time (usually a few seconds), then expectorating it. If for no other reason, its unparalleled simplicity has made this the most widely used chemosensory procedure. The chief experimental advantage of sip-and-spit is that it stimulates a large region of the anterior oral cavity, which has made it attractive for making inferences about normal perception during eating or drinking.

1.3.1.1 Problems of Stimulus Control

The sip-and-spit procedure has inherent limitations related to stimulus control. First and foremost, the spatial extent of stimulation during sipping is inexact and variable. The stimulus spreads unpredictably throughout the mouth, depending on how an individual sips and manipulates the solution. Tipping the head down after ingestion and keeping the tongue still can limit but not eliminate this spread. Variability in spatial stimulation is a serious disadvantage when the objective is to measure sequential interactions between stimuli (e.g., desensitization) because stimulus repetitions cannot be assumed to strike the same areas of the mouth.

Another difficulty with spatial control is that stimuli are spread to the lips during expectoration. This is not a problem for gustatory research, but can be a very serious one for chemesthetic research. Because they are sensitive to chemical irritants,[34] incidental exposure of the lips can, at a minimum, complicate the subject's task of detecting or rating irritation in the mouth. Inasmuch as sensory irritations at different body sites can interact in an inhibitory fashion,[35] stimulation of the lips has the potential to reduce perception of irritation in the mouth. This problem is even greater with another, less-common procedure for stimulating the tongue called the flow procedure, in which stimuli are flowed over the extended tongue and collected in a funnel below. With this procedure, solutions inevitably run from the tongue onto the lower lip and chin, which for chemical irritants would result in unacceptable perioral irritation.

Another limitation of the sip-and-spit procedure is that, although often referred to as a whole-mouth procedure, it fails to stimulate the entire mouth. Stimuli do not reliably reach the back of the tongue, and only reach the pharynx if they are spread there inadvertently when subjects swallow between trials. Thus, the sip-and-spit procedure cannot be used to test the perceptual contribution of the glossopharyngeal and vagus nerves. This is an important point. A recent study showed that when subjects swallowed rather than expectorated the stimulus, sensory irritation was reported to be as strong or stronger in the throat than in the front of the mouth.[36] Testing with the sip-and-spit procedure is therefore almost certain to underestimate the full experience of sensory irritation during normal consumption, which can only be captured when stimuli are swallowed.

Swallowing stimuli presents another set of problems, however, including the possibility of post-ingestive effects and the difficulty of clearing or rinsing the palate and throat between stimuli. For irritants the former problem is greater than the latter, which for lipophilic substances exists even in the sip-and-spit procedure. Stimuli that are swallowed should therefore be restricted to very small volumes (e.g., 5 ml), and should be delivered using protocols that minimize the possibility of sensitization and desensitization over trials.

1.3.1.2 Problems of Stimulus Preparation

Preparation of sip-and-spit solutions can be problematic when stimuli are strongly hydrophobic. For such chemicals (e.g., capsaicin, piperine, zingerone, and menthol), the use of solvents and surfactants becomes necessary to get them and keep them in aqueous solution. Aqueous capsaicin solutions provide a good example. Because it is virtually insoluble in water, capsaicin must first be dissolved in ethanol. But because ethanol is itself a sensory irritant, it should not be present in test solutions at concentrations greater than about 5%. This problem can be solved by dissolving capsaicin in a volume of ethanol that will total less than 5% in the final test solution. In the author's experience, a 4% solution has been adequate for this purpose. Four-percent ethanol is too low to keep capsaicin in solution, however, so the surfactant Tween 80 is added as an emulsifying agent before bringing the solution to volume with H_2O. (Note that the need for an emulsifier may not be apparent from a visual inspection of solution. Irritants are often prepared in such low concentrations that precipitates are not visible to the naked eye.) The addition of 1% Tween 80 has been found sufficient to maintain capsaicin and menthol in solution. Finally, because 1% Tween 80 tastes bitter to most people, sucrose (0.25 to 0.5 *M*) can be added to counteract the bitterness.

It is important to recognize that use of a vehicle can affect the rate at which stimuli are delivered to the mucosa. Surfactants can slow the movement of chemicals out of solution, and sucrose can reduce the irritancy of capsaicin.[37] The likelihood that vehicles affect perception makes it essential to use the same vehicle whenever the sensory characteristics of different chemicals are being compared.

1.3.2 Techniques for Localized Testing

The two most common approaches used to test chemesthesis on small areas of the mouth or tongue are the filter paper and swab techniques. The filter paper technique involves the application of small amounts of stimulus to filter paper disks (usually <1 cm diameter) that are then placed on the tongue or other oral site (using forceps) and left in contact for several seconds or longer. The swab technique simply involves wetting a cotton-tipped applicator (e.g., a Q-tip) with the stimulus and swabbing it briefly onto a target site. In principle, the ability to apply stimuli to specific locations on the tongue or oral mucosa makes these procedures useful for measuring spatial variations in sensory irritation and for studying temporal phenomena such as sensitization and desensitization that depend on repeated stimulation of the same receptors.

1.3.2.1 Issues of Stimulus Control

In practice, positioning filter papers accurately with forceps can be difficult, and the ability to swab small oral areas with precision and consistency is limited by the manual dexterity of the experimenter. Even when placement is precise, there is risk that the stimulus will spread to nearby areas, especially to opposing oral surfaces. Reducing excess stimulus on the papers or swabs before applying them to the mucosa (e.g., by drawing them along the edge of a beaker or cup) can reduce but not eliminate this problem. A good practice when using filter paper is to pipette the stimulus onto the disks in a fixed volume that has been determined in pilot tests to thoroughly wet the papers without producing runoff.

1.3.2.2 Advantages of Stimulus Preparation

Another advantage of filter papers and swabs over the sip-and-spit procedure is that some hydrophobic stimuli can be prepared without the need for a complex vehicle. For example, stock solutions of capsaicin can be made in 100% ethanol, then pipetted in precise volumes onto filter papers or swabs and allowed to dry (10 to 15 min). Drying liberates the ethanol but leaves the relatively nonvolatile capsaicin in the paper or cotton. The filter papers or swabs are then re-wetted with water (deionized or distilled) just prior to application to the mucosa. The primary limitation of this procedure is the volatility of the stimulus molecule relative to ethanol. A volatile chemical such as menthol will continue to dissipate unless the swabs or papers are tightly wrapped in plastic or parafilm. Even so, fresh stimuli should be made before each testing session or, at a minimum, each day.

1.4 ADAPTING STANDARD PSYCHOPHYSICAL METHODS TO THE UNIQUE REQUIREMENTS OF CHEMESTHESIS

The temporal characteristics of oral chemesthesis impose serious limitations on psychophysical methodology. Both the persistence of sensations after stimulation has ended and the potential for sensitization at short ISIs place limits on the rate of stimulation, and hence on the number of stimuli that can be applied in a testing session. On the other hand, lengthening the ISI to avoid persistence and sensitization increases the likelihood of desensitization. To avoid or at least minimize the effects of these temporal constraints, experimenters must be prepared to modify methods that have been developed for use in other sensory systems.

1.4.1 Threshold Measurement

1.4.1.1 Choosing a Method

Measuring the threshold for detection is among the most challenging of all psychophysical tasks. Unfortunately, the most sophisticated and sensitive of the available methods, those based on the theory of signal detection (TSD), require more stimulus presentations than can reasonably be administered with chemical irritants. Even the more economical and less-sensitive method of limits, in which subjects report the presence or absence of a stimulus as it is alternately increased and decreased in intensity, cannot be used when the stimulus under test can cause desensitization. In these cases, the chemesthetic researcher is left with only a couple of options: using very long ISIs or limiting multiple stimulus presentations in each session to a single ascending concentration series.

Increasing the ISI enough to avoid desensitization is usually impractical. Even stimuli that have relatively short desensitization periods, such as zingerone[29] and menthol,[38] still require ISIs in excess of 5 min. Instead, an ascending concentration series, essentially a unidirectional method of limits, is usually the safer choice. This procedure minimizes desensitization because weaker stimuli tend not to desensitize (or adapt) the response to stronger ones.[13] Pilot tests must be conducted to select a range of stimuli that begins with at least one concentration too low to be detected, and ends with at least one concentration certain to be perceived. Thresholds are then calculated by administering several (at least three) replicates of the series on separate days and taking the average (or median) of the lowest concentration detected in each series.

The principal weaknesses of this method are that it is vulnerable to guessing and it is likely to underestimate sensitivity. The guessing problem can be somewhat reduced by collecting repeated responses to each stimulus over a period of 3 to 5 min. Here, the slow rise and decay of sensation can work to the experimenter's advantage by giving subjects several chances to perceive the stimulus during each observation period. The decision as to whether detection has occurred can then be based on more than one response per stimulus. In addition, the method can be used in conjunction with an intensity scaling method (see below) to obtain information about perceived intensity when sensations are reported. This information can then be used to establish an intensity-based criterion for detection, for example, the lowest concentration consistently rated above barely detectable throughout the observation period.

The tendency toward underestimating sensitivity arises because of the well-known proclivity of subjects to wait to respond until they are certain they perceive the stimulus. Because the opposite tendency occurs with descending series (i.e., subjects tend to wait longer to make certain the sensation has disappeared), the practice of alternating ascending and descending series in the classical method of limits enables experimenters to counterbalance the judgmental biases by averaging across the two kinds of estimates. Lacking this opportunity in chemesthesis, thresholds obtained with ascending series must be assumed to underestimate sensitivity. This does not, however, compromise the utility of such thresholds as relative measures of sensitivity across stimuli, conditions, or subjects.

1.4.1.2 Use of Vehicle Controls and Paired Comparisons

Whenever stimuli are presented in a vehicle other than water, it is essential that a vehicle-only condition be included in the design. It should never be assumed that the vehicle is imperceptible, as even small changes in rheology can sometimes produce subtle sensory cues. In addition, the problem of false positives in detection tasks can be counteracted to some extent by the use of blanks (which give an indication of the probability of false positives), or by simultaneously presenting the vehicle to another (usually contralateral) site. A paired comparison task of this sort, in which the subject must decide which of two sites contains the stimulus (or elicits the stronger sensation,[39]) reduces the criterion problem by changing the task to a difference judgment rather than an absolute detection task. That is, subjects need not perceive a clearly identifiable sensation of irritation; they need only perceive a difference between sites. This is the basis of the oft-used two-alternative forced-choice task (2AFC), which is a so-called criterion-free method based on TSD.[40]

1.4.2 Suprathreshold Intensity Measurement

1.4.2.1 Choosing a Method: The Labeled Magnitude Scale

Suprathreshold psychophysics appears deceptively simple. A variety of methods are available that make it easy to collect a lot of data on perceived intensity in a short period of time. But like any form of measurement, the precision, reliability, and validity of the data depend on the quality of the measurement instrument chosen and the skill with which it is applied. A detailed discussion of the pros and cons of different methods of intensity scaling is beyond the scope of this chapter. The author focuses instead on one method that has been designed specifically for measuring oral sensory irritation and contrast its features with other methods that are in common use.

The labeled magnitude scale, or LMS, is a semantic scale of sensation intensity.[41,42] As shown in Figure 1.1, the intensity labels of the LMS are spaced nonlinearly at locations that were determined empirically by semantic scaling (see also

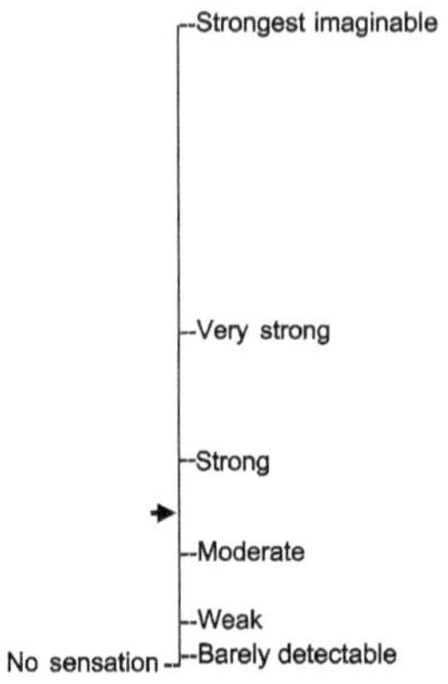

FIGURE 1.1 The Labeled Magnitude Scale is shown as it appears on the computer screen during an experiment. The subject uses a mouse to move the arrow-shaped cursor to the location on the scale that corresponds to the perceived intensity of the sensation, then clicks the mouse to enter the response. *Strongest imaginable* refers to the most intense sensation that a subject can imagine experiencing.

Reference 43). It is this unique spacing, and the fact that the scale's upper limit is the strongest imaginable sensation, that enables the LMS to yield data equivalent to what can be obtained with the unconstrained method of numerical magnitude estimation.[44] Thus, in contrast to typical equal-interval category scales, the LMS can be used to estimate the relative strength of sensations produced by different stimuli and/or different conditions. It is possible to establish, for example, that stimulus X produces sensations 3 times stronger than stimulus Y, or that a particular treatment causes a 20% decrease in perceived intensity. Moreover, because its semantic labels are associated with meaningful sensation magnitudes, the LMS provides much richer semantic information about sensations than do typical scales, whose category boundaries are arbitrarily spaced at equal distances along the scale. A researcher using the LMS can therefore conclude, for example, that one stimulus produces a sensation a little less than strong, while another produces a sensation just above weak. Finally, because the LMS is a fixed scale that all subjects are instructed to use in the same way, inferences can be made about individual differences in perception based on differences in intensity ratings. These inferences must be tempered, however, by the caveat that despite careful instructions and practice, not all people can use the scale in exactly the same way, particularly if their sensory experiences outside the laboratory are different.[45]

1.4.2.2 The Importance of Practice

Subjects who are inexperienced in intensity scaling must master two new tasks: thinking about sensations analytically and quantitatively, and translating their perceptions onto an external scale. It is important that they become comfortable with both tasks before data collection starts. This can be accomplished by beginning each experiment with a combined training and practice session in which subjects are

BOX 1.2
Instructions for Using the Labeled Magnitude Scale

You will be asked to rate the intensity of a variety of real and remembered sensations by indicating where they lie on a scale of all possible sensations. The scale contains commonly used terms like weak and strong, and the top of the scale is the strongest sensation of any kind that you can imagine experiencing.

When you make your ratings, you should use the terms just as you would in daily life. But do not limit your ratings to the terms themselves. A good strategy is to first decide which term most closely describes the strength of a sensation, then fine-tune your rating by moving the cursor to the appropriate place on the scale between that descriptor and the next most appropriate one. For example, if you think a sensation is about moderate, but a little bit stronger, you should move the cursor to the appropriate place just above moderate. If you think another sensation is more than just barely noticeable but less than weak, you should move the cursor to the appropriate place between barely noticeable and weak, etc.

Do you have any questions about how to use the scale?

given instructions on how to use the scale and then given practice in the scaling procedure. In the case of the LMS, its unique form and the paramount importance of understanding that it encompasses the full range of possible sensations make it wise to give subjects clear instructions and then gauge their understanding before proceeding further. Generic instructions that enable the LMS to be used to collect data without reference to a particular sensory modality or sensation quality are shown in Box 1.2. The experimenter should read the instructions to each subject, then ask the subject to use the scale to rate the intensity of several commonly encountered or imagined sensations. The examples of common or imagined sensations shown in Box 1.3 were designed to cover a range of perceptual intensities in oral chemesthesis, somesthesis and taste, so that subjects would learn to rate qualitatively different sensations on a common scale. The experimenter then observes the subject's response to ensure that sensations having very different intensities (e.g., the burn of cinnamon gum and biting the tongue) are rated in a reasonable way relative to one another and to the semantic labels. The experimenter must never dictate where on the scale a sensation should be rated, but can follow an unusual response with a neutral question, such as, "In your experience, is the burn of cinnamon gum stronger than the pain from biting your tongue?" The answer to a question of this kind can usually reveal whether a rating reflects a misunderstanding of the scaling task or only carelessness.

Once the experimenter is convinced the subject understands the task, the subject should then be given a chance to rate actual sensations similar to those he/she will encounter in the experiment. There is good reason to assume that without such practice the first several responses collected in the experiment will include a learning

BOX 1.3
Imagined and Remembered Sensations for Practice with the Labeled Magnitude Scale

- Sweetness of a cherry lifesaver
- Coldness of an ice cube in the mouth
- Burning your mouth with a hot pizza
- Warmth of warm bread
- Saltiness of a potato chip
- Pain from biting one's tongue
- Sourness of a dill pickle
- Coldness of ice cream
- Saltiness of soup broth
- Heat of hot tea
- Bitterness of celery
- Saltiness of a pretzel
- Sweetness of a banana
- Burn of cinnamon gum
- Carbonation of a cola drink
- Sweetness of cotton candy
- Sourness of sourdough bread
- Sweetness of a red apple
- Spiciness of a ginger cookie
- Coldness of ice water
- Bitterness of black coffee
- Sourness of a lemon
- Burn of yellow mustard
- Hottest burn ever experienced from a spicy food

component as subjects become increasingly comfortable scaling sensations to which they typically pay little attention in daily life. Practice stimuli need not be the same as the test stimuli; and in cases where the order of presentation of stimuli in an experiment is critical, it may be advantageous to use a different stimulus. For example, practice for an experiment with capsaicin or a similar irritant that causes a predominately burning or stinging sensation could involve rating a wide range of concentrations of ethanol (e.g., 10 to 50%) applied to the oral area to be studied. Ethanol is an attractive practice stimulus because it is easy to prepare, inexpensive, does not produce lasting desensitization, and its sensation dissipates quickly to allow several stimulus presentations in a relatively short period of time. However, ethanol is not suitable for use in a sip-and-spit procedure because the high concentrations needed to produce irritation are too readily absorbed into the bloodstream through the oral mucosa.

Another value of a practice session is that it can be used to screen subjects for sensitivity. The problem of outliers is particularly troublesome in chemesthesis, where large individual differences in sensitivity and in the response to stimuli over time are often encountered.[23,27,46] Screening subjects prior to data collection can avoid the complications and potential for biases that come with eliminating outliers after the fact, when all the data are in and hypotheses are about to be tested statistically. Sensitivity screening can be accomplished by presenting the test stimulus (or a subset) in the practice session, then setting a criterion level of response for inclusion in the experiment. By giving every subject a common set of (ascending) concentrations, the practice session can also be used to find which concentration produces the desired sensation intensity (e.g., moderate on the LMS) for each subject.

1.4.2.3 Sensation Quality

Because chemesthesis arises from stimulation of neurons sensitive to temperature, mechanical stimulation, or both,[9,11,14,15] sensory irritation can be qualitatively complex. This fact must be taken into account when designing experiments to measure perceived intensity. Is the datum of interest the strength of the overall sensation, or is it important to know which qualities of irritation are present and what their intensities are? While there may be reasons for wanting to know only about total intensity, it is more often important to acquire data on the strength of individual sensation qualities. Among other things, data of this sort can potentially reveal the extent to which different types of sensory receptors (e.g., cold fibers vs. nociceptors) contribute to perception of an irritant.

When the strength of individual qualities is of interest, it is wise to obtain ratings for each of the qualities that are likely to be present. Research on taste and smell has shown that subjects tend to overestimate the strength of a single quality when they do not have the opportunity to rate the intensity of other qualities that accompany it.[47] In addition, it is difficult for subjects to understand what is meant by irritation if individual qualities are not clearly defined; and once they are defined, it makes sense to obtain intensity ratings for each of them. Box 1.4 contains a list of the sensory qualities and their definitions that are frequently encountered with sensory irritants. It is recommended that the experimenter read the list of definitions to subjects rather than trusting them to do so carefully on their own, and that the list be left in view for reference throughout the experiment.

In practice, making independent judgments of sensation qualities can be difficult. Some combinations, like burning and cold, may be relatively easy to discriminate, while others, like burning and stinging, may not be. Ideally, it would be helpful to present subjects with a stimulus prototype for each sensory quality. But unlike taste, where prototypes for the basic tastes are readily available, chemesthesis lacks stimuli that are truly qualitatively pure. A few chemicals do, however, have sufficiently distinct qualities to be of limited use in this regard. In low concentrations, menthol is primarily cooling, but becomes burning or stinging at higher concentrations. In moderate and high concentrations, capsaicin is usually reported to produce only burning, but at low concentrations it is sometimes described as warm or tingling. High concentrations of citric acid produce stinging, but stinging is accompanied by sourness on gustatory areas. Experimenters must therefore exercise care in using these stimuli as standards.

BOX 1.4
Definitions of Common Oral Chemesthetic Sensations

- Burning:
  The sensation that commonly results from very high temperatures (i.e., thermal burns) or certain chemical irritants (e.g., alcohol; may or may not be accompanied by a temperature sensation.
- Stinging/Pricking:
  Sharp, uncomfortable sensations similar to those produced by an insect sting or a pinprick; may be constant (stinging) or brief (pricking).
- Tingling:
  A lively pins-and-needles sensation.
- Numbness:
  A diffuse (e.g., fuzzy) sensation like that experienced during the onset or offset of local anesthesia (e.g., from Novocain). NOT the absence of sensation.
- Warm:
  The sensation of mild heating.
- Hot:
  The sensation that exceeds warmth, but is not yet painful.
- Cool/Cold:
  Sensations of mild (cool) or extreme (cold) cooling.
- Pain:
  Any sensation that hurts.

1.5 SUMMARY

On first impression, psychophysical research in oral chemesthesis appears easy to do. Its procedures are technically simple, and elaborate or expensive equipment is usually not required. However, it should be clear from the foregoing discussion that measuring chemesthesis is in fact a complicated business. Because it is a nascent research area with few tried-and-true methods, each new experiment is likely to bring with it a host of methodological challenges. Attempting to meet these challenges by simply borrowing methods from taste and other senses is ill-advised because the nature of chemesthesis places unique limitations and demands on experimental procedures. Instead, researchers must begin by identifying the method that best suits the aims of an experiment and then adapt the method as needed to accommodate the sensory characteristics of chemesthesis. To do this successfully requires more than just a familiarity with chemosensory psychophysics. The multimodal nature of chemesthesis requires that experimenters also have a basic understanding of the neurophysiological and psychophysical properties of somatosensation. This chapter cannot serve as a substitute for either of these requirements. It is intended instead as an introductory guide to help researchers new to the field avoid the most serious methodological pitfalls of this complex and intriguing chemical sensibility.

REFERENCES

1. Parker, G.H., The relations of smell, taste, and the common chemical sense in vertebrates, *J. Acad. Natl. Sci. Phila.,* 15, 221, 1912.
2. Walker, J.C., Tucker, D., and Smith, J.C., Odor sensitivity mediated by the trigeminal nerve in the pigeon, *Chem. Senses Flav.,* 4, 107, 1979.
3. Silver, W.L., Arzt, A.H., and Mason, J.R., A comparison of the discriminatory ability and sensitivity of the trigeminal and olfactory systems to chemical stimuli in the tiger salamander, *J. Comp. Physiol. A,* 164, 55, 1988.
4. Jancso, N., Role of the nerve terminals in the mechanism of inflammatory reactions, *Bull. Millard Fillmore Hosp.,* 7, 53, 1960.
5. Keele, C.A. and Armstrong, D., *Substances Producing Pain and Itch,* Edward Arnold, Ltd., London, 1964.
6. Bessou, P. and Perl, E.R., Response of cutaneous sensory units with unmyelinated fibers to noxious stimuli, *J. Neurophysiol.,* 32, 1025, 1969.
7. Dash, M.S. and Deshpande, S.S., Human skin nociceptors and their chemical response, *Adv. Pain Res. Ther.,* 1, 47, 1976.
8. Adriaensen, H., Gybels, J., Handwerker, H.O., and Van Hees, J., Latencies of chemically evoked discharges in human cutaneous nociceptors and of the concurrent subjective sensations, *Neurosci. Lett.,* 20, 55, 1980.
9. Szolcsanyi, J., Sensory receptors and the antinociceptive effects of capsaicin, in *Tachykinin Antagonists,* Hakanson, R. and Sundler, F., Eds., Elsevier Science Publishers B.V., Amsterdam, 1985, 45-54.
10. Green, B.G., Sensory interactions between capsaicin and temperature, *Chem. Senses,* 11, 371, 1986.
11. Harada, S., Maeda, S., and Kasahara, Y., Trigeminal nerve fiber responses to the temperature, tactile, and chemical stimulation applied on the rat tongue, in *Proc. 21st Japanese Symp. Taste and Smell,* Sato, T., Ed., 1987, 165-168.
12. Green, B.G., Mason, J.R., and Kare, M.R., *Chem. Senses, Vol. 2: Irritation*, Marcel Dekker, New York, 1990, v-vi.
13. Rentmeister-Bryant, H. and Green, B.G., Perceived irritation during ingestion of capsaicin or piperine: Comparison of trigeminal and non-trigeminal areas, *Chem. Senses,* 22, 257, 1997.
14. Caterina, M.J., Schumacher, M.A., Tominaga, M., Rosen, T.A., Levine, J.D., and Julius, D., The capsaicin receptor: a heat-activated ion channel in the pain pathway, *Nature,* 389, 816, 1997.
15. Nagy, I. and Rang, H., Noxious heat activates all capsaicin-sensitive and also a subpopulation of capsaicin-insensitive dorsal root ganglion neurons, *Neuroscience,* 88, 995, 1999.
16. Melzack, R. and Wall, P.D., Pain mechanisms: a new theory, *Science,* 150, 971, 1965.
17. Sandkuhler, J., The organization and function of endogenous antinociceptive systems, *Prog. Neurobiol.,* 50, 49, 1996.
18. Green, B.G., Spatial and temporal factors in the perception of ethanol irritation on the tongue, *Percept. Psychophys.,* 44, 108, 1988.
19. Whitehead, M.C. and Kachele, D.L., Development of fungiform papillae, taste buds, and their innervation in the hamster, *J. Comparative Neurol.,* 340, 515, 1994.
20. Dixon, A.D., Sensory nerve terminations in the oral mucosa, *Arch. Oral Biol.,* 5, 105, 1961.
21. Lawless, H.T., Oral chemical irritation: psychophysical properties, *Chem. Senses,* 9, 143, 1984.

22. Green, B.G. and Lawless, H.T., The psychophysics of somatosensory chemoreception in the nose and mouth, in *Smell and Taste in Health and Disease,* Getchell, T.V., Doty, R.L., Bartoshuk, L.M., and Snow, J.B., Eds., Raven Press, New York, 1991, 235-253.
23. Green, B.G., Capsaicin desensitization and stimulus-induced recovery on facial compared to lingual skin, *Physiol. Behav.,* 65, 517, 1998.
24. Green, B.G., Capsaicin sensitization and desensitization on the tongue produced by brief exposures to a low concentration, *Neurosci. Lett.,* 107, 173, 1989.
25. Green, B.G., Temporal characteristics of capsaicin sensitization and desensitization on the tongue, *Physiol. Behav.,* 49, 501, 1991.
26. Green, B.G., Rapid recovery from capsaicin desensitization during recurrent stimulation, *Pain,* 68, 245, 1996.
27. Cliff, M.A. and Green, B.G., Sensitization and desensitization to menthol and capsaicin in the oral cavity: interactions and individual differences, *Physiol. Behav.,* 59, 487, 1996.
28. Prescott, J., The generalizability of capsaicin sensitization and desensitization, *Physiol. Behav.,* 66, 741, 1999.
29. Prescott, J. and Stevenson, R.J., Desensitization to oral zingerone irritation: effects of stimulus parameters, *Physiol. Behav.,* 60, 1473, 1996.
30. Sizer, F. and Harris, N., The influence of common food additives and temperature on threshold perception of capsaicin, *Chem. Senses,* 10, 279, 1985.
31. Green, B.G., Oral chemesthesis: the importance of time and temperature for the perception of chemical irritants, in *Sensory Science Theory and Applications in Foods,* Lawless, H.T. and Klein, B.P., Eds., Marcel Dekker, New York, 1991, 107-123.
32. Green, B.G., Menthol modulates oral sensations of warmth and cold, *Physiol. Behav.,* 35, 427, 1985.
33. Wahren, L.K., Torebjork, E., and Jorum, E., Central suppression of cold-induced C fibre pain by myelinated fibre input, *Pain,* 38, 313, 1989.
34. Lawless, H.T. and Stevens, D.A., Responses by humans to oral chemical irritants as a function of locus of stimulation, *Percept. Psychophys.,* 43, 72, 1988.
35. Green, B.G., Interactions between chemical and thermal cutaneous stimuli: inhibition (counterirritation) and integration, *Somatosens. Mot. Res.,* 8, 301, 1991.
36. Rentmeister-Bryant, H. and Green, B.G., Perceived irritation during ingestion of capsaicin or piperine: comparison of trigeminal and non-trigeminal areas, *Chem. Senses,* 22, 257, 1997.
37. Prescott, J., Allen, S., and Stephens, L., Interactions between oral chemical irritation, taste and temperature, *Chem. Senses,* 18, 389, 1993.
38. Green, B.G. and McAuliffe, B.L., Menthol desensitization of capsaicin irritation. Evidence of a short-term anti-nociceptive effect, *Physiol Behav.,* 68, 631, 2000.
39. Dressier, J.M., Nguyen, N., Sieffermann, J.M., Carstens, E., and O'Mahony, M., Oral irritant properties of piperine and nicotine: psychophysical evidence for asymmetrical desensitization effects, *Chem. Senses,* 24, 405, 1999.
40. Gescheider, G., *Psychophysics: Method, Theory and Application*, 2nd ed. L. Erlbaum Associates, Hillside, NJ, 1985.
41. Green, B.G., Shaffer, G.S., and Gilmore, M.M., Derivation and evaluation of a semantic scale of oral sensation magnitude with apparent ratio properties, *Chem. Senses,* 18, 683, 1993.
42. Green, B.G., Dalton, P., Cowart, B.J., Shaffer, G.S., Rankin, K.M., and Higgins, J., Evaluating the 'Labeled Magnitude Scale' for measuring sensations of taste and smell, *Chem. Senses,* 21, 323, 1996.

43. Borg, G., A category scale with ratio properties for intermodal and interindividual comparisons, in *Psychophysical Judgment and the Process of Perception,* Geissler, H.-G. and Petxold, P., Eds., VEB Deutscher Verlag der Wissenschaften, Berlin, 1982, 25-34.
44. Stevens, S.S., *Psychophysics,* Wiley, New York, 1975.
45. Lawless, H.T., Horne, J., and Spiers, W., Contrast and range effects for category, magnitude and labeled magnitude scales in judgments of sweetness intensity, *Chem. Senses,* 25, 85, 2000.
46. Prescott, J. and Swain-Campbell, N., Responses to repeated oral irritation by capsaicin, cinnamaldehyde and ethanol in PROP tasters and non-tasters, *Chem. Senses,* 25, 239, 2000.
47. Frank, R.A. and Van der Klaauw, N.J., The contribution of chemosensory factors to individual differences in reported food preferences, *Appetite,* 22, 101, 1994.

2 Behavioral Methods in Olfactory Research With Rodents

Burton Slotnick and Heather Schellinck

CONTENTS

2.1 INTRODUCTION

The past decade has seen a striking increase in research on olfaction. Interest has been driven both by discoveries in molecular biology and by behavioral studies demonstrating that rats can use olfactory cues to perform complex learning tasks.

0-8493-2329-0/02/$0.00+$1.50

The molecular biological studies have generated hypotheses concerning odor quality coding that can be examined at the level of behavior. The learning studies have made olfaction amenable to analysis by sensory psychologists, molecular biologists, and cognitive neuroscientists. Each research area requires methods for generating, controlling, and delivering odor stimuli and for testing the ability of animals to detect and discriminate odors. This chapter is largely confined to studies of olfactory behavior in rodents and describes methods for achieving control of odor stimuli, techniques for training rats and mice on odor detection and discrimination tasks, and some of the results obtained using these methods.

A fundamental problem common to all studies of olfaction concerns the control and delivery of odor stimuli. With the exception of encapsulated odors used in scratch-and-sniff odor identification tests with humans,[30,37,38] there is no standardized or widely accepted method for producing odors or presenting them to test subjects. This remains true despite the publication of several dozen types of odor generator systems and odor delivery devices, including those specifically designed for humans,[163,173] dogs,[107] rodents,[148] and insects,[119] as well as general designs adaptable to most species. In addition, and in contrast to research in most other sensory modalities, no standard stimulus against which other stimuli can be compared has been established; nor is there general agreement on what metric should be used to measure odor concentration.

In part, the absence of such standards is because vapor stimuli are difficult to control and the olfactory system of macrosmatic animals will respond to an exceedingly large number of qualitatively different stimuli that vary along several physical dimensions. The armamentarium of equipment needed to control odor generation and flow and to measure odor concentration is rarely available in the laboratory of most investigators. Further, measuring instruments, when present, are typically difficult to employ. Unlike light meters, sound meters, or easily duplicated and calibrated stimuli used to study somatic sensation, there is no commercially available odor meter. Physical devices that reflect molecular concentration of gases, such as gas chromatographs or mass spectrophotometers, present their own unique difficulties and demands for expertise in standardization, maintenance, and analysis.

It is also the case that the demands for precise control of odor stimuli vary widely among studies. In some, the question of primary concern is whether an animal can detect an odor while others require accurate psychophysical measures of odor sensitivity and discrimination. As a result, a wide variety of odor presentation methods and test procedures have been used. The most often used methods are, unfortunately, those that exert little or no control of odor vapors and provide relatively crude measures of behavior. Such methods may prove useful in studies such as those concerned with habituation and odor aversion learning, but are inadequate for detailed sensory analyses. In contrast, automated olfactometry combined with instrumental or operant training procedures allow excellent control of the stimulus and stimulus control of behavior.

2.2 STATIC PRESENTATION OF ODORS

The most common, although least satisfactory, method of presenting odor stimuli is to allow vapors to freely diffuse from a point source. This simple technique has been widely used, for example, to test olfaction in food-finding tasks, assessing odor preference, avoidance of odors, habituation to odors, and in studies of odor aversion learning. This chapter section considers both the benefits and problems associated with these methods.

2.2.1 Food-Finding Tests

Food-finding, a popular test for olfaction in rodents, requires that a hungry animal find a scented or natural food object in a substrate of sand or the animal's normal bedding material.[1] Finding the food reinforces search behavior and the time taken to do so provides the dependent measure of olfaction. Typically, only one or a few trials are given daily and, of course, the position of the target is varied randomly over trials. Hungry rats, mice, and hamsters learn to search the substrate and, generally, after 5 to 10 trials are able to find the food object in 5 to 30 sec, depending on the size of the test arena and distinctiveness of the target odor. The method has been validated by the demonstration that olfactory bulbectomized rats show little or no improvement in performance over trials and frequently fail to locate the target in a 5-min test.[59] Also, response latencies are generally longer in rats with damage to the olfactory system such as that produced by zinc sulfate or an olfactotoxin.[1,55] Speed of learning, asymptotic levels of performance, and the magnitude of an effect produced by some experimental treatments are all undoubtedly influenced by level of deprivation, size of the test arena, competing odor components of the substrate, and odor properties of the target. These factors vary widely among studies, making cross-study comparisons difficult at best for this very simple test of olfaction. Even if such parameters were standardized, food-finding tasks have inherent disadvantages and are distinctly inferior to other tests for olfactory function. For example, in food-finding and other tests using static presentation of odors, there is little or no control of vapor diffusion or odor gradients within the test arena. Thus, the subject is required to detect the vapor of one substance (the target) against the background vapors of the substrate and other potential ambient odors. The required search behavior may also make demands on spatial memory and, importantly, the dependent measure, time, does not index accuracy. Thus, food-finding is not an appropriate test for odor detection and, apart from it being simple, there seems little reason to sacrifice control of the stimulus and a measure of response accuracy when better methods are available. These shortcomings are particularly evident when the test is used to assess if a particular treatment (e.g., intranasal irrigation of zinc sulfate) has produced anosmia. The failure of rats or mice in such studies to find a hidden food object in a brief test has been taken as evidence for loss of smell. It is possible, however, that increased response latencies may reflect difficulties in making the requisite odor discrimination, in following odor gradients, decreased motivation, or other factors including not engaging in control levels of search behavior.[142,152,180]

2.2.2 Odor Preference Tests

In another widely used measure of olfaction, the so-called preference test, two or more odors are placed in a test arena and the frequency and/or duration of a subject's approach to and investigation of each odor is recorded. Preference tests have been used most often to assess the response to social odors such as those denoting sexual status[17] and kin identity.[62,63] More recently, they have been used to assess the effectiveness of a prior associative learning experience.[132,133]

Testing of animals in the home cage environment is generally not recommended because rodents tend to sniff indiscriminately at all novel objects introduced into the home cage.[74] Thus, tests are generally conducted in an open field or a maze.[46,69] Preference tests, like food-finding tests, have a number of serious shortcomings. The method as generally practiced provides little or no control over odor dispersion and, when two or more odors are presented simultaneously, a mixture of their vapors could increase the difficulty in discriminating between them. Moreover, because the odors will reach the subject over time regardless of whether the animal investigates them, there is a possibility that an animal need not approach an odor directly to show a preference. Also, the subject may habituate to the odor before showing a measurable response.

A variety of factors that influence activity, including the subject's circadian rhythm, can also affect test outcomes. In practice, rats and hamsters spend relatively little time investigating most test odors. Apparently, odors that have no reinforcing consequences maintain little attentive behavior and decisions concerning preferences are often based on small although generally reliable differences in sample time between one odor and another. While there are exceptions, notably the vigorous investigatory behavior by male hamsters toward hamster vaginal discharge (HVD), Coppola and O'Connell[25] found that the opportunity to contact this attractant provided only marginal support of operant responding. Their results suggest that pheromonal stimuli do not have strong reinforcing properties. The tendency of rodents to habituate quickly to odors that have no further behavioral consequences may contribute to this outcome. Test-retest reliability for preference tests has not been reported but, given the many factors affecting test outcome, it is unlikely to be high. As described below, clearer and more consistent results are obtained in a learning paradigm where preference is assessed for odors that had previously been paired with reinforcement.

Because of their limited motor abilities, special problems are encountered in adapting odor preference tests for neonates. Generally, a simple two-choice apparatus is used. The pups are placed on a mesh covering and the odors diffuse through the mesh from containers placed below the chamber. A neutral zone can be created in the middle of the chamber to minimize the possibility that inaccurate measures may result from chance movements.[26,165,177] As in the food-finding task, parameters used in neonatal preference tests vary widely among studies, as do criteria defining a preference response.[29,50,63,70,100] In addition, the influence of temperature in poikleothermic neonates has largely been ignored in many of these studies. Warm ambient temperatures would increase the ability of subjects to move about but could also reduce the motivation of the animals to do so. For testing odor preference in juveniles, a variety of mazes have been designed[73,109] but, again, inadequate control of odors in a maze can degrade choice behavior and test results may be difficult to interpret.

The natural preference or aversion for specific odors can also influence the outcome of preference tests that are designed to assess appetitive and aversive conditioning.[3,20] In some instances, for example, both neonates and juveniles preferred no odor to peppermint, lemon, or orange odors; however, when tested against each other, they showed no difference in preference for any of the odors.[3] Finally, and perhaps most importantly, it is unclear how the outcome of preference tests, particularly those that use natural odors, should be interpreted. They may reflect an absolute or relative preference for one of the test odors, the relative novelty of each odor, the detectability of the odors, or aversion to the least sampled odor, etc. A failure to display a preference can also result from both odors being equally attractive or aversive.

2.2.3 Habituation Tests

Habituation-dishabituation and habituation-discrimination tests generally yield more easily interpretable results and are reasonably well-suited to static stimulus presentation methods. Habituation is said to occur if successive presentations of the same stimulus odor results in a decreased behavioral response to that odor. Rodents will generally investigate a novel odor but quickly habituate to repeated presentations, even when those presentations are separated by several days.[6] Indeed, habituation can be a major confound in studies of odor preference.

A habituation-discrimination test involves a two-phase process: first, the response to the odor is habituated and then the initial stimulus and a novel odor are presented simultaneously. Responding differentially to the two odors provides evidence for discrimination. The habituation-dishabituation procedure is similar except that only one odor is presented in the post-habituation test. The test session usually consists of brief consecutive trials in one daily session. Generally, each test begins with a no-odor adaptation period during which each subject is presented with the odor vehicle.[19] After a series of habituation trials, samples of a second odor are presented. If the subject dishabituates, it shows an increase in sniffing on the first presentation of the novel odor. These methods have proven useful for examining odor discrimination in rodents, primates, and other mammals.[75,76,80,98,110,167] As noted by Martin and Beauchamp,[98] special care is necessary when habituation tests are used to assess identification of individual subjects. Unless the odors are equally familiar prior to testing, the animal may respond on the basis of familiar vs. nonfamiliar, rather than discriminating between individuals.

Habituation tests have proven quite useful in a variety of studies and Sundberg et al.[166] have devised a standard method of scoring the development of habituation. This method has been used extensively by Brown and colleagues[18,131,136] to provide evidence that congenic strains of rats, that differ only in the genes of the major histocompatibility complex, produce discriminably different urinary odors. Their results demonstrate the typical pattern of changes produced by habituation and dishabituation (Figure 2.1). This task has recently been successfully adapted for use with mice.[118] Because mice are frequently quite active during the tests, videotaping the session may be required to accurately score behavioral responses. Johnston and colleagues[75,168,171] have also used the habituation method to advantage in studies of social behavior in hamsters to assess recognition of individual members of a group, including kin recognition, odor memory, and the effect of gonadectomy on scent production.

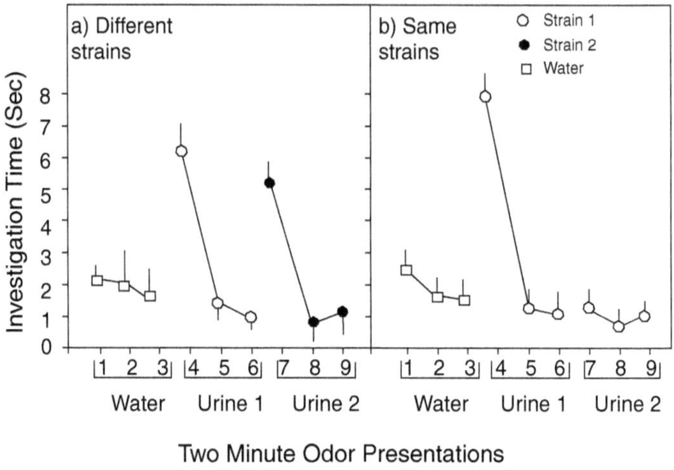

FIGURE 2.1 Mean (SEM) time spent rearing and sniffing at urine odors from (a) individuals from two different strains of rats, and (b) two individuals from the same strain. In (a), dishabituation occurred on Trial 7, indicating that the odors from two strains could be discriminated. In (b), the rats did not dishabituate when presented with a second urine sample from an individual of the same strain. (Redrawn from Figure 3 in Brown, R.E., Singh, P.B., and Roser, B., *Physiol. Behav.,* 40, 65-73, 1987.)

The principal benefits of both the habituation-discrimination and the habituation-dishabituation tests come from their simplicity of design, speed, and ease of testing. The occurrence of dishabituation in the test phase provides evidence for the discriminability of the odors. Sequential presentation of test materials eliminates the problems of animals being presented with a mixture of their vapors. As with the preference test, the main drawback to habituation tests is in the interpretation of a null outcome and that habituation itself precludes or sharply limits the ability to obtain repeated measures with the same subject.

2.2.4 Conditioned Odor Aversions

A more complex but potentially powerful olfactory test using static odor presentation is conditioned odor aversion. Typically, in these studies, thirsty animals are allowed to drink water while in the presence of a novel odor (conditioning stimulus). Although the odorant is generally placed near the drinking tube, it probably vaporizes to fill the test chamber. Ingestion is followed by administration of an illness-producing agent (e.g., lithium chloride) and, at some future time, the subject is retested in the presence of the same odor. Assuming appropriate controls, inhibition of intake provides evidence for odor detection.

Tests of odor aversion learning have not been widely adopted, perhaps as a result of the uncertainty concerning the strength and longevity of conditioned odor aversions. In support of their sensory-and-gate channeling analysis of aversion learning,[53,54] Garcia and associates provided evidence that odors were weak cues for illness but their strength could be potentiated if the odor was presented with a taste as a compound conditioned stimulus. In fact, the evidence concerning the strength of simple odor aversion learning is mixed and some investigators have demonstrated strong single-trial odor aversion learning in the absence of potentiation by taste.[10,115,156] When odors are presented orally, conditioned odor aversions have the same properties that characterize conditioned taste aversions. The aversion is learned in a single pairing of the odor and illness inducing agent. It is learned when there are long delays between administration of the odor and the unconditioned stimulus, and conditioning strength varies as a function of stimulus concentration.[156] Generally, strong long-lasting aversions result and can be demonstrated in adults,[156] neonates,[128,157] and even in the fetus.[159]

Taste aversion learning has proven to be a particularly valuable method for examining taste classification or taste generalization.[112] In such studies, the subject, conditioned to one tastant, is presented with different tastants in post-conditioning tests. The extent to which the novel tastant inhibits fluid intake provides one measure of perceptual similarity between the conditioning stimulus and the test stimulus. Such methods may prove useful for assessing stimulus generalization among odors.

There are a number of disadvantages associated with the use of odor aversion tests to study learning. In contrast to the extensive literature on taste aversion learning, the optimal parameters for achieving robust conditioning (e.g., odor concentration, length of exposure, and mode of exposure) have not been studied systematically and it is not known whether all or most odors would be effective conditioned stimuli.[115] Nevertheless, aversion learning has unique properties that could be exploited for a better understanding of odor perception in animals.

2.2.5 Odor-Cued Taste Avoidance

Training animals to use an odor cue to avoid a bitter-tasting solution provides a simple yet reasonably well-controlled method to test odor detection and discrimination.[27,78] Thirsty rats are given repeated, short-duration (~30 s) exposures to a water tube. On the conditioning trial, the water is replaced with an odor solution (e.g., 0.1% amyl acetate) containing a bitter-tasting substance (e.g., quinine, denatonium benzoate, cycloheximide). Generally, after one or two exposures to the conditioning stimulus, rats carefully sniff at the tube opening and lick at it if the fluid is water (or an odor that had not been paired with a bitter taste), but avoid licking if the conditioning odor is detected. Learning is rapid (occurs in one to three exposures to the bitter taste) and long-lasting. Stimulus concentration can be varied to assess odor sensitivity and each trial generates a response latency measure. The method is simple, requires no special apparatus, produces rapid acquisition and, in contrast to food finding and other static stimulus presentation methods, allows for reasonably good control of the stimulus and provides a measure of response accuracy. The method may also be useful for field tests of non-standard laboratory animals. For example, Hunt, Croft, and Slotnick,[68] used odor-cued taste avoidance to examine odor detection and discrimination in kangaroos maintained in a semi-natural environment.

2.2.6 Conditioned Odor Preference and Associative Learning

Other designs using static presentation of odors have proven useful in studying a variety of associative learning phenomena. For example, it has been demonstrated that rearing young pups in an artificially scented nest or with the odors of a foster strain or species can alter later odor preference.[21,41,93,94,103] It is unclear, however, whether these outcomes represent associative learning or simply the effects of stimulus familiarity.[22,50,69] In some instances, this early experience was ineffective in altering preference behavior probably because the conditioning odors were constantly present and consequently not consistently associated with reinforcing stimuli.[57] Johanson and Teicher[71] provided a more convincing demonstration of conditioned odor preference in neonates by pairing an odor with oral infusions of milk in a warm environment.

Another related approach for investigating olfactory learning in neonatal rats involves the pairing of tactile or unconditioned stimulation (US) and an odor or conditioning stimulus (CS). In this paradigm, animals are exposed to an odor while receiving perineal stroking with a sable brush. The stimulation mimics maternal contact. Following one or more trials, the success of the conditioning is assessed in a preference test using a Y-maze. Sullivan, Leon, and colleagues[72,164,165,178] have used this method extensively to examine the neurobiology of olfactory learning.

Schellinck and co-workers,[47,48,132] have developed a simple discrimination task to assess odor learning in adult mice. Initially, mice are given the opportunity to dig in a small dish in which woodchips are diffused with an odor from a lower level. One odor is paired with sugar reinforcement in the woodchips (CS+ stimulus) and a second odor is presented alone (CS– stimulus). In a subsequent preference test, the mice are presented with both odors but no sugar. They do not need to be food restricted during training[47] and the use of digging rather than sniffing provides an easily scored and objective measure of preference. In the post-training test, mice dig consistently and almost exclusively in the woodchips containing the CS+ odor. The animals rapidly learn the task and remember the discrimination for at least 90 days.[132]

The socially transmitted food preference task as first described by Galef and Wigmore[49] is becoming increasingly popular as a method of assessing memory in rodents.[79] The rationale for the test is based on the demonstration that rats and mice develop a preference for food that they have smelled on the breath of conspecifics.[51,52] There are three phases to this test. First, a demonstrator animal is fed a scented food; then the demonstrator interacts with an observer animal. Finally, the food-deprived observer is given the choice of two scented foods to eat: the scented food that the demonstrator has eaten and a non-familiar scented food. Preference is measured in terms of the consumption of the food the demonstrator has eaten. This task provides consistent results and the subjects require only one trial to form a memory for the demonstrator odor.

Having rats dig in scented sand has been used by Eichenbaum and co-workers to examine the neuroanatomical basis of more cognitive-based olfactory learning tasks. The test procedures provide a good example of the fact that olfactometric methods (as described below) do not lend themselves to the solution of all olfactory learning tasks and that, in some cases, being clever is more important than being automated. Two examples from a series of studies by Eichenbaum and associates illustrate this point. In one study, rats were first trained to dig through a cup filled with scented sand to obtain a food reward.[45] They were then trained to sample two simultaneously presented cups and were rewarded only if they responded to the correct one. The stimulus pairs can be designated as A+/B–, B+/C–, C+/A–, where + and – indicate the reinforced and non-reinforced stimuli in each pair. Note that in this task, correct responding depends on the configuration of the stimuli and not on associations between individual stimuli and reinforcement. Dusek and Eichenbaum[45] found that rats readily learned the task and performed at high levels of accuracy in sessions where all stimulus pairs were presented in random order.

A modification of this task was successful in assessing transitive inference.[44] Rats were first trained to discriminate between odor pairs A+/B–, B+/C–, C+/D–, and D+/E–. In the final test sessions, they were given occasional presentations of odor pair B & D. Because stimulus B was positive when paired with C, and stimulus D was negative when paired with C, choosing B over D in the B & D pairings would provide evidence for transitive inference. Rats readily learned the four initial discriminations and on the probe trials they demonstrated transitive inference by choosing B in the B & D pairings. The results of these two studies suggest that rats have a robust capacity to learn the relational organization among a series of odor items. The digging in scented sand test has also been adapted successfully for use in mice.[101,102]

2.2.7 Other Methods

A wide variety of other methods have been used to determine if a subject can detect an odor or discriminate between odors.[16,23,24,84] These tests range from using species-typical responses (e.g., odor-dependent sex and maternal behaviors, and nipple search in neonates) to conditioned odor avoidance[117] and various types of odor-cued maze acquisition tasks. Yamazaki and colleagues[11,181] have used a Y-maze to demonstrate that mice can learn to discriminate the odors of groups of mice congenic for the major histocompatibility complex (MHC), thus, implicating the MHC in individual recognition in mice. In another instance, Staubli et al.[161] and Thanos and Slotnick[170] examined basic memory storage mechanisms by training rats in a series of odor discriminations in a simple maze. Stevens[162] described several potential confounds that must be controlled when using mazes to study olfactory discrimination. These include preventing the odor of a reinforcement serving as a cue. Rats have been shown to discriminate between the odor of a 45-mg food pellet and clean air in a T-maze[160] and ensuring that the experimenter does not cue the subject in any fashion. There is no standard maze for odor discrimination tasks and designs vary greatly among studies. Despite the use of fans to direct air flow in some studies, precise control of odors is difficult to achieve in a maze. Often, odors from the arms of a maze can commingle at choice points. Such methodological problems may have contributed to the inability of Reid and Morris[125] to replicate prior learning set and reversal learning experiments using an olfactometer.[137,138,144,154,155]

In general, studies that employ idiosyncratic methods that are unlikely to be widely adopted provide only existential demonstrations of olfactory function. One clear exception is the long-term program of research by Hudson and co-workers on olfaction in neonatal rabbits. These investigators have developed well-controlled standard methods for assessing olfaction in neonatal rabbits using nipple search behavior.[2,65,66,67,77]

2.3 OLFACTOMETRY

The most widely accepted alternative to static presentation of odors is the use of a dynamic airflow system in which the experimenter exerts control over the purity, concentration, timing, and presentation of the odor. Mazes and runways that use fans to direct odor flows to choice points are sometimes referred to as olfactometers, but, at best, they only grossly approximate desired levels of odor control achieved by more sophisticated systems. Olfactometers, devices used to generate and deliver a preset concentration of a vapor for a fixed duration, have evolved over the years from the relatively crude devices of Zwaardemaker, Sternberg, and others (reviews in References 31, 40, 107, 124) to sophisticated computer-controlled systems suitable for animal psychophysics. Several types of olfactometers for laboratory studies in rats and mice are now commercially available (e.g., http://www.chemsenses.com).

Pfaffman, Goff, and Bare[120] were perhaps the first to combine aspects of operant conditioning and olfactometric control of odor stimuli. Their design was simple but incorporated features that are now standard components of small-animal olfactometers. In their system, pressurized charcoal-filtered air was divided into clean and odorized air streams, each controlled by needle valves and monitored by in-line flowmeters. Hand-operated stopcocks allowed the experimenter to add air from an odorant-containing bottle to a clean air stream connected to the test chamber. Rats were trained in a glass wind tunnel to respond for intermittent water reinforcement and to inhibit responding upon presentation of the odor stimulus. This system was modified by Braun, Wermuth, and Haberly[13] by the introduction of electrically operated solenoid valves to control the onset and offset of odors, and was used by Braun and Marcus[12] in what was perhaps the first detailed study of odor discrimination in rats using automated olfactometry combined with operant conditioning. The system was simple and effective and the Braun and Marcus report provided a convincing demonstration that olfaction could be studied in rats using reasonably well-controlled methods.

Several designs for automated olfactometers and animal training methods were published in the decade of the Braun and Marcus work and, while each proved useful in examining one or another aspect of olfactory controlled behavior, these appeared to have generated little enthusiasm for highly controlled studies of olfaction in animals. Thus, odor generator systems described by Braun and Marcus,[12] Bennett,[7,8] and Sakellaris[129] were each used in one or two published studies and, apparently, not thereafter. At that time, few laboratories were devoted to behavioral sensory studies of olfaction, and the apparatus, often illustrated by diagrams that provided little guidance for duplicating the system, appeared complex.

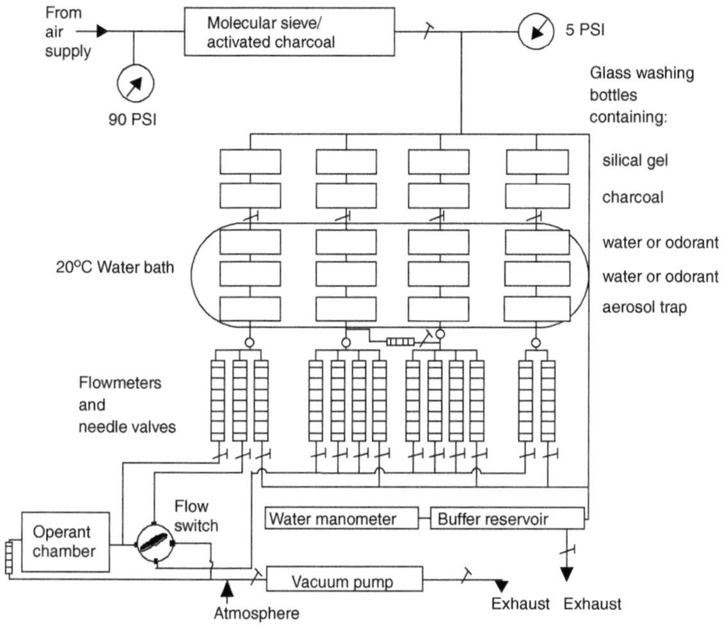

FIGURE 2.2 Air dilution olfactometer used in initial psychophysical studies by Tucker and colleagues.[81,113]

2.3.1 Air Dilution Olfactometers

It was the work of Moulton and Tucker, two investigators dedicated to the development of animal models to study olfactory processes, that set the standards for automated air dilution odor generators suitable for animal psychophysical studies. The odor generators from each investigator's laboratory were somewhat different but both incorporated basic principles of chemical engineering. They paid special attention to air filtration, temperature regulation, efficient odor saturators, turbulent mixture of air streams, regulation and measurement of airflows, and the use of high-grade Teflon body valves to control airflows. Their systems were relatively complex and, except for the most demanding work, were probably somewhat over-engineered. The initial studies of Tucker, Moulton, and their associates provided detailed psychophysical analysis of odor detection thresholds in rats, pigeons, and dogs.[81,95,107,108,121,122] This data and the systems used still set the standard for quantitative olfactometry in animals.

Typical of these odor generator systems is the dynamic air dilution system developed by Tucker and used in electrophysiological and behavioral studies.[60,61,81,113,121,172] As shown in Figure 2.2, the system was complex and used multiple components and numerous lines and flow controllers. Pressure-regulated compressed air was dried, filtered for oil and particulates, and cleaned using a molecular sieve and activated charcoal before being divided into two odor lines, a clean air wash, and clean air dilution streams. Odor saturation was achieved by bubbling air through a liquid odorant contained in two sequential gas-washing bottles. Aerosols produced

of glass with tight-fitting Teflon connectors. A hand-operat
operator to switch between an air wash line and the odor lir
was diluted within the flow switch and by a continuous flow
line connecting the valve to the test chamber. The volume of od
and measured by a series of flowmeters and needle valves con
pressures upstream and downstream of the saturators and press
lines from the animal test chamber were carefully regulate
changes in pressure or volume of air could not be used as a
menter switched from a clean airline to an odor line.

The performance accuracy of the system was assessed us
phy. Chromatogram peak areas of air samples taken from the
were nearly perfectly related to dilutions calculated from air
determined by flowmeter settings.[121]

A different and somewhat simpler odor generator design
and associates.[108] It employed multiple air dilution stages and
valves for controlling stimulus samples. The design may ha
the control in the Tucker system but, as determined by peak
ionization detector, it generated reasonably accurate odor dilu

The Moulton system was further modified by Slotnick ar
their system is described in detail because it is not only usef
studies but was sufficiently simple to be easily reproduced b
The system, illustrated diagrammatically in Figure 2.3, met t
Dravnieks[40] for generating a saturated vapor, producing air
airflows, and delivering the stimulus to the subject. It used wide
although some components had to be made by a glassblower. A
odor generator except for the compressor and refrigerant drye
1.3-m wide board attached to a wall.

Air was supplied from an 80-gal tank, which was pressuriz
80 pounds per square inch (PSI) by an oil-free Thomas compre
the tank was dehydrated using a small refrigerant dryer and deod
of three canister-type charcoal filters (Cole-Parmer). Line press
PSI by a stainless steel pressure regulator (Porter) and the result
to each of the clean airlines (Figure 2.3). Glass tubing (8 and
(OD)) was used for all connections and glass tubes were joined b
(for 8-mm glass) or AWG#7 (for 4-mm glass) standard wall
Palmer). High-capacity Porter or Key Instrument flowmeters (40
to monitor clean air flows and lower capacity flowmeters (20
odorized airflows. Glass and Teflon needle valves (Kontes) serv
to flowmeters. Each flowmeter was calibrated in place using a s

The odor saturator was a 25-mm diameter, 300-mm long
half filled with liquid odorant and immersed in a water bath
refrigerated water bath was used. This proved unnecessary
cooling of a sufficiently large water bath served to maintain the

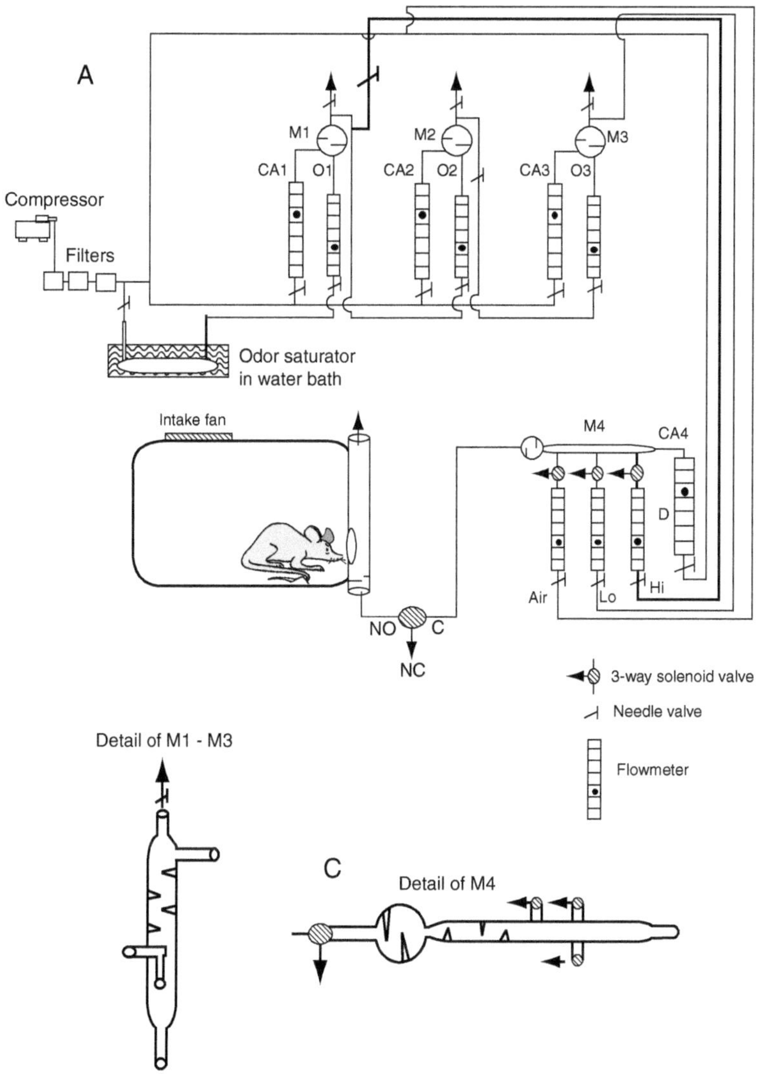

FIGURE 2.3 A relatively simple air dilution olfactometer used by Slotnick and colleagues.[4,5,9,148,150,151] Arrows indicate exhaust paths. The odor saturator is maintained at ~1°C below ambient temperature by evaporative cooling. Odor-saturated air in line O1 is diluted with clean air (line CA1) in manifold M1. The relatively high odor concentrations in M1 are controlled and monitored via the needle valve and flowmeter labeled Hi. This line is closed when lower concentrations are used. Additional dilutions made in manifolds M2 and M3 are controlled and monitored via the needle valve and flowmeter labeled Lo. A final dilution of stimulus flows (Air, Lo, and Hi) and clean air (CA4) occurs in manifold 4 (M4). Note that stimuli are being continuously generated and exhausted via the common and normally open ports of three-way solenoid valves connecting output flowmeters (Air, Lo, and Hi) to M4. Operation of a solenoid valve manifolds the stimulus flow with clean air in M4. As described in the text, brief operation of the final valve (FV) at the beginning of a trial can be used to signal the onset of a stimulus, reduce the ascending concentration phase produced by mixture of the stimulus and clean air flows in M4, and mask non-odor cues produced by operation of stimulus control valves.

below the ambient temperature of the room and this was effective in avoiding condensation of the odor in downstream components. As determined by peak height output of a flame ionization detector, air flows of 10 to 100 ml/min over the surface of the odorant were >95% saturated at the outport of the saturator tube.

Dilution of this saturated vapor was achieved by four sequential dilutions with clean air. As shown in Figure 2.3, the first dilution occurred by mixing clean air from channel CA1 with odor-saturated air in channel O1 in manifold M1. The clean air flowmeters could be reliably set in the range of 500 to 4000 ml/min and the odor flowmeters could be reliably set from 20 to 200 ml/min. Thus, the first dilution (odor flow into manifold M1/total flow in manifold M1) could be varied from approximately 4 to 0.25% of vapor saturation. A fraction of the odorized air in manifold M1 was further diluted by mixing it with clean air in manifold M2 and, in a similar manner, a subsequent dilution could be made in manifold M3. To minimize contamination of the second and third dilution stages when they were not needed, the relatively high concentrations of the odor used in initial training (the output from manifold M1) was fed to a separate output flowmeter (labeled Hi in Figure 2.3) and valve via the darker line in Figure 2.3. Clean air, high concentration and lower concentrations of odorized air were connected to separate final output flowmeters (Figure 2.3, Air, Hi, and Lo). The outports of these flowmeters were connected via the common and normally closed ports of separate three-way Teflon body solenoids (Parker) to side arms of manifold M4. The input of this manifold was connected to a clean air source (Figure 2.3, CA4) and the output was connected to the animal's sampling port via the common and normally open ports of a three-way Teflon body solenoid (Mace). Manifolds M1 to M3 were 100-mm long, 25-mm wide tubes with push-ins to create turbulence and ensure mixture of the air and odor streams. Manifold M4 was a 75-mm long, 12-mm diameter tube with 4-mm side tubes that connected to the solenoids. The final dilution of the stimulus occurred in manifold M4 and flow rates in lines connected to that manifold were kept constant and independent of flows in upstream dilution channels. The four dilution steps of this system could generate any odor concentration from 4% to approximately $4 \times 10^{-6}\%$ of vapor saturation. This range of dilutions is more than adequate for determining the absolute detection threshold of most odorants. The range could be extended by using additional dilution steps, higher capacity flowmeters in the clean air channels, lower capacity meters in the odor channels, or by lowering the temperature of the odor saturator water bath.

The system was washed in place by pressure injection of approximately 2 liters of 95% ethanol and then air-dried. After washing and at periodic intervals, glass-to-Teflon connections were checked for air leaks using a soap solution.

There are a number of advantages and a few disadvantages to this design. On the plus side, the system is relatively easy to construct and maintain. Preliminary gas chromatographic analyses in our laboratory and detailed GC analyses of an earlier version of this system[97] and similar systems[107,108] demonstrated that the concentrations produced did not vary significantly from calculations based on flowmeter settings; i.e., the dynamic air dilution system performed as intended. Further, it produced the high odor concentrations needed for initial training and sufficiently low concentrations for determining detection threshold. Also, a simple modification of the output flowmeters allowed intensity difference thresholds to be determined.[148] Finally, the system provided a highly controlled environment for generating and automatically dispensing odor samples, the purity of which should be limited only by the purity of the source odorant and air used for dilution.

A disadvantage of this and similar air dilution systems arises with the need to change stimulus concentration. This requires adjustments of each needle valve to alter flow rates and to balance backpressures in each manifold. Thus, rapid changes in odor concentration cannot be made within a test session. However, once set, flows are reasonably stable but do require small adjustments from time to time. These small deviations from the initial set points could be avoided using digitally controlled mass flowmeters but the cost of using multiple such controllers is prohibitive. Further, even the use of electronic flow controllers may not immediately produce the desired change in odor concentration. This is because odorant molecules will be absorbed and adsorbed onto the walls of the ductwork and, when flows are set for a lower concentration, that concentration can be achieved only slowly because desorption occurs. Indeed, very low concentrations may not be achieved at all if, after using high-concentration odors, desorption results in a higher concentration than that desired. The problems of adsorption, absorption, and desorption of odorant molecules from ductwork are ones common to all odor generator systems. The speed at which odorant is cleared or reduced in the system, or the speed at which saturation of connecting lines occurs such that the intended concentration is being delivered to the subject, will vary as a function of properties of the odorant molecule, flow rates, odor concentration, length of ductwork, duct material, and temperature. There are various solutions for ensuring desaturation of the system when a lower odor concentration is desired. These range from air-washing the ductwork for an extended period of time to washing ductwork with a hot liquid solvent such as ethanol and air-drying the system at an elevated temperature.[108] The potential confound in odor concentration from desorption may be significant at near-threshold levels and a failure to control for this factor may account, in part, for the variability in absolute detection thresholds reported among studies.[148,150,175]

There do not appear to be any systematic studies of odor saturation or desorption rates in air dilution olfactometers and, most often, the investigator must make this determination for him/herself. In unpublished studies, Slotnick and Schoonover monitored the performance of the system illustrated in Figure 2.3 by tracking the response of a flame ionization detector that was in-line with a fraction of the flow from the downstream port of the final mixing manifold (Figure 2.3). For isoamyl acetate, and at the flows used, saturation of the system for relatively high concentrations (0.5 to 0.001% of vapor saturation) occurred within 8 to 13 min, while approximately 20 min was needed to achieve a steady state in generating much lower concentrations. Desaturation to undetectable levels was achieved within 15 to 20 min after the stopcocks controlling airflow through the odor saturator tube were turned off. However, these times may vary considerably for different odors. In a related study, Slotnick and Schoonover[148] used changes in performance accuracy to estimate odor clearance time in their olfactometer.

Another solution to this problem is an olfactometer that simultaneously generates multiple odor concentrations. Walker and O'Connell[174] describe one such system

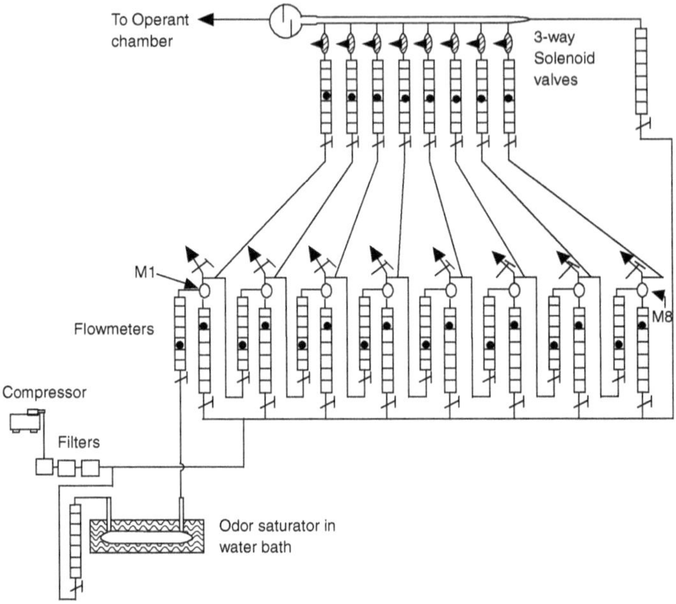

FIGURE 2.4 Design for a multiple concentration air dilution olfactometer similar to that used by Doty and colleagues.[32,35] A fraction of each of eight dilutions produced in manifolds M1 through M8 is controlled by separate needle valves and flowmeters. Operation of one of the three-way solenoid valves introduces that concentration to a manifold whose output is connected to the animal chamber.

but a somewhat simpler and more flexible one would be a modification of that employed by Doty and colleagues.[32,35] The system is similar in principle to that in Figure 2.3 except that it contains more dilution stages and a measured portion of the diluted vapor from each stage provides an input to a separate three-way solenoid (Figure 2.4). Air streams can be adjusted to produce sequential dilutions and the experimenter can easily switch between concentrations and hence use a method of limits or staircase psychophysical procedure to determine thresholds within a session. The system is more complex than that illustrated in Figure 2.4 but it makes available multiple concentrations of an odor, any one of which can be delivered to the subject by operation of the appropriate control valve.

Despite the advantages of a system that simultaneously produces multiple dilutions of a single odor, it is important to appreciate that stable psychophysical functions with animals generally require extended training at each dilution step and that only a finite number of trials can be given before satiation or fatigue occurs and the subject is no longer motivated to perform. Thus, it is probably the case that only after extended preliminary training over a series of concentrations could one expect to produce a reliable psychophysical function within a single session. With these provisions in mind, the multiple concentration odor generator shown in Figure 2.4 should prove useful for assessing the action of short-acting drugs and other manipulations on odor sensitivity.[32-38]

A separate issue in the construction of automated odor generators concerns the selection of electrically operated solenoid valves. A variety of solenoid valves are available but few of these are suitable for olfactometry. Inexpensive valves contain materials that are easily contaminated[125] and cannot be cleaned with strong solvents (e.g., ethanol or acetone). Even Teflon body valves (valves in which air flow contacts only Teflon) can become contaminated, especially when exposed over many cycles of operation to high concentrations of odors. While the Teflon block of such valves can be cleaned with most solvents, over time odors may be adsorbed and, even after a solvent wash, provide an odor cue. Dedicating separate valves to the control of high and low odor concentrations, as in the system illustrated in Figure 2.3, minimizes this problem. Valve contamination can be eliminated entirely by the use of pinch valves. These act by pinching closed a soft tubing material. Operation of the valve retracts the actuator (in normally closed valves), thus unpinching the tubing. When a new odor or a different concentration of an odor is used, this piece of tubing can be easily replaced. Pinch valves might be an ideal solution to control odor flow if an odorless and trouble-free soft tubing were available but, at present, it is unclear that it is. Small pinch valves that are usable in olfactometry require tubing with a durometer rating of about 50 or lower. Latex, silicone, and C-flex tubing are usable and widely available. Latex has an unacceptable odor, as do lower grades of silicone such as that used as aquarium tubing. Platinum-cured silicone and C-flex tubing work well in pinch valves, are essentially odorless, and can be autoclaved. A minor problem with pinch valves is that the interior walls of silicone and C-flex tubing adhere when pinched closed for many hours and, before use, this bond must be broken by gently rubbing the tubing. Also, the proper operation of the solenoid actuator depends on it being slightly depressed by the tubing and, as the compressed tubing loses its resilience, valve operation may fail. Thus, pinch valves require more attention from the experimenter than do traditional flow-through valves but these mechanical concerns are easily resolved by tubing replacement. To the authors' knowledge, pinch valves have not yet been employed in air dilution olfactometers designed for psychophysical tests but they have been used in multiple liquid dilution olfactometers of the type described below.

Another potential solution to the problem of valve contamination is a valveless olfactometer. The unit contains valves but they control only flows to exhaust lines. The critical details of the design are shown in Figure 2.5. As illustrated, the line connecting the output flowmeter to manifold D has a side arm connected to a vacuum pump via a normally open solenoid valve. The vacuum is adjusted to ensure that all air from the flowmeter is led to an exhaust line. Operation of the valve closes this line and forces the air stream into the manifold. Valve contamination is not a concern because all air passing through the valve is exhausted. As far as we know, this system has not been used in any published studies and, while we have no data concerning its performance, it should not be difficult to implement and evaluate.

2.3.2 Dual Air Dilution Odor Generators

Studies of simple odor discrimination, discrimination of odor mixtures, odor masking, odor memory, and related olfactory functions require a system that can generate

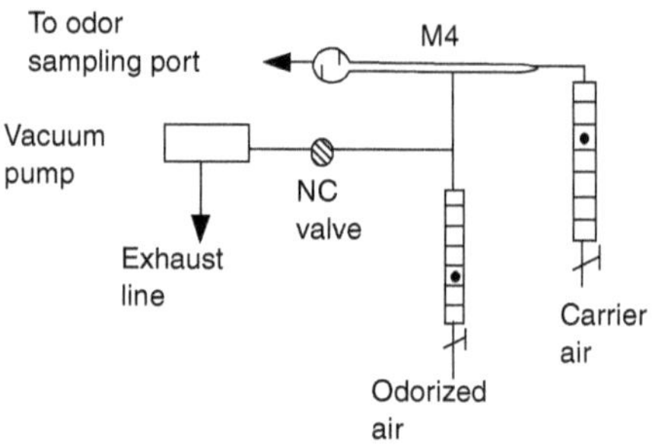

FIGURE 2.5 Design for a valveless olfactometer. By placing the control valve between the stimulus line (Odorized air) and a vacuum exhaust, the valve can control the stimulus without potentially contaminating the stimulus line.

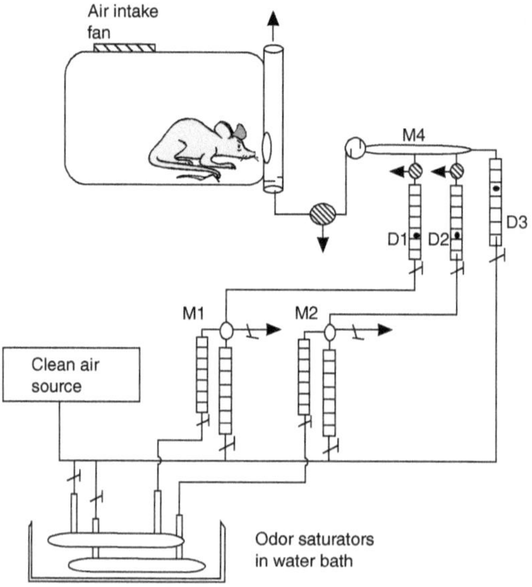

FIGURE 2.6 A simple dual stimulus air dilution system. See Figure 2.3 for details.

two or more odors. As one can readily imagine, incorporating two or more odor generators of the type illustrated in Figure 2.3 in a single system would be unacceptably cumbersome. Fortunately, for most such studies, it is unnecessary to produce concentrations in the threshold range and, thus, fewer dilution stages are required. An example of a simple air dilution dual odor generator is that used by Slotnick and Schoonover.[148,149] As shown in Figure 2.6, the system contains two identical, two-stage dilution systems. Using flows of 20 to 100 ml/min in the odor channels and a maximum flow of 4000 ml/min in the clean air channels, the concentration in each system can be varied from approximately 4 to 0.015% of vapor saturation. For most odors, the lower concentrations would be well above detection threshold but still safely below threshold for trigeminal stimulation. Also, odor concentrations could be varied over a sufficiently wide range to assess odor masking,[83] to provide a variety of vapor phase mixtures of the two odors[88,89] and to test whether discriminative responding was based on intensity or qualitative differences between odors. The advantages and disadvantages of the system are similar to those described for the single odor psychophysical system (Figure 2.3). Of course, when new odors are to be used, the entire system must be cleaned with an appropriate solvent and air-dried. Unless pinch valves or exhaust-only valves are used, it may be necessary to have valves dedicated to each odor employed.

2.3.3 Flow Rates and Control Procedures

Although flowmeters can be used to monitor flows over a wide range, there are constraints on what flows can be used in generating odors and in sequential air dilutions. For over-the-surface saturators of the type illustrated in Figure 2.3, vaporization from a pure liquid odorant follows Ficks diffusion law.[40] The degree of saturation at the exit port of the saturator will be a function of the vapor pressure of the odorant, the rate of flow over the odorant surface and the surface area exposed to that flow. With sufficiently slow flows and sufficiently long path length, saturation can be achieved for most odorant materials. As discussed by Dravnieks,[40] for most odorants, air flowing at rates up to 100 ml/min over a 15-cm long and 1.25-cm wide surface of an odorant will be 95 to 98% saturated at the exit port.

Flow rates at subsequent air dilution stages are governed only by flowmeter capacity. However, it is difficult to produce accurate flows at rates under approximately 10 ml/min and, while very high flow rates can be used to dilute odor flows, the precision in generating such flows decreases as the flow rate increases. This means that, in practice, it is difficult to accurately achieve dilutions of more than two orders of magnitude at any one dilution stage.

The flow rate of the odor sample delivered to the subject should be relatively high to ensure that the sample reaches the subject quickly after operation of valves controlling the odor. Using an early version of the system illustrated in Figure 2.3, Slotnick and Nigrosh[145] found optimal odor detection performance at delivery rates of approximately 3 l/min. In subsequent work and with olfactometers with shorter path lengths between the final dilution stage and the subject, we find that flow rates of 2 l/min are adequate for rats and mice. These determinations were made by examining the relationship between stimulus sampling time (as described below) and flow rate.

A number of control procedures have been employed to ensure that discriminative performance of rats trained in olfactometers is entirely dependent on vapor cues. Rats trained on odor detection and visual detection tasks and then olfactory bulbectomized continue to perform well on the visual task but show no retention or ability to relearn the odor detection task.[149] Other control tests, including attempts to train animals when the same stimulus is generated in each of the olfactometer channels or using odor-free channels after washing the system should be routinely employed to assess whether non-vapor cues are available as discriminative stimuli.

Perhaps the most convincing evidence that discriminative performance is based on odor cues and that the olfactometer system is performing as expected comes from psychophysical studies demonstrating a decrease in performance accuracy as the stimulus concentration is lowered.[81,96,97,148,150,151]

There are a number of potential non-olfactory cues that may differentially signal S+ and S– trials. These include changes in flow rate or pressure pulses when odor or air is added to the carrier stream, as well as the production of auditory cues by valves or by the sounds generated by air flowing through stimulus channels when these streams are added to the carrier stream. The systems designed by Tucker and by Moulton included special provisions to avoid extraneous cues at the onset of odor presentation. The Tucker olfactometer used a hand-operated, specially constructed Teflon-body valve to introduce the odor stimulus. The Moulton unit used doors to separate the subject from the odor sampling port between trials and when stimulus control valves were operated. A quite different approach to the problem was used in the Slotnick and Schoonover air dilution system[148] and in the eight-channel liquid dilution system described below. In these systems, the onset of a new trial was made quite explicit by operation of a final valve (Figure 2.3). This valve served to shunt all airflow from the subject odor sampling port into an exhaust path. The valve operated simultaneously with operation of the stimulus control valve and was relaxed 0.8 to 1.2 s later. The stimulus control valve was de-energized 1 to 5 s (in different studies) after termination of the final valve. Two functions are served by diverting the air stream into an exhaust path at the onset of a trial.[139,145] First, it acts as a warning signal to the animal that a stimulus is about to be delivered, and second, it allows the stimulus to mix with the carrier stream and thus reduces or eliminates the ascending concentration phase of the odor before it is presented to the animal. Allowing a brief period for the odor and carrier streams to mix would be particularly important in studies of intensity difference thresholds and when vapor phase mixtures of two or more odors are used.[83,151] Also, the noise of the final valve helps mask potential differences in the sound of stimulus valves. In the authors' experience, the get-ready signal and abrupt onset of the stimulus produced by brief operation of the final valve at the beginning of a trial clearly enhances odor detection and discrimination performance.

2.3.4 Liquid Dilution Systems for Controlling Multiple Odors

If an experimenter needs to present many different odors within a test session, then a liquid dilution based olfactometer may be required. The advantage of liquid dilutions is that the desired concentration can be produced in a single container, thus simplifying the odor generator by eliminating the need for multiple flow controllers, flowmeters, associated connections, and tubing paths. A simple liquid dilution system for controlling eight different odors is shown in Figure 2.7. Generating odors in this system differs from air dilution systems in a number of ways. Odors are not continuously generated; instead, samples of the headspace above the odorant material are taken as needed. As shown in Figure 2.7, airflow to each odor saturator bottle is controlled by separate upstream and downstream normally closed valves. Operating these valves manifolds a fraction of the saturator headspace with a clean air stream connected to the test chamber. A virtually unlimited number of odors can be

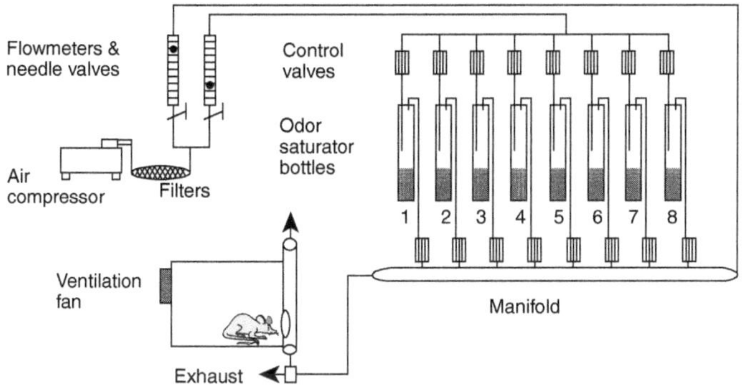

FIGURE 2.7 An eight-channel liquid dilution olfactometer used in multiple odor discrimination tasks. Odorants previously diluted in an appropriate solvent are stored in odor saturator bottles 1 to 8. The upstream and downstream lines of each saturator are controlled by normally closed pinch valves. Operation of these valves introduces a sample of the saturator headspace into the manifold. Note that the clean air and the stimulus lines are each controlled and monitored by a single needle valve and flowmeter.

thus controlled although, most commonly, an eight-odor system has been employed. This type of system has proven quite useful in a variety of studies of simple odor discrimination, discrimination of odor mixtures, for assessing intensity difference thresholds, odor memory, olfactory matching to sample and other tasks that require presentation of multiple odors.[9,89,90,134,139,147,152-155,180]

However, the simplicity and convenience of a liquid dilution system involves significant compromises in control of odor concentration and some constraints on which odors can be used. This is because there is no universal odorless solvent and most odor-solvent combinations fall far short of forming ideal mixtures. Without a physical analysis, it is difficult at best to specify the odor concentration in the headspace of a mixture. The problem is further complicated because generation of an odor from the headspace requires the addition of odor-free air, thus diluting the headspace concentration each time the odor sample is produced. The time required for the headspace to re-saturate will be a function of the vapor pressure of the odorant, the partial pressure of the odorant in the mixture, and the amount of headspace replaced by clean air. If the time required for re-saturation is greater than the intervals between generating odor samples, then each sequential odor sample will be reduced in concentration. Re-saturation times can be minimized in a number of ways: by bubbling the input air through the liquid odor; by ensuring that the headspace volume is largely relative to the volume replaced by clean air when an odor sample is generated, or by using long, horizontally oriented saturator tubes to increase the surface area of the odorant. However, air sparging may produce aerosols and require aerosol filters on the output line of each saturator tube. Maximizing the headspace and/or surface area of the odorant is probably a better solution. In any case, if knowledge of the precise odor concentration delivered to the subject is needed, a gas chromatographic analysis would be required to evaluate the performance of the system. In practice, we have used 20-ml volumes of odorant material held in 200-ml bottles as odor saturators and replaced the odorant material after each 200-trial session. Odor samples were generated by passing 50 ml/min of clean air into the saturator for 3 to 4 s on each trial, thus displacing approximately 3% of the headspace. As judged by human observers, there was no apparent decrease in odor concentration for a wide variety of odors diluted with either water or odorless mineral oil over a series of trials spaced at 30-s intervals.

2.3.5 Test Chamber

Operant conditioning studies using air dilution or liquid dilution olfactometers for generating odor samples have used either a wind-tunnel type test chamber or a more traditional chamber containing a tube or small flow chamber for presenting odors. A wind tunnel has the advantage of ensuring that the animal makes contact with the odor stimulus: indeed, on each trial it is literally surrounded by the stimulus. However, the wind tunnel itself must be specially fabricated and washed and dried after each test session. A discrete tube with an opening for sampling the air stream is more easily made and can be attached to virtually any type of chamber (e.g., Figure 2.3). The time spent sampling the stimulus can be determined on each trial, by monitoring snout insertions into the port by a photobeam or other detector. Training animals to insert their snout into an odor tube, to keep it there until the stimulus is presented, and then to respond on a manipulandum takes somewhat more time than training them in a wind tunnel but, still, training can be completed in one to three 30-min sessions. In our initial studies, a wind-tunnel chamber was used but the desire to measure stimulus-sampling time and to use a more easily constructed chamber led us to change to the discrete tube configuration. For rats, we use a 25-mm diameter, 100-mm long glass tube that contains a 20-mm diameter hole for snout insertions. The air inlet end tapers to 4 mm and the outlet end is connected by a flexible exhaust line coupled to a 20-CFM fan. A second 20-CFM intake fan is used to maintain positive pressure in the chamber, thus ensuring that odors passing through the sampling tube do not leak into the chamber. A 14-gauge stainless steel tube located on the same wall as the stimulus delivery tube is used to deliver liquid reinforcement via a solenoid-controlled, gravity-fed reservoir. Licks on this tube, detected by a touch-sensitive circuit, serve as the operant response. The location of the response tube is not critical; rats quickly learn to inhibit responding even when the tube is located within the odor sampling port.[86,99] Other configurations are possible, including the use of multiple odor sampling tubes, each associated with a separate manipulandum or a single sampling tube with separate response place or manipulanda for each odor presented.[183]

2.3.6 Training Methods

Two quite different methods of training animals in olfactometers have been used and both are successful in producing good stimulus control of behavior. Tucker and other members of the Tallahassee (Florida) group have long used a conditioned suppression method in a variety of animal psychophysical studies.[61,81,113,158] In odor detection tasks, this involves making the presentation of an odor predictive of a brief electrical shock. Animals, trained on a bar-press task, are reinforced with food or water on a variable-interval schedule. The odor-shock pairings result in a suppression of responding when the odor is detected. For psychophysical studies, odor serves as the negative stimulus (the stimulus predictive of shock) and clean air as the positive stimulus. The procedure, although quite effective, has several shortcomings: shock intensity must be adjusted for each individual animal; and because approximately 20-s time periods are needed to determine response rate before, during, and after odor presentation, relatively few trials can be given in each training session. Problems in maintaining stimulus control can also occur with stimuli that are at or below detection threshold because animals can receive shocks after stimulus presentations they may not be able to detect. Despite these various problems, the method proved successful for assessing odor detection and discrimination.

Most contemporary studies have used the more efficient go, no-go discrete trials discrimination training method. After a fixed or variable inter-trial interval, attending behavior (e.g., insertion of snout into the odor sampling port) produces an S+ or S– odor. A single response or a fixed number of responses made after presentation of the S+ stimulus is rewarded, but meeting the response criterion after presentation of the S– stimulus is not rewarded and may be punished by extending the length of the next intertrial interval. Reinforcement is asymmetrical; only correct responses to the S+ stimulus are rewarded. Symmetrical reinforcement can be achieved using a response choice method: two or more spatially separated manipulanda are used and, after presentation of a stimulus, responding on the one designated as correct is reinforced. However, outcomes using such choice procedures are distinctly inferior to the go, no-go method.[145] Rats and mice require extensive training before they reliably perform well using a two-stimulus, two-response choice method. This is because strong position preferences for responding occur, are idiosyncratic, and must be overcome by correction procedures tailored to the requirements of individual animals. Even after stable and accurate discrimination behavior is obtained, position habits may again predominate when new odors are used or when discrimination difficulty is increased. Such problems would be exponentially multiplied as the number of stimuli and responses are increased. In contrast, rats and mice very quickly learn to inhibit responding to S– stimuli in a go, no-go procedure, and initial training on the operant task can generally be completed in two or three 30-min sessions. Correction procedures are not required and when new stimuli are used, subjects soon learn not to respond to the presentation of the non-reinforced stimulus. Training failures or loss of stimulus control after training occur in a significant number of animals using the two-manipulandum choice method but are seldom encountered with the go, no-go method. Not being reinforced for responding appears to be a sufficiently strong punishment and, indeed, rats and mice appear to be biologically "prepared" not to respond to stimuli that are not reinforced.

With well-motivated animals and appropriate sized reinforcements (1 to 5 µl for mice and 20 to 50 µl for rats), 200 — and as many as 600 — trials can be given in a daily session. S+ and S– trials are presented in quasi-random order to ensure that there are an equal number of each type within blocks of 20 to 40 trials and that one type of trial does not occur more than three times in succession.

2.4 STIMULUS CONTROL OF BEHAVIOR ACHIEVED USING OLFACTOMETRY

2.4.1 ODOR SAMPLING BEHAVIOR

Ensuring attention to the relevant stimulus and measuring attentive behavior are major concerns in all animal discrimination studies. For olfactory studies, these problems are considerably simplified when controlled stimulus presentations are used in combination with a discrete odor sampling port.

Youngentob and co-workers[184] provided a detailed analysis of sniff patterns exhibited by rats trained to detect different concentrations of an odor. Odors were presented in a small chamber that was effectively closed when a rat inserted its snout into the chamber port. Sniffing, recorded using a pneumotachograph, could be characterized by 52 different measures, including inspiration and expiration volume, duration, average flow rate, frequency, etc. Among the many interesting findings in this work was that the volume and average flow rate of individual inspiratory sniffs increased with decreasing stimulus concentration and that the maximal inspiratory and expiratory sniffs tended to occur in the middle or at the end of a bout of sniffing. In general, Youngentob's results indicate that rats can and do adjust their sniffing strategies under different stimulus conditions. Thus, the perceptual features of an odor can be determined not only by its physiochemical properties but also by the mode of stimulus sampling used by the subject.

Slotnick[139] and Slotnick et al.[154] provided a simple index of odor sampling by measuring how long rats kept their snouts in an odor stream during a brief (1 or 2 s) presentation of the stimulus on each S+ and S– trial of a session. Rats were required to sample the stimulus for at least 0.1 s and failure to do so aborted the trial. At the beginning of a session, when discrimination accuracy is at chance levels, there are no differences in sampling times between S+ and S– trials and rats generally sample the stimulus for 0.4 to 0.8 s (Figure 2.8). As performance accuracy increases, stimulus sampling time on S+ trials does not change or decreases slightly, but sampling time on S– trials increases. Indeed, in many cases, an increase in S– sample times precedes an increase in response accuracy (Figure 2.8B) and this suggests that sampling time may provide a particularly sensitive measure of detection or discrimination. This difference in S+ and S– sampling time is characteristic of both odor detection and odor discrimination tasks and occurs even when an S– odor is clearly more intense than the S+ odor. In psychophysical tests, the S– (air) stimulus was sampled longer than the S+ (odor) stimulus, except at near-threshold concentrations of the S+ stimulus.[139] As the stimulus approached threshold values (e.g., when the odor stimulus was relatively difficult to detect), rats increased their sampling time on S+ trials and, at near-threshold values of the stimulus, S+ sampling time exceeded S– sampling time. However, even at near-threshold values of the stimulus, sampling time rarely exceeded 0.8 s. Short or minimal S+ sampling time is probably governed by reinforcement contingencies; once the rat detects or, in the case of a two-odor discrimination, recognizes the S+ stimulus, it removes its snout from the sampling port and begins to respond on the water delivery tube. However, as

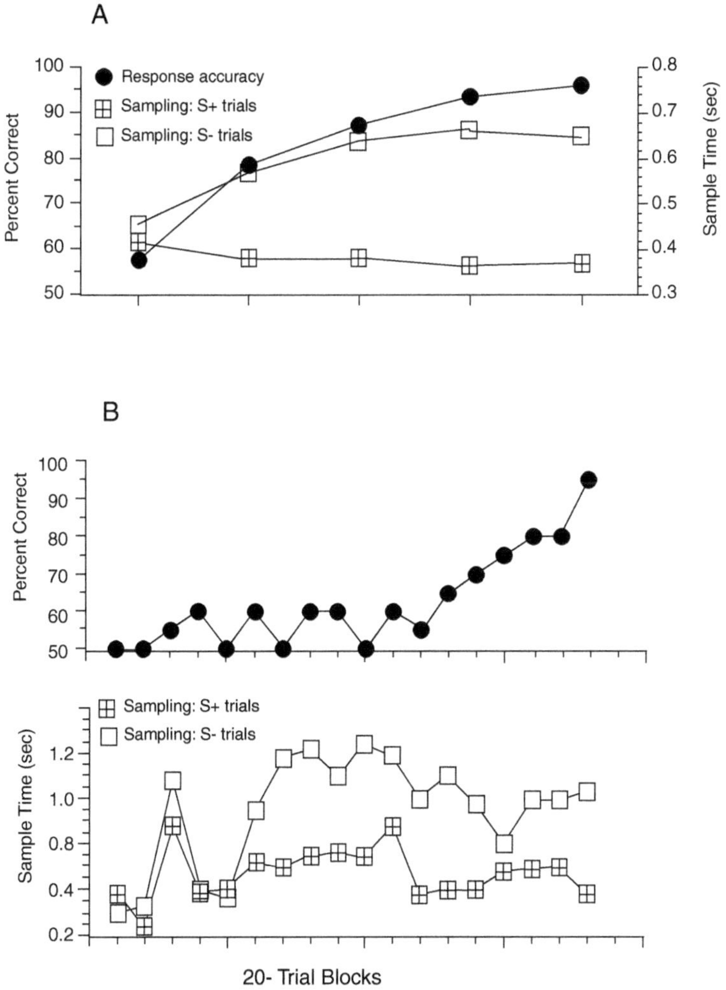

FIGURE 2.8 Mean response accuracy and stimulus sampling time on blocks of 20 S+ and S– trials. A: acquisition of a simple odor detection task. Note that an increase in S– sample time is closely related to response accuracy. B: acquisition of a relatively difficult odor mixture task. The S+ stimulus was 0.1% amyl acetate and the S– stimulus was a vapor-phase mixture of 0.1% amyl acetate and 0.05% propyl acetate. Note that an increase in S– sample time precedes an increase in response accuracy. See discussion in text.

previously noted,[139] this explanation does not account for why, at near-threshold values of a detection task, S+ sampling time would exceed S– sampling time. That is, the rat should require no more time to determine that the stimulus is not present than it does to determine that it is present. One explanation for this difference in sampling time at near-threshold values of a stimulus is that, even in odor detection tasks, sampling behavior is governed by odor recognition: the rat samples not just to detect that a stimulus is present, but also to recognize that stimulus and, at very low concentrations of the stimulus, recognition requires a longer sample than does detection. In any case, in addition to response accuracy, the difference in S+ and S– sampling times provides a useful and sensitive measure of the extent to which the animal's behavior is under the control of the stimulus.

2.4.2 Psychophysics

Absolute detection and intensity difference thresholds have been determined for a number of odorants in dogs, rabbits, rats, and mice using the Tucker or Moulton type olfactometer.[4,5,9,15,56,81,95-97,105,108,116,122,146,148,152,174,185] In general, these studies have produced orderly psychophysical functions but the absolute values reported vary considerably. Comparisons among species or even between studies using the same species is complicated by differences in odors used, methods used for training, how odors were generated, and by criteria for determining threshold values. Walker and Jennings[175] provide a thoughtful review of these issues. For amyl acetate, the odor used most often in psychophysical studies, the detection threshold is reported to be relatively high in the macrosmatic human and pigeon but 1 to 3 orders of magnitude lower in the rat and mouse (Slotnick and Schoonover, unpublished studies).

Absolute detection threshold and other measures of odor sensitivity have been used to assess odor masking and potential deficits after experimental manipulations of the olfactory system. The later studies have provided quantitative measures of residual function after transection of the lateral olfactory tract and other lesions of the olfactory system,[8,87-89,113,143,147-151,153] partial destruction of the epithelium with olfactotoxins,[28,55,135,180,185] and zinc sulfate,[142,152] and after treatment with drugs and other agents.[5,15,32-36,39,58,122,129] In general, these studies have demonstrated relatively subtle effects or savings in sensory capacity that, undoubtedly, would not have been seen using the much less sensitive static presentation methods described above. For example, based on the results of food-finding tasks, it has been commonly assumed that intranasal application of zinc sulfate in rats produces a more or less short-lived anosmia. However, tests using olfactometric methods revealed considerable olfactory function in zinc sulfate-treated rats.[142,152]

2.4.3 Salience of Olfactory Stimuli

Automated odor generators combined with operant conditioning provided the tools not only for detailed psychophysical analyses of olfactory function, but also for examining the efficacy of odors in controlling learned behavior and the ability of rats to discriminate among stimuli. Initial studies showing that rats quickly acquire both simple and complex olfactory tasks and have an exceptionally good memory for odors generated a more widespread interest in olfactory learning. The efficiency of the go, no-go training method as well as the salience of odor cues was demonstrated in several early reports of olfactory learning using an automated olfactometer.[111,145] After preliminary training on the S+ stimulus, rats made few responses toward the S– stimulus when discrimination training was initiated (Figure 2.9). This

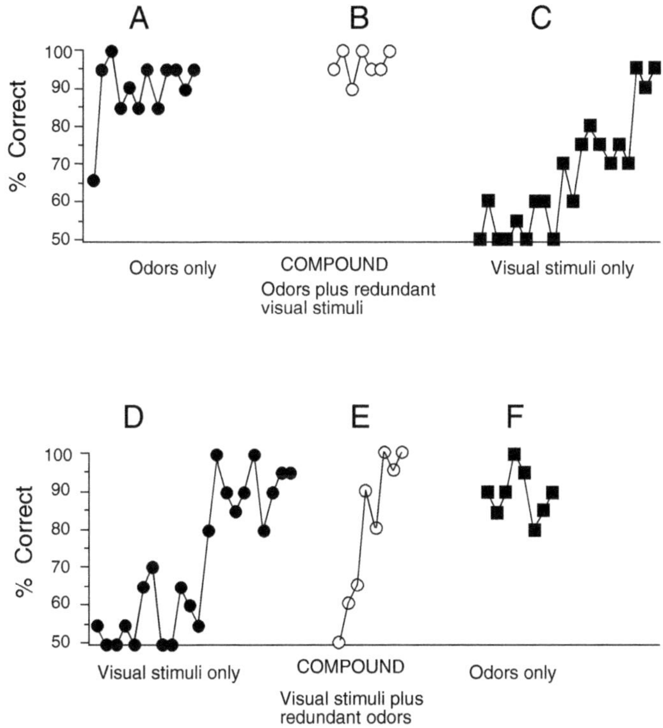

FIGURE 2.9 Stimulus control by odor and visual stimuli. A, B, and C show the complete discrimination training for one rat. All trials were given in a single continuous session. A shows rapid acquisition of a two-odor discrimination task (S+, amyl acetate; S–, ethyl acetate). In B, a redundant visual cue (S+, flashing light; S–, steady light) was provided for the next 160 trials. In C, the odor stimulus was turned off. D, E, and F show the results for a different rat first trained on the visual discrimination (D), given redundant odor cues (E), and tested after the visual cues were turned off (F). Curves C and F reflect learning of the redundant cue in conditions B and E, respectively. (Redrawn from Nigrosh, B., Slotnick, B., and Nevin, J., *J. Comparative Physiological Psychol.,* 80, 285, 1975.)

virtually errorless learning with abrupt introduction of the S– stimulus is seldom obtained in discrimination studies unless special procedures for fading in the S– stimulus are used.[169] Clear evidence that rats preferentially attended to odor stimuli was provided by tests of overshadowing and cue competition. In each test, stimulus control by odor cues was distinctly superior to control exerted by visual or auditory cues (Figure 2.9). Olfactory cues were also superior in more complex forms of learning. Thus, when the significance of the S+ and S– stimuli were reversed in each daily training session, rats trained with odors quickly learned to respond correctly in each session, while those trained with visual, auditory, or taste cues made many errors on each reversal.[42,111,138,141] Odor cues also proved at least as effective as taste cues in learning an illness-induced, classically conditioned aversion.[156]

2.4.4 Complex Olfactory Discrimination Tasks

The exceptional ability of rats to acquire odor discrimination tasks was soon confirmed in other laboratories using variants of the odor generator and training methods described above. These methods, combined with selective lesions of the olfactory bulb and olfactory pathways, proved useful in examining the functional neurology of the olfactory system and the role of olfactory-related structures such as the amygdala, entorhinal cortex, and hippocampus in various forms of odor memory. Because the olfactory system has well-defined projections to prefrontal cortex and limbic structures and because rats quickly acquired a variety of complex odor-based tasks, several investigators have suggested that the rat olfactory system may provide a useful model for studies in cognitive neuroscience.[44,45,91,114,137,186,187] In support of this contention were studies demonstrating that, when provided with olfactory cues, rats were able to acquire learning-set and matching to sample problems, tasks on which rats had proven distinctly inferior when tested with stimuli from other modalities. The performance on learning-set tasks was particularly impressive. In learning-set studies, animals are typically presented with a novel two-stimulus discrimination task in each session. Performance generally improves over the problem series and, with additional training, inter-problem transfer becomes nearly perfect such that after the first "information" trial on a new problem (in which a response to one of the novel stimuli is reinforced or not reinforced), no or few errors are made in subsequent trials. In general, it is assumed that this level of performance can be achieved only if the subject has adopted a strategy or abstracted the rules governing the solution to the task (e.g., win-stay, lose-shift). Many primates trained over a series of several hundred visual discrimination problems achieve this level of performance, but species of "lower" phyletic rank generally perform much more poorly.[176] However, the rate of inter-problem transfer shown by rats trained on a series of two-odor discrimination rivaled that of primates trained with visual stimuli.[144,154] This rather remarkable level of inter-problem transfer occurs even when rats are presented with eight novel stimuli in each successive discrimination task (Reference 155; Figure 2.10) and also extends to successive reversal learning tasks.[111,138,143]

Additional evidence for the superiority of odor learning comes from the ability of rats to learn a matching to sample or non-matching to sample task. The task involves the rat sampling one stimulus (the sample stimulus) and then reporting by responding or not responding whether the second stimulus presented (the comparison stimulus) was identical to the first. In general, rats perform rather poorly on matching or non-matching to sample tasks when visual or auditory stimuli are used.[43,123,127,130] However, they rather quickly acquire such tasks when odors are used as discriminative stimuli and achieve high levels of performance. By imposing a delay between the sample and comparison stimulus, the task provides a convenient and repeatable within-session measure of short-term olfactory memory. For example, Otto and Eichenbaum[114] used a successive presentation method in which responding to an odor was reinforced only if that odor was different from the last odor presented. Interestingly, performance on this delayed non-matching to sample task obeyed several of the rules governing verbal learning: performance improved when inter-item

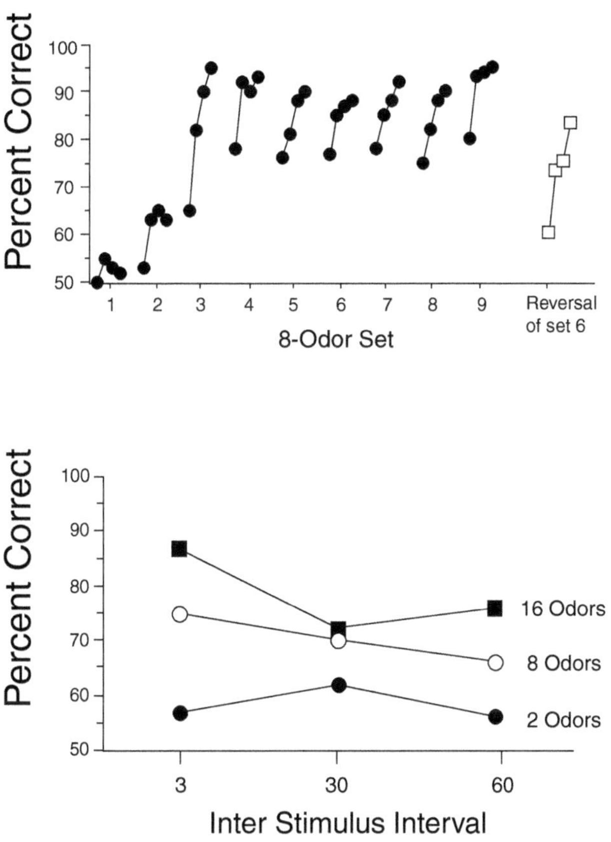

FIGURE 2.10 Top graph shows acquisition of nine sequential eight-odor tasks. In each problem, all eight odors were novel; four severed as S+ stimuli and four as S– stimuli. To test odor memory, after training on Set 9, rats were tested on a reversal of odors used in Set 6 (odors used previously as S+ now served as S– stimuli and those used previously as S– now served as S+ stimuli). Memory for odors in Set 6 is reflected by the initial low accuracy scores in this task. (Redrawn from Slotnick, B., Kufera, A., and Silberberg, A.M., *Physiol. Behav.,* 50, 555, 1991.) Bottom graph shows accuracy scores in a continuous non-matching to sample task as a function of the number of odors used and the inter stimulus interval. (Redrawn from Otto, T. and Eichenbaum, H., *Behav. Neurosci.,* 106, 762, 1992.)

interference was decreased by the use of many stimulus elements and was decreased as the delay between the presentation of individual items was increased (Figure 2.10). This task has been used to assess the role of brain structures in odor learning and odor memory[114] and to evaluate cognitive decline in normal aging.[187]

Odor matching to sample tasks is also readily learned. Thus, Lu et al.[90] using a discrete trials method, allowed rats to sample one of three odors. The sample odor was followed a few seconds later by a second odor and, if that odor was identical to the first, then responding on a key was rewarded. Rats not only learned this task in just a few training sessions but, when tested over a series of such problems, each using a novel set of odors, they showed positive inter-problem transfer, eventually achieving near-errorless acquisition of a new task; e.g., they could learn to match to sample. Performance accuracy in these tests was unaffected by a 10-s delay between the sample and comparison stimulus or by presenting a strong irrelevant odor during the delay period.

The rat's long-term memory for odors is also impressive. In one study, rats were trained to discriminate among nine separate, eight-odor problems.[155] In each session, all eight odors of a set were presented in random order. Performance accuracy improved in each sequential set and rats made few errors in learning the last few sets. To test their memory, the odors used in set 6 were presented again but with the significance of each odor reversed; those that had served as positive odors were now negative and those that were negative were now positive. If rats had little or no memory for these odors, then one would expect the set to be acquired with few errors. However, rats made many errors, especially in initial trials of the session, thus demonstrating that, despite ample opportunity for anterograde and retrograde interference, they remembered these odors and their prior positive and negative assignments. Other studies demonstrated that both rats and mice retain essentially perfect memory for a set of eight odors learned many (up to 30) days previously (Reference 9; Bodyak and Slotnick, unpublished results). The excellent performance in these studies suggests that such tasks were not difficult for the rat and that their capacity to learn and to remember different odors may be quite large.

2.4.5 Odor Quality Perception

A fundamental issue in olfactory research concerns the factors that determine the perceptual quality of an odor.[85,92,104,126] Odor quality can be assessed using verbal reports with human subjects but it is less clear whether animals can be trained to report on their perceptual experience. In an interesting departure from traditional operant conditioning procedures used in the studies reviewed above, Youngentob et al.[182,183] devised a discrimination task to assess the perceptual similarities among odors. Rats were trained to sample an odor in a central arena from which extended five short runways that were highly discriminable from one another. On each trial, one of five odors was presented and the odor served as a cue for which runway should be traversed to obtain reinforcement. Of particular interest were the patterns of errors made because these reflected the extent to which the rats confused one odor with another. The outcomes can be expressed as a confusion matrix, a form of analysis that has proven useful in studying human olfaction[82,179] and taste.[64] After training and when odor concentration was relatively high, rats made few errors. More errors were made at reduced odor concentration and confusion matrix analysis revealed that among the five odors (amyl acetate, propyl acetate, acetic acid, phenol alcohol, and anethole), the most likely confusions were between amyl acetate and propyl acetate. The task has been expanded to ten odors and ten separate type of responses[182] and, potentially, could yield data revealing the perceptual similarity of odors along several dimensions. As Youngentob et al.[182,183] suggest, this confusion matrix task could prove useful in evaluating changes in odor perception that might result from experimentally induced alterations in the olfactory system. It should also be noted that the performance of rats in this task represents a form of cross-modality matching, the demonstration of which is relatively rare in the rodent learning literature.

A somewhat different approach to assess odor perception is that of Lu and Slotnick.[88] The question addressed was whether removal of a part of the olfactory bulb identified in 2-deoxyglucose (2-DG) studies as potentially important in mediating responses to propionic acid would alter the perceptual quality of the odor. The training procedure was designed to have rats learn the rule "respond to any odor providing it is not propionic acid or does not contain propionic acid." It was assumed that if the rats learned this rule and if olfactory bulb lesions significantly altered the perception of propionic acid, then, in postoperative tests, the rats would treat propionic acid as a novel or at least different odor and, hence, respond to it. To instill the rule, rats were trained using propionic acid as the S– stimulus and a variety of 54 other odors as S+ stimuli. In preoperative training, rats learned to respond to any novel odor while continuing to inhibit responses to propionic acid. In postoperative odor recognition tests, propionic acid and other novel odors were used and, to avoid the confound of reinforcement as an instructional variable, responses to novel stimuli were not reinforced in initial trials of a session. The results demonstrated that both control and experimental rats responded to novel odors and continued to inhibit to propionic acid and odor mixtures that contained propionic acid. The outcome suggested that lesions of the propionic acid focal area identified by 2-DG did not significantly alter the perceptual quality of the odor.

Both the Youngentob et al. and the Lu and Slotnick methods for assessing odor quality perception in rats are potentially useful but both methods require extensive training and could yield results difficult to interpret. Another approach to this problem is to assess odor classification. Odor classification provides an approximation to an odor generalization gradient and may prove useful when, as in the case of olfaction, the physical parameters determining perceptual similarity are not well-defined. An example of this approach is the Slotnick and Boren study[140] in which rats were first trained to discriminate between exemplars of odors from the bedding of male and female rats. Then, in non-reinforced test trials, they were presented with odors from the bedding of novel male and female rats. Rats responded appropriately with regard to the gender source of the test stimuli, indicating that there was an odor common to sex type and that rats may have a concept of male and female based on odors. Obviously, this general approach can be expanded to examine other potential odor classifications schemes. Thus, for example, if there is a perceptual commonality among fatty acid odors that distinguishes them from a commonality among, say, aldehydes,[106] rats could be trained to discriminate between exemplars of each odor class and then tested with novel odors from each group. Recently, a modification of this approach was reported by Linster and Hasselmo,[85] who attempted to assess whether odor similarity among aliphatic aldehydes could be predicted on the basis of carbon chain length. Rats were trained to respond to one aldehyde and then tested with other aldehydes. The results were partially successful; rats responded to aldehydes that differed by one carbon atom (e.g., rats trained on *n*-butylaldehyde responded more strongly to *n*-vereraldehyde than to *n*-heptylaldehyde), but responses were not graded as a function of carbon length among other members of the homologous series. It is unclear whether the failure to demonstrate a generalization gradient in this study was because odor similarity was, in fact, not determined by carbon length chain among these homologous odors or because Linster and Hasselmo used a static and unconventional method for presenting odors that may not have provided adequate control of odor sampling and odor concentration. However, their hypothesis and other hypotheses concerning a physical dimension that might determine odor quality are clearly amenable to olfactometric methods using odor classification or similar test procedures.

2.5 CONCLUSION

This brief review reveals that a wide variety of methods have been used to assess olfaction in rodents. Many simple tests using static presentation of stimuli are effective in demonstrating odor detection or discrimination and may provide optimal methods for assessing some olfactory behaviors (e.g., measures of habituation and some forms of associative learning). However, such tests do not provide sufficient control of the stimulus needed to assess odor sensitivity, anosmia, hyposmia, or factors governing the ability to discriminate among stimuli. Excellent control of stimulus concentration, duration, mode of presentation, and sampling behavior can be achieved with relatively simple odor generators. These, coupled with operant conditioning, provide a powerful method for detailed analysis of olfactory sensory functions, including the application of psychophysical methods to determine odor sensitivity, studies of odor discrimination, odor masking, odor memory, and the learning of complex or cognitive-based tasks.

REFERENCES

1. Alberts, J.R. and Galef, Jr., B.G. (1971). Acute anosmia in the rat: a behavioral test of a peripherally-induced olfactory deficit, *Physiol. Behav.,* 6, 619.
2. Allingham, K., Brennan, P.A., Distel, H., and Hudson, R. (1999). Expression of c-*Fos* in the main olfactory bulb of neonatal rabbits in response to garlic as a novel and conditioned odor, *Behav. Brain Res.,* 104, 157.
3. Amiri, L., Dark, T., Noce, K.M., and Kirstein, C.L. (1998). Odor preferences in neonatal and weanling rats, *Developmental Psychobiol.,* 33, 157.
4. Apfelbach, R., Russ, D., and Slotnick, B.M. (1991). Olfactory development and sensitivity to odors in rats, *Chem. Senses,* 16, 209.
5. Apfelbach, R., Weiler, E., Asselbergs, W.F., Polak, E.H., and Slotnick, B. (1999). Selective and reversible reduction of odor sensitivity in the rat by concanavalin A, *Physiol. Behav.,* 65, 513.
6. Beauchamp, G.K. and Wellington, J.L. (1984). Habituation to individual odors occurs following brief, widely-spaced presentations, *Physiol. Behav.,* 32, 511.
7. Bennett, M.H. (1972). Effects of asymmetrical mucosal stimulation upon two-odor discrimination in the rat, *Physiol. Behav.,* 9, 301.
8. Bennett, M.H. (1968). The role of the anterior limb of the anterior commissure in olfaction, *Physiol. Behav.,* 3, 507.

9. Bodyak, N. and Slotnick, B. (1999). Performance of mice in an automated olfactometer: odor detection, discrimination and odor memory, *Chem. Senses,* 24, 637.
10. Bouton, M.E., Jones, D.L., McPhillips, S.A., and Swartzentruber, D. (1986). Potentiation and overshadowing in odor-aversion learning: role of method of odor presentation, the distal-proximal cure distinction and the conditionability of odor, Learning and Motivation, 17, 115.
11. Boyse, E.A., Beauchamp, G.K., Bard, J., and Yamazaki, K. (1991). Behavior and the major histocompatibility complex of the mouse, in *Psychoneuroimmunology,* Ader, R. and Felten, D., Eds., Academic Press, San Diego, 1218.
12. Braun, J.J. and Marcus, J. (1969). Stimulus generalization among odorants by rats, *Physiol. Behav.,* 4, 245.
13. Braun, J.J., Wermuth, B., and Haberly, L. (1967). An olfactory discrimination apparatus: modification of the basic wind-tunnel design, *Psychonomic Sci.,* 9, 515.
14. Brennan, P.A., Schellinck, H.M., DeLa Riva, C., Kendrick, K.M., and Keverne, K.B. (1998). Changes in neurotransmitter release in the main olfactory bulb following an olfactory conditioning procedure in mice, *Neuroscience,* 87, 583.
15. Brosvic, G.M., Risser, J.M., Mackay-Sim, A., and Doty, R.L. (1996). Odor detection performance in hypothyroid and euthyroid rats, *Physiol. Behav.,* 59, 117.
16. Brown, R.E. (1979). Mammalian social odors: a critical review, *Adv. Study of Behavior,* 10, 103.
17. Brown, R.E. (1977). Odor preference and urine-marking scales in male and female rats: effects of gonadectomy and sexual experience on responses to conspecific odors, *J. Compar. Physiological Psychol.,* 91, 1190.
18. Brown, R.E., Roser, B., and Singh, P.B. (1989). Class I and class II regions of the major histocompatability complex both contribute to individual odors in congenic inbred strains of rats, *Behav. Genet.,* 19, 659.
19. Brown, R.E., Singh, P.B., and Roser, B. (1987). The major histocompatibility complex and the chemosensory recognition of individuality in rats, *Physiol. Behav.,* 40, 65-73.
20. Brown, R.E. and Willner, J.A. (1983). Establishing an effective scale for odor preferences of infant rats, *Behav. Neural Biol.,* 38, 251.
21. Carter, C.S. and Marr, J.N. (1970). Olfactory imprinting and age variables in the guinea-pig, *cavia porcellus, Animal Behav.,* 18, 238.
22. Caza, P.A. and Spear, N.E. (1984). Short-term exposure to an odor increases its subsequent preference in preweanling rats: a descriptive profile of the phenomenon, *Developmental Psychobiol.,* 17, 407.
23. Cheal, M.L. and Sprott, R.L. (1971). Social olfaction: a review of the role of olfaction in a variety of animal behaviors, *Psycholo. Rep.,* 29, 195.
24. Cheal, M.L. (1975). Social olfaction: a review of the ontogeny of olfactory influences on vertebrate behavior, *Behav. Biol.,* 15, 1.
25. Coppola, D.M. and O'Connell, R.J. (1988). Are pheromones their own reward?, *Physiol. Behav.,* 44, 811.
26. Cornwall, C.A. (1975). Golden hamster pups adapt to complex rearing odors, *Behav. Biol.,* 14, 175.
27. Darling, F.M.C. and Slotnick, B.M. (1994). Odor-cued taste avoidance: a simple and efficient method for assessing olfaction in rats, *Physiol, Behav.,* 55, 817.
28. Dhong, H.J., Chung, S.K., and Doty, R.L. (1999). Estrogen protects against 3-methylindole-induced olfactory loss, *Brain Res.,* 824, 312.
29. Doane, H.M. and Porter, R.H. (1978). The role of diet in mother-infant reciprocity in the spiny mouse, *Developmental Psychobiol.,* 11, 271.

30. Doty, R.L., Shaman, P., Kimmelman, C.P., and Dann, M.S. (1984). University of Pennsylvania Smell Identification Test: a rapid quantitative olfactory function test for the clinic, *Laryngoscope,* 94, 176.
31. Doty, R.L. (1995). Introduction and historical perspective, in *Handbook of Olfaction and Gustation,* Doty, R.L., Ed., Marcel Dekker, New York, 1.
32. Doty, R.L. and Ferguson-Segall, M. (1987). Odor detection performance of rats following d-amphetamine treatment: a signal detection analysis, *Psychopharmacology,* 93, 87.
33. Doty, R.L. and Risser, J.M. (1989). Influence of the D-2 dopamine receptor agonist quinpirole on the odor detection performance of rats before and after spiperone administration, *Psychopharmacology,* 98, 310.
34. Doty, R.L., Bagla, R., and Kim, N. (1999). Physostigmine enhances performance on an odor mixture discrimination test, *Physiol. Behav.,* 65, 801.
35. Doty, R.L., Ferguson-Segall, M., Lucki, I., and Kreider, M. (1988). Effects of intrabulbar injections of 6-hydroxydopamine on ethyl acetate odor detection in castrate and non-castrate male rats, *Brain Res.,* 44, 95.
36. Doty, R.L., Li, C., Bagla, R., Huang, W., Pfeiffer, C., Brosvic, G.M., and Risser, J.M., (1988). SKF 38393 enhances odor detection performance, *Psychopharmacology,* 136, 75.
37. Doty, R.L., Li, C., Mannon, L.J., and Yousem, D.M. (1999). Olfactory dysfunction in multiple sclerosis: relation to longitudinal changes in plaque numbers in central olfactory structures, *Neurology,* 53, 880.
38. Doty, R.L., McKeown, D.A., Lee, W.W., and Shaman, P. (1995). A study of the test-retest reliability of ten olfactory tests, *Chem. Senses,* 20, 645.
39. Doty, R.L., Risser, J.M., and Brosvic, G.M. (1991). Influence of adrenalectomy on the odor detection performance of rats, *Physiol. Behav.,* 49, 1273.
40. Dravnieks, A. (1975). Instrumental aspects of olfactometry, in *Methods in Olfactory Research,* Moulton, D.G., Turk, A., and Johnston, J.W., Eds., Academic Press, New York.
41. D'Udine, B., and Partridge, L. (1981). Olfactory preferences of inbred mice (*Mus musculus*) for their own strain and for siblings: effects of strain, sex and cross-fostering, *Behavior,* 78, 314.
42. Duncan, H. and Slotnick, B. (1990). Olfactory and visual reversal learning in the pigeon, *Chem. Senses,* 15, 59.
43. Dunnett, S.B. and Martel, F.L. (1990). Proactive interference effects on short-term memory in rats: I. Basic parameters and drug effects, *Behav. Neurosci.,* 104, 655.
44. Dusek, J.A. and Eichenbaum, H. (1997). The hippocampus and memory for orderly stimulus relations, *Proc. Natl. Acad. Sci.,* 94, 7109.
45. Dusek, J.A. and Eichenbaum, H. (1998). The hippocampus and transverse patterning guided by olfactory cues, *Behav. Neurosci.,* 112, 762.
46. Ferkin, M.H. and Rutka, T.F. (1990). Mechanisms of sibling recognition in meadow voles, *Can. J. Zool.,* 68, 609.
47. Forestell, C.A., Schellinck, H.M., Dreumont, S., and LoLordo, V.M. (2001). Effect of food restriction on the acquisition and expression of a conditioned odor discrimination task, *Physiol. Behav.,* 72, 559.
48. Forestell, C.A., Schellinck, H.M., Lolordo, V.M., Brown, R.E., and Wilkinson, M. (1999). Olfactory conditioning and immediate early gene expression in mice, *Chem. Senses,* 24, 618.
49. Galef, B.G. Jr. and Wigmore, S.W. (1983). Transfer of information concerning distant food: a laboratory investigation of the information-centre hypothesis, *Animal Behav.,* 31, 748.

50. Galef, B.G., and Kaner, H.C. (1980). Establishment and maintenance of preference for natural and artificial olfactory stimuli in juvenile rats, *J. Comparative Physiolog. Psychol.,* 94, 588.
51. Galef, B.G. Jr. and Whiskin, E.E. (1998). Determinants of the longevity of socially learned food preferences of Norway rats, *Animal Behav.,* 55, 967.
52. Galef, B.G. Jr., Mason, J.R., Preti, G., and Bean, N.J. (1988). Carbon disulfide: a semiochemical mediating socially induced diet choice in rats, *Physiol. Behav.,* 42, 119.
53. Garcia, J. and Rusiniak, K.W. (1980). What the nose learns from the mouth, in *Chemical Signals: Vertebrates and Aquatic Invertebrates,* Muller-Schwarze, D. and Silverstein, R.M. Eds. Plenum Press, New York, 141.
54. Garcia, J., Lasiter, P.S., Bermudez-Rattoni, F., and Deems, D.A. (1985). A general theory of aversion learning, *Ann. N.Y. Acad. Sci.,* 443, 8.
55. Genter, B.M., Owens, D.M., Carlone, H.B., and Crofton, K.M. (1996). Characterization of olfactory deficits in the rat following administration of 2,6-dichlorobenzonitrile (dichlobenil), 3,3′-iminodipropionitrile, or methimazole, *Fundamental Appl. Toxicol.,* 29, 71.
56. Goff, W.R. (1961). Measurement of absolute olfactory sensitivity in rats, *Am. J. Psychol.,* 384.
57. Goldblatt, A. (1978). Effects of early olfactory experience on a tilt table odor preference, *Behav. Biol.,* 22, 269.
58. Greiner, R.S., Moriguchi, T., Hutton, A., Slotnick, B., and Salem, N. (1999). Rats with low levels of brain docosahexaenoic acid show impaired performance in olfactory-based and spatial learning tasks, *Lipids,* 34, Suppl:S239.
59. Hendricks, K.R., Knott, J.N., Lee, M.E., Gooden, M.D., Evers, S.M. and Westrum, L.E. (1994). Recovery of olfactory behavior. I. Recovery after a complete olfactory bulb lesion correlates with patterns of olfactory nerve penetration, *Brain Res.,* 648, 121.
60. Henton, W.W. (1969). Conditioned suppression to odorous stimuli in pigeons, *J. Exp. Anal. Behav.,* 12, 175.
61. Henton, W.W., Smith, J.C., and Tucker, D. (1966). Odor discrimination in pigeons, *Science,* 1138.
62. Hepper, P.G. (1987). The discrimination of different degrees of relatedness in the rat: evidence for a genetic identifier?, *Animal Behav.,* 35, 549.
63. Hepper, P.G. (1987). The amniotic fluid: an important priming role in kin recognition, *Animal Behav.,* 35, 1343.
64. Hettinger, T.P., Gent, J.F., Marks, L.E., and Frank, M.E. (1999). A confusion matrix for the study of taste perception, *Perception and Psychophysics,* 61, 1510.
65. Hudson, R. (1985). Do newborn rabbits learn the odor stimuli releasing nipple-search behavior?, *Developmental Psychobiol.,* 18, 575.
66. Hudson, R. and Distel, H. (1983). Nipple location by newborn rabbits: behavioural evidence for pheromonal guidance, *Behaviour,* 85, 260.
67. Hudson, R. and Distel, H. (1998). Induced peripheral sensitivity in the developing vertebrate olfactory system, *Ann. N.Y. Acad. Sci.,* 855, 109.
68. Hunt, M., Croft, D., and Slotnick, B. (1999). Olfaction in the red kangaroo, *Physiol. Behav.,* 67, 365.
69. Janus, C. (1989). The development of olfactory preferences for artificial odors briefly experienced by the precocial spiny mouse young, *Behav. Neural Biol.,* 52, 430.
70. Janus, C. and Holman, S. (1989). Development of sex differences in the response of spiny mouse pups to adult male odors, *Physiol. Behav.,* 46, 895.

71. Johanson, I.B. and Teicher, M.H. (1980). Classical conditioning of an odor preference in three-day old rats, *Behav. Neural Biol.,* 29, 132.
72. Johnson, B.A. and Leon, M. (1996). Spatial distribution of [^{14}C]2-deoxyglucose uptake in the glomerular layer of the rat olfactory bulb following early odor preference learning, *J. Comparative Neurol.,* 376, 557.
73. Johnson, B.A., Woo, C.C., Duon, H., Nguyen, V., and Leon, M. (1995). A learned odor evokes an enhanced Fos-like glomerular response in the olfactory bulb of young rats, *Brain Res.,* 699, 192.
74. Johnston, R.E. (1981). Attraction to odors in hamsters: an evaluation of methods, *J. Compar. Physiological Psychol.,* 95, 951.
75. Johnston, R.E. (1983). Mechanisms of individual discrimination in hamsters, in *Chemical Signals in Vertebrates,* Vol. 3, Mueller-Schwarze, D. and Silverstein, R.M., Eds., Plenum, New York, 245.
76. Johnston, R.E., Derzie, A., Chiang, G., Jernigan, P., and Lee, H. (1993). Individual scent signatures in golden hamsters: evidence for specialization of function, *Animal Behav.,* 45, 1061.
77. Keil, W., Von Stralendorff, F., and Hudson, R. (1990). A behavioral bioassay for analysis of rabbit nipple-search pheromone, *Physiol. Behav.,* 47, 525.
78. Kimura, Y., Miwa, T., Furukawa, M., and Umeda, R. (1991). Effects of topical application of steroids on olfactory disturbance in mice, *Chem. Senses,* 16, 297.
79. Kogan, J.H., Frankland, P.W., Blendy, J.A., Coblentz, J., Marowitz, Z., Schutz, G., and Silva, A.J. (1997). Spaced training induces normal long-term memory in CREB mutant mice, *Curr. Biol.,* Jan 1;7.1.
80. Krames, L. (1970). Responses of female rats to the individual body odors of male rats, *Psychonomic Sci.,* 20, 274.
81. Krestel, D., Passe, D., Smith, J.C., and Johsson, L. (1984). Behavioral determination of olfactory thresholds to amyl acetate in dogs, *Physiol. Behav.,* 8, 169.
82. Kurtz, D.B., White, T.L., Hornung, D.E., and Belknap, E. (1999). What a tangled web we weave: discriminating between malingering and anosmia, *Chem. Senses,* 24, 697.
83. Laing, D.L., Panhuber, H., and Slotnick, B.M. (1989). Odor masking in the rat, *Physiol. Behav.,* 45, 689.
84. Langdon, P.E., Harley, C.W., and McLean, J.H. (1997). Increased B adrenoreceptor activation overcomes conditioned olfactory learning deficits induced by serotonin depletion, *Developmental Brain Res.,* 102, 291.
85. Linster, C. and Hasselmo, M.E. (1999). Behavioral responses to aliphatic aldehydes can be predicted from known electrophysiological responses of mitral cells in the olfactory bulb, *Physiol. Behav.,* 66, 497.
86. Lovelace, C. and Slotnick, B. (1995). Memory for brief, widely spaced odor presentations in the rat, *Chem. Senses,* 20, 183.
87. Lu, X.C. and Slotnick, B. (1990). Acquisition of an olfactory learning-set in rats with lesions of the thalamic mediodorsal nucleus, *Chem. Senses,* 15, 713.
88. Lu, X.C. and Slotnick, B. (1994). Recognition of propionic acid vapor after removal of the olfactory bulb area associated with high 2-DG uptake, *Brain Res.,* 639, 26.
89. Lu, X.M. and Slotnick, B. (1998). Olfaction in rats with extensive lesions of the olfactory bulbs: implications for odor coding, *Neuroscience,* 84, 849.
90. Lu, X.M., Slotnick, B.M., and Silberberg, A.M. (1993). Odor matching and odor memory in the rat. *Physiol. Behav.,* 53, 795.

91. Lynch, G., Granger, R., and Larson, J. (1989). Some possible functions of simple cortical networks suggested by computer modeling, in *Neural Models of Plasticity: Experimental and Theoretical Approaches,* Byrne, J.H. and Berry, W.O., Eds., Academic Press, San Diego, 329.
92. Malnic, B., Hirono, J., Sato, T., and Buck, L.B. (1999). Combinatorial receptor codes for odors, *Cell,* 96, 713.
93. Marr, J.N. and Lilliston, L.G. (1969). Social attachment in rats by odor and age, *Behaviour,* 33, 277.
94. Marr, J.N. and Gardener, L.E. (1965). Early olfactory experience and later social behavior in the rat: preference, sexual responsiveness, and the care of young, *J. Genet. Psychol.,* 107, 167.
95. Marshall, D.A., Blumer, L., and Moulton, D.G. (1981). Odor detection curves for n-pentanoic acid in dog and man, *Chem. Senses,* 6, 445.
96. Marshall, D.A. and Moulton, D.G. (1981). Olfactory sensitivity to alpha-ionone in humans and dogs, *Chem. Senses,* 6, 53.
97. Marshall, D.A., Doty, R.L., Lucero, D.P., and Slotnick, B. (1981). Odor detection thresholds in the rat for the vapors of three related perfluorocarbons and ethylene glycol dinitrate, *Chem. Senses,* 6, 421.
98. Martin, I.G. and Beauchamp, G.K. (1982). Olfactory recognition of individuals by male cavies (*Cavia apera*), *J. Chem. Ecol.,* 8, 1241.
99. McBride, S.A. and Slotnick, B. (1997). The olfactory thalamocortical system and odor reversal learning examined using an asymmetrical lesion paradigm, *Behav. Neurosci.,* 111, 1273.
100. McLean, J.H., Darby-King, A., and Hodge, E. (1996). 5-HT2 receptor involvement in conditioned olfactory learning and memory in the neonate rat pup, *Behav. Neurosci.,* 110, 1426.
101. Mihalick, S.M., Langlois, J.C., Krienke, J.D., and Dube, W.V. (2000). An olfactory discrimination procedure for mice, *J. Exp. Anal. Behav.,* 73, 305.
102. Mihalick, S.M., Langlois, J.C., and Krienke, J.D. (2000). Strain and sex differences on olfactory discrimination learning in C%&BL/6J and DBA/2J inbred mice (*Mus musculus*), *J. Comp. Psych.,* 114, 365.
103. Miller, J.S., Jagielo, J.A., and Spear, N.E. (1989). Changes in the retrievability of associations to elements of the compound CS determine the expression of overshadowing, *Animal Learning and Behavior,* 18, 157.
104. Mobarts, P., Wang, F., Dulac, C., Chai, S.K., Nemes, A., Mendelsohn, M., Edmondson, J., and Axel, R. (1996). Visualizing an olfactory sensory map, *Cell,* 87, 675.
105. Moncrieff, R.W. (1967). The *Chem. Senses,* CRC Press, Cleveland.
106. Mori, K and Yoshihara, Y. (1995). Molecular recognition and olfactory processing in the mammalian olfactory system, *Prog. Neurobiol.,* 45, 585.
107. Moulton, D.G. and Marshall, D.A. (1976). The performance of dogs in detecting alpha-ionone in the vapor phase, *J. Compar. Physiol.,* 110, 287.
108. Moulton, D.G., Celebri, G., and Fink, R.P. (1970). Olfaction in mammals — two aspects: proliferation of cells in the olfactory epithelium and sensitivity to odours, *Ciba Foundation Symposium on Taste and Smell in Vertebrates,* Wolsenholme, G.E.W., Knight, J., and Churchill, A., Eds., Ciba-Geigy, London, 227.
109. Nelson, E. and Panksepp, J. (1996). Oxytocin mediates acquisition of maternally associated odor preferences in preweanling rat pups, *Behav. Neurosci.,* 110, 583.
110. Newman, K.S. and Halpin, Z.T. (1979, 1988). Individual odours and mate recognition in the prairie vole, *Microtus ochrogaster, Animal Behav.,* 36.

111. Nigrosh, B., Slotnick, B., and Nevin, J. (1975). Reversal learning and olfactory stimulus control in rats, *J. Compar. Physiological Psychol.,* 80, 285.
112. Nowlis, G.H., Frank, M.E., and Pfaffmann, C. (1980). Specificity of acquired aversions to taste qualities in hamsters and rats, *J. Compar. Physiological Psychol.,* 94, 932.
113. Oley, N., DeHan, R.S., Tucker, D., Smith, J.C., and Graziadei, P.P. (1975). Recovery of structure and function following transection of the primary olfactory nerves in pigeons, *J. Compar. Physiological Psychol.,* 88, 477.
114. Otto, T., and Eichenbaum, H. (1992). Complementary roles of the orbital prefrontal cortex and the perirhinal-entorhinal cortices in an odor-guided delayed-nonmatching-to-sample task, *Behav. Neurosci.,* 106, 762.
115. Panhuber, H. (1982). Effect of odor quality and intensity on conditioned aversion learning in the rat, *Physiol. Behav.,* 28, 149.
116. Passe, D.H. and Walker, J.C. (1985). Odor psychophysics in vertebrates, *Neurosci. Biobehav. Rev.,* 9, 431.
117. Peele, D.B., Allison, S.D., Bolon, B., Prah, D., Jensen, K.F., and Morgan, K.T. (1991). Functional deficits produced by 3-methylindole-induced olfactory mucosal damage revealed by a simple olfactory learning task, *Toxicol. Appl. Pharmacol.,* 107, 191.
118. Penn, D., and Potts, W.K. (1998). Untrained mice discriminate MHC-determined odors, *Physiol. Behav.,* 63, 235.
119. Perritt, D.W., Couger, G., and Barker, R.W. (1993). Computer-controlled olfactometer system for studying behavioral responses of ticks to carbon dioxide, J. Med. Entomol., 30, 571.
120. Pfaffmann, C., Goff, W.R., and Bare, J.K. (1958). An olfactometer for the rat, *Science,* 128, 1007.
121. Pierson, S. (1974). Conditioned suppression to odorous stimuli in the rat, *J. Compar. Physiological Psychol.,* 86, 708.
122. Pietras, R.J. and Moulton, D.G. (1974). Hormonal influences on odor detection in rats: changes associated with the estrous cycle, pseudopregnancy, ovariectomy, and administration of testosterone propionate, *Physiol. Behav.,* 12, 475.
123. Pontecorvo, M.J. (1983). Effects of proactive interference on rats, continuous non-matching-to-sample performance, Animal Learn. Behav., 11, 356.
124. Prah, J.D., Sears, S.B., and Walker, J.C. (1995). Modern approaches to air dilution olfactometry, in *Handbook of Olfaction and Gustation,* Doty, R.L., Ed., Marcel Dekker, New York, 227.
125. Reid, I.C. and Morris, R.G.M. (1992). Smells are no surer: rapid improvement in olfactory discrimination is not due to the acquisition of a learning set, *Proc. R. Soc. London, Ser. B,* 247, 137.
126. Ressler, K.J., Sullivan, S.L. and Buck, L.B. (1994). Information coding in the olfactory system: Evidence for a stereotyped and highly organized epitope map in the olfactory bulb, *Cell,* 79, 1245.
127. Rothblat, L.A. and Hayes, L.L. (1987). Short-term object recognition memory in the rat: nonmatching with trial-unique junk stimuli, *Behav. Neurosci.,* 101, 587.
128. Rudy, J.W. and Cheatle, M.D. (1979). Ontogeny of associative learning: acquisition of odor aversions by neonatal rats, in *Ontogeny of Learning and Memory,* Spear, N.E. and Campbell, B.A., Eds., John Wiley & Sons, Hillsdale, NJ, 157.
129. Sakellaris, P.C. (1987, 1972). Olfactory thresholds in normal and adrenalectomized rats, *Physiol. Behav.,* 9, 495.
130. Sakurai, Y., Rat's auditory working memory tested by continuous non-matching-to-sample performance, *Psychobiology,* 15, 277.

131. Schellinck, H.M., Rooney, E., and Brown, R.E. (1995). The individual odors of germ-free mice are not discriminated by rats in a habituation-dishabituation procedure, *Physiol. Behav.,* 57, 1005.
132. Schellinck, H.M., Forestell, C., and LoLordo, V. (in press). A simple and reliable test of olfactory learning and memory in mice, *Chem. Senses* (in press).
133. Schellinck, H.M., Forestell, C., Dill, P., LoLordo, V., and Brown, R.E. (2000). A simple test for olfactory learning in mice, in *Chem. Signals in Vertebrates,* Vol. IX.
134. Schellinck, H.M., Slotnick, B.M. and Brown, R.E. (1997). Odors of individuality originating from the major histocompatibility complex are masked by diet cues in the urine of rats, *Animal Learn. Behav.,* 25, 193.
135. Setzer, A.K. and Slotnick, B. (1998). Disruption of axonal transport from olfactory epithelium by 2-methylindole. *Physiol. Behav.,* 65, 479.
136. Singh, P.B., Herbert, J., Roser, B., Arnott, L., Tucker, D.K., and Brown, R.E. (1990). Rearing rats in a germ-free environment eliminates their odours of individuality, *J. Chem. Ecol.,* 16, 1667.
137. Slotnick, B. (1994). The enigma of olfactory learning revisited, *Neuroscience,* 58, 1.
138. Slotnick, B. (1984). Olfactory stimulus control in the rat, *Chem. Senses,* 9, 157.
139. Slotnick, B. (1990). *Olfactory Perception in Animals, Comparative Perception,* Vol. 1: Basic Mechanisms, Stebbins, W. and Berkley, M., Eds., John Wiley & Sons, New York, 155.
140. Slotnick, B. and Born, W.S. (1979). Transfer of sex-odor discrimination in rats, *Physiol. Behav.,* 23, 589.
141. Slotnick, B. and Brosvic, G. (1987). Failure of the rat to acquire a taste reversal learning set, *Chem. Senses,* 12, 333.
142. Slotnick, B. and Guttman, L. (1977). Evaluation of intranasal zinc sulfate treatment on olfactory discrimination in rats, *J. Compar. Physiological Psychol.,* 91, 942.
143. Slotnick, B. and Kaneko, N. (1981). Role of mediodorsal thalamic nucleus in olfactory discrimination learning in rats, *Science,* 214, 91.
144. Slotnick, B. and Katz, H. (1974). Olfactory learning-set formation in rats, Science, 185, 796.
145. Slotnick, B. and Nigrosh, B.J. (1974). Olfactory stimulus control evaluated in a small animal olfactometer, *Perceptual and Motor Skills,* 39, 583.
146. Slotnick, B. and Ptak, J.E. (1977). Olfactory intensity-difference thresholds in rats and humans, *Physiol. Behav.,* 19, 795.
147. Slotnick, B. and Risser, J. (1990). Odor memory and odor learning in rats with lesions of the lateral olfactory tract and mediodorsal thalamic nucleus, *Brain Res.,* 529, 23.
148. Slotnick, B. and Schoonover, F.W. (1984). Olfactory thresholds in normal and unilaterally bulbectomized rats, *Chem. Senses,* 9, 325.
149. Slotnick, B. and Schoonover, F.W. (1992). Olfactory pathways and the sense of smell, *Neurosci. Biobehav. Rev.,* 16, 453.
150. Slotnick, B. and Schoonover, F.W. (1993). Olfactory sensitivity of rats with transection of the lateral olfactory tract, *Brain Res.,* 616, 132.
151. Slotnick, B., Bell, G.A., Panhuber, H., and Laing, D.G. (1997). Detection and discrimination of propionic acid after removal of its 2-DG identified major focus in the olfactory bulb: a psychophysical analysis, *Brain Res.,* 762, 89.
152. Slotnick, B., Glover, P., and Bodyak, N. (2000). Does intranasal application of zinc sulfate produce anosmia in the rat?, *Behav. Neurosci.,* 114, 814.
153. Slotnick, B., Graham, S., Laing, D.G., and Bell, G.A. (1987). Detection of propionic acid vapor by rats with lesions of olfactory bulb areas associated with high 2-DG uptake, *Brain Res.,* 417, 343.

154. Slotnick, B., Hanford, S., and Hodos, W. (2000). Can rats acquire an olfactory learning set ?, *J. Exp. Psychol.: Animal Behav. Processes,* 26, 399.
155. Slotnick, B., Kufera, A., and Silberberg, A.M. (1991). Olfactory learning and odor memory in the rat, *Physiol. Behav.,* 50, 555.
156. Slotnick, B., Westbrook, F. and Darling, F.M.C. (1997). What the rat's nose tells the rat's mouth: long delay aversion conditioning with aqueous odors and potentiation of taste by odors, *Animal Learn. Behav.,* 25, 357.
157. Smith, F.J., Charnock, D.J., and Westbrook, R.F. (1983). Odor-aversion learning in neonate rat pups: the role of duration of exposure to an odor, *Behav. Neural Biol.,* 37, 284.
158. Smith, J. (1970). *Conditioned Suppression as an Animal Psychophysical Technique,* Stebbins, W.C., Ed., Appleton-Century-Crofts, Educational Division, Meredith Corporation, New York, 125.
159. Smotherman, W.P. (1982). Odor aversion learning by the rat fetus, *Physiol. Behav.,* 29, 769.
160. Southall, P.F. and Long, C.J. (1969). Odor cues in a maze discrimination, *Psychonomic Sci.,* 16, 126.
161. Staubli, U., Fraser, D., Kessler, M., Lynch, G. (1986). Studies on retrograde and anterograde amnesia of olfactory memory after denervation of the hippocampus by entorhinal cortex lesions, *Behav. Neural Biol.,* 46, 432.
162. Stevens, D.A. (1975). Laboratory methods for obtaining olfactory discrimination in rodents, in *Methods in Olfactory Research,* Moulton, D.G., Turk, A., and Johnston, A.W. Jr., Eds., Academic Press, New York, 375.
163. Stone, H., Pryor, G., and Steinmetz, G. (1969). The design and operation of an improved olfactometer for behavioral and physiological investigation, *Behav. Res. Meth. Instrument.,* 1, 153.
164. Sullivan, R.M., McGaugh, J.L., and Leon, M. (1991). Norepinephrine-induced plasticity and one-trial olfactory learning in neonatal rats, *Development. Brain Res.,* 60, 219.
165. Sullivan, R.M., Wilson, D.A., Lemon, C., and Gerhardt, G.A. (1994). Bilateral 6-OHDA lesions of the locus coeruleus impair associative olfactory learning in newborn rats, *Brain Res.,* 643, 306.
166. Sundberg, H., Doving, K., Novikov, S., and Ursin, H. (1982). A method for studying responses and habituation to odors in rats, *Behav. Neural Biol.,* 34, 113.
167. Swaisgood, R.R., Lindburg, D.G., and Zhou, X. (1999). Giant pandas discriminate individual differences in conspecific scent, *Animal Behav.,* 57, 1045.
168. Syau-Cheng, L. and Johnston, R.E. (1994). Individual odors in Djungarian hamsters (*Phodopus campbelli*), *Ethology,* 96, 117.
169. Terrace, H.S. (1963). Errorless transfer of a discrimination across two continua, *J. Exp. Anal. Behav.,* 6, 223.
170. Thanos, P. and Slotnick, B. (1997). Short-term odor memory: effects of posterior transection of the lateral olfactory tract, *Physiol. Behav.,* 61, 903.
171. Todrank, J., Heth, G., and Johnston, R.E. (1998). Kin recognition in golden hamsters: evidence for kinship odours, *Animal Behav.,* 55, 377.
172. Tucker, D. (1963). Physical variables in the olfactory stimulation process, *J. Gen. Physiol.,* 46, 453.
173. Vigouroux, M., Viret, P, and Duchamp, A. (1998). A wide concentration range olfactometer for delivery of short reproducible odor pulses, *J. Neurosci. Meth.,* 24, 57.
174. Walker, J.C. and O'Connell, R.J. (1986). Computerized odor psychophysical testing in mice, *Chem. Senses,* 11, 439.

175. Walker, J.C. and Jennings, R.A. (1992). Comparisons of odor perception in humans and animals, in *The Human Sense of Smell,* Laing, D.G., Doty, R.L., and Breiphol, W., Eds., Springer-Verlag, Berlin, 261.
176. Warren, J.M. (1973). Learning in vertebrates, in *Comparative Psychology: A Modern Survey,* Dewsbury, D.A. and Rethlingshafer, D.A., Eds., McGraw-Hill, New York, 471.
177. Weldon, D.A., Fedorcik, G.C., LoRusson, C.M., Tiburzi, M.J., and Lenoci, J.M. (1997). Olfactory conditioning impairment following posttraining NMDA receptor blockade in neonatal rats, *Neurobiol. Learn. Memory,* 67, 34.
178. Wilson, D.A. and Sullivan, R.M. (1994). Neurobiology of associative learning in the neonate: early olfactory learning, *Behav. Neural Biol.,* 61, 1.
179. Wright, H.N. (1987). Characterization of olfactory dysfunction, *Arch. Otolaryngol. Head Neck Surg.,* 113, 163.
180. Xu, W. and Slotnick, B. (1999). Olfaction and peripheral olfactory connections in methimazole-treated rats, *Behav. Brain Res.,* 102, 41.
181. Yamazaki, K., Yamaguchi, M., Baranoski, L., Bard, J., Boyse, E.A., and Thomas, L. (1979). Recognition among mice: evidence from the use of a Y-maze differentially scented by congenic mice of different histocompatibility types, J. Exp. Med., 150, 755.
182. Youngentob, S.L., Markert, L.M., Hill, T.W., Matyas, E.P., and Mozell, M.M. (1991). Odorant identification in rats: an update, *Physiol. Behav.,* 49,1293.
183. Youngentob, S.L., Markert, L.M., Mozell, M.M., and Hornung, D.E. (1990). A method for establishing a five odorant identification confusion matrix task in rats, *Physiol. Behav.,* 47, 1053.
184. Youngentob, S.L., Mozell, M.M., Sheehe, P.R. and Hornung, D.E. (1987). A quantitative analysis of sniffing strategies in rats performing odor detection tasks, *Physiol. Behav.,* 41, 59.
185. Youngentob, S.L., Schwob, J.E., Sheehe, P.R., and Youngentob, L.M. (1997). Odorant threshold following methyl bromide-induced lesions of the olfactory epithelium, *Physiol. Behav.,* 62, 1241.
186. Zagreda, L., Goodman, J., Druin, D.P., McDonald, D., and Diamond, A. (1999). Cognitive deficits in a genetic mouse model of the most common biochemical causes of human mental retardation, *J. Neurosci.,* 19, 6175.
187. Zyzak, D.R., Otto, T., Eichenbaum, H., and Gallagher, M. (1995). Cognitive decline associated with normal aging in rats: a neuropsychological approach, *Learn. Memory,* 2, 1.

Section 2

Molecular and Cell Biology of Taste and Olfaction

3 Recordings from Vertebrate Olfactory Receptor Neurons: From Isolated Cells to Intact Epithelial Preparations

Minghong Ma and Gordon M. Shepherd

CONTENTS

3.1 BACKGROUND

As the first neuronal component in the olfactory system, olfactory receptor neurons (ORNs) are responsible for detecting and discriminating among thousands of odors and sending the information directly to the olfactory bulb in the brain. Considerable efforts have been made to develop preparations from which odor-induced responses and membrane properties of ORNs can be studied by different physiological approaches.

0-8493-2329-0/02/$0.00+$1.50

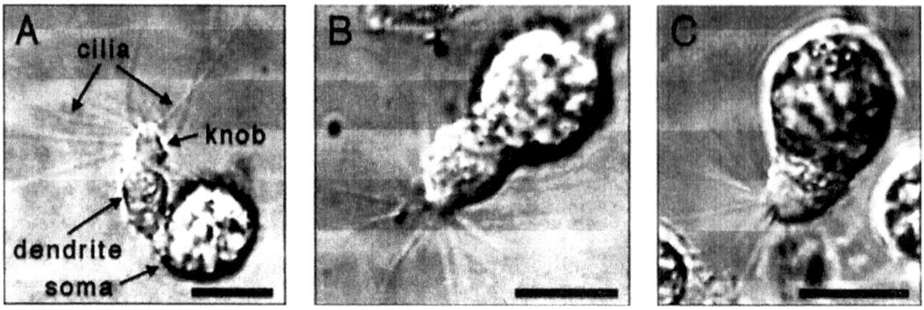

FIGURE 3.1 Phase contrast images of isolated salamander ORNs. (Adapted from Leinders-Zufall, T. et al., *J. Neurosci.*, 17, 4136, 1997.)

Odor-induced responses in the olfactory epithelium *in vivo* have been primarily obtained via electro-olfactogram (EOG) and single-unit recordings. EOG recordings, which summate the activities of many receptor neurons,[1] have been widely used to study the overall responsiveness and the spatial organization of the epithelium in responding to odors.[2-5] Single-unit extracellular recordings have been used to record the activity of individual ORNs *in vivo*. These studies have provided essential information on the peripheral coding of olfaction by analyzing the impulse pattern changes induced by odor stimulation (described in detail by Duchamp-Viret in Chapter 4 of this book).

The properties of ORNs and odor-induced responses in the epithelial sheet *in vitro* have been studied by a number of techniques, including extracellular recordings,[6,7] intracellular recordings,[8] and voltage-sensitive dye imaging.[9-11]

This chapter focuses on the methods for recording from isolated ORNs and *in situ* ORNs in the intact epithelium.

3.2 RECORDING FROM ISOLATED OLFACTORY RECEPTOR NEURONS

Isolated ORNs have become the most commonly used *in vitro* preparations in the past 10 to 15 years after two powerful techniques — patch clamp and calcium imaging — became available. Individual ORNs can be dissociated by cutting the olfactory epithelium into small pieces, followed by gentle trituration using a fire-polished Pasteur pipette. Dissociation is often facilitated by enzymatic treatment and/or by the use of low-divalent media. This procedure has produced viable ORNs in a variety of species, including catfish,[12] salamander,[13] frog,[14] newt,[15] rat,[16] and mouse.[17] An ORN can be easily identified by its characteristic morphology, i.e., a large cell body, a single thick dendrite, and a number of cilia (5 to 15) emanating from the dendritic knob (see Figure 3.1).[18]

Studies of isolated ORNs by patch clamp and calcium imaging techniques have provided critical functional evidence regarding olfactory signal transduction mechanisms.[12-16,19-30] A major pathway of signal transduction begins with the binding of odor molecules to specific receptors, which initiates a cascade of events leading to cAMP formation and opening of Ca^{2+}-permeable cyclic nucleotide-gated (CNG)

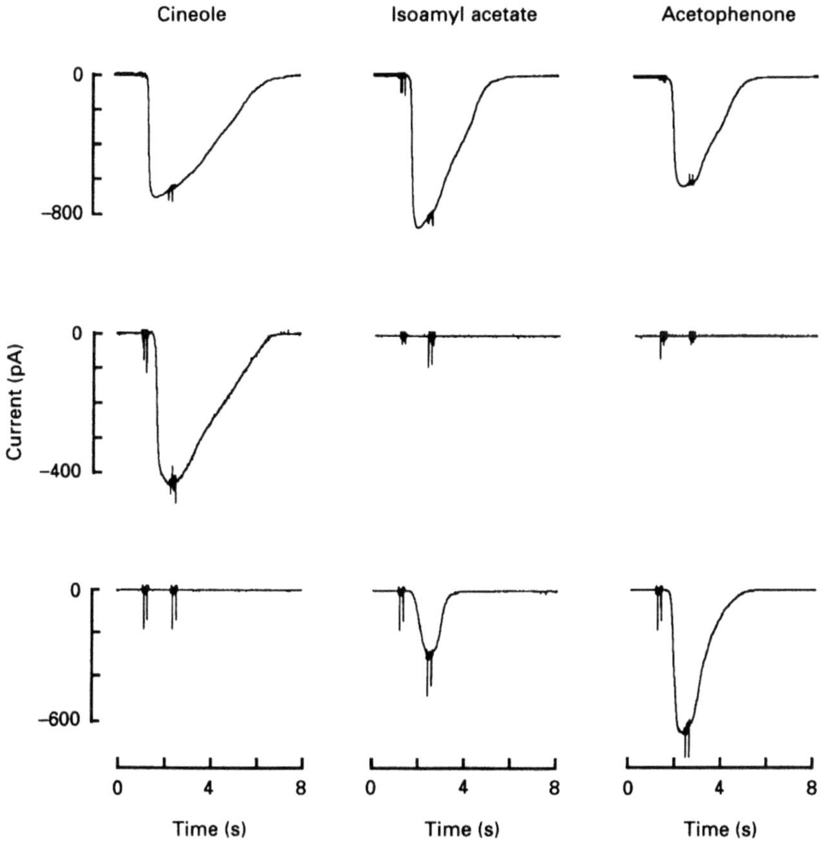

FIGURE 3.2 ORNs respond to odors by generating inward currents. Current responses to the indicated odors at 5×10^{-4} *M* applied for 1.2 s in three different cells. Holding potential is –55 mV. (Adapted from Firestein, S., Picco, C., and Menini, A., *J. Physiol., London,* 468, 1, 1993.)

channels. ORNs respond to odor stimulation by generating inward currents under voltage clamp (see Figure 3.2),[21] and by showing increased calcium concentrations in the cilia, dendritic knob, dendrite, and soma (see Figure 3.3).[31]

A technique called tissue printing has been developed to achieve simultaneous recordings of calcium increases in dissociated neighboring ORNs in frog (see Figure 3.4).[32] Sato et al.[33] extended this technique to dissociated mouse ORNs, from which they successfully recorded odor-induced calcium changes and studied the tuning properties of these neurons to specific odors.

In summary, isolated preparations have been very effective for studies of the membrane properties of vertebrate ORNs and of cellular mechanisms underlying olfactory signal transduction. Isolated ORNs can be easily identified under a phase contrast microscope, and all the compartments, such as the soma, dendrite, knob, and even the cilia, can be patched. These neurons are also suitable for calcium imaging recordings by offering easy loading of calcium-sensitive dyes via acetoxymethyl (AM) esters and a high signal-to-noise ratio.

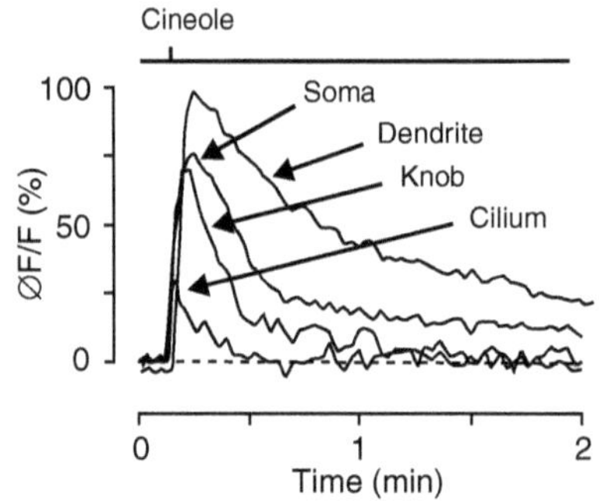

FIGURE 3.3 ORNs respond to odors by showing increased calcium concentrations. Optical imaging reveals transient Ca^{2+} elevations in cilia, dendritic knob, dendrite, and soma after the application of a single pulse (1 s) of odor. (Adapted from Leinders-Zufall, T. et al., *J. Neurosci.*, 18, 5630, 1998.)

3.3 RECORDING FROM OLFACTORY RECEPTOR NEURONS IN THE INTACT EPITHELIAL PREPARATIONS

Compared with amphibian ORNs, odor-induced responses from rodent ORNs are more difficult to record because of the small size of these neurons and the relatively poor yield during dissociation. However, recent progress in molecular studies in olfaction using rodents as model animals makes it essential to reliably record from rodent ORNs. There remains a lack of information about physiological properties underlying odor coding at the epithelial level. For example, knowledge of the odor response spectra of individual ORNs expressing a particular receptor gene is limited,[34,35] and little is known about the overlap in response spectra between ORNs expressing different receptor genes and the spatial distribution of ORNs with distinct response spectra. To address these questions, an epithelial preparation is needed in which odor responses can be tested at the resolution of individual ORNs while the spatial organization is maintained.

To meet this need, the authors have developed an intact olfactory epithelial preparation from both rat and mouse. Infrared differential interference contrast (DIC) video microscopy allows one to visualize and record from individual dendritic knobs of ORNs. This preparation offers the advantages of approximating *in vivo* physiological conditions, while furnishing an opportunity to map individual neuron responses in the epithelium in a spatially defined manner, using both electrophysiological and optical imaging methods.

3.3.1 Intact Epithelial Preparations

The animals used in the authors' studies are Sprague-Dawley rats (2 to 4 weeks old) and c57bl/6 mice (4 to 12 weeks old), both male and female. An animal is anesthetized by intraperitoneal injection of an overdose of ketamine (300 mg/kg body weight) and decapitated. The head is immediately put into icy artificial cerebrospinal fluid

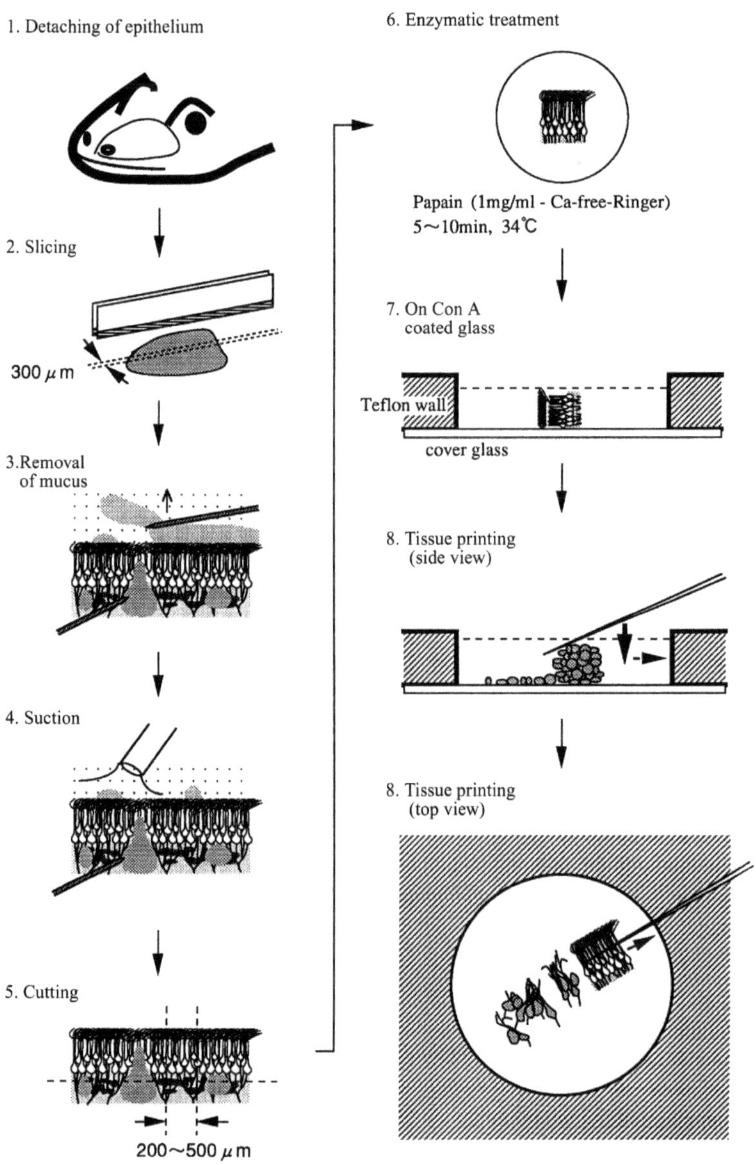

FIGURE 3.4 Tissue printing method for isolation of olfactory receptor neurons. (Adapted from Hirono, J. et al., *J. Neurosci. Methods,* 42, 185, 1992.)

(ACSF), which contains (in m*M*) NaCl (124), KCl (3), $MgSO_4$ (1.3), $CaCl_2$ (2), $NaHCO_3$ (26), NaH_2PO4 (1.25), glucose (15); pH 7.6, 305 mOsm. The skin covering the skull, the eyes, and the lower jaw is removed. Two coronal cuts, one in the skull across the olfactory bulbs and the other in the anterior end of the nostril, are made to dissect out the nasal cavity *en bloc*. The entire nasal cavity is then transferred into a culture dish and examined under a dissecting microscope (Figure 3.5A). The

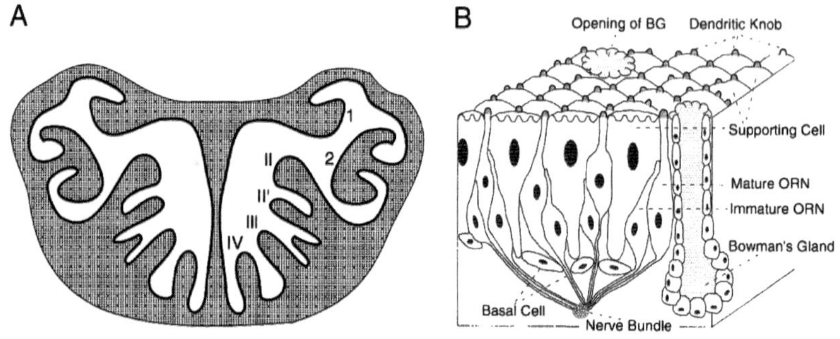

FIGURE 3.5 Structure of rodent olfactory epithelium. A: Schematic diagram of the nasal cavity illustrating the epithelia attached to the septum and the turbinates. Ectoturbinates: 1, 2; endoturbinates: II, II′, III, and IV. B: Schematic drawing of the structure of olfactory epithelium. ORN, olfactory receptor neuron. BG, Bowman's gland. 10 to 15 cilia emanating from each knob are not drawn. (Modified from Ma, M., Chen, W.R., and Shepherd, G.M., *J. Neurosci. Methods,* 92, 31, 1999; Ma, M. and Shepherd, G.M., *Proc. Natl. Acad. Sci. U.S.A.,* 97, 12869, 2000.)

palate is removed to expose the ventral part of the nasal cavity. Two sagittal cuts, one on each side of the septal wall, are made to expose the epithelia. The turbinates on each side are freed from the surrounding tissues and bones. The epithelia that are still attached to the septal wall or to the turbinates are kept in ACSF bubbled with 95% O_2 and 5% CO_2 at room temperature for up to 8 h prior to use.

A swatch of epithelium up to several millimeters across is then peeled off from the septal wall or from the turbinates and transferred to a recording chamber under a dissecting microscope and stabilized by a customized net with the mucus layer facing up. The recording chamber is continuously perfused with the oxygenated ACSF throughout the experiments at a rate of about 3 ml/min.

3.3.2 Visualization of Dendritic Knobs of ORNs

Our setup for visualizing the dendritic knobs of rodent ORNs consists of an upright infrared DIC microscope (Olympus BX50WI) equipped with a 40× water-immersion objective and a cooled CCD-300-RC camera (Dage-MTI) connected to a TV monitor. An extra 4× magnification is achieved using a magnification changer (Luigs & Neumann GmbH, Germany).

When the superficial (mucus) surface of epithelium is examined under the microscope, the olfactory epithelium is distinguished from the respiratory epithelium by a sharp border with several structural features. The respiratory epithelium is characterized by the moving microvilli of the epithelial cells, whereas the olfactory epithelium is characterized by the evenly distributed openings of Bowman's glands and numerous dendritic knobs of ORNs stretching out from the gaps between supporting cells (Figure 3.5B). The density of the dendritic knobs varies among different areas, ranging from 5 to 10 knobs/100 μm^2 for rats and 10 to 20 knobs/100 μm^2 for mice (Figure 3.6). The origins of 10 to 15 cilia emanating from a single knob can sometimes be seen by adjusting the focus.

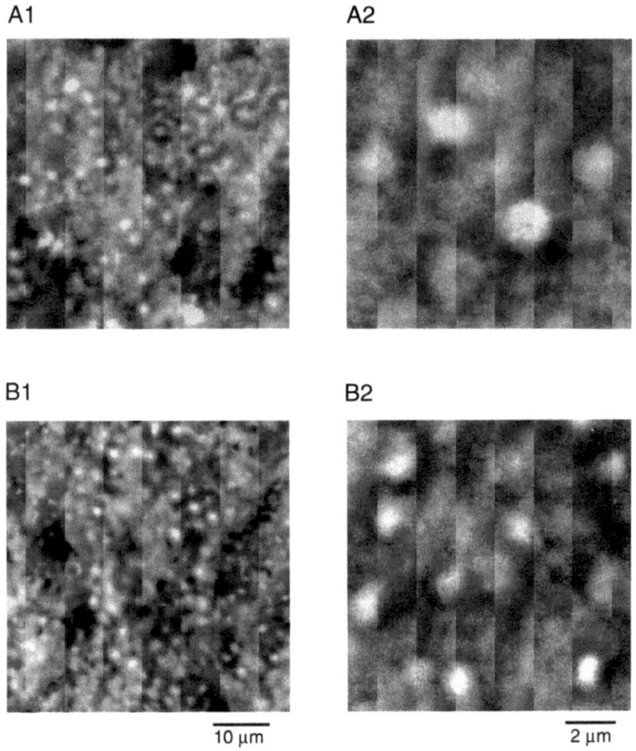

FIGURE 3.6 Visualization of ORN dendritic knobs in olfactory epithelium from rat (A) and mouse (B). A1, B1: the dendritic knobs of rat and mouse ORNs are seen under video microscopy with a 40× objective. A2, B2: an extra 4× is achieved by a magnification changer. Patch-clamp is preformed under this magnification. (Modified from Ma, M., Chen, W.R., and Shepherd, G.M., *J. Neurosci. Methods,* 92, 31, 1999.)

3.3.3 Patch Clamp of Dendritic Knobs of ORNs

Our setup for electrophysiological recordings contains an EPC-9 patch clamp amplifier combined with the Pulse/Pulsefit software package (HEKA Electronic, Germany) running on a Macintosh computer.

A dendritic knob is patched while both the knob and the electrode are visualized on the monitor screen with the highest magnification (Figures 3.6 and 3.7A). The resistance of the electrodes varies from 8 to 12 MΩ when filled with the internal solution for whole-cell recording, which contains (in m*M*) KCl (130), $MgCl_2$ (5), $NaHCO_3$ (1), $MgSO_4$ (1), HEPES sodium salt (5), K_2ATP (2.5); pH 7.2, 290 mOsm. The electrode is given a positive pressure (10 cm H_2O) while approaching to the cell to prevent blockage of the electrode and clean the surface of the membrane. Once the electrode touches the membrane, a small negative pressure is given through suction to obtain gigaohm seal. Two methods have been tried to gain electrical access to the cell cytoplasm. In whole-cell recording, pulses of negative pressure along with zap signals (600 mV, 0.1 msec) are used to rupture the membrane. In perforated

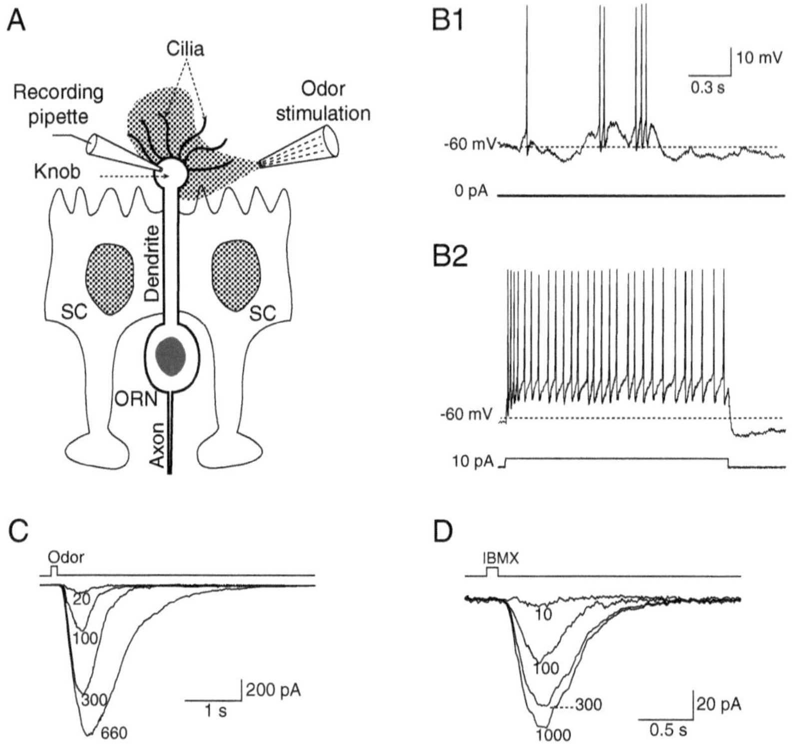

FIGURE 3.7 Electrophysiological properties of rat ORNs. A: Schematic diagram illustrating patch-clamp recording performed on the dendritic knobs and the delivery of odor stimuli by multiple-barrel pipettes. ORN: olfactory receptor neuron; SC, supporting cell. B: Impulse firing patterns of rat ORNs under current-clamp mode. Perforated patch-clamp with amphotericin B is performed on the dendritic knob of a rat ORN. B1: The cell spontaneously fires irregular action potentials. B2: Action potentials are elicited by injecting a depolarizing current of 10 pA into the cell. C: Odor responses in an ORN are elicited by applying 100-msec odor pulses at different concentrations (in μ*M*). Odor stimulation is a mixture of the four following odors at equal concentrations: acetophenone, n-amyl acetate, benzaldehyde, and cineole. D: Responses in a different ORN are elicited by applying 100-msec IBMX pulses at different concentrations. The holding potential in both cells shown in C and D is –60 mV. (Modified from Ma, M., Chen, W.R., and Shepherd, G.M., *J. Neurosci. Methods,* 92, 31, 1999.)

patch clamp, 260 μ*M* amphotericin B or nystatin are used to permeablize the membrane of a cell-attached patch.[36,37] Amphotericin B or nystatin are dissolved in the following solution (in m*M*): KCl (17.7), KOH (105.3), methanesulfonic acid (82.3), EGTA (5.0), HEPES (10), sucrose (70); pH 7.2 (KOH), 310 mOsm. Because the dendritic knobs are such small structures (the diameter of the knobs ranges from 2 to 3.5 μm for rat ORNs and 1 to 2 μm for mouse ORNs), perforated patch clamp is more practical than whole-cell recordings.

The electrophysiological properties of rodent ORNs are characterized under the whole-cell configuration. Most ORNs recorded from the intact epithelial preparation fire spontaneous action potentials (Figure 3.7B1). When stimulated by depolarizing current injections, nearly half of the ORNs fire repetitive action potentials (Figure 3.7B2) and the mean firing rates increase with the intensity of injected currents in a linear manner.

3.3.4 Odor Responses of Individual ORNs: Patch Clamp Recording

The odor stimuli are delivered by pressure ejection using Picospritzer II (General Valve Co., Fairfield, NJ) through multiple-barrel pipettes that are placed within 5 to 10 μm downstream of the dendritic knobs. All odors are prepared as 0.4-*M* stock solutions in dimethyl sulfoxide (DMSO) and diluted to the final concentrations by adding ACSF. 3-Isobutyl-1-methyl-xanthine (IBMX) is prepared in a 10-m*M* stock solution containing 5% DMSO and diluted to the final concentrations with ACSF.

Odor response properties of rodent ORNs are tested by puffing odors directly onto the cilia and the knobs of ORNs (Figure 3.7A). Under voltage-clamp mode, odor stimuli induce up to 1.5 nA inward currents in a dose-dependent manner (Figure 3.7C). IBMX, a potent inhibitor of phosphodiesterase, also induces inward currents in a dose-dependent manner, but with a shorter delay — presumably by increasing the intracellular cAMP level directly, which in turn open the CNG channels (Figure 3.7D).

3.3.5 Odor Responses of ORN Populations: Calcium Imaging Recording

Our calcium imaging system consists of a SensiCam Super VGA camera (PCO, Germany) mounted on an upright microscope (BX50WI) equipped with a fluorescence attachment (Olympus BX-FLA), a calcium green filter set (XF22, Omega Optical Inc.), a 150-W Xenon short arc lamp (USHIO Inc., Japan), a PCI interface card, and Imaging Workbench 2.1 (Axon Instrument Inc., CA) as software running on a PC machine (Dell, XPS R400). Using a 40× objective and a binning factor of 2, the camera can view an image size of 215 μm by 172 μm with a resolution of 0.33 μm/pixel.

The intact epithelial preparation allows one to simultaneously visualize large populations of the dendritic knobs of mouse ORNs, which makes it possible to monitor odor responses from a group of mouse ORNs with their spatial relations intact by using calcium imaging techniques. The calcium changes in the dendritic knobs have been recorded in dissociated rodent ORNs.[27,29] It has been noted that the time course of calcium changes in the dendritic knob is closer to that in the cilia, whereas calcium changes in the dendrite and soma are slower and longer lasting.[18]

Before the epithelial preparation is transferred to the recording chamber, it is loaded with calcium-sensitive dye (calcium green-1/AM, Molecular Probes, Eugene, OR). The dye is prepared as a 5-m*M* stock solution in DMSO with 20% Pluronic F-127 (Molecular Probes) and then diluted to a loading solution with a final concentration of 10 to 20 μ*M* by adding ACSF. Olfactory epithelia attached to the septal wall are placed in a 1.5-ml tube containing the oxygenated loading solution for a period of 20 to 40 min, followed by a washing-out period of at least 20 min. To reduce the disturbance of the bath during recording, odor stimulation is applied through bath perfusion.

Under fluorescence illumination, up to 400 dendritic knobs of mouse ORNs can be seen as bright spots in one image; no supporting cells are labeled (Figure 3.8). Calcium imaging data are acquired at 0.5 to 1 Hz, with five frames as baseline before the onset of odor delivery during each trial. The fluorescence changes are calculated for individual dendritic knobs as follows:

$$\Delta F/F = [(F_1 - B_1) - (F_0 - B_0))/(F_0 - B_0)$$

where F_1 and B_1 are fluorescence in the knob and background, respectively, at any given time point; and F_0 and B_0 are fluorescence at the beginning of the experiment, which are averaged from the first five control frames. In the preparation shown in Figure 3.8, 30 out of 216 dendritic knobs show increased fluorescence intensity responding to an odor mixture. Our studies thus far demonstrate that imaging odor-induced responses of ORN populations in the intact epithelium may provide an efficient way for functional analysis of olfactory coding at the epithelial level.[38,39]

3.4 CONCLUSION

The intact rodent epithelial preparation is suitable for providing functional mapping of responsive ORNs *in situ.* The ORNs in the preparation are closer to *in vivo* physiological conditions than isolated ORNs. Most of them have well-protected cilia, intact dendrites, and relatively long axons, with the mucus, supporting cells, and lamina propria maintained in their normal configuration. Presumably, only mature ORNs are examined and recorded here because the preparation is viewed from the surface.[40] The recordings are made from the dendritic knob, which is close to the cilia where the olfactory signal transduction takes place. An attractive application of the preparation is to simultaneously monitor the activities of hundreds of ORNs at the resolution of individual cells via calcium imaging. Physiological studies can potentially be carried out on these preparations with known locations in different olfactory receptor expression zones with maintained spatial relations among neighboring ORNs. This is especially useful in using transgenic mice as model animals in which molecular markers are introduced into the olfactory epithelium. This preparation, combined with optical imaging technique, can enhance the study of peripheral coding of olfactory information by providing physiological data with greater efficiency than would be possible to obtain by other methods.

ACKNOWLEDGMENTS

This work was supported by grants to G.M.S. from NIDCD (R01); NIDCD, NIA, NASA and NIMH (Human Brain Project); and ARO (MURI); and a PHS training grant to M.M.

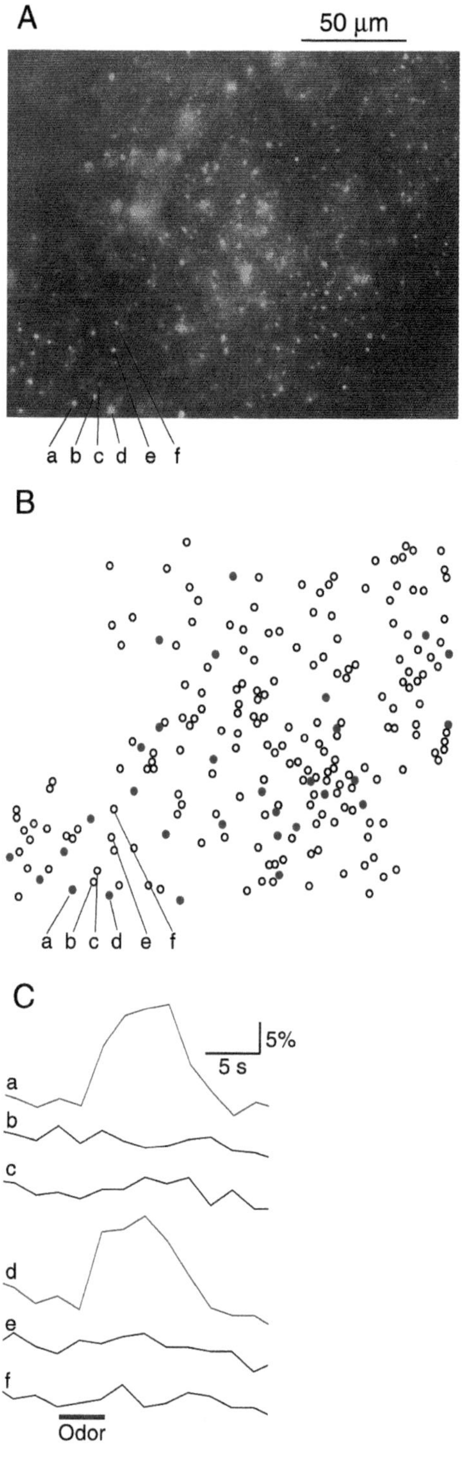

FIGURE 3.8 (see color insert following page 238) Mapping odor responses of mouse ORNs in the intact epithelial preparation by calcium imaging. A: Many dendritic knobs can be seen as bright spots under fluorescence illumination in an *en face* view, after a swatch of epithelium is loaded with 15 μ*M* calcium green-1 AM. B: 216 dendritic knobs are mapped for further analysis. A circle represents a single cell. Red-filled circles (30 out of 216) stand for cells that respond to a 4-s odor stimulus, a mixture containing the following six odors at 10 μ*M*: acetophenone, anisaldehyde, n-amyl acetate, benzaldehyde, cineole, and 3-heptanone. C: The fluorescence intensity changes (ΔF/F) induced by an odor stimulus are measured and calculated for individual dendritic knobs; the traces corresponding to cells "a–f" are shown. (Modified from Ma, M. and Shepherd, G.M., *Proc. Natl. Acad. Sci. U.S.A.*, 97, 12869, 2000.)

REFERENCES

1. Ottoson, D., Analysis of the electrical activity of the olfactory epithelium, *Acta Physiol. Scand. Suppl.*, 122, 6, 1956.
2. Mackay-Sim, A. and Kesteven, S., Topographic patterns of responsiveness to odorants in the rat olfactory epithelium, *J. Neurophysiol.*, 71, 150, 1994.
3. Scott, J.W., Davis, L.M., Shannon, D., and Kaplan, C., Relation of chemical structure to spatial distribution of sensory responses in rat olfactory epithelium, *J. Neurophysiol.*, 75, 2036, 1996.
4. Scott, J.W., Shannon, D.E., Charpentier, J., Davis, L.M., and Kaplan, C., Spatially organized response zones in rat olfactory epithelium, *J. Neurophysiol.*, 77, 1950, 1997.
5. Belluscio, L., Gold, G.H., Nemes, A., and Axel, R., Mice deficient in G(olf) are anosmic, *Neuron*, 20, 69, 1998.
6. Frings, S. and Lindemann, B., Single unit recording from olfactory cilia, *Biophys. J.*, 57, 1091, 1990.
7. Frings, S. and Lindemann, B., Current recording from sensory cilia of olfactory receptor cells *in situ*. I. The neuronal response to cyclic nucleotides, *J. Gen. Physiol.*, 97, 1, 1991.
8. Masukawa, L.M., Hedlund, B., and Shepherd, G.M., Electrophysiological properties of identified cells in the *in vitro* olfactory epithelium of the tiger salamander, *J. Neurosci.*, 5, 128, 1985.
9. Kent, P.F. and Mozell, M.M., The recording of odorant-induced mucosal activity patterns with a voltage-sensitive dye, *J. Neurophysiol.*, 68, 1804, 1992.
10. Kent, P.F., Mozell, M.M., Murphy, S.J., and Hornung, D.E., The interaction of imposed and inherent olfactory mucosal activity patterns and their composite representation in a mammalian species using voltage-sensitive dyes, *J. Neurosci.*, 16, 345, 1996.
11. Youngentob, S.L., Kent, P.F., Sheehe, P.R., Schwob, J.E., and Tzoumaka, E., Mucosal inherent activity patterns in the rat: evidence from voltage-sensitive dyes, *J. Neurophysiol.*, 73, 387, 1995.
12. Restrepo, D., Miyamoto, T., Bryant, B.P., and Teeter, J.H., Odor stimuli trigger influx of calcium into olfactory neurons of the channel catfish, *Science*, 249, 1166, 1990.
13. Firestein, S. and Werblin, F., Odor-induced membrane currents in vertebrate-olfactory receptor neurons, *Science*, 244, 79, 1989.
14. Nakamura, T. and Gold, G.H., A cyclic nucleotide-gated conductance in olfactory receptor cilia, *Nature*, 325, 442, 1987.
15. Kurahashi, T., Activation by odorants of cation-selective conductance in the olfactory receptor cell isolated from the newt. *J. Physiol. (London)*, 419, 177, 1989.
16. Lowe, G. and Gold, G.H., Nonlinear amplification by calcium-dependent chloride channels in olfactory receptor cells, *Nature*, 366, 283, 1993.
17. Maue, R.A. and Dionne, V.E., Preparation of isolated mouse olfactory receptor neurons, *Pflügers Arch.*, 409, 244, 1987.
18. Leinders-Zufall, T., Rand, M.N., Shepherd, G.M., Greer, C.A., and Zufall, F., Calcium entry through cyclic nucleotide-gated channels in individual cilia of olfactory receptor cells: spatiotemporal dynamics, *J. Neurosci.*, 17, 4136, 1997.
19. Firestein, S., Shepherd, G.M., and Werblin, F.S., Time course of the membrane current underlying sensory transduction in salamander olfactory receptor neurones, *J. Physiol. (London)*, 430, 135, 1990.
20. Firestein, S., Darrow, B., and Shepherd, G.M., Activation of the sensory current in salamander olfactory receptor neurons depends on a G protein-mediated cAMP second messenger system, *Neuron*, 6, 825, 1991.

21. Firestein, S., Picco, C., and Menini, A., The relation between stimulus and response in olfactory receptor cells of the tiger salamander, *J. Physiol. (London)*, 468, 1, 1993.
22. Kurahashi, T. and Yau, K.W., Co-existence of cationic and chloride components in odorant-induced current of vertebrate olfactory receptor cells, *Nature*, 363, 71, 1993.
23. Kurahashi, T. and Menini, A., Mechanism of odorant adaptation in the olfactory receptor cell, *Nature*, 385, 725, 1997.
24. Lowe, G. and Gold, G.H., The spatial distributions of odorant sensitivity and odorant-induced currents in salamander olfactory receptor cells, *J. Physiol. (London)*, 442, 147, 1991.
25. Nakamura, T., Tsuru, K., and Miyamoto, S., Regulation of Ca^{2+} concentration by second messengers in newt olfactory receptor cell, *Neurosci. Lett.*, 171, 197, 1994.
26. Okada, Y., Teeter, J.H., and Restrepo, D., Inositol 1,4,5-trisphosphate-gated conductance in isolated rat olfactory neurons, *J. Neurophysiol.*, 71, 595, 1994.
27. Restrepo, D., Okada, Y., and Teeter, J.H., Odorant-regulated Ca^{2+} gradients in rat olfactory neurons, *J. Gen. Physiol.*, 102, 907, 1993.
28. Schild, D., Lischka, F.W., and Restrepo, D., InsP3 causes an increase in apical $[Ca^{2+}]_i$ by activating two distinct current components in vertebrate olfactory receptor cells, *J. Neurophysiol.*, 73, 862, 1995.
29. Tareilus, E., Noe, J., and Breer, H., Calcium signals in olfactory neurons, *Biochim. Biophys. Acta*, 1269, 129, 1995.
30. Zufall, F. and Leinders-Zufall, T., Identification of a long-lasting form of odor adaptation that depends on the carbon monoxide/cGMP second-messenger system, *J. Neurosci.*, 17, 2703, 1997.
31. Leinders-Zufall, T., Greer, C.A., Shepherd, G.M., and Zufall, F., Imaging odor-induced calcium transients in single olfactory cilia: specificity of activation and role in transduction, *J. Neurosci.*, 18, 5630, 1998.
32. Hirono, J., Sato, T., Tonoike, M., and Takebayashi, M., Simultaneous recording of $[Ca^{2+}]_i$ increases in isolated olfactory receptor neurons retaining their original spatial relationship in intact tissue, *J. Neurosci. Methods*, 42, 185, 1992.
33. Sato, T., Hirono, J., Tonoike, M., and Takebayashi, M., Tuning specificities to aliphatic odorants in mouse olfactory receptor neurons and their local distribution, *J. Neurophysiol.*, 72, 2980, 1994.
34. Zhao, H., Ivic, L., Otaki, J.M., Hashimoto, M., Mikoshiba, K., and Firestein, S., Functional expression of a mammalian odorant receptor, *Science*, 279, 237, 1998.
35. Malnic, B., Hirono, J., Sato, T., and Buck, L.B., Combinatorial receptor codes for odors, *Cell*, 96, 713, 1999.
36. Horn, R. and Marty, A., Muscarinic activation of ionic currents measured by a new whole-cell recording method, *J. Gen. Physiol.*, 92, 145, 1988.
37. Rae, J., Cooper, K., Gates, P., and Watsky, M., Low access resistance perforated patch recordings using amphotericin B, *J. Neurosci. Methods*, 37, 15, 1991.
38. Ma, M., Chen, W.R., and Shepherd, G.M., Electrophysiological characterization of rat and mouse olfactory receptor neurons from an intact epithelial preparation, *J. Neurosci. Methods*, 92, 31, 1999.
39. Ma, M. and Shepherd, G.M., Functional mosaic organization of mouse olfactory receptor neurons, *Proc. Natl. Acad. Sci. U.S.A.*, 97, 12869, 2000.
40. Farbman, A.I., *Cell Biology of Olfaction*, Cambridge University Press, Cambridge, 1992, 58.

4 Recordings from Olfactory Receptor Neurons in the Rat *in Vivo*

M.A. Chaput, P. Duchamp-Viret, and A. Duchamp

CONTENTS

4.1 INTRODUCTION

Electrophysiological recordings of single units and global responses (electro-olfactogram or EOG) from the olfactory epithelium *in vivo* lead to direct assessment of odor reactivity of the peripheral level of the olfactory pathways. In the past, such recordings have been extensively utilized to investigate odor coding in amphibians (Duchamp-Viret and Duchamp, 1997, for a review). In the frog, the sensitivity of

0-8493-2329-0/02/$0.00+$1.50

single olfactory receptor neurons (ORNs) to pure odorants was systematically studied in our laboratory for a hundred odor molecules (Duchamp et al., 1974; Revial et al., 1978a, b, 1982, 1983). Individual ORNs were found to respond to different chemical groups of pure odorant compounds such as camphor, terpene, and aromatic molecules and were thus considered poorly selective. This led us to propose the concept of qualitative olfactory space where odors are grouped according to the similitude of the response profiles they elicited across the ORNs' population. According to this poor selectivity, odor quality coding was assumed to be based on an across-neuron pattern constituted by a unique combination of ORNs for each odor molecule.

More recently, molecular olfactory receptors (ORs) where odor molecules interact with ORNs have been identified (Buck and Axel, 1991). From the convergent results of genetic and molecular biology studies (Buck, 1993; Kishimoto et al., 1994; Ressler et al., 1993; Vassar et al., 1993; Zhao et al., 1998), ORs have been shown to be G protein-coupled seven transmembrane proteins and each ORN was hypothesized to be likely endowed with a single OR. Thus, following these lines of evidence, only one odor or a small group of odors having a similar molecular structure was assumed to bind to a single OR, and rat ORNs were hypothesized to be highly selective, in contrast with frog ORNs.

Given this state of knowledge, the gap between electrophysiological and molecular data obtained respectively in the frog and rat made it necessary to study the electrophysiological response properties of individual ORNs to odors in the rat *in vivo*. Such experiments, by offering direct access to the ORN odor reactivity in physiological conditions, revealed that rat ORNs are as poorly selective as those of the frog (Duchamp-Viret et al., 1999; 2000). This important result has been confirmed through the molecular biology approach of Malnic and colleagues (1999). This chapter describes the methods utilized to record EOG and single-unit ORN activities in the rat *in vivo* and illustrates the main results obtained.

4.2 METHODS

4.2.1 ANIMAL

Experiments presented in this chapter were carried out in accordance with the European Communities Council Directive for the care and use of laboratory animals. Adult Wistar rats (250 to 300 g) purchased from IFFA-CREDO (L'Arbresle, France) were anesthetized by an intraperitoneal injection of equithesine at the initial dose of 3 ml/kg. Anesthetic was supplemented as necessary to maintain a deep level of anesthesia, as determined by the lack of withdrawal reflex of the leg in response to a moderately intense toe pinch. Rectal temperature was maintained at 37 ± 0.5°C by a homeothermic blanket (Harvard apparatus, U.S.A.) and surgical wounds of the animals were regularly infiltrated with 2% procaine. To obtain 250 ml equithesine, 10.625 g chloral hydrate were dissolved in 24.7 ml ethanol. They were added to 40.5 ml pentobarbital at 6×10^{-3} dilution, 99 ml propylene glycol (propane diol), and 5.315 g magnesium sulfate previously dissolved in distilled water. The volume was adjusted to 250 ml with distilled water.

4.2.1.1 Tracheotomy

In some experiments, animals were tracheotomized to limit the movement of the turbinates due to the passage of the respiratory flow and to reduce the risk of olfactory mucosa drying. Tracheotomy facilitated recordings. It was nevertheless possible to record EOG and single units in freely breathing rats, i.e., without bypassing the alternation of the air flows coming in and going out of the nasal cavity during inhalation and exhalation periods, respectively. In the latter condition, the respiratory activity influenced the EOG and single-unit responses, and multi-peak responses synchronized on inhalation were observed as shown in Figure 4.2.

4.2.1.2 Access to the Mucosa

For recordings, anesthetized animals were secured in a stereotaxic apparatus. Access to the olfactory mucosa was gained by removing the nasal bones and then gently slipping aside the dorsal recess underlying these bones. A midline rostro-caudal incision was first made in the skin covering the head. Then the nasal bone was gently drilled while paying careful attention not to damage the underlying mucosa because blood on the recording site will prevent the odor molecules to reach ORs. The bone was first drilled rostro-caudally following the suture between the two nasal bones and the suture between the nasal and incisivium bones, before joining the ends of the previously drilled lines (Figure 4.1). The delimited piece of bone was gently lifted with a blunt spatula and removed so as to expose the underlying mucosa. This mucosa was rostro-caudally incised with the tip of a small scalpel blade and then gently pulled over the edges of the wound so as to expose the endoturbinate II in the depth of the nasal cavity.

4.2.1.3 Recording Sites

The opening of the nasal cavity permitted access to the endoturbinate II and to a portion of the septal wall (Figure 4.1). EOG and single-unit recordings could be performed on the dorsal and lateral parts of this turbinate and on the septal wall. On the dorsal endoturbinate II, units were generally isolated at a depth ranging from 0 to 350 μm.

4.2.2 Electrodes

4.2.2.1 EOG

The EOG signal was recorded with glass micropipettes (Glass reference : H15/10 from Jencons Scientific, Cherrycourt Way, Stanbridge Road, Leighton Buzzard, Bedfordshire, LU7 8UA, England) pulled with a vertical puller (Sutter Instrument Company; 51 Digital Drive, Novato, CA 94949, http://www.sutter.com/home.html). Their tips were broken to about 50-μm OD and they were filled with a saline solution (NaCl 0.9% or Ringer solution) thickened with 5% of agar-agar or gelatin to prevent the recording electrolyte from flowing out of the micropipette.

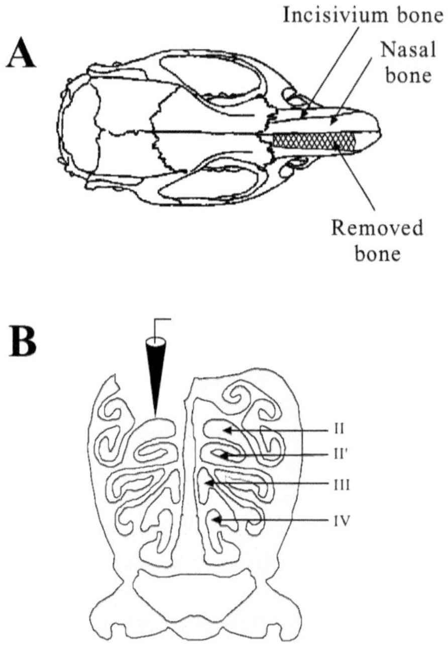

FIGURE 4.1 A: Dorsal view of the rat skull showing the position of the removed bone to gain access to the olfactory mucosa. B: Schematic cross-section of the rat nose, indicating the position of the recording sites and their relationships to the turbinate bones. Abbreviations: II and II′, dorsal and ventral lamella of endoturbinate II, respectively; III, endoturbinate III; IV, endoturbinate IV.

4.2.2.2 Single Units

ORN action potentials were recorded using metal-filled glass micropipettes, the properties and manufacturing of which have been fully described by Gesteland and collaborators (1959). Briefly, these microelectrodes were made with H15/10/137 glass tubes from Jencons Scientific pulled with a microelectrode puller from Sutter Instrument. Their tips were broken off under direct vision through a microscope so as to obtain an OD of about 2 to 3 μm. Then, a small piece of laboratory-made alloy rod composed of a mixture of indium (20%) and wood metal (80%) was inserted into the shank of the micropipette and pushed to its tip using a stainless steel rod just thin enough to fit into the micropipette shank. This operation was performed in front of a hot air stream just warm enough to melt the alloy (melting point ~70°C) while the stainless steel rod was constantly pushed to completely fill the micropipette tip. Electrodes were then checked under a microscope to verify that they had not been broken by the filling operation and that they were completely and perfectly filled (no bubble in the alloy). Then the stainless steel rod was replaced by a copper wire hot-welded at the back of the electrode.

Electrode impedance was measured at 1000 Hz. It was adjusted by plugging the microelectrode tip into an bath containing an electrolyte composed of 1% platinum tetrachloride, 0.01% of lead acetate, and 2% viscous gelatin and by plateing it with a platinum ball by passing a 0.1-μA current through the electrode. The best ORN recordings were obtained with microelectrodes having an impedance ranging from 2 to 4 MΩ. Microelectrodes were often re-used several times in the same day and even for several days. Between two penetrations, the plateing operation should be repeated after having the electrode tip cleaned between two humidified fingers. Although delicate and time-consuming to manufacture, such metallic microelectrodes are, in our opinion, the best we have ever utilized to record individual ORNs in the frog and rat in terms of the signal-to-noise ratio and the stability and selectivity of the recordings. Indeed, it is also possible to record ORNs with micropipettes of about 4 to 5 μm OD at the tip, filled with 3 *M* NaCl and 10 g/l gelatin and having an impedance of 7 to 10 MΩ. In that case, a single electrode can be utilized to monitor both EOG and single spikes (Adamek and Gesteland, 1984), but we found such microelectrodes very noisy and few that were stable, especially for ORN single-unit recordings.

4.2.3 Recordings and Data Acquisition

EOG and single units were recorded with laboratory-made amplifiers, the former with a DC amplifier with a high impedance input (>100 MΩ) and the latter with an AC passband amplifier (300 to 3000 Hz) with a high-impedance input probe (> 500 MΩ). Data were stored on a Data Tape Recorder (Bio-Logic — Science Instruments SA, http://www.bio-logic.fr) 1 rue de l'Europe, F-38640 — CLAIX — France. During experiments, the single-unit nature of the recordings was controlled online by triggering the recorded cells near the background noise on the screen of a storage oscilloscope. This allowed one to control the characteristics of the polyphasic spike of the studied cell to ensure that the same cell was recorded throughout the experimental procedure. The single-unit activity and EOG signal were sampled offline at 15 kHz and 200 Hz, respectively, using a CED-1401 data acquisition system (Cambridge Electronic Design Ltd., http://www.ced.co.uk) Science Park, Milton Road, Cambridge CB4 0FE, England, connected to a computer. They were processed offline using the programming facilities offered by the Spike2 language (Cambridge Electronic Design Ltd., United Kingdom). Spikes suspected to belong to the same cell were first selected by triggering the unitary signal with respect to a trigger level adjusted on the computer screen, and then by verifying by visual inspection the consistency of the shape of the sorted spikes on the computer screen. The EOG signal was analyzed so as to determine its peak amplitude and its latency with respect to stimulation onset.

4.2.4 Odor Stimulation

Pure odor compounds (see Duchamp-Viret et al., 2000 for the exhaustive list of compounds) were applied directly near the surface of the turbinate using a dynamic multistage olfactometer described elsewhere (Vigouroux et al., 1988) that ensured precise control of the concentration range and allowed the delivery of 12 discrete concentrations. Stimuli consisted of odor pulses of 2-s duration delivered at 200 ml/min. Depending on their saturated vapor pressure (SV), compounds were delivered at concentrations ranging from 3×10^{-8} or 5×10^{-7} *M* (SV/562) to 2×10^{-5} or 3×10^{-4} *M* (SV).

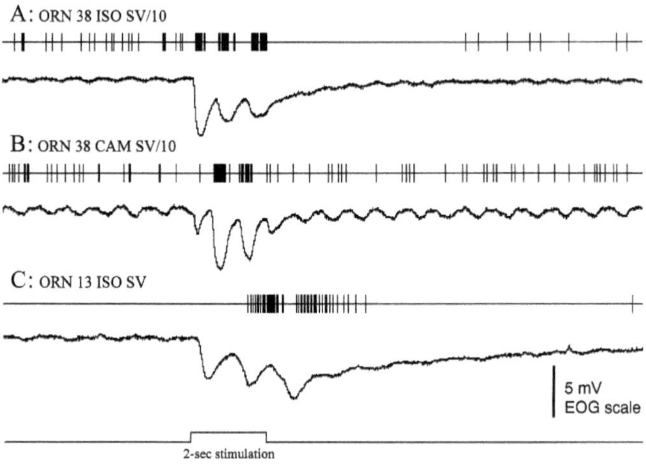

FIGURE 4.2 Examples of ORN (top traces) and EOG (bottom traces) responses recorded in freely breathing rats. In A and B are shown the EOG and single-unit responses of the same ORN to isoamyl acetate (ISO) and camphor (CAM) at SV/10. In C, another ORN was stimulated with ISO delivered near the saturated vapor pressure. The stimulus was delivered during 2 s. Respiration was not recorded per se, but was often visible on the EOG trace, as in B.

4.3 ELECTROPHYSIOLOGICAL RECORDINGS

4.3.1 Response Characteristics in Freely Breathing Rats

EOG signals and single-unit firing activities recorded before, during, and after odor presentation in two different freely breathing rats are shown in Figure 4.2. This figure illustrates the main types of EOG and single ORN responses observed in breathing animals. In the two animals, isoamyl acetate (ISO) and camphor (CAM) elicited a multiphasic EOG composed of successive waves synchronized with respiration. EOG amplitude decreased from the first to the third wave in Figure 4.2A, and from the second to the third wave in Figure 4.2B. This reflects a slight adaptation. However, EOG waves were found to persist during long-lasting odor presentations (Chaput, M.A. unpublished observations). In Figure 4.2B, the first EOG wave was smaller than successive ones because the stimulus arrived during inhalation. In Fig 2C, the EOG amplitude clearly increased from the first to the third waves. This illustrates the progressive recruitment of an increasing number of ORNs.

Single-unit responses could be clearly matched to the EOG signal recorded simultaneously. For example, the ORN shown in Figure 4.2C was recruited during the second EOG wave. Given that EOG mirrors ORN population recruitment, this ORN may be assumed to belong to a supplementary population of ORNs recruited during the second wave. In contrast, the ORN in Figure 4.2A started to discharge as soon as the early phase of the first EOG wave and then fired one burst per EOG wave. At the same time, the amplitude of the second and third waves decreased, indicating that some ORNs adapted and thus did not further participate in the EOG.

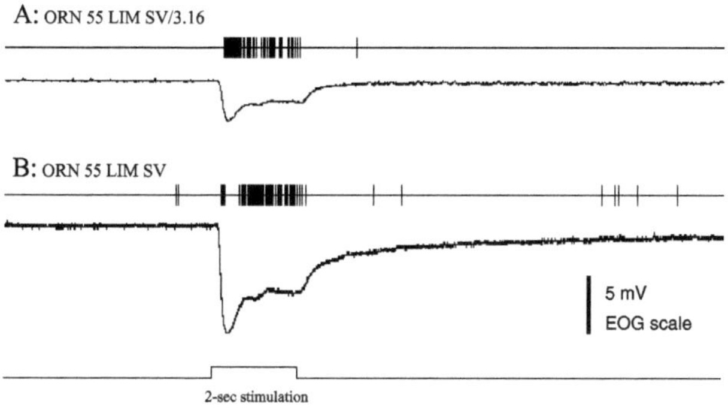

FIGURE 4.3 Examples of EOG and single-unit responses of the same ORN to limonene (LIM) at SV/3.16 (A) and SV (B) recorded in tracheotomized rat.

Finally, the ORN in Figure 4.2B was not recruited until the second EOG wave. This is likely due to the lack of synchronization of the stimulus delivery on the respiratory activity evoked above, an argument that is coherent with the observation of the small first EOG wave.

To summarize, EOG and single-unit responses are generally synchronized with respiration in freely breathing rats. Single units, having reached their threshold, have their spike discharge time-locked on successive EOGs and deliver one burst per wave. This time-locking may also express through phasic frequency increases that punctuate a tonic response. Moreover, according to the cells and odors, the phase-locking of the spike bursts over the time course of EOG can differ.

4.3.2 Response Characteristics in Tracheotomized Rats

Figure 4.3 exemplifies EOG and single-unit odor responses recorded in tracheotomized animals. When the respiratory flow is suppressed in the nasal cavity, the EOG response is monophasic and resembles EOG responses recorded in the frog or the salamander. It shows a unique depolarization phase, followed by a plateau that persists for the entire duration of stimulus delivery. The repolarization begins immediately after the odor is switched off, as evidenced by the track break-off. ORNs respond by a temporal firing pattern that is time-locked to the stimulus and EOG duration. Depending on stimulus intensity, it consists of a tonic burst (Figure 4.3A) or in phasic bursts followed by a rebound (Figure 4.3B).

4.3.3 Single Unit Temporal Response Patterns and Changes with Odor Concentration

In our opinion, the fact that different odors induce different temporal patterns can be ascribed almost exclusively to the fact that they induce different levels of ORN depolarization. In other words, differences in temporal patterns do not specify odor quality. In contrast, they ensure odor intensity coding. These assertions are based not only on our recent results in the rat, but also on our previous observations in the frog (Duchamp-Viret, 1988). In the ORN population, the standard profile of evolution that we observed with concentration is the following: at threshold and near-threshold concentrations, ORN response patterns are generally simple and consist of a tonic response. Then, the response latency decreases, the discharge frequency increases, while the response duration decreases. At high concentrations, the pattern is a phasic early burst with spikes of decreasing amplitudes: the sustained depolarization reduces the driving force for Na^+ and K^+ for high depolarization levels. This decremental response is often followed by a rebound of activity with spikes of increasing amplitudes that mirrors the return of the cell membrane potential to its resting value. In theory, if stimuli could be delivered over a continuous concentration range, all stages of the standard evolution profile described above should be observed. As we delivered discrete concentrations sampled over a concentration range, only some of the possible patterns of a given cell were observed.

Figure 4.4 illustrates the fact that ORN temporal response patterns do not specify odor quality. Indeed, the same cell showed very similar response patterns to very different odors. It was excited by the eight odorants. Anisole (ANI), methyl amyl ketone (MAK), isoamyle acetate (ISO), limonene (LIM), and cyclodecanone (CDN) evoked a tonic excitation at the lowest concentration, for which the cell threshold was clearly surpassed. In contrast, excitations were less pronounced for cyclohexanone (XON), cineole (CIN), and vanillin (VAN) because they likely corresponded to threshold responses. When considering these responses with respect to the saturated vapor pressures of the odorants, the most efficacious stimulus in terms of the ratio between the cell depolarization level and the odorant molecular concentration was ANI, followed by MAK and ISO, XON, CDN, CIN, and VAN, respectively.

The evolution of ORN temporal response patterns as described above are illustrated through the concentration/response series shown in Figure 4.5. This figure gives, in parallel, a typical example of the evolution of the response of a single ORN and the EOG response recorded in the same region of the mucosa. As evidenced by comparing the single-unit traces, the cell displayed large response dynamics ranging over more than a two-logarithm concentration scale. At the lowest delivered concentration (SV/562), the response consisted of a sustained tonic discharge that corresponded to a clear supra-threshold response. It evolved toward an initial phasic decremental discharge, followed by an incremental rebound when concentrations were increased (as shown by the traces obtained from SV/100 to SV/10). Frequencies in the initial discharge continuously increased from threshold to maximum while latencies progressively shortened.

Through typical examples of ORN responses to odorants, Figure 4.5 also illustrates the ambiguity in determining the response type of a cell without having the possibility to expand the time scale of the recorded trace. Indeed, the cell presented in this figure might be suspected to respond to SV/31.6 and SV/10 by an inhibition followed by a later and later off effect. However, when looking at the initial part of the response on an expanded time scale (Figure 4.6), it becomes evident that the initial response was not an inhibition, but a high-frequency excitatory response composed of decremental spikes. Thus, the mode of response of the cell is coherent throughout the concentration range: from threshold to maximum, it was an excitation.

The presence of decremental and incremental spikes is a well-known phenomenon, already observed through extracellular (Revial et al., 1982) as well as intracellular

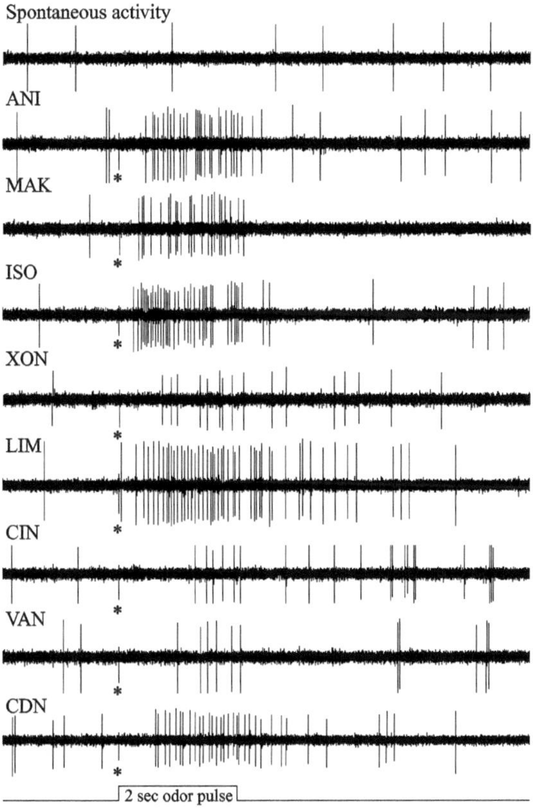

FIGURE 4.4 Spontaneous activity (top trace) and representative responses of the same ORN to different odor stimuli delivered at the lowest available concentration that induced a response. Anisole (ANI), methyl amylketone (MAK), and ISO were delivered at SV/562, the lowest concentration allowed by the olfactometer. They induced sustained responses. Their thresholds were thus lower than 10^{-7} *M*. Cyclohexanone (XON) induced a response at SV/100 (i.e., 2.5 × 10^{-6} *M*). LIM, cineole (CIN), and vanillin (VAN) induced a response at SV/10 and cyclodecanone (CDN) at SV/5. The asterisk indicates the stimulation artifact. (From Duchamp-Viret, P., Chaput, M.A., and Duchamp, A., *Science,* 284, 2171-2174, 1999. With permission.)

(Trotier and Mac Leod, 1983) recordings performed in different species. When discharge frequencies are very high, spike amplitudes can be reduced in proportion to the amplitude of the odor-evoked depolarization, as shown by Trotier and Mac Leod (1983) through intracellular recordings in the salamander. When repolarizing, ORNs give incremental discharges that represent in some ways an off effect of the initial response.

4.4 CONCLUSION

Until recently there were few physiological data on ORNs in mammals. This chapter demonstrates that it is possible to record the spontaneous activity and odor responses

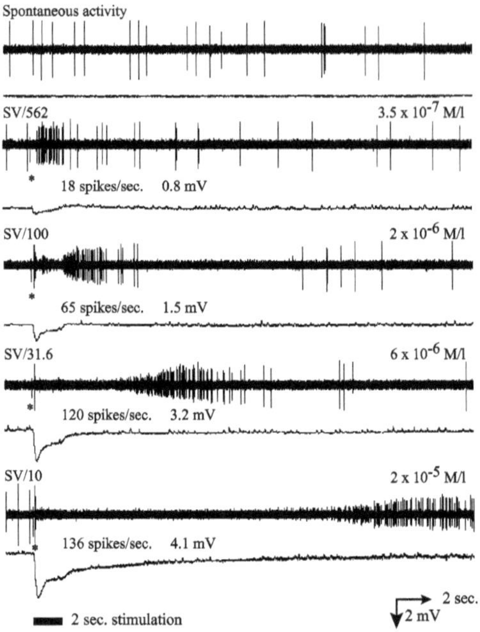

FIGURE 4.5 Spontaneous (two top traces) and odor-evoked single-unit responses and EOGs (pairs of lower traces) of the same ORN to increasing concentrations of ANI. Firing frequencies in the initial response burst and EOG amplitudes are given above the corresponding pairs of recordings. Concentrations are given in proportions of SV and in mole/liter on the left and right side of the figure, respectively. Stars below single-unit recordings indicate the artifact at the beginning of the stimulation. (From Duchamp-Viret et al., *J. Neurosci.,* 20(6), 2383–2390, 2000. With permission.)

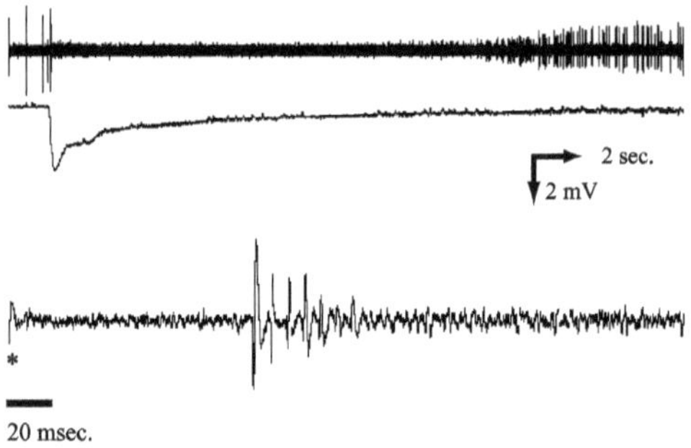

FIGURE 4.6 Example of decremental single-unit response (EOG is given in parallel) to ANI at SV/10 (top traces) and the initial burst of the same response at an expanded time scale (bottom trace). It clearly appears that the initial response was a high-frequency burst, and not a suppression, as it seems when observing the upper single-unit trace.

of a single ORN *in vivo* in freely breathing as well as in tracheotomized adult rats. The preparation described allows one to obtain such recordings from the endoturbinate II and the septal wall. Access to other turbinates by passing through the frontal bone is impossible because it requires the removal of endoturbinate II, a bloody operation that cannot be performed in living animals. It is possible through other surgical routes (Scott et al., 1997; Scott and Brierly, 1999), but requires one to kill the animals. However, we observed that when rats died, EOG turned to a square form and ORNs rapidly ceased to be functional; spontaneous activity and odor responsiveness disappeared within few minutes.

Regarding the single-unit recording technique, it is always tedious to isolate a single ORN with a high signal-to-noise ratio, although the use of the metallic electrodes described above facilitates the task. Patience and know-how are necessary to succeed in ORN recordings. When descending a microelectrode into the mucosa, spikes can be picked up as soon as the tip reaches the surface of the epithelius. The difficulty is to find and isolate a single ORN that emits large spikes and then to estimate the stability of the recordings.

To obtain insight into the coding mechanisms underlying olfactory discrimination, electrophysiological experiments offer the advantage of direct characterization of the odor response properties of individual ORNs. A fruitful approach would be to combine physiological recordings performed in mammals *in vivo* with molecular biology research so as to further analyze the binding properties of the molecular olfactory receptors carried by a single ORN.

REFERENCES

Adamek, A. and Gesteland, R.C. (1984). Saline filled pipette that records olfactory receptor cell activity, *Chem. Senses*, 9, 121-126

Buck, L. (1993). Identification and analysis of a multigene family encoding odorant receptors: implications for mechanisms underlying olfactory information processing, *Chem. Senses,* 18, 203-208.

Buck, L. and Axel, R. (1991). A novel multigene family may encode odorant receptors — a molecular basis for odor recognition, *Cell*, 65, 175-187.

Chaput, M.A., Duchamp-Viret, P., and Duchamp, A. (2000). Peripheral odor coding in the rat and frog: quality and intensity specification, *J. Neurosci.,* (in press).

Duchamp, A., Revial, M.F., Holley, A. and Mac Leod, P. (1974). Odor discrimination by frog olfactory receptors, *Chem. Senses,* 1, 213-233.

Duchamp, A. and Sicard, G. (1984). Influence of stimulus intensity on odour discrimination by olfactory bulb neurons as compared with receptor cells, *Chem. Senses*, 8, 355-366.

Duchamp-Viret, P. (1988). Le codage intensitif du stimulus olfactif: rôle des neurorécepteurs et de deutoneurones, étude électrophysiologique chez la grenouille, Thesis, Université Claude Bernard, Lyons.

Duchamp-Viret, P. and Duchamp, A. (1997). Odor processing in the frog olfactory system, *Prog. in Neurobiol.,* 53, 561-602.

Duchamp-Viret, P., Chaput, M.A., and Duchamp, A. (1999). Odor response properties of rat olfactory receptor neurons, *Science,* 284, 2171-2174.

Duchamp-Viret, P., Duchamp, and A. Chaput, M. (2000). Peripheral odor coding in the rat and frog: quality and intensity specification, *J. Neurosci.,* 20(6), 2383-2390.

Gesteland, R.C. (1959). Comments on microelectrodes, *Proc. Inst. Radio Eng.,* 47, 1856-1862.

Kishimoto, J., Cox, H., Keverne, E.B., and Emson, P.C. (1994). Cellular localization of putative odorant receptor mRNAs in olfactory and chemosensory neurons: a nonradioactive *in situ* hybridization study, *Mol. Brain Res.,* 23, 33-39.

Malnic, B., Hirono, J., Sato, T., and Buck, L.B. (1999). Combinatorial receptor codes for odors, *Cell.* 96,713-723.

Ressler, K.J., Sullivan, S.L., and Buck, L.B. (1993). A zonal organization of odorant receptor gene expression in the olfactory epithelium, *Cell,* 73, 597-609.

Revial, M.F., Duchamp, A., Holley, A., and Mac Leod, P. (1978a). Frog olfaction: odour groups acceptor distribution and receptor categories, *Chem. Senses,* 3, 23-33.

Revial, M.F., Duchamp, A., and Holley, A. (1978b). Odour discrimination by frog olfactory receptors: a second study, *Chem. Senses,* 3, 7-21.

Revial, M.F., Sicard, G., Duchamp, A., and Holley, A. (1982). New studies on odour discrimination in the frog's olfactory receptor cells. I. Experimental results, *Chem. Senses,* 7, 175-190.

Revial, M.F., Sicard, G., Duchamp, A., and Holley, A. (1983). New studies on odour discrimination in the frog's olfactory receptor cells. II. Mathematical analysis of electrophysiological responses, *Chem. Senses,* 8, 179-194.

Scott, J.W., Shannon, D.E., Charpentier, J., Davis, L.M., and Kaplan, C. (1997). Spatially organized response zones in rat olfactory epithelium, *J. Neurophysiol.,* 77, 1950-1962.

Scott, J.W. and Brierly, T. (1999). A functional map in rat olfactory epithelium, *Chem. Senses,* 24, 679-690.

Trotier, D. and Mac Leod, P. (1983). Intracellular recordings from salamander olfactory receptor cells, *Brain Res.,* 268, 225-237.

Vassar, R., Ngai, J., and Axel, R. (1993). Spatial segregation of odorant receptor expression in the mammalian olfactory epithelium, *Cell,* 74, 309-318.

Vigouroux, M., Viret, P., and Duchamp, A. (1988). A wide concentration range olfactometer for delivery of short reproducible odor pulses, *J. Neurosci. Methods,* 24, 57-63.

Zhao, H. Ivic, I., Otaki, J.M., Hashimoto, M., Mikoshiba, K., and Firestein, S. (1998). Functional expression of a mammalian odorant receptor, *Science,* 279, 237.

5 Voltage-Sensitive and Calcium-Sensitive Dye Imaging of Activity in the Olfactory Bulb: Presynaptic Inhibition, Maps of Receptor Cell Input, and Oscillations

Matt Wachowiak, Ying-Wan Lam, Lawrence B. Cohen, and Michal R. Zochowski

CONTENTS

0-8493-2329-0/02/$0.00+$1.50

5.1 INTRODUCTION

An optical measurement using voltage-sensitive or calcium-sensitive dyes as indicators of activity can be beneficial in several circumstances. This chapter describes three examples in the study of olfactory processing. Some details of the methods used to make the measurements are provided for each example. In addition, because the optical signals are small (fractional intensity changes ($\Delta I/I$) of between 5×10^{-2} and 10^{-4}), optimizing the signal-to-noise ratio in the measurements is important. The chapter concludes with a general discussion of optical recording methods, including the choice of dyes, sources of noise, light sources, optics, and cameras. The general approach is three pronged. First, test dyes to find the dye with the largest signal-to-noise ratio. Second, reduce the extraneous sources of noise (vibrations, line frequency noise, etc.). Third, maximize the number of photons measured to reduce the relative shot noise (noise arising from the statistical nature of photon emission and detection).

One advantage of an optical measurement is the possibility of recording from processes that are too small or fragile for electrode recording. We utilize this advantage when recording from the nerve terminals of the olfactory receptor axons in studies of presynaptic inhibition and in determining the maps of the input to the bulb from the olfactory receptor neurons. The second advantage of an optical measurement is the ability to make simultaneous measurements from many locations. This was important in the study of odor-induced oscillations in the bulb and in making maps of the input from receptor neurons.

The voltage-sensitive dyes that we have used are membrane-bound chromophores that change their fluorescence in response to changes in membrane potential. In a model preparation, the giant axon from a squid, these fluorescence signals are fast (following membrane potential with a time constant of <10 μsec) and their size is linearly related to the size of the change in potential. Thus, these dyes provide a direct, fast, and linear measure of the change in membrane potential of the stained membranes. There are other optical signals from membrane-bound dyes (e.g., absorption and birefringence) and another class of dyes sense membrane potential by redistribution; these topics are discussed elsewhere (Cohen and Salzberg, 1978). Similarly, the evidence that pharmacological effects and photodynamic damage resulting from the voltage-sensitive dyes are manageable can be found in earlier reviews (Cohen and Salzberg, 1978; Waggoner, 1979; Salzberg, 1983; Cohen and Lesher, 1986; Grinvald et al., 1988).

The calcium-sensitive dyes we have used are thought to be located in the axoplasm of the neurons and they change their fluorescence in response to changes in the intracellular free calcium. However, the relationship of the dye signals to calcium concentration is generally nonlinear and the dye response often lags behind the change in calcium. In addition, the dyes add to the calcium buffering in the cytoplasm (Neher, 2000). Thus, calcium-sensitive dyes provide a measure for the calcium concentration in the axoplasm but one that must be interpreted with care. Furthermore, in our measurements of the map of the input to the olfactory bulb from olfactory receptor neurons, we are using the calcium signal as a measure of neuron activity; clearly, this measure is indirect.

Voltage-sensitive and calcium-sensitive dyes might be expected to have signals with differing localization even if they were distributed equally over the area of a neuron. A voltage-sensitive dye is expected to have a signal everywhere the action potential exists, in the cell body, along the axon, and in the nerve terminals. On the other hand, a calcium-sensitive dye is expected to have a signal relatively restricted to the nerve terminal because the calcium influx is largest at the nerve terminal.

Two kinds of cameras have been used in our experiments; both have frame rates faster than 1000 ft/sec. One camera is a photodiode array with 464 pixels and the second a cooled, back-illuminated CCD camera with 6400 pixels. Although the spatial resolution of the two cameras differs rather dramatically, the most important difference is in the range of light intensities over which they provide an optimal signal-to-noise ratio. The CCD camera is optimal at low light levels and the photodiode array at high light levels.

In the experiments described below, we used two basic staining protocols. For the voltage-sensitive dyes, we bathed the lobster olfactory lobe or turtle olfactory bulb with a solution of the dye for tens of minutes. In contrast, for the calcium-sensitive dye, we infused the dye solution (containing a low concentration of detergent) into the nose of the turtle and then waited several days for the dye to be transported to the nerve terminals in the olfactory bulb. These two staining procedures are described first. We then show how optical recording of activity in the olfactory bulb can be used to address three different aspects of sensory processing. First, we describe the use of both voltage- and calcium-sensitive dyes to study presynaptic inhibition onto olfactory receptor terminals in the glomeruli. Second, we describe the use of calcium-sensitive dyes to determine the map of receptor cell input to the glomeruli. And, third, we describe the use of voltage-sensitive dyes to study the oscillations that occur in the bulb in response to odor presentation. In all three measurements, a large number of neurons and/or processes are imaged onto each pixel of the camera. Thus, the signals report the average of the changes that occur in this population.

All of the experimental protocols were approved by the Yale Animal Care and Use Committee and by the Marine Biological Laboratory Institutional Animal Care and Use Committee.

5.2 STAINING PROCEDURES

5.2.1 VOLTAGE-SENSITIVE DYES

For the *in vitro* experiments on presynaptic inhibition (Section 5.3), lobster brain tissue was stained with the styryl dye, di-2-ANEPHQ (JPW-2081; synthesized by J. P. Wuskell and L.M. Loew, University of Connecticut Health Center; Farmington,

CT). Turtle olfactory bulb tissue was stained by bath application of either of two styryl dyes, di-4-ANEPPS (0.05 mg/ml) or RH-414 (0.05 mg/ml; Grinvald et al., 1994) (both from Molecular Probes, Eugene, OR) for 30 min. The dye solutions were then replaced by saline and the experiments carried out on dye that remained bound to the preparation. The dye appeared to wash out from the lobster olfactory lobe over a period of about 1 h. The washout time for the turtle experiments was substantially longer.

For the *in vivo* measurements of oscillations in the bulb (Section 5.5), the turtles were first anesthetized by placing them in ice for 2 h. A craniotomy was performed over the olfactory bulb. The dura and arachnoid matter were then carefully removed to facilitate staining. The exposed olfactory bulb was stained by covering it with dye solution (RH414, 0.02 to 0.2 mg/ml) for 60 min. Excess dye was then washed away with turtle saline.

5.2.2 Calcium-Sensitive Dyes

The olfactory receptor axon terminals in the turtle (Sections 5.3 and 5.4) were selectively labeled by anterograde transport of the dextran-conjugated calcium indicator Calcium Green-1 dextran (10 kD molecular weight; Molecular Probes) applied to the olfactory epithelium using a protocol adapted from Friedrich and Korsching (1997). Turtles were first chilled on ice for 1 to 2 h, then placed upside down with the mouth held open and 25 to 60 μl of a 2% dye solution dissolved in 0.1 *M* NaCl plus 0.5% Triton-X 100 was injected into each naris. The pharyngeal opening of each naris was plugged to prevent leakage, and the epithelium was stained for 10 to 20 min, after which the plugs were removed and the mouth closed, allowing the animal to expel the dye from the nose. The turtles were held at room temperature for 4 to 18 days before recording. The nerve terminal labeling is first apparent after about 4 days but appeared most intense after 8 to 10 days. This level of intensity persisted for at least 18 days (the longest interval attempted). The nares and olfactory bulb of experimental animals revealed a relatively intense fluorescence in the olfactory cell body layer of the epithelium and a much less intense staining of the olfactory nerve and glomerular layers of the olfactory bulb. While glomeruli could be clearly distinguished by the labeled fibers, no fluorescence was visible in subglomerular layers (Figure 5.1). We observed no differences in the gross appearance of the olfactory epithelium (postmortem examination) or in the size and sensitivity of electroolfactogram (EOG) recordings between stained and unstained animals.

5.3 PRESYNAPTIC INHIBITION OF PRIMARY OLFACTORY AFFERENTS IS MEDIATED BY DIFFERENT MECHANISMS IN LOBSTER AND TURTLE

Little is known about presynaptic inhibition in the olfactory system. Until recently, the evidence for its existence has been indirect (Nickell et al., 1994; Keller et al., 1998). In other parts of the nervous system, the known functional roles of presynaptic afferent inhibition include preventing habituation, sharpening the receptive fields of

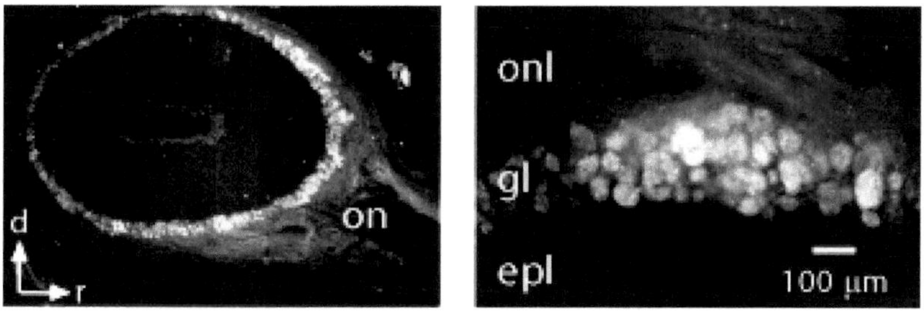

FIGURE 5.1 Calcium green-1 dextran labeling of turtle olfactory receptor neurons. Sagittal olfactory bulb section showing pattern of dextran labeling. Left: uniform labeling of glomeruli in all olfactory bulb regions. Right: specific labeling of olfactory nerve axons and their terminal branches within glomeruli. Absence of labeling in the external plexiform layer (epl) indicates that the dye is confined to olfactory receptor neurons. Abbreviations: on, olfactory nerve; d, dorsal; r, rostral; l, lateral; onl, olfactory nerve layer; gl, glomerular layer. (Wachowiak, M. and Cohen, L. B., modified from *J. Neurosci.,* 19, 8808-8817, 1999.)

afferent fibers, and serving as a mechanism for gain control of sensory input to the CNS (Watson, 1992). We investigated presynaptic regulation of transmission at the first olfactory synapse by selectively imaging axon terminals of receptor neurons in *in vitro* preparations of the lobster olfactory lobe and turtle olfactory bulb. In both species, action potential propagation into axon terminals following olfactory nerve stimulation was measured using voltage-sensitive dyes. In addition, in the turtle, calcium influx into terminals was measured by selectively labeling receptor neurons with a dextran-conjugated calcium indicator dye. Optical signals were recorded from the olfactory lobe of the spiny lobster, *Panulirus argus*, using a perfused, isolated brain preparation and the olfactory bulb of the box turtle, *Terrepene carolina*, using a superfused hemisected bulb preparation.

A 464-element photodiode array camera (NeuroPlex, RedShirtImaging, LLC, Fairfield, CT) was used to record the optical signals. For the voltage-sensitive dyes, the preparation was illuminated with light from a tungsten-halogen lamp passed through a 520 ± 45 nm interference filter, reflected from a 590-nm long-pass dichroic mirror and focused on the preparation using a 25-mm, 0.95 f, CCTV lens (Kleinfeld and Delaney, 1996). For the Calcium Green signals, a 460 ± 30 nm interference filter and a 515-nm dichroic mirror were used. Each photodiode received light from a square region of the object plane approximately 170 μm on a side. The optics were optimized to increase the number of photons in the measurement of epifluorescence at low magnification. In epifluorescence, the intensity reaching the objective image plane is proportional to the fourth power of the objective numerical aperture (NA)(Inoue, 1986). Because conventional microscope optics have small numerical apertures at low magnifications, we assembled a microscope (macroscope) based on a 25-mm focal length, 0.95 f, C-mount, camera lens (used with the C-mount end facing the preparation) (Salama, 1988; Ratzlaff and Grinvald, 1991; Kleinfeld and Delaney, 1996). With a magnification of 4×, the intensity reaching the photodetector was 100 times larger with this lens than with a conventional 4×, 0.16 NA, microscope lens. The same optics were used for the experiments in Sections 5.4 and 5.5.

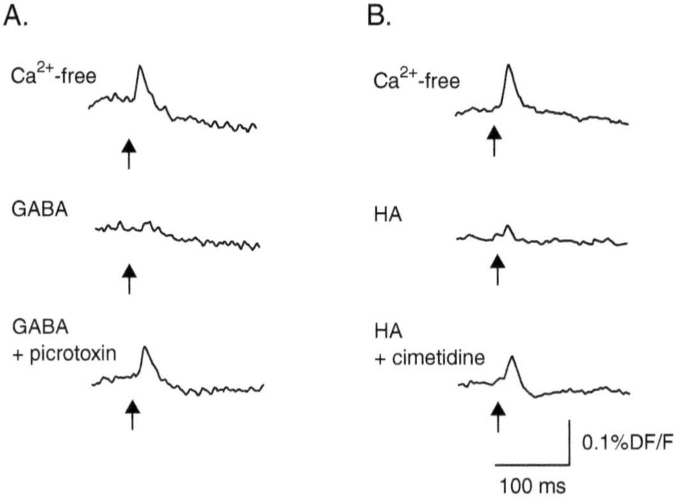

FIGURE 5.2 The effect of GABA, histamine, and two antagonists on the lobster olfactory lobe response evoked by a shock to the olfactory nerve in Ca^{2+}-free saline to eliminate postsynaptic components of the response. A: GABA (100 μ*M*) nearly eliminates the evoked response and the effect of GABA is blocked in the presence of 100 μ*M* picrotoxin. B: A different preparation than in A. Histamine (30 μ*M*) nearly eliminates the evoked response, and this effect is blocked by 5 m*M* cimetidine. Digitally filtered using low- and high-pass cutoffs of 5.7 and 60 Hz. (Modified from Wachowiak, M. and Cohen, L.B., *J. Neurosci.,* 19, 8808-8817, 1999.)

Because all of the neurons and processes in the imaged part of the bulb are directly stained with the voltage-sensitive dye, and because of the efforts to maximize the light intensity, the intensity reaching the image plane in this measurement was about 10^8 photons/msec, far above the intensity that would saturate the CCD camera (see discussion in Section 5.6). For additional details of the methods, see Lam et al., 1999 and Wachowiak and Cohen, 1998 and 1999.

5.3.1 Lobster Olfactory Lobe

Optical signals arising selectively from lobster olfactory receptor cell axon terminals were isolated by delivering electrical shocks to the olfactory (antennular) nerve using a preparation perfused with Ca^{2+}-free/high-Mg^{2+} saline to eliminate all post-synaptic signals initiated by the receptor transmitter. It was known that the transmitters GABA and histamine both mediate inhibition in the lobster olfactory lobe via distinct receptors and that the histamine receptor is weakly antagonized by cimetidine while the GABA receptor is antagonized by picrotoxin (Wachowiak and Ache, 1997). Therefore, we tested the ability of the two transmitters to mediate inhibition of olfactory afferents. Both transmitters significantly and reversibly reduced the amplitude of the olfactory lobe spike signal in Ca^{2+}-free saline (Figure 5.2). These effects of GABA and histamine were selectively blocked by their respective receptor antagonists (Figure 5.2).

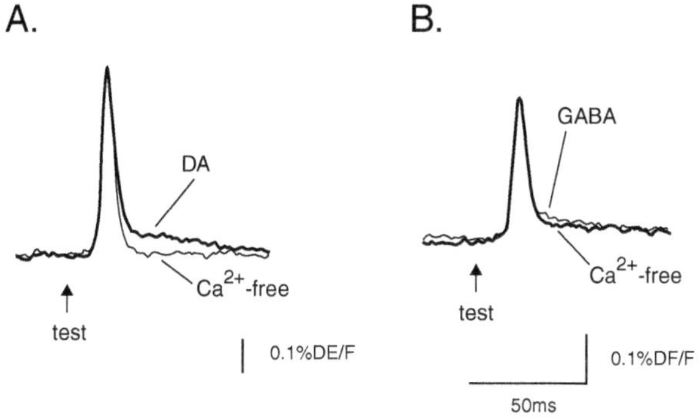

FIGURE 5.3 GABA and dopamine suppress postsynaptic activity but do not affect receptor cell action potentials in turtle olfactory bulb. Effect of GABA and dopamine (1 m*M*) on the test response in Ca^{2+}-free ringers. Control and drug traces are superimposed. Dopamine eliminates the remaining slow component of the evoked response but does not affect the compound action potential amplitude. GABA has little effect on the slow component and no effect on action potential amplitude. Digital filter cutoffs of 1.4 to 200 Hz. (Modified from Wachowiak, M. and Cohen, L.B., *J. Neurosci.,* 19, 8808-8817, 1999.)

In other experiments, the effect of GABA on the olfactory lobe response was antagonized by picrotoxin, after which histamine, still in a picrotoxin background, was applied. Histamine greatly reduced the response amplitude, indicating that GABA and histamine mediate suppression of receptor cell activity via distinct receptor pathways. Likewise, cimetidine (5 m*M*) reduced the effect of 30 μ*M* histamine by 60% but did not change the effect of GABA (data not shown).

The results do not discriminate whether presynaptic inhibition is mediated by a reduced action potential amplitude or by conduction failure in the terminal region. Both phenomena are documented in sensory afferent terminals of crustaceans and insects (Boyan, 1988; Burrows and Laurent, 1993; Clarac and Cattaert, 1996). Because the terminal branching regions of sensory afferents in the arthropod CNS appear to be largely passive, shunting conductances such as those mediated by ionotropic GABA and histamine receptors could have a large effect on action potential amplitude at the axon terminal and, subsequently, on transmitter release.

5.3.2 Turtle Olfactory Bulb

In the vertebrate olfactory bulb, the transmitters GABA and dopamine are known to mediate inhibition, and $GABA_B$ and dopamine D_2 receptors are concentrated in and around glomeruli (Nickell et al., 1994; Bowery et al., 1987). With voltage-sensitive dye recordings in Ca^{2+}-free ringers (to eliminate postsynaptic signals), neither GABA, dopamine, nor the D_2 receptor antagonist quinpirole affected the amplitude or time-course of the receptor axon compound action potential (fast component; Figure 5.3). The mean amplitude of the response in Ca^{2+}-free ringers was 103 ± 3% of control values when GABA was applied and 102 ± 6% for dopamine application. Thus, neither transmitter affects action potentials in the receptor axon

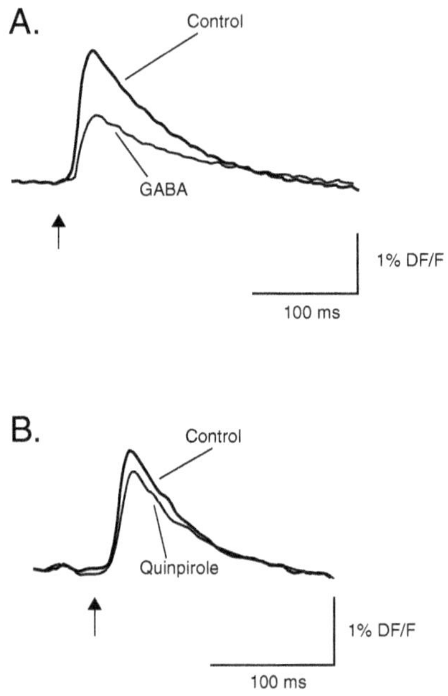

FIGURE 5.4 Effect of the transmitter GABA and the dopamine D_2 agonist quinpirole on the evoked calcium influx in the turtle. A: Superimposed traces in normal turtle ringers and 1 m*M* GABA. B: Superimposed traces in normal ringers and 100 μ*M* quinpirole. Digital filter cutoffs of 1.4 and 30 Hz. (Modified from Wachowiak, M. and Cohen, L.B., *J. Neurosci.,* 19, 8808-8817, 1999.)

terminals. Furthermore, action potential propagation in turtle olfactory receptor axons does not show the paired-pulse suppression observed in the lobster (data not shown).

We used calcium dye recordings to test whether presynaptic inhibition might be mediated by processes subsequent to action potential invasion of the axon terminal. These experiments were carried out in normal turtle ringers. In contrast to the voltage-sensitive dye results, GABA (1 m*M*) significantly and reversibly reduced the amplitude of the calcium influx resulting from stimulating the olfactory nerve to 60% of its initial value (Figure 5.4). The D_2 agonist quinpirole (100 μ*M*; n = 5) also affected calcium influx, reducing the response to 80% of the initial amplitude (Figure 5.4). 50 μ*M* baclofen, a $GABA_B$ agonist, strongly inhibited evoked calcium influx, reducing the response amplitude to 30% of its original amplitude (data not shown). Thus, it was concluded that, in the turtle, GABA and dopamine mediate presynaptic inhibition not by affecting action potentials directly, as in the lobster, but by reducing calcium influx via $GABA_B$ and dopamine D_2 receptors. While mediated by different cellular mechanisms, presynaptic regulation of olfactory input to the CNS, via dual synaptic pathways, appears to be a feature common to vertebrates and invertebrates.

Presynaptic inhibition mediated by inhibitory interneurons innervating several adjacent glomeruli could directly shape and refine spatial patterns of input across glomeruli. This pre-glomerular level of processing would be in addition to glomerular and subglomerular processing steps already partially characterized in the vertebrate olfactory bulb (review: Nickell and Shipley, 1992). Alternatively, presynaptic inhibition could have a modulatory role, regulating sensitivity to olfactory input by controlling the effectiveness of transmission from primary afferents to second-order neurons. Additional experiments are needed to distinguish between this and other possible roles in olfactory coding.

5.4 MAPS OF INPUT TO THE OLFACTORY BULB FROM OLFACTORY RECEPTOR NEURONS

To understand how odorants are distinguished and how one odorant is recognized as the same across a concentration range of several orders of magnitude, we measured the spatial pattern of the olfactory receptor neuron input to the olfactory bulb in the three-toed box turtle, *T. triunguis*.

To monitor the input to the bulb, we labeled the nerve terminals of olfactory receptor neurons with Calcium Green-1 dextran 10 kD using the method described above in Section 5.2. Because the amount of dye that reaches the nerve terminals is small, the light intensity in such a measurement is low and a better signal-to-noise ratio is obtained with a cooled CCD camera (see discussion in Section 5.6). We formed a magnified (4×) image of the bulb on an 80 × 80 CCD camera. For additional details of the methods, see Wachowiak et al., 2001.

We recorded the changes in Calcium Green fluorescence that resulted from a 2-s odorant pulse delivered to the nose. The signals measured had approximately the same time-course everywhere in the bulb, and we therefore characterized the response by the amplitude of the signal as a function of its position on the bulb.

We tested a few different odorants on the same animal. In every instance, the maps were different for different odorants. In some cases, there was substantial overlap in the activated area but at higher magnification, the patterns were easily distinguished (data not shown). Thus, the maps of input from the olfactory receptor neurons could be used for odor identification.

Figure 5.5 shows three pseudo-color representations of activity in response to 2-hexanone at two concentrations. Red represents a large signal in each measurement and blue represents a signal 30% as large. The left-hand image shows the response to hexanone at a concentration that was 0.3% of saturation. The largest signal in the response was colored red (relative scaling). Both right-hand images show the response to 10% hexanone using different scaling procedures. The bottom image shows the response using the same scale as that used for the response to 0.3% hexanone (absolute scaling). This image is qualitatively different from the 0.3% image. In contrast, the top image shows the response using relative scaling; again, the largest signal was colored red. This image is very similar to the image on the left although the concentration of odorant differed by a factor of 30. We have observed similar results over a concentration range of up to 600 (data not shown). Thus, relative maps of input to the olfactory bulb appear to be approximately concentration invariant. We hypothesize that concentration invariant odorant identification

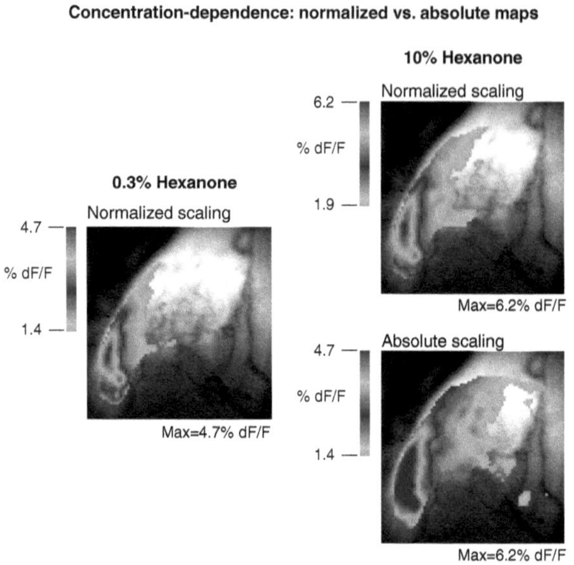

FIGURE 5.5 (see color insert following page 238) Normalized maps of receptor neuron input to the turtle olfactory bulb are concentration invariant. The left panel shows a pseudocolor map of the response to a 0.3% dilution of saturated vapor of 2-hexanone. The map is normalized to the maximum signal amplitude for this trial. The right panels show pseudocolor maps of a response to a 10% dilution of 2-hexanone. The map on the top is normalized to its maximum signal amplitude. The map on the bottom (absolute scaling) shows the same data using the same scaling as for the 0.3% trial. The figure shows a concentration-dependent increase in the number of glomeruli activated above a given absolute level, but shows a concentration invariance in the relative levels of input to all glomeruli activated by an odorant. (4× image magnification; the field of view is approximately 4 mm × 4 mm.)

could be achieved if higher olfactory centers read the relative maps of the input to the olfactory bulb.

5.5 OSCILLATIONS IN THE OLFACTORY BULB IN RESPONSE TO ODORS

Odor stimuli have long been known to induce stereotyped local field-potential responses consisting of sinusoidal oscillations of 10 to 80 Hz riding on top of a slow DC signal. Since its first discovery in the hedgehog (Adrian, 1942), odor-induced oscillations have been seen in phylogenetically distant species including locust, (Laurent and Naraghi, 1994), turtle (Beuerman, 1975), and monkey (Hughes and Mazurowski, 1962). We measured the voltage-sensitive dye signal that accompanies these oscillations in the box turtle. Because the optical measurements have a spatial resolution that is about 25 times better than local field-potential measurements (Zochowski et al., 2000), we obtained a more detailed visualization of the spatiotemporal characteristics of the oscillations.

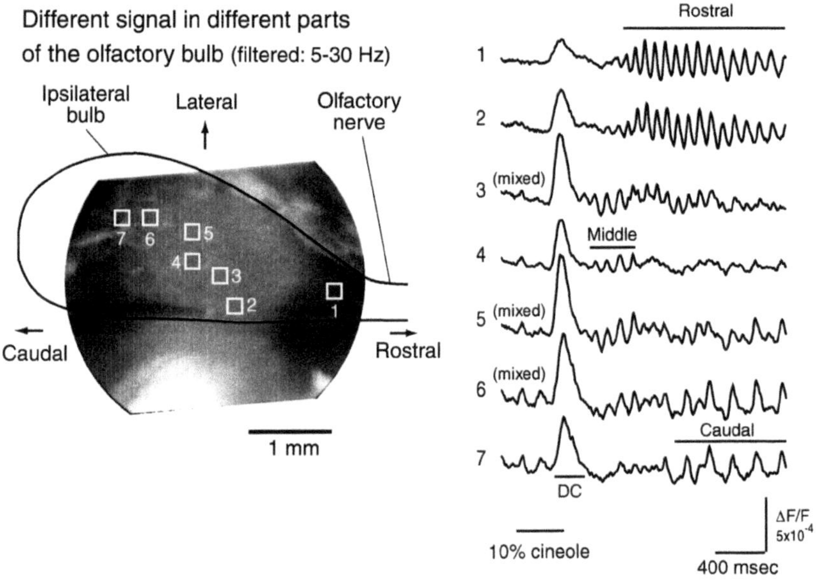

FIGURE 5.6 Simultaneous optical recordings from seven different areas of an olfactory bulb. An image of the olfactory bulb is shown on the left. Signals from seven selected pixels are shown on the right. The positions of these pixels are labeled with squares and numbers on the image of the bulb. All seven signals have a filtered version of the DC signal at the time indicated by the bar labeled DC. The oscillation in the rostral region has a high frequency and relatively long latency and duration (detectors 1 and 2). The oscillation from the middle region has a high frequency and short latency and duration (detector 4). The oscillation from the caudal region has a lower frequency and the longest latency (detector 7). The signal from detectors between these regions (3, 5 and 6) appear to contain a mixture of two components. The horizontal line labeled "10% cineole" indicates the time of the command pulse to the odor solenoid. The data are filtered by a high-pass digital RC (5 Hz) and low-pass Gaussian (30 Hz) filters. (Modified from Lam et al., 1999.)

The direct staining with the voltage-sensitive dye results in a high light intensity; therefore, we used the photodiode array camera (see discussion in Section 5.6). For additional details of the methods, see Lam et al., 1999.

5.5.1 Multiple Components of the Odor-Induced Response

We found three different oscillations in the bulb in response to odorants. These oscillations differed in their amplitude, location, frequency, latency, shape, and duration but were similar for the two odors tested, isoamyl acetate and cineole. In Figure 5.6, the recordings from seven selected diodes in a single trial are shown. The location of these diodes is indicated on the image in the left panel by the numbered squares. In rostral locations (detectors 1 and 2), we found a relatively large, long-lasting oscillation with a relatively high frequency (~15 Hz). On a diode from a middle location (detector 4), there was a relatively small, brief, short-latency oscillation (~14 Hz). On detector 7 from the caudal bulb, the oscillation had a lower

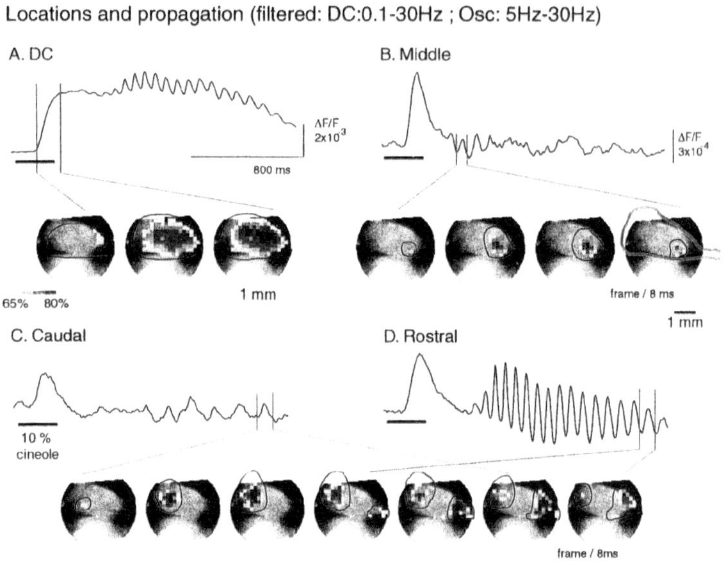

FIGURE 5.7 (see color insert following page 238) The locations and propagation of the four components from the trial shown in Figure 5.5. Multiframe pseudo-color displays of the signals are overlaid on the image of the olfactory bulb. The red color and the black contour lines label the areas where the signals are larger than 80 percent and 20 percent of the peak. The DC component in this animal covers almost the entire bulb. The other three panels show the position and propagation of one cycle (indicated by the red and green lines) of the three different oscillations. The black horizontal bars indicate the time of the odor command-pulse. The data are filtered by a high-pass digital RC (5 Hz) and low-pass Gaussian (30 Hz) filters. (Modified from Lam, Y. et al., *J. Neurosci.,* 20, 749-762, 1999.)

frequency (~7 Hz) and longer latency. We named the oscillations for the regions of the dorsal bulb in which they occurred: rostral, middle, and caudal. In areas between two regions, the recorded oscillations were combinations of two signals; i.e., rostral/middle in detector 3 and middle/caudal in detectors 5 and 6. The three oscillations also had different shapes; the rostral and caudal oscillations had relatively sharp peaks while the middle oscillation was more sinusoidal. In addition to the three oscillations, a DC signal, which appears as a single peak after high-pass filtering in Figure 5.6, was observed over most of the ipsilateral olfactory bulb. Figure 5.7A shows the time-course of an unfiltered recording from the rostral region. Following the start of the odor pulse, the optical signal rose to a plateau and then continued for a period of seconds. After a delay, the rostral oscillations appeared on the DC response.

Figure 5.7 shows the time-courses of the signals from four detectors from this trial, together with multiple-frame images indicating the position and propagation during one cycle. The rostral, middle, and caudal oscillations are seen more clearly after the DC signal was reduced with a high-pass filter (Figure 5.7B, C, D). In these multi-frame images, the red color and area enclosed by the black line indicate the areas where the signals are larger than 80% and 20% of the maximum signal. The

DC signal covered most of the olfactory bulb. The rostral signal (D) propagated in the caudal direction, the middle signal (B) did not appear to propagate, and the caudal signal (C) propagated in a lateral-caudal direction.

Oscillations are not restricted to the olfactory system. Despite their ubiquity, the roles and functions of oscillation are not well understood. However, in view of their ubiquity and the fact that they involve a very large number of neurons, it is reasonable to speculate that oscillations may have an important role in odor processing. Our data shows that the odor-induced oscillations in the olfactory bulb are substantially more complicated than had been anticipated and that multiple functional population domains are processing olfactory input in parallel.

5.6 OPTICAL RECORDING METHODS

The three examples given above involved fractional intensity changes ($\Delta I/I$) that ranged from 10^{-4} to 6×10^{-2}. To measure these signals, the noise in the measurements had to be an even smaller fraction of the resting intensity. In the chapter sections that follow, some of the considerations necessary to achieve such a low noise are outlined. The general approach is three-pronged. First, test dyes to find the dye with the largest signal-to-noise ratio. Second, reduce the extraneous sources of noise (vibrations, line frequency noise, etc.). Third, maximize the number of photons measured to reduce the relative shot noise (noise arising from the statistical nature of photon emission and detection).

5.6.1 Choosing Voltage-Sensitive Dyes

Several voltage-sensitive dyes (see e.g., Figure 5.8) have been used to monitor changes in membrane potential in a variety of preparations. This figure illustrates four different chromophores. For each chromophore, approximately 100 analogs have been synthesized in an attempt to optimize the signal-to-noise ratio that can be obtained in a variety of preparations. This screening was made possible by synthetic efforts of three laboratories: Jeff Wang, Ravender Gupta, and Alan Waggoner then at Amherst College; Rina Hildesheim and Amiram Grinvald at the Weizmann Institute; and Joe Wuskell and Leslie Loew at the University of Connecticut Health Center.

For each of the four chromophores illustrated in Figure 5.8 there were 10 or 20 dyes that gave approximately the same signal size on squid axons (Gupta et al., 1981). However, dyes that gave nearly identical signals on squid axons gave very different responses on other preparations, and thus tens of dyes usually have to be tested to obtain the largest possible signal. Often, dyes that worked well in squid did poorly in other preparations because they did not penetrate through connective tissue or along intercellular spaces to the membrane of interest. Others dyes appeared to have a relatively low affinity for neuronal vs. non-neuronal tissue. Finally, in some cases, the dye penetrated well and the staining appeared to be specific, but the signals were nonetheless small.

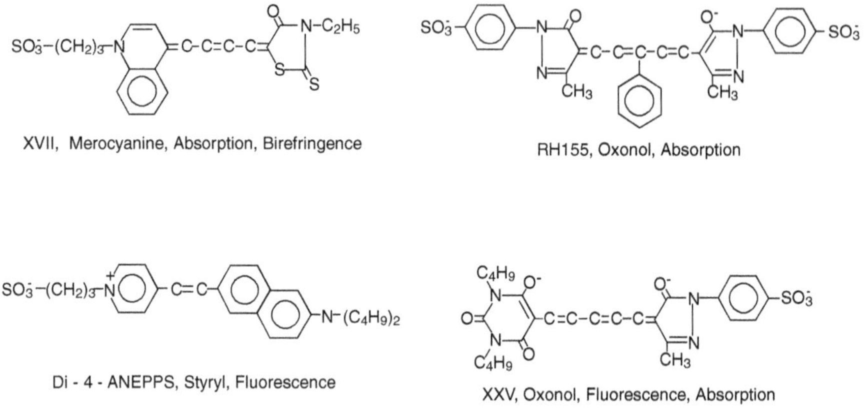

FIGURE 5.8 Examples of four different chromophores that have been used to monitor membrane potential. The merocyanine dye XVII (WW375) and the oxonol dye RH155 are commercially available as NK2495 and NK3041 from Nippon Kankoh-Shikiso Kenkyusho Co. Ltd., Okayama, Japan. The oxonol XXV (WW781) and styryl di-4-ANEPPS are available commercially as dye R-1114 and D-1199 from Molecular Probes, Junction City, OR.

5.6.2 Measuring Technology

5.6.2.1 Three Kinds of Noise

5.6.2.1.1 Shot Noise

The limit of accuracy with which light can be measured is set by the shot noise arising from the statistical nature of photon emission and detection. The root-mean-square deviation in the number emitted is the square root of the average number emitted (I) over a long measuring period. Therefore, the signal in a light measurement will be proportional to I and the noise in that measurement will be proportional to the square root of I. Thus, the signal-to-noise ratio (S/N) is proportional to the square root of the number of measured photons; more photons measured means a better signal-to-noise ratio. The S/N is also inversely proportional to the square root of the bandwidth of the photodetection system (Braddick, 1960; Malmstadt et al., 1974). The basis for this square root dependence on intensity is illustrated in Figure 5.9. In Figure 5.9A the result of using a random number table to distribute 20 photons into 20 time windows is shown. In Figure 5.9B, the same procedure was used to distribute 200 photons into the same 20 bins. Relative to the average light level, there is more noise in the top trace (20 photons) than in the bottom trace (200 photons). On the right side of Figure 5.9 the measured signal-to-noise ratios are listed and we show that the improvement from A to B is similar to that expected from the above square root relationship. This square root relationship is indicated by the dotted line in Figure 5.10 which plots the light intensity divided by the noise in the measurement (S/N) vs. the light intensity. At high light intensities, this ratio is large and thus small changes in intensity can be detected. For example, at 10^{10} photons/msec a fractional intensity change of 0.1% can be measured with a signal-to-noise ratio of 100. On the other hand, at low intensities, the ratio of intensity

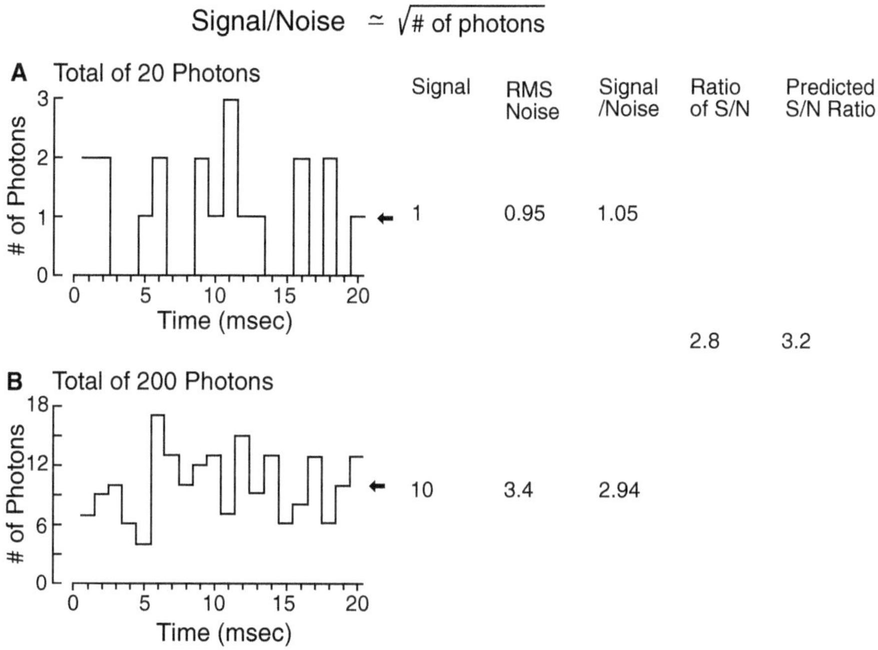

FIGURE 5.9 Plots of the results of using a table of random numbers to distribute 20 photons (top, A) or 200 photons (bottom, B) into 20 time bins. The result illustrates the fact that when more photons are measured, the signal-to-noise ratio is improved. On the right, the signal-to-noise ratio is measured for the two results. The ratio of the two signal-to-noise ratios was 0.43. This is close to the ratio predicted by the relationship that the signal-to-noise ratio is proportional to the square root of the measured intensity. (Redrawn from Wu, J.Y. and Cohen, L.B., *Fluorescent and Luminescent Probes for Biological Activity,* W.T. Mason, Ed., Academic Press, London, 389-404, 1993.)

divided by noise is small and only large signals can be detected. For example, at 10^4 photons/msec, the same fractional change of 0.1% can be measured with a signal-to-noise ratio of 1 only after averaging 100 trials. In a shot-noise limited measurement, improvement in the signal-to-noise ratio can only be obtained by (1) increasing the illumination intensity, (2) improving the light-gathering efficiency of the measuring system, or (3) reducing the bandwidth.

A tungsten filament light source emits an average of 10^{16} photons/msec with an rms noise of 10^8 photons/msec. Because only a tiny fraction of the 10^{16} photons/msec emitted by a tungsten filament will be measured, a signal-to-noise ratio of 10^8 (see above) cannot be achieved. A partial listing of the light losses follows. A 0.9-NA lamp collector lens would collect 0.1 of the light emitted by the source. Only 0.2 of that light is in the visible wavelength range; the remainder is infrared (heat). Limiting the incident wavelengths to those that have the signal means that only 0.1 of the visible light is used. Thus, the light reaching the preparation might typically be 10^{13} photons/msec. Only a fraction (say 0.1) of the incident photons is absorbed by the fluorophores, only a fraction of the absorbed photons appears as emitted

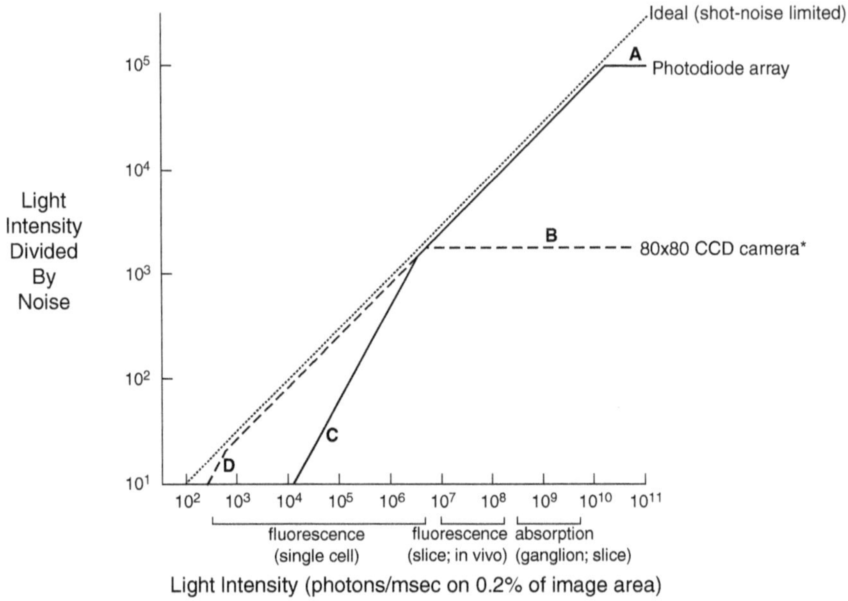

FIGURE 5.10 The ratio of light intensity divided by the noise in the measurement as a function of light intensity in photons/msec/0.2% of the object plane. The theoretical optimum signal-to-noise ratio (dotted line) is the shot-noise limit. Two camera systems are shown, a photodiode array with 464 pixels (solid lines) and a cooled, back-illuminated, 2-kHz frame rate, 74 × 74 pixel CCD camera (dashed lines). The photodiode array provides an optimal signal-to-noise ratio at higher intensities, while the CCD camera is better at lower intensities. The approximate light intensity per detector in fluorescence measurements from a single neuron, fluorescence measurements from a slice or *in vivo* preparation, and in absorption measurements from a ganglion or a slice is indicated along the *x*-axis. The signal-to-noise ratio for the photodiode array falls away from the ideal at high intensities (A) because of extraneous noise, and at low intensities (C) because of dark noise. The lower dark noise of the cooled CCD allows it to function at the shot-noise limit at lower intensities until read noise dominates (D). The CCD camera saturates at intensities above 5×10^6 photons/msec/0.2% of the object plane.

photons (say a quantum efficiency of 0.1), and only a fraction of the emitted photons is collected by the objective (a 0.9-NA objective would collect 0.1 of the emitted photons.) If the camera has a quantum efficiency of 1.0, then a total of 10^{10} photoelectrons/msec will be measured. With a camera of 1000 pixels, there will be 10^7 photoelectrons/msec/pixel. The shot noise will be 3×10^3 photoelectrons/msec/pixel. Thus, the best that can be expected is a relative noise 3×10^{-4} of the resting light (a signal-to-noise ratio of 70 dB).

Figure 5.10 compares the performance of two particular camera systems: a photodiode array (solid lines) and a cooled CCD camera (dashed lines) with the shot-noise ideal. The photodiode array approaches the shot-noise limitation over the range of intensities from 3×10^6 to 10^{10} photons/msec. This is the range of intensities obtained in absorption measurements and fluorescence measurements on *in vitro* slices and intact brains. On the other hand, the cooled CCD camera approaches the shot noise limit over the range of intensities from 3×10^2 to 5×10^6 photons/ms. This is the range of intensities obtained from fluorescence experiments on branches of individual cells and neurons. In the discussion that follows, we will indicate the aspects of the measurements and the characteristics of the two camera systems that cause them to deviate from the shot-noise ideal. The two camera systems we have chosen to illustrate in Figure 5.10 have outstanding dark noise and saturation characteristics; other commercially available cameras would be dark-noise limited at higher light intensities and would saturate at lower intensities.

5.6.2.1.2 Extraneous Noise

A second type of noise, termed extraneous or technical noise, is more apparent at high light intensities where the sensitivity to this kind of noise is high because the fractional shot noise and dark noise are low. One type of extraneous noise is caused by fluctuations in the output of the light source (see below). Two other sources of extraneous noise are vibrations and movement of the preparation. A number of precautions for reducing vibrational noise have been described (Salzberg et al., 1977, London et al., 1987). The pneumatic isolation mounts on many vibration isolation tables are more efficient in reducing vertical vibrations than in reducing horizontal movements. One solution is air-filled soft rubber tubes (Newport Corp., Irvine, CA). Minus K Technology sells Biscuit benchtop vibration isolation tables with very low resonant frequencies. They provide outstanding vibration isolation. Nevertheless, it has been difficult to reduce vibrational noise to less than 10^{-5} of the total light. For this reason, the performance of the photodiode array system is shown reaching a ceiling in Figure 5.10 (segment A, solid line).

5.6.2.1.3 Dark Noise

Dark noise will degrade the signal-to-noise ratio at low light levels. Because the CCD camera is cooled and the photosensitive area (and capacitance) is small, its dark noise is substantially lower than that of the photodiode array system. The excess dark noise in photodiode array systems accounts for the fact that segment C in Figure 5.10 is substantially to the right of segment D.

5.6.2.2 Light Sources

Three kinds of sources have been used. Tungsten filament lamps are a stable source, but their intensity is relatively low, particularly at wavelengths less than 480 nm. Arc lamps are somewhat less stable but can provide more intense illumination. Measurements made with laser illumination have been substantially noisier (Dainty, 1984).

5.6.2.2.1 Tungsten Filament Lamps

It is not difficult to provide a power supply stable enough so that the output of the bulb fluctuates by less than 1 part in 10^5. In absorption measurements, where the fractional changes in intensity are relatively small, only tungsten filament sources have been used. On the other hand, fluorescence measurements often have larger fractional changes that will better tolerate light sources with systematic noise, and the measured intensities are lower, making improvements in signal-to-noise ratio from brighter sources attractive.

5.6.2.2.2 Arc Lamps

Opti-Quip, Inc., provides 150- and 250-watt xenon power supplies, lamp housings, and arc lamps with noise that is in the range 1 to 3 parts in 10^4. The 150-watt bulb yielded 2 to 3 times more light at 520 ± 45 nm than a tungsten filament bulb and, in turn, the 250-watt bulb was 2 to 3 times brighter than the 150-watt bulb. The extra intensity is especially useful for fluorescence measurements from single neurons or from weakly stained nerve terminals. If the dark noise is dominant, then the signal-to-noise ratio will improve linearly with intensity; and if the shot noise is dominant, it will improve as the square root of intensity.

5.6.2.3 Optics

5.6.2.3.1 Numerical Aperture

The need to maximize the number of measured photons is a dominant factor in the choice of optical components. In epifluorescence, both the excitation light and the emitted light pass through the objective, and the intensity reaching the photodetector is proportional to the *fourth* power of numerical aperture (Inoue, 1986). Therefore, the numerical aperture of the objective is a crucial factor.

5.6.2.3.2 Depth of Focus

Salzberg et al. (1977) determined the effective depth of focus for a 0.4-NA objective lens by recording an optical signal from a neuron when it was in focus and then moving the neuron out of focus by various distances. They found that the neuron had to be moved 300 μm out of focus to cause a 50% reduction in signal size. Using 0.5-NA optics, 100 μm out of focus led to a reduction of 50% (Kleinfeld and Delaney, 1996).

5.6.2.3.3 Light Scattering and Out-of-Focus Light

Light scattering can limit the spatial resolution of an optical measurement. London et al. (1987) measured the scattering of 705-nm light in *Navanax* buccal ganglia. They found that inserting a ganglion in the light path caused light from a 30-μm spot to spread so that the diameter of the circle of light that included intensities greater than 50% of the total was roughly 50 μm. The spread was greater, to about 100 μm, with light of 510 nm. Figure 5.11 illustrates the results of similar experiments carried out on the salamander olfactory bulb. The top panel indicates that when no tissue is present, essentially all of the light (750 nm) from a small spot falls on one detector. The bottom panel illustrates the result when a 500-μm-thick slice of olfactory bulb is present. The light from the small spot is spread to about 200 μm. Mammalian cortex appears to scatter more than the olfactory bulb. Thus, light scattering will cause considerable blurring of signals in adult vertebrate preparations. In the measurements from olfactory nerve terminals in glomeruli (Section 5.4), we are imaging through 50 to 100 μm of tissue and thus the blurring from light scattering should be small.

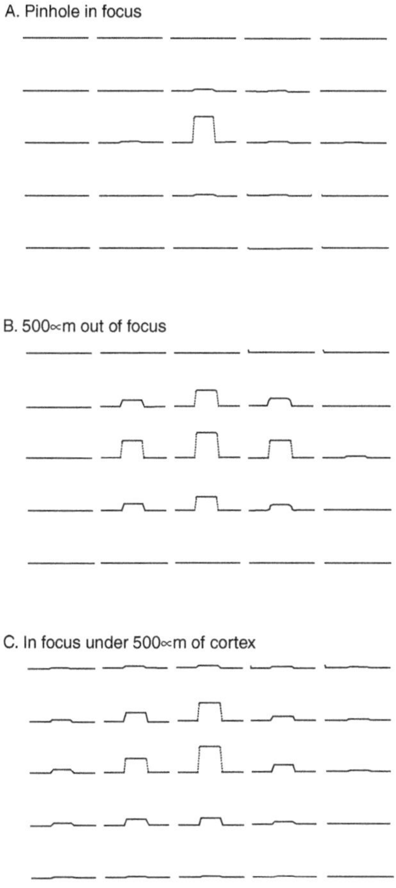

FIGURE 5.11 Effects of focus and scattering on the distribution of light from a point source onto the array. A: A 40-μm pinhole in aluminum foil covered with saline was illuminated with light at 750 nm. The pinhole was in focus. More than 90% of the light fell on one detector. B: The stage was moved downward by 500 μm. Light from the out-of-focus pinhole was now seen on several detectors. C: The pinhole was in focus but covered by a 500-μm slice of salamander cortex. Again, the light from the pinhole was spread over several detectors. A 10 × 0.4-NA objective was used. Kohler illumination was used before the pinhole was placed in the object plane. The recording gains were adjusted that so the largest signal in each of the three trials would be approximately the same size in the figure. (Redrawn from Orbach, H.S. and Cohen, L.B., *J. Neurosci.,* 3, 2251-2262, 1983.)

A second source of blurring is signal from regions that are out of focus. For example, if the active region is a cylinder (a column) perpendicular to the plane of focus, and the objective is focused at the middle of the cylinder, then the light from the in-focus plane will have the correct diameter at the image plane. However, the light from the regions above and below are out of focus and will have a diameter that is too large. The middle panel of Figure 5.11 illustrates the effect of moving the small spot of light 500 μm out of focus. The light from the small spot is spread to about 200 μm. Thus, in preparations with considerable scattering or with out-of-focus signals, the actual spatial resolution may be limited by the preparation and not by the number of pixels in the imaging device.

5.6.2.3.4 *Confocal Microscopes*

The confocal microscope (Petran and Hadravsky, 1966) substantially reduces both the scattered and out-of-focus light that contributes to the image. A recent modification using two-photon excitation of the fluorophore further reduces out-of-focus fluorescence and photobleaching (Denk et al., 1995). With both types of microscope, one can obtain images from intact vertebrate preparations with much better spatial resolution than was achieved with ordinary microscopy. However, at present, the sensitivity of these microscopes is relatively poor and many milliseconds are required to record the image from a single *x-y* plane. The kinds of problems that can be approached using a confocal microscope are limited by these factors. In addition, the very high *z*-dimension resolution of confocal microscopy can be a drawback if only a very thin section of the preparation is recorded in each frame.

5.6.2.4 Cameras

Because the signal-to-noise ratio in a shot-noise limited measurement is proportional to the square root of the number of photons converted into photoelectrons (see above), quantum efficiency is important. Silicon photodiodes have quantum efficiencies approaching the ideal (1.0) at wavelengths where most dyes absorb or emit light (500 to 900 nm). In contrast, only specially chosen vacuum photocathode devices (phototubes, photomultipliers, or image intensifiers) have a quantum efficiency as high as 0.15. Thus, in shot-noise limited situations, a silicon diode may have a signal-to-noise ratio that is substantially larger.

Many factors must be considered in choosing an imaging system. Two important considerations are the requirements for spatial and temporal resolution. Because the signal-to-noise ratio in a shot-noise limited measurement is proportional to the square root of the number of measured photons, increases in either temporal or spatial resolution will reduce the signal-to-noise ratio. Our discussion considers systems that have frame rates near 1 kHz.

5.6.2.4.1 *Silicon Diode Arrays*

Parallel readout arrays. Photodiode arrays with 256 to 1020 elements are now in use in several laboratories (e.g., Iijima et al., 1989; Zecevic et al., 1989; Nakashima et al., 1992; Hirota et al., 1995). These arrays are designed for parallel readout; each detector followed by its own amplifier whose output can be digitized at frame rates of 1 kHz. While the need to provide a separate amplifier for each diode element limits the number of pixels in parallel readout systems, it contributes to the very large (10^5) dynamic range that can be achieved with these systems. A discussion of amplifiers has been presented (Wu and Cohen, 1993). Two parallel readout array systems are commercially available: Argus-50 (256 pixels), Hamamatsu Photonics K.K. (www.hpk.co.jp) and NeuroPlex-II (464 pixels), RedShirtImaging, LLC (www.redshirtimaging.com).

TABLE 5.1
Characteristics of Fast CCD Camera Systems (As Reported by the Manufacturer)

	Frame Rate (Hz) Full Frame	Well Size (× 1000 e)	Read Noise (Electrons)	Back Illum.	Bits a-to-d	Pixels
MiCAM 01[a]	1333	—	—	No	12	92 × 64
Dalsa CA-D1-0128[b]	756	300	360	No	12	128 × 128
LSR, NeuroCAM UltraPix[c]	500	300	5 @100 Hz	Yes	12	80 × 80
SciMeasure[d]	1100	300	8 3 @ 100 Hz	Yes	14	74 × 74
TILL Photonics, IMAGO[e]	140	35	14	No	12	160 × 120

[a] www.scimedia.co.jp
[b] www.dalsa.com
[c] www.astrocam.co.uk/cameras/ultrapixpreci.htm
[d] www.redshirtimaging.com
[e] www.TILL-photonics.com

Serial readout arrays. By using a serial readout, the number of amplifiers is greatly reduced. In addition, it is simpler to cool CCD chips to reduce the dark noise. However, because of saturation, presently available CCD cameras cannot be used with the higher intensities available in some neurobiological experiments (Figure 5.10). The high-intensity limit of the CCD camera is set by the light intensity that fills the electron wells on the CCD chip. This accounts for the bending over of the CCD camera performance at segment B in Figure 5.10. A dynamic range of even 10^3 is not easily achieved with presently available CCD cameras. Thus, these cameras will not be optimal for measurements of absorption or for fluorescence measurements where the staining intensity is high. The light intensity would have to be reduced, with a consequent decrease in signal-to-noise ratio. On the other hand, CCD cameras are close to ideal for measurements from branches of individual neurons stained with internally injected dyes. Table 5.1 compares several CCD cameras with frame rates near 1 kHz.

5.7 FUTURE DIRECTIONS

Because the light measuring apparatus is already reasonably optimized (see above), any improvement in the sensitivity of these optical measurements will need to come from the development of better dyes and/or investigating signals from additional optical properties of the dyes. The dyes in Figure 5.8 and the vast majority of those synthesized are of the general class named polyenes (Hamer, 1964). It is possible that improvements in signal size can be obtained with new polyene dyes (see Waggoner and Grinvald, 1977 and Fromherz, 1991 for a discussion of maximum possible fractional changes in absorption and fluorescence). On the other hand, the fractional change on squid axons has not increased in recent years (Gupta et al., 1981; L. B. Cohen, A. Grinvald, K. Kamino, and B. M. Salzberg, unpublished results), and most improvements (e.g., Grinvald et al., 1982; Momose-Sato et al., 1995; Tsau et al., 1996; Antic and Zecevic, 1995) have involved synthesizing analogs that work well on new preparations.

The best of the styryl and oxonol polyene dyes have fluorescence changes of 10 to 20%/100 mV in situations where the staining is specific to the membrane whose potential is changing (Grinvald et al., 1982; Loew et al., 1992; Rohr and Salzberg, 1994). Recently, Gonzalez and Tsien (1995) introduced a new scheme for generating voltage-sensitive signals using two chromophores and energy transfer. While these fractional changes were also in the range of 10%/100 mV, more recent results are about 30% (J. Gonzalez and R. Tsien, personal communication). However, because one of the chromophores must be very hydrophobic and does not penetrate into brain tissue, it has not been possible to measure signals with this pair of dyes in intact tissues (T. Gonzalez and R. Tsien, personal communication; A. Obaid and B. M. Salzberg, personal communication).

5.7.1 Neuron-Type Specific Staining

An important new direction is the development of methods for neuron-type specific staining. Three quite different approaches have been tried. First, the use of retrograde staining procedures has recently been investigated in the embryonic chick and lamprey spinal cords (Tsau et al., 1996). An identified cell class (motoneurons) was selectively stained. In lamprey experiments, spike signals from individual neurons were sometimes measured (Hickie et al., 1996). Further efforts at optimizing this staining procedure are needed. The second approach is based on the use of cell-type specific staining developed for fluorescein by Nirenberg and Cepko (Nirenberg and Cepko, 1993). It might be possible to use similar techniques to selectively stain cells with voltage-sensitive or ion-sensitive dyes. Third, Siegel and Isacoff (1997) constructed a genetically encoded combination of a potassium channel and green fluorescent protein. When introduced into a frog oocyte, this molecule had a (relatively slow) voltage-dependent signal with a fractional fluorescence change of 5%. Neuron-type specific staining would make it possible to determine the role of specific neurons types in generating the input-output function of a brain region.

Optical recordings already provide unique insights into brain activity and organization. Clearly, improvements in sensitivity or selectivity would make these methods more powerful.

ACKNOWLEDGMENTS

The authors are indebted to their collaborators Vicencio Davila, Chun Falk, Amiram Grinvald, Kohtaro Kamino, David Kleinfeld, Les Loew, Bill Ross, Guy Salama,

Brian Salzberg, Alan Waggoner, Jian-Young Wu, and Joe Wuskell for numerous discussions about optical methods. Supported by NINDS Grant, NS08437, a Brown-Coxe Fellowship from the Yale University School of Medicine, and an NRSA Fellowship, DC 00378.

REFERENCES

Adrian, E.D. (1942). Olfactory reactions in the brain of the hedgehog, *J. Physiol.,* 100, 459-473.

Antic, S. and Zecevic, D. (1995). Optical signals from neurons with internally applied voltage-sensitive dyes, *J. Neurosci.,* 15, 1392-1405.

Beuerman, R.W. (1975). Slow potentials of the turtle olfactory bulb in response to odor stimulation of the nose, *Brain Res.,* 97(1), 61-78.

Bowery, N.G., Hudson, A.L., and Price, G.W., (1987). $GABA_A$ and $GABA_B$ receptor site distribution in the rat central nervous system, *J. Neurosci.,* 20, 365-383.

Boyan, G.S. (1988). Presynaptic inhibition of identified wind-sensitive afferents in the cercal system of the locust, *J. Neurosci.* 8, 2748-2757.

Braddick, H.J.J. (1960). Photoelectric photometry, *Rep. Prog. Physics,* 23, 154-175.

Bullock, T.H. and McClune, M.C. (1989). Lateral coherence of the electrocorticogram: a new measure of brain synchrony, *Electroencephalogr. Clin. Neurophysiol.,* 73, 479-498.

Burrows, M. and Laurent, G. (1993). Synaptic potentials in the central terminals of locust proprioceptive afferents generated by other afferents form the same sense organ, *J. Neurosci.,* 13, 808-819.

Clarac, F. and Cattaert, D., (1996). Invertebrate presynaptic inhibition and motor control, *Exp. Brain Res.,* 112: 163-180.

Cohen, L.B. and Lesher, S. (1986). Optical monitoring of membrane potential: methods of multisite optical measurement, *Soc. Gen. Physiol. Ser.,* 40, 71-99.

Cohen, L.B. and Salzberg, B.M. (1978). Optical measurement of membrane potential, *Rev. Physiol. Biochem. Pharmacol.,* 83, 35-88.

Dainty, J.C. (1984). *Laser Speckle and Related Phenomena,* Springer-Verlag, New York.

Denk, W., Piston, D.W., and Webb. W. (1995). Two-photon molecular excitation in laser-scanning microscopy, in *Handbook of Biological Confocal Microscopy,* J.W. Pawley, Ed., Plenum Press, New York, 445-458.

Friedrich, R. and Korsching, S. (1997). Combinatorial and chemotopic odorant coding in the zebrafish olfactory bulb visualized by optical imaging, *Neuron,* 18, 737-752.

Fromherz, P., Dambacher, K.H., Ephardt, H., Lambacher, A., Muller, C.O., Neigl, R., Schaden, H., Schenk, O., and Vetter, T. (1991). Fluorescent dyes as probes of voltage transients in neuron membranes: progress report, *Ber. Bunsenges. Phys. Chem.,* 95, 1333-1345.

Gonzalez, J.E. and Tsien, R.Y. (1995). Voltage sensing by fluorescence energy transfer in single cells, *Biophysic. J.,* 69, 1272-1280.

Grinvald, A., Hildesheim, R., Farber, I.C., and Anglister, L. (1982). Improved fluorescent probes for the measurement of rapid changes in membrane potential, *Biophys. J.,* 39, 301-308.

Grinvald, A., Lieke, E.E., Frostig, R.D., and Hildesheim, R. (1994). Cortical point-spread function and long-range lateral interactions revealed by real-time optical imaging of Macaque monkey primary visual cortex, *J. Neurosci.,* 18, 9977-9988.

Grinvald, A., Frostig, R.D., Lieke, E., and Hildesheim, R. (1988). Optical imaging of neuronal activity, *Physiol. Rev.,* 68, 1285-1366.

Gupta, R.K., Salzberg, B.M., Grinvald, A., Cohen, L.B., Kamino, K., Lesher, S., Boyle, M.B., Waggoner, A.S., and Wang, C.H. (1981). Improvements in optical methods for measuring rapid changes in membrane potential, *J. Memb. Biol.,* 58, 123-137.

Hamer, F.M. (1964). *The Cyanine Dyes and Related Compounds,* Wiley, New York.

Hickie, C., Wenner, P., O'Donovan, M., Tsau, Y., Fang, J., and Cohen, L.B. (1996). Optical monitoring of activity from individual and identified populations of neurons retrogradely labeled with voltage-sensitive dyes, *Abs. Soc. Neurosci.,* 22, 321.

Hirota, A., Sato, K., Momose-Sato, Y., Sakai, T., and Kamino, K. (1995). A new simultaneous 1020-site optical recording system for monitoring neural activity using voltage-sensitive dyes, *J. Neurosci. Meth.,* 56, 187-194.

Hughes, J.R. and Mazurowski, J.A. (1962). Studies on the supracallosal mesial cortex of unanesthetized, conscious mammals. II. Monkey. B. Responses from the olfactory bulb. *Electroencephalograph. Clin. Neurophys.,* 14, 635-645.

Iijima, T., Ichikawa, M., and Matsumoto, G. (1989). Optical monitoring of LTP and related phenomena, *Abst. Soc. Neurosci.,* 15, 398.

Inoue, S. (1986). *Video Microscopy.* Plenum Press, New York, 128.

Keller, A., Yagodin, S., Aroniadou-Anderjaska, V., Zimmer, L., Ennis, M., Sheppard, N., and Shipley, M. (1998). Functional organization of rat olfactory bulb glomeruli revealed by optical imaging, *J. Neurosci.,* 18, 2602-2612.

Kleinfeld, D. and Delaney, K. (1996). Distributed representation of vibrissa movement in the upper layers of somatosensory cortex revealed with voltage-sensitive dyes, *J. Comp. Neurol.,* 375, 89-109.

Lam, Y.-W., Cohen, L.B., Wachowiak, M., and Zochowski, M.R. (1999). Odors elicit three different oscillations in the turtle olfactory bulb, *J. Neurosci.,* 20, 749-762.

Laurent, G. and Naraghi, M. (1994). Odorant-induced oscillations in the mushroom bodies of the locust, *J. Neurosci.,* 14(5 Pt. 2), 2993-3004.

Loew, L.M., Cohen, L.B., Dix, J., Fluhler, E.N., Montana, V., Salama, G., and Wu, J.Y. (1992). A napthyl analog of the aminostyryl pyridinium class of potentiometric membrane dyes shows consistent sensitivity in a variety of tissue, cell, and model membrane preparations, *J. Memb. Biol.,* 130, 1-10.

London, J.A., Zecevic, D., and Cohen, L.B. (1987). Simultaneous optical recording of activity from many neurons during feeding in *Navanax, J. Neurosci.,* 7, 649-661.

Malmstadt, H.V., Enke, C.G., Crouch, S.R., and Harlick, G. (1974). *Electronic Measurements for Scientists,* Benjamin, Menlo Park, CA.

Momose-Sato, Y., Sato, K., Sakai, T., Hirota, A., Matsutani, K., and Kamino, K. (1995). Evaluation of optimal voltage-sensitive dyes for optical measurement of embryonic neural activity, *J. Memb. Biol.,* 144, 167-176.

Nakashima, M., Yamada, S., Shiono, S., Maeda, M., and Sato, F. (1992). 448-detector optical recording system: development and application to *Aplysia* gill-withdrawal reflex, *IEEE Trans. on Biomed. Eng.,* 39, 26-36.

Neher, E. (2000). Some quantitative aspects of calcium fluorimetry, in *Imaging Neurons: A Laboratory Manual,* Ed. by R. Yuste, F. Lanni, and A. Konnerth. Cold Spring Harbor Laboratory Press, Cold Spring Harbor, ME, 31.1.

Nickell, W.T. and Shipley, M. (1992). Neurophysiology of the olfactory bulb, in *Science of Olfaction,* Serby, M. and Chobor, K., Eds., Springer-Verlag, New York, 172-212.

Nickell, W.T., Behbehani, M.M., Shipley, and M.T. (1994). Evidence for $GABA_B$-mediated inhibition of transmission from the olfactory nerve to mitral cells in the rat olfactory bulb, Brain Res. Bull., 35, 119-123.

Nirenberg, S. and Cepko, C. (1993). Targeted ablation of diverse cell classes in the nervous system *in vivo, J. Neurosci.,* 13, 3238-3251.

Orbach, H.S. and Cohen, L.B. (1983). Optical monitoring of activity from many areas of the *in vitro* and *in vivo* salamander olfactory bulb: a new method for studying functional organization in the vertebrate central nervous system, *J. Neurosci.,* 3, 2251-2262.

Petran, M. and Hadravsky, M. (1966). Czechoslovakian patent 7720.

Ratzlaff, E.H. and Grinvald, A. (1991). A tandem-lens epifluorescence microscope: hundred-fold brightness advantage for wide-field imaging, *J. Neurosci. Meth.,* 36, 127-137.

Rohr, S. and Salzberg, B.M. (1994). Multiple site optical recording of transmembrane voltage in patterned growth heart cell cultures: assessing electrical behavior, with microsecond resolution, on a cellular and subcellular scale, *Biophys. J.,* 67, 1301-1315.

Salama, G. (1988). Voltage-sensitive dyes and imaging techniques reveal new patterns of electrical activity in heart and cortex, *SPIE Proc.,* 94, 75-86.

Salzberg, B.M. (1983). Optical recording of electrical activity in neurons using molecular probes, in *Current Methods in Cellular Neurobiology,* J.L. Barker and J.F. McKelvy, Eds., Wiley, New York, 139-187.

Salzberg, B.M., Grinvald, A., Cohen, L.B., Davila, H.V., and Ross, W.N. (1977). Optical recording of neuronal activity in an invertebrate central nervous system: simultaneous monitoring of several neurons, *J. Neurophysiol.,* 40, 1281-1291.

Siegel, M.S. and Isacoff, E.Y., (1997). A genetically encoded optical probe of membrane voltage, *Neuron,* 19, 735-741.

Tsau, Y., Wenner, P., O'Donovan, M.J., Cohen, L.B., Loew, L.M., and Wuskell, J.P. (1996). Dye screening and signal-to-noise ratio for retrogradely transported voltage-sensitive dyes, *J. Neurosci. Meth.,* 70, 121-129.

Wachowiak, M. and Ache, B.W. (1997). Dual inhibitory pathways mediated by GABA and histaminergic interneurons in the lobster olfactory lobe, *J. Comp. Physiol. A,* 180, 357-372.

Wachowiak, M. and Cohen, L.B. (1998). Presynaptic afferent inhibition of lobster olfactory receptor cells: Reduced action potential propagation into axon terminals, *J. Neurophysiol.,* 80, 1011-1015.

Wachowiak, M., Cohen, L.B., and Zochowski, M. (2001). Spatial representations of odorant quality and intensity by receptor neuron input to turtle olfactory bulb glomeruli, submitted for publication.

Wachowiak, M. and Cohen, L.B. (1999). Presynaptic inhibition of primary olfactory afferents mediated by different mechanisms in lobster and turtle, *J. Neurosci.,* 19, 8808-8817.

Waggoner, A.S. (1979). Dye indicators of membrane potential, *Annu. Rev. Biophys. Bioeng.,* 8, 47-68.

Waggoner, A.S. and Grinvald, A. (1977). Mechanisms of rapid optical changes of potential sensitive dyes, *Ann. N.Y. Acad. Sci.,* 303, 217-241.

Watson, A.H.D. (1992). Presynaptic modulation of sensory afferents in the invertebrate and vertebrate nervous system. *Comp. Biochem. Physiol. A. Physiol.,* 103, 227-239.

Wu, J.Y. and Cohen, L.B. (1993). Fast multisite optical measurements of membrane potential, in *Fluorescent and Luminescent Probes for Biological Activity,* W.T. Mason, Ed., Academic Press, London, 389-404.

Zecevic, D., Wu, J.Y., Cohen, L.B., London, J.A., Hopp, H.P., and Falk, C.X. (1989). Hundreds of neurons in the *Aplysia* abdominal ganglion are active during the gill-withdrawal reflex, *J. Neurosci.,* 9, 3681-3689.

Zochowski, M., Wachowiak, D.M., Falk, C.X., Cohen, L.B., Lam, Y.-W., Antic, S., and Zecevic, D. (2000). Imaging membrane potential with voltage-sensitive dyes, *Biological Bull.,* 198: 1-21.

6 Gustatory System Development: New Experimental Approaches in Amphibian and Mammalian Embryos

Linda A. Barlow and Thomas E. Finger

CONTENTS

0-8493-2329-0/02/$0.00+$1.50

6.1 INTRODUCTION

Embryonic development of the peripheral taste system involves the genesis of receptor cells and sensory neurons, which ultimately connect to form a functional sensory system. Typically, taste buds form only in the epithelium of the mouth and pharynx, and these receptor organs become innervated by sensory axons of the facial, glossopharyngeal, and vagal nerves. Taste buds possess a readily identifiable multicellular morphology, are found in a restricted region of the head, and are innervated by only a small subset of nerves. Thus, the developing taste periphery provides a more easily studied microcosm of the developing peripheral nervous system; by studying taste system development, one then hopes to elucidate general cellular and molecular mechanisms driving peripheral nervous system (PNS) development. Certainly, many of the problems faced by the taste periphery are common to the PNS. How are remote populations of cells (i.e., receptor organs and sensory neurons) generated, and their development temporally coordinated? What determines the pattern of receptor organs, in this case, taste buds? And how do sensory neurons find their taste bud targets such that proper and specific synaptic contacts are made and maintained?

Much of what has been learned about the development of this system has come from descriptive studies in a variety of vertebrates species (catfish: Landacre, 1907; rat: Farbman, 1965; Farbman and Mbiene, 1991; Mistretta, 1972; Oakley et al., 1991; hamster: Miller and Smith, 1988; Whitehead and Kachele, 1994; chicken: Ganchrow et al., 1995; salamanders: Barlow and Northcutt, 1995). Typically, using normal histological criteria, taste buds are only apparent late in embryonic development, either at birth or hatching. This is well after the epithelium has been innervated. This observation, combined with numerous demonstrations of the role of nerves in the maintenance of adult taste buds, has resulted in the view that early in-growing nerves induce the subsequent formation of taste buds (Barlow, 2000a; Oakley, 1993a). Nonetheless, in these studies in general, the potential inductive power of gustatory nerves had not been tested explicitly.

One exception is the early work of Leon Stone. In a series of studies using amphibian embryos, Stone found that taste buds would develop when the pharyngeal region of one embryo was grafted ectopically to the flank of another embryo (Stone, 1932; 1940). These manipulations were carried out prior to innervation of the pharynx, which was critical if the inductive capacity of nerves was to be tested. Stone concluded that nerves were not necessary for the induction of taste buds. However, because of the limited technical repertoire of the times, he did not assess the degree to which these ectopic taste buds were innervated, and possibly induced, by spinal nerves.

In contrast to the work in amphibians, the majority of studies designed to test the inductive capacity of nerves in mammals have focussed on postnatal development, and have been done in rodents. Unfortunately, at these postnatal stages, the epithelium has been innervated for over a week, and taste buds are already evident in many taste papillae. Thus, disrupting this intact innervation has not addressed the issue of taste bud induction by nerves. Rather, these studies have clearly demonstrated that an intact nerve supply is necessary for maintenance and differentiation of taste buds (Hosley et al., 1987a; 1987b; Zalewski, 1976). Even studies of embryonic development of taste buds (Farbman, 1968; 1971; 1972a) were done after papillae have become innervated (Farbman, 1965; Farbman and Mbiene, 1991) and thus represent interruption of innervation, rather than prevention.

In the last 5 years, results from a series of new approaches, including molecular biology, molecular genetics, and advances in tissue culture methodology, have changed our sense of how the taste periphery develops. Rather than viewing the formation of taste buds as a late embryonic event that is triggered by nerves, we now recognize that the onset of taste bud development occurs much earlier in embryogenesis, prior to innervation of the epithelium, and is independent of nerve contact. This conclusion has been reached based on experimental studies in mammals and amphibians, as well as gene expression patterns in mammals. The methods and results of these studies form the bulk of this chapter.

6.2 WHAT ARE THE EMBRYONIC ORIGINS OF TASTE BUDS?

The predominant view of the origins of taste buds has been that they are derived from local epithelial cells within the mouth and pharynx (Cook and Neal, 1921; Johnston, 1910; Landacre, 1907; and Farbman, 1972b; Oakley, 1993a for review). However, over the years, there have been several suggestions that taste buds arise from neurogenic cells that migrate into the epithelium early in development. One idea was that a subset of cells from the migratory neural crest contributed to taste buds (Reutter, 1978), and the other notion was that cells from the neurogenic epibranchial placodes might give rise to taste bud progenitor cells (Gans and Northcutt, 1983). Both neural crest and epibranchial placodes contribute sensory neurons to the gustatory cranial ganglia (VII, IX, and X) (D'Amico-Martel and Noden, 1983; Landacre, 1910; Narayanan and Narayanan, 1980; Stone, 1922). Because placodes and neural crest produce these sensory neurons early on, both of their putative contributions to taste buds were hypothesized also to occur early, well before the epithelium became innervated by these sensory neurons. For this reason, this early migration of neural crest or placodal cells into the epithelium might have been missed by researchers examining the taste bud-bearing epithelium later in embryonic development. In view of these disparate ideas of taste bud origins, it was clear that in order to understand the mechanisms governing taste bud development, it was critical to identify where the cells that give rise to taste receptor organs come from.

This can be accomplished by labeling specific regions of early embryos and following those cells through development to assess their ultimate fate. This fate mapping approach has been used in amphibian embryos for nearly a century (e.g., Spemann, 1921; Vogt, 1929). However, recent advances in technology have allowed two important improvements in fate mapping techniques. First, classical fate mapping in amphibians typically was only feasible for brief portions of the entire embryonic period. This temporal restriction was due to both the aqueous, and therefore soluble nature of the vital dyes used, as well as the dilution of the dye applied to the embryo surface with progressive cell divisions during embryogenesis (Barlow and Northcutt, 1994). Modern approaches have both circumvented the dilution of label problems and amply lengthened the period over which cell fate can be tracked in amphibian and avian embryos. These current methods have allowed elucidation of taste bud origins in amphibians, and are discussed below. Second, early fate mapping was primarily restricted to amphibian embryos, and the degree to which amphibian development was representative of development in other vertebrates, such as mammals and birds, was unknown. With advances in fate mapping in cultured mammalian embryos, as well as the development of *in ovo* mapping techniques in birds, we now know that the fates of embryonic tissues are conserved across vertebrates (Gilbert, 1997). While long-term fate mapping with more conventional techniques has proved to be a challenge in mammalian embryos, molecular genetic approaches have been successfully used to track the fates of embryonic tissues in mice. In particular, molecular genetics have revealed the embryonic origins of taste buds in mice, as discussed below.

6.2.1 Fate Mapping with the Lipophilic Dye DiI

One way to track cells in embryos is to label their membranes with a lipophilic fluorescent dye. This can be accomplished through the use of a pair of long-chain-length (typically C18) carbocyanine dyes, conventionally known as DiI and DiO. DiI is 1,1′-dioctadecyl-3,3,3′,3′- tetramethylindocarbocyanine perchlorate, and fluoresces in the rhodamine range, whereas DiO, 3,3′-dioctadecyloxacarbocyanine perchlorate is excited and emits in the fluorescein range (Molecular Probes, Eugene OR; http://www.probes.com/media/pis/mp00282.pdf). For labeling of amphibian embryos, DiI is typically dissolved in DMSO or ethanol at a concentration of 0.5 to 1%, and then diluted 1:10 with a sterile 0.3 *M* sucrose solution for microinjection with a small caliber micropipet (Collazo et al., 1993). A small volume (picoliter range) of dissolved DiI solution is then focally injected into an embryonic region of interest using a picoinjector. The small bolus of dye diffuses into the membranes of cells at the injection site, thus labeling their surfaces. Alternatively, a micropipet filled with DiI solution can be allowed to weep when low-level backpressure is applied through a picoinjector. The weeping pipet is placed against the surface of the embryo for a minute or more to allow DiI labeling of a tiny surface region (Eagleson et al., 1995; Eagleson and Harris, 1989).

The microinjection approach was used, in part, to determine the embryonic origins of taste buds in amphibians. The epithelium that lines the majority of the pharynx in salamanders is derived from endoderm, and the pharyngeal endoderm arises from a set of cells that reside at the dorsal lip of the blastopore at the onset of gastrulation (Cleine and Slack, 1985; Pasteels, 1942; Vogt, 1929). If taste buds arise from the local endodermal epithelium, as was suspected, then injecting the dorsal lip with DiI should result in the labeling of both the pharyngeal epithelium and its resident taste buds at the end of the embryonic development. This was indeed the case. Brightly fluorescent DiI was found in the membranes of both general epithelial cells and the fusiform cells of taste buds (Barlow and Northcutt, 1995).

A comparable labeling approach has been used in cultured mouse and rat embryos to track the fates of various embryonic tissues, including the neural crest and the somitic mesoderm (Osumi-Yamashita et al., 1994; Trainor and Tam, 1995; Trainor et al., 1994). In general terms, the method is very similar to that described for amphibians. DiI is dissolved in either DMSO or dimethylformamide, and diluted to a concentration of ~0.25% (w/v) in either sterile 0.3 *M* sucrose or dimethylformamide. A small volume of this solution is then microinjected into an embryonic tissue of interest to follow the migration and development of the labeled cells. This approach has been used to track cells in either early, intact mouse and rat embryos (Osumi-Yamashita et al., 1994; Trainor and Tam, 1995; Trainor et al., 1994) or in dissected subregions of embryos (Imai et al., 1996), both raised in *ex utero* culture. To date, the culture period for rodent embryos is approximately 60 h (Imai et al., 1996), whereas explants of the mandible may survive for 8 days *in vitro*. This approach of fate mapping cells in cultured mammalian embryos or explants has yet to be applied to investigations of the taste periphery.

6.2.2 Fate Mapping with Labeled Grafts

6.2.2.1 Cell Tracking Based on Endogenous Markers

An alternative and frequently used approach to fate mapping in amphibian and avian embryos has been the grafting of labeled or otherwise identifiable sets of embryonic cells into unlabeled or heterospecific host embryos. The earliest and possibly most well-known example of this approach is the experiments performed by Spemann and Mangold to demonstrate the organizing power of the dorsal lip of the blastopore during amphibian gastrulation (Spemann and Mangold, 1924). These researchers microsurgically removed the dorsal lip from the embryo of one salamander species and grafted it into a host embryo of another salamander species. Based on pigmentation differences between the two species, Spemann and Mangold were able to follow cells of the graft and assess their fates. A similar technique is to take advantage of albino and pigmented strains of a single species of salamander, such as the axolotl (Barlow and Northcutt, 1994). Again, by transplanting a small region from a pigmented embryo into an albino host embryo (which completely lacks melanin), the fates of the grafted cells can be readily tracked. This approach has been used to assess if some taste buds arise from the ectoderm that lines the most anterior regions of the pharynx in axolotls (Figure 6.1B). In these experiments, the presumptive stomadeal ectoderm of an albino embryo was replaced with ectoderm from the same region of a pigmented donor embryo. The resultant pigmentation chimera possessed taste buds both with and without pigment containing cells; those with pigment were ectodermal in origin and those without were likely endodermal (Barlow, 2000b). While this type of approach based on the existence of intrinsic markers is readily

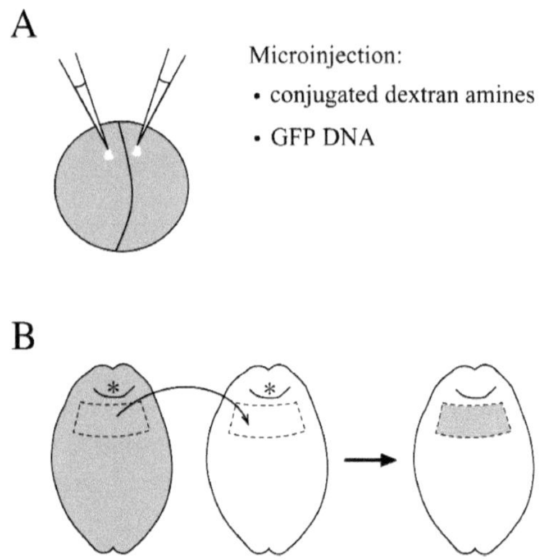

FIGURE 6.1 Schematic diagram of fate mapping approaches using microinjection and homotopic grafting with lineage tracer substances. A: A fine micropipet is used to pressure-inject a small volume of lineage tracer into each cell of a two-cell stage amphibian embryo. Either conjugated dextran amines (e.g., fluorescein dextran, rhodamine dextran, biotinylated dextran) or DNA encoding GFP can be injected in this fashion. B: Homotopic grafting is used here to fate map cells of the stomadeal ectoderm of a late neurula stage axolotl embryo. In this view of the ventral surfaces of a pigmented (left) and an albino (middle) embryo, the stomadeal ectoderm (dashed outline) located just posterior to the developing forebrain (*; anterior is up) is moved from the pigmented donor and placed in the same location in an albino host. The resultant chimeric embryo is shown on the right. This same type of grafting experiment can be performed using stomadeal ectoderm from an embryo that had been injected with dextran amine or GFP DNA at the two-cell stage (as in A). In this case, ectoderm from a labeled embryo would be grafted homotopically into an unlabeled host embryo.

applicable in amphibian and avian embryos (Le Douarin and Kalcheim, 1999), comparable methods in rodent embryos have yet to be developed.

6.2.2.2 Cell Tracking Using Exogenous Markers

An alternative method to follow grafted cells is to label them prior to grafting. This can be accomplished in amphibians by injecting label into all cells of early embryos (at the 1-, 2- or 4- cell stage) and then allowing the injected embryos to develop until the appropriate embryonic tissue has developed and can be then grafted into uninjected host embryos (Figure 6.1). A set of compounds, the dextran amines, has been used extensively for this purpose. These are large molecular weight, hydrophilic polysaccharides, varying from 3 to 200 kD, and can be tagged with specific fluorescent molecules to allow their detection in living embryos (Molecular Probes, Eugene, OR: http://www.probes.com/media/pis/mp01800.pdf). Typically, 10 to 70 kD dextran amines have proven optimal for labeling of amphibian embryos, as the larger molecular weight dextrans are cytotoxic and the smaller molecular weight dextrans pass through some gap junctional contacts (Delarue et al., 1992; Gimlich and Gerhart, 1984; O'Rourke and Fraser, 1986). Dextrans are water soluble and diffuse readily in the cytoplasm of injected cells. Thus, by injecting all cells of early embryos with these tagged polysaccharides, the entire embryonic cytoplasm is quickly labeled. This method has a distinct advantage over the pigment-to-albino grafts mentioned above. Whereas pigment granules are restricted to cell bodies (Barlow and Northcutt, 1995; Northcutt et al., 1995), dextrans will diffuse down axons and thus label the peripheral processes of neurons (O'Rourke and Fraser, 1986). Certainly, this type of approach would be informative in studies of the development of gustatory sensory neurons. While not yet used to investigate taste bud development per se, dextran-labeled grafts have been employed to examine the embryonic origins of discrete gut regions, including the pharynx of *Xenopus laevis*, the African clawed frog (Chalmers and Slack, 2000).

One problem that arises with the use of dextran-labeled cells is that with subsequent cell divisions, the dextran within each cell is serially diluted. Thus, if the labeled cells belong to an extensively dividing population, such as an epithelium, the dextran may not be detectable at the end of the experimental period (L. Barlow, personal observation). To get around this technical difficulty, dextrans can be conjugated with biotin so that, upon fixation, the dextran signal can be amplified using avidin-HRP/DAB or avidin-FITC visualization. However, another, newer approach is now commonly used to avoid the reduction in labeling intensity encountered with dextrans.

Green fluorescent protein (GFP) occurs naturally in the jellyfish *Aequorea victoria* and confers bioluminescence to that creature (Chalfie et al., 1994). The nucleotide sequence encoding this protein can be introduced into non-fluorescing cells of mammals, amphibians, flies, and yeast (to name only a few taxa), causing these cells to fluoresce when GFP is expressed as protein (Chalfie, 1995; Conn, 1999; Misteli and Spector, 1997). Thus, similar to that described above for dextran amine labeling in fate mapping, early amphibian embryos can be microinjected with DNA encoding GFP, which is driven by a constitutively active promoter (Zernicka-Goetz et al., 1996; Figure 6.1). The exogenous sequence is transcribed and translated into protein by the host cells machinery, and these cells then fluoresce within a few hours of microinjection. The exogenous DNA does not become incorporated into the host genome, however, the extrachromosomal GFP DNA is replicated to some degree along with the host DNA (Cleaver and Kreigstein, 1999; Ruiz i Altaba et al., 1993). In this way, the dilution of label problem is obviated. Cells that possess GFP DNA will replicate it as they prepare for cytokinesis and, therefore, resultant daughter cells will continue to produce new GFP protein.

As with the dextran grafting approach, GFP-labeled embryonic tissue can then be grafted into an unlabeled host embryo, and the fates of those labeled cells can be followed (Figure 6.1). However, these grafts can be mosaic in terms of GFP label (Cleaver and Kreigstein, 1999; Ruiz i Altaba et al., 1993); not all cells within the grafted region will express GFP. This problem can be traced back to the viscous nature of nucleotide solutions. When DNA is injected into early amphibian embryos, unlike dextrans, this solution is thick and does not instantly diffuse throughout the cytoplasm. As a consequence, depending on the plane of early cell divisions with respect to the bolus of DNA solution, not all cells will receive a portion of the GFP DNA and mosaic embryos will be created. Thus, GFP-expressing grafts can be used to assess only the fates of labeled cells, and not necessarily the fates of all cells within an embryonic tissue graft (as is the case with the dextran labeling).

6.2.3 Determining Embryonic Origins Through Chimeric Analysis

An entirely different approach to determining the embryonic origins of complex tissues is by chimeric or mosaic analysis. This method relies on the use of an animal composed of two genetically distinct populations of cells. This can be accomplished either by fusion of two genetically different 4- to 8-cell embryos (Lo, 1986) or by use of strains that carry an inheritable marker on one allele; such as a higher level of enzyme activity (e.g., glucuronidase; Herrup and Mullen, 1979), or a transgenic marker such as lacZ under control of a constitutive promoter (Tan and Breen, 1993). In amphibian chimeras (Jacobson and Hirose, 1978), coarse mosaics can form, where large areas of the animal arise from one or the other cell populations. In mammals, the different embryonic populations intermingle extensively, forming a fine-grain mosaic. Despite the extensive intermingling of cells early in embryogenesis, the mosaic/chimeric animals are useful in addressing questions such as:

- Do taste buds contain a population of cells that migrate into the developing epithelium, or do taste buds contain only cells from adjacent epithelial areas?
- Do taste buds arise from single progenitor cells; i.e., is there a mother cell for each taste bud?
- Do the different types of cells in taste buds originate from different progenitors?

The basic principle underlying mosaic/chimeric analysis is that in the early stages of embryogenesis, there is extensive intermingling and migration of proliferative cell populations. However, later in embryogenesis, after organogenesis is underway, epithelial cells form sheets and tend to maintain their position relative to surrounding epithelial cells. A given epithelial progenitor will give rise to neighboring daughter cells, which inhabit a patch of epithelium (see Figure 6.2). Thus, the epithelium is a patchwork of clonally related cells.

If taste buds contained cells that migrate into epithelium (e.g., from placode or neural crest) then the phenotype of the migratory cells would be determined independently of that of the surrounding epithelium. Accordingly, the migratory cells would match the surrounding epithelium only 50% of the time. If, on the other hand, taste buds arise from local epithelial elements, then the phenotype of the taste buds would always match that of the surrounding epithelium. Stone et al. (1995) found that taste buds always match the surrounding epithelium and therefore must arise entirely from local epithelial elements; there is no contribution from migratory populations of either the neural crest or epibranchial placodes.

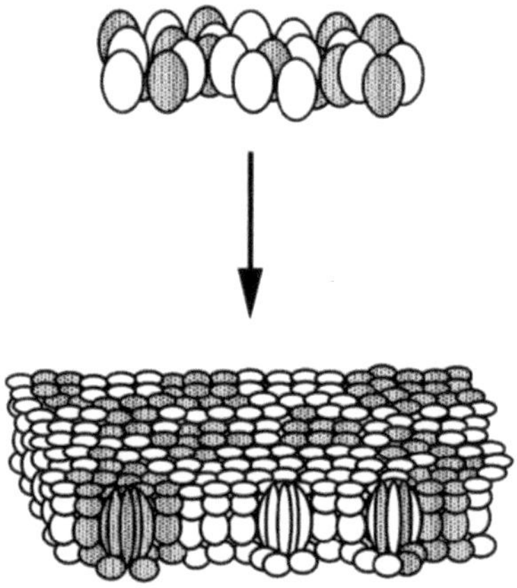

FIGURE 6.2 Chimeric analysis of epithelium. The embryonic epithelium consists of two phenotypic populations of cells (grey and white) distributed randomly within the epithelium. As the epithelium matures, each progenitor cell gives rise to a patch of epithelial cells that retains the phenotypic character of the progenitor. Taste buds in the mature epithelium can either reside entirely within a patch of one type or another (TB1 and TB2), or else on the border between two phenotypically different patches of epithelium (TB3). When taste buds reside in homogeneous epithelium (TB1 and TB2), they always match the phenotype of the surrounding epithelium, and hence are derived from local epithelium (Stone et al., 1995). Taste buds on borders contain mixed populations of cells and thus must arise from multiple progenitors.

Mosaic/chimeric analysis can also be used to determine the approximate number of progenitors contributing to a tissue compartment (Herrup, 1986). Taste buds residing on the border of two epithelial phenotypic patches potentially may have progenitors from either side. If taste buds were to arise from a single progenitor cell, then all the cells of the taste bud would be of one phenotype or the other. If multiple progenitors contribute to a taste bud, then taste buds would sometimes contain a mixed population of cells. Stone et al. (1995) found that taste buds contain mixed populations and therefore arise from multiple progenitors. Further, the percentage of labeled cells in the mixed taste buds was quite variable, indicating either that taste buds arise from a large (>4) progenitor population or that progenitors contribute unequally to the mature endorgan. The mixed-population taste buds can also be examined to determine whether particular types of taste cells are related by lineage. Preliminary accounts indicate this to be the case (Stone et al., 1996).

6.3 WHAT DEVELOPMENTAL MECHANISMS ARE RESPONSIBLE FOR TASTE BUD FORMATION?

The conventional view of taste bud development has been that these receptor organs are induced to form by signals they receive from ingrowing nerve fibers. As discussed earlier, until recently, all of the experimental studies using rodents that were intended to test the inductive role of nerves were performed on embryonic or postnatal tongue epithelium that was *already* innervated (Farbman, 1968; 1971; 1972a; Hosley et al., 1987a; 1987b; Oakley, 1993b; Torrey, 1940). Once an epithelium has become innervated, disrupting that innervation, even at early embryonic stages, is not the same experiment as preventing initial nerve contact of a naïve epithelium. The former type of experiment is likely assessing the role of innervation in the maintenance of continued taste bud development, whereas the latter experiment asks directly if nerve contact changes the fates of cells within the epithelium. Thus, to test if nerves are inducing taste buds (i.e., nerves change the fate of cells that otherwise would develop as epithelial) nerves must be blocked from contacting the epithelium.

6.3.1 Tests of the Neural Induction Model

6.3.1.1 Explant Culture Using Amphibian Embryos

One way to prevent the epithelium from becoming innervated is to remove it from the embryo prior to innervation. This can be readily accomplished in amphibians. Explants of amphibian embryos can survive for extended periods in extremely simple culture media, (e.g., Chien et al., 1993; Gurdon et al., 1999; Henry et al., 1996; Holtfreter, 1996; and see Willier et al., 1955, for a review of early studies). Rather than relying on direct maternal supply of nutrition, as do mammals, or on a single large yolk reserve, as do birds for example, the nutritive yolk of amphibians is distributed in granules that are provisioned in every cell of the embryo. Even explants of frog and salamander embryos possess sufficient endogenous nutrients to power them through embryogenesis at a pace comparable to intact controls (Graveson et al., 1997). Thus, it is possible to remove the presumptive taste bud-bearing epithelium from early salamander embryos, prior to innervation, and assess if these explants will generate taste buds with a time-course comparable to control embryos. In these experiments, the embryonic epithelium was removed microsurgically from host embryos using flame-etched tungsten needles, rinsed through a series of washes with sterile 60% Liebowitz-15 medium supplemented with 0.1% BSA and antibiotics (L-15), and then raised in individual wells of a 24-well plate in L-15 for up to 10 days (Barlow et al., 1996). In this serum-free medium, the isolated epithelium from neurula or early tailbud stage salamander embryos gave rise to taste buds whose morphology and immunoctyochemical profile were similar to those of normal taste buds (Barlow et al., 1996). In addition, when these explants were probed immunocytochemically for neurons, no such cells were encountered. This experiment represents a clear test, and rejection, of the neural induction hypothesis, and is not confounded by the issues of neural maintenance encountered with nerve interruption experiments in rodents discussed above.

6.3.1.2 Explant Cultures Using Rodent Embryos

Organ culture systems for mammals have not been utilized as effectively as those for amphibia for the study of taste bud development, but the methodologies exist. Two laboratories have independently developed organ culture methods (Mbiene et al., 1997; Morris-Wiman et al., 2000). The former entails culture of either the tongue primordium or lower jaw with tongue starting just after fusion of the branchial arches (i.e., before nerves have reached the developing epithelium). This culture system involves removal of the entire lower jaw or tongue from gestational day 12 mice (Hall et al., 1999a) or day 13–14 rats, and placement of the tissue at an air-medium interface. This system permits elaborate development of fungiform papillae, albeit lacking in differentiated taste buds.

The second culture method entails roller culture of branchial arches either with or without their innervating ganglia (Morris-Wiman et al., 2000). The tissues are taken from the embryo a day earlier than in the previous method (e.g., mouse gestational day 11). The latter method has the advantage of permitting assessment of the importance of innervation to the process under study. In contrast to the interface culture system, in the roller culture system, the circumvallate papilla develops well and exhibits the beginning of differentiated taste buds, whereas fungiform development is poor.

Both culture systems employ enriched media, i.e., media containing sera. The drawback of this supplementation is that the chemical composition of the medium is not defined and is likely to contain numerous growth and trophic factors, which may or may not be present during normal development.

These *in vitro* experiments from mammalian preparations all confirm that papillary morphogenesis can proceed without innervation but that formation of fully differentiated taste buds requires innervation. This has been interpreted by some to mean that nerves induce formation of taste buds in a competent epithelium; however, molecular studies reveal that molecular markers of taste bud progenitors (e.g., BDNF) are present in cultured tongues although the taste buds do not undergo cytodifferentiation (Nosrat et al., 1998). Thus, taste nerves cannot be said to induce taste buds in an undifferentiated epithelium; rather, in mammals, the taste nerves may permit differentiation of an already specified taste bud progenitor population.

6.3.2 Examination of Nerve-Independent Mechanisms of Taste Bud Genesis

Given the results described above in both mammals and amphibians, there is a growing consensus that much of taste bud development in vertebrate embryos occurs prior to and is independent of innervation, and may be intrinsic to the epithelium. However, it has been indirectly proposed that taste bud development may rely on signaling between the epithelium and the adjacent mesenchyme (Mistretta, 1998). In particular, a close relationship between epithelium and mesenchyme exists in the developing gustatory papillae of mammals (Farbman, 1965; Mistretta, 1972; Mistretta and Haus, 1996), and signaling between these embryonic cell populations may be essential for formation of papillae. It has also been suggested, based on gene expression data presented below, that papillae represent, in part, taste bud primordia. Thus, epithelial-mesenchymal interactions may be necessary for papillae as well as taste bud development. Once again, a direct test of this idea can be performed using explants of appropriate epithelium from amphibian embryos.

6.3.2.1 Testing the Role of Mesenchyme in the Induction of Taste Buds

To examine if mesenchymal contact of the epithelium is essential for taste bud genesis, the presumptive taste bud-bearing epithelium must be removed and cultured prior to the migration of mesenchymal cells to the pharyngeal region. Because mesenchymal migration is complete well before the epithelium is innervated (Northcutt and Brändle, 1995), the epithelium must be removed quite early and then cultured (as described above) for up to 12 days, when taste buds have developed in intact controls (Barlow and Northcutt, 1995). Nonetheless, likely because of intrinsic yolk stores, even these early explants develop at a pace similar to intact embryos. Further, these explants, while devoid of embryonic mesenchyme, differentiate taste buds, which are morphologically and immunocytochemically indistinguishable from control buds (Barlow and Northcutt, 1997). Therefore, at least in amphibians, taste bud development occurs independently of mesenchymal cell contact.

6.3.2.2 Candidate Molecule Approach to Assess Early Events in Taste Bud Development

Just as early studies of embryogenesis entailed descriptive studies of the appearance and sequence of reorganization of tissues, one way of studying molecular events during embryogenesis is to undertake descriptive studies of the timing and distribution of developmentally important molecules. Several studies have identified a variety of developmentally regulated molecules that appear in areas of the epithelium where taste buds will appear but prior to cytological evidence for taste bud primordia (Bitgood and McMahon, 1995; Nosrat et al., 1996; Hall et al., 1999b; Jung et al., 1999; Jeffery, 2001).

The common method for assessing the expression of developmentally regulated genes is through *in situ* hybridization. Because many of these gene products are either secreted or present at low levels, detection through immunocytochemistry is difficult. Even with *in situ* hybridization, detection of low-abundance mRNA is difficult. Nonetheless, the use of *in situ* hybridization has uncovered several families of developmentally regulated molecules that may be involved in taste bud or taste papillary development. These include BDNF (Nosrat et al., 1996; Nosrat and Olson, 1995), sonic hedgehog (Shh) and its signaling partners (Bitgood and McMahon, 1995; Hall et al., 1999b; Jung et al., 1999), and bone morphogenetic proteins (BMPs; Jung et al., 1999). For Shh and BMP-2, tissue expression is widespread as fusion of the arches first makes a tongue primordium. Slightly later in development — at a time before nerve fibers reach the epithelium — expression of these molecules becomes limited to small loci of about 5 to 10 cells that presage the locus of fungiform papillae. This molecular differentiation is approximately concurrent with the earliest signs of cytodifferentiation, but prior to epithelial innervation (Farbman and Mbiene, 1991). More careful, multiple label studies are necessary to determine the sequence of molecular specification of the fungiform epithelium in these early, preneural stages of development.

Another method for studying expression of developmentally regulated molecules is to utilize transgenic animals in which the lacZ coding sequence replaces some of the coding sequence for the molecules of interest. For example, BDNF-lacZ mice (where the lacZ sequence replaces the BDNF coding sequence and therefore falls under the control of the BDNF regulatory elements) can be used to describe the exact timing of expression and cell populations expressing BDNF early in taste development (Yee et al., 2000).

6.3.3 Examination of the Role of Neurotrophins in Taste Development

Neurotrophins are secreted signaling molecules that are generated by target tissues during development and are required for the survival of the neurons destined to innervate that tissue. Those nerve cells whose processes reach the correct target have access to the neurotrophin and can survive, while those that fail to reach the target undergo apoptotic cell death. Each particular type of sensory endorgan (e.g., Merkel cells, muscle spindles, and hair follicle mechanoreceptors) produces a particular type or sequence of neurotrophin matching the neurotrophin dependency of the innervating ganglion cells. Neurotrophins may also serve as stop signals to inhibit further extension of the growing neurites.

The family of neurotrophins includes NGF, BDNF, NT3, and NT4/5. The cognate receptors comprise trkA (for NGF), trkB (for BDNF, NT4/5 and to a lesser extent, NT3), trkC (for NT3), and the low-affinity nonspecific receptor, p75. The usual approaches to studying the role of neurotrophins in sensory systems include studies of the timing and location of expression of the neurotrophins and their receptors, mis-expression of the neurotrophins or their receptors, and knockouts of neurotrophins and receptors.

6.3.3.1 Expression

The timing and pattern of expression of the neurotrophins and their receptors can be studied by either immunocytochemistry and/or *in situ* hybridization. Because neurotrophins are produced at relatively low levels and are rapidly secreted, immunocytochemical detection of neurotrophins is fraught with the likelihood of false-negatives, i.e., the inability to detect the low levels of protein in developing cells. Further complicating the situation, secreted neurotrophins may bind to the low-affinity p75 receptors produced by cells at some distance from the site of generation. Thus, even an immunocytochemically positive cell might not be producing the neurotrophin detected on its surface.

In situ hybridization has been widely utilized to detect even low-abundance mRNA in various tissues. Signal amplification can be achieved either by using long exposures with radioactive probes (Nosrat et al., 1996) or by enzyme amplification (e.g., tyramide-based amplification) of nonradioactively tagged probes (e.g., digoxigenin or fluorescein; Yang et al., 1999). Radioactive *in situ* hybridization was used by Nosrat and Olson (1995) to demonstrate that BDNF mRNA is produced by taste bud primordia while NT3 is produced by the perigemmal epithelium (Nosrat et al., 1996). This led to the suggestion that BDNF is the neurotrophin underlying taste-specific innervation of taste buds. BDNF is present in the right cell population (patches of lingual epithelium) at the right time (just prior to or at the time of innervation).

For BDNF to function effectively as the gustatory neurotrophin, its cognate receptor, trkB, should be present in the gustatory ganglia (geniculate, petrosal, nodose) at the same, or earlier, stages of development. Nosrat (1997) and Pirvola et al. (1994) have reported the presence of non-truncated (i.e., potentially functional) trkB transcripts at E15-18 during rat development. This finding is consistent with the role of BDNF as a specific gustatory neurotrophin during development.

6.3.3.2 Knockouts

Knockout mice have been utilized to test whether a particular neurotrophin is required in a developing system. If the hypothesis is correct that BDNF produced by developing taste buds primordia sustains the trkB-producing neurons of the gustatory ganglia, then elimination of either BDNF or trkB should result in wholesale apoptotic death of the gustatory ganglion cells resulting in failure to innervate presumptive taste buds. The lack of innervation should result in secondary failure of differentiation of taste bud primordia. Thus, significantly fewer taste buds should form in the knockout animals. Although the BDNF and trkB knockouts do exhibit pronounced deficits in taste bud formation (Cooper and Oakley, 1998; Fritzsch et al., 1997; Mistretta et al., 1999; Nosrat et al., 1997; Oakley et al., 1998), the effect is not absolute. BDNF knockout mice exhibit significantly fewer gustatory ganglion cells and taste buds, but some taste buds persist, especially at the anterior end of the tongue. Those that do persist, however, are insufficient to drive avoidance behaviors to bitter and thus may not be fully functional (Nosrat et al., 1997). It seems likely that the remnant taste buds can be sustained morphologically by the trigeminal innervation (see also Kinnman and Aldskogius, 1988), although the trigeminal innervation is incapable of conveying gustatory quality information.

The persistence of taste buds in BDNF knockout mice may be partially accounted for by the presence of NT4/5 in the developing gustatory epithelium. NT4/5 is a suitable ligand to activate the trkB receptor, and NT4/5 knockout mice exhibit subtle losses of the gustatory periphery (Liebl et al., 1999).

Curiously, trkB knockout mice exhibit a remarkable persistence of fairly well differentiated taste buds, although in significantly lower numbers than in wild-type animals. Gustatory ganglion cells in these animals undergo apoptotic cell death prior to the time at which they innervate the epithelium (Fritzsch et al., 1997). Thus, the presumptive taste buds in these animals undergo growth and differentiation despite the absence of the specific gustatory innervation. This finding largely undercuts the hypothesis that gustatory nerve fibers are necessary to induce the differentiation of taste buds from already competent primordia.

The use of knockout mice to test hypotheses regarding neurotrophin dependency is compromised by several features of this signaling system. First, all of the neurotrophins can interact, albeit with lesser efficiency, with most neurotrophin receptors.

Thus, some rescue of function can be attributed to non-specific interactions that would not play a significant role in rescue of ganglion cells during normal development (Conover and Yancopoulos, 1997). Second, the system may alter expression of the remaining neurotrophins or trk receptors in response to the elimination of one of the family members (Rice et al., 1998). Thus, compensatory changes in neurotrophin expression may rescue populations that would normally be entirely dependent on the ablated neurotrophin. Finally, many of the neurotrophin knockout animals do not survive long after birth (e.g., BDNF null mice: Jones et al., 1994; Nosrat et al., 1997). Thus, these animals are of limited utility in the study of the taste system, which undergoes extensive postnatal development and change.

6.3.3.3 Neurotrophin Mis-expression

Another way to study the function of neurotrophins is by causing their expression in the wrong tissues or tissue compartments. This overabundance of neurotrophins can have a variety of effects over and above increasing the survival of particular classes of sensory neurons (Rice et al., 1998). For example, neurotrophins can serve as signals to cease process elongation and to begin elaboration of sensory terminals (Rice et al., 1998). Thus, expression of neurotrophins in foreign tissues is likely to result in both survival of excess neurons as well as misdirection or elaboration of their terminal processes.

In a transgenic line of mice, BDNF is misexpressed in tissues of the tongue. In one, BDNF was placed under the control of the nestin promoter (Ringstedt et al., 1999). In these animals, BDNF is expressed by muscle as well as by numerous neurons. Thus, outgrowing gustatory nerves encounter high levels of BDNF in the muscle of the tongue long before the fibers reach gustatory epithelium. As might be expected, the excess BDNF permits survival of excess gustatory ganglion cells. The presence of high levels of BDNF deep in the tongue appears to deflect the gustatory axons from their normal sites of termination and thus the taste epithelium is deficient in gustatory innervation. This results in abnormally formed, small taste buds and fungiform papillae (Ringstedt et al., 1999).

6.4 DEVELOPMENT OF TASTE BUD INNERVATION

Taste buds are innervated by branches of the VIIth, IXth, and Xth cranial nerves. In all vertebrates, each of these nerves has a specific topographic innervation territory within the mouth and pharynx. VIIth nerve fibers innervate anterior taste buds (located on fungiform papillae in mammals), while IXth nerve fibers innervate intermediately positioned taste buds (of the foliate papillae in mammals), and the Xth nerve innervates the most posterior taste buds (of the circumvallate papillae in mammals) (Kinnamon, 1987; Northcutt et al., 2000). During embryogenesis, the arrival of these nerves at the epithelium occurs well before taste buds differentiate (Barlow et al., 1996; Farbman and Mbiene, 1991; Hall et al., 1999b; Landacre, 1907; Whitehead and Kachele, 1994). Interestingly, in some studies in rodents, it appears as if nerve branches, rather than exploring the epithelium broadly, grow directly to developing papillae (Farbman and Mbiene, 1991; Whitehead and Kachele, 1994).

These data suggest that ingrowing axons may be using discretely localized chemical cues to find their targets. Additionally, earlier data from studies of the development of the trigeminal innervation of the maxilla indicated that the target tissue likely emitted a long-range chemoattractant that guided trigeminal axons from a distance (Davies, 1988; Lumsden and Davies, 1983). Recently, an *in vitro* approach was developed to test if geniculate neurons (gVII) are guided to the developing tongue by long-range chemoattractants (Rochlin and Farbman, 1998; Rochlin et al., 2000). The results of these studies have provided some surprising insights.

6.4.1 *In Vitro* Development of Gustatory Ganglia

6.4.1.1 Culture of Gustatory Ganglia

In rat embryos, neurons of the cranial ganglia of the VIIth, IXth, and Xth nerves appear between embryonic days 11 and 14 (E11 and 14) (Altman and Bayer, 1982), and begin to send out nerve branches beginning at E12 (Altman and Bayer, 1982; Rochlin and Farbman, 1998; Rochlin et al., 2000). Ganglia can be dissected from rat embryos at these early stages and at later stages, and cultured in a collagen gel matrix overlaid with culture medium (Davies and Lindsay, 1985; Lumsden and Davies, 1983; Rochlin and Farbman, 1998; Rochlin et al., 2000). While neurons survive under these conditions, collagen alone, however, is not permissive for axonal outgrowth, and thus the gel matrix must be supplemented with small quantities of specific neurotrophins (Arumae et al., 1993; Lumsden and Davies, 1983; Rochlin and Farbman, 1998; Rochlin et al., 2000). Axon outgrowth from cultured geniculate ganglia is supported by BDNF, NT-4/5, and GDNF at early stages (E12). Sensitivity to BDNF and NT-4/5 persists in older cultured ganglia (E14), but responsiveness to GDNF is quickly lost (by E13) (Rochlin et al., 2000). Comparable temporal changes in neurotrophin selectivity have been demonstrated in trigeminal ganglia (Arumae et al., 1993; Enokido et al., 1999; Rochlin and Farbman, 1998). One caveat common to virtually all of these *in vitro* studies, however, is that the ganglia are removed at a stage when neurons have already begun to send out axons. Thus, dissecting out the ganglion involves axotomizing any neurons whose processes have already left the ganglion. One way to avoid this potential confound is to culture instead the embryonic precursors of the ganglia so that neurons are generated and send out axons *de novo in vitro*.

Sensory neurons of vertebrate cranial ganglia arise from two distinct embryonic sources: the cranial neural crest and the epibranchial placodes (D'Amico-Martel and Noden, 1983; Landacre, 1910; Narayanan and Narayanan, 1980; Northcutt and Brändle, 1995; Stone, 1922; Yntema, 1937; 1943; 1944). A subset of the cranial neural crest migrates from its early location, the dorsal neural tube, to contribute both neurons and glia (satellite glia and Schwann cells) to the developing ganglia, whereas epibranchial placodes give rise only to neurons. Epibranchial placodes, in contrast, are epithelial thickenings of columnar cells that appear in the embryonic head ectoderm just dorsal to the forming branchial pouches, hence their name. The placodes are believed to generate immature neurons that coalesce into a ganglion structure. Depending on the vertebrate species and the ganglion in question, these placodally derived aggregates may remain as discrete ganglia (as in chicken: D'Amico-Martel and Noden, 1983; Narayanan and Narayanan, 1980)), or may form fused ganglia with neurons derived from neural crest (as in fishes and amphibians: Landacre, 1910; Northcutt and Brändle, 1995). Regardless of their final position, it is believed that these placodally derived neurons, rather than those derived from neural crest, likely innervate taste buds (Landacre, 1907).

Epibranchial placodes from chick embryos will generate sensory neurons *in vitro* (Vogel and Davies, 1993). And we have recently found, using amphibian embryos, that ectoderm destined to give rise to epibranchial placodes also produces neurons *in vitro* (Gross and Barlow, 2000; Gross and Barlow, personal observation). Presumptive placodal ectoderm can be removed from axolotl neurula stage embryos before placodes have formed (Northcutt and Brändle, 1995), and placed in a commercially available collagen gel that has been minimally supplemented with NGF, several growth factors, and ECM components (GFR-Matrigel, Becton-Dickinson, Franklin Lakes, NJ; http://www.bdbiosciences.com/labware/Library/ecm/pdf/p128-129-110.pdf). The gel containing the ectoderm is then immersed in serum-free L-15 medium with BSA (Barlow et al., 1996). After 4 days *in vitro*, neurons, which have developed *de novo* from placodes, begin to send axons out into the gel matrix. This outgrowth continues at a pace comparable to that described for axolotl sensory neurons *in vivo* (Barlow et al., 1996; Northcutt and Brändle, 1995). As is the case with cultured rat neurons, this growth is not observed when placodes are cultured in collagen matrix without supplementation (Gross and Barlow, unpublished observations). The advantage of developing this system in axolotls is that the pharyngeal epithelium will generate taste buds under culture conditions comparable to those that allow placodal neurogenesis to occur (Barlow and Northcutt, 1997). Further, by culturing neurons *de novo* from epibranchial placodes, not only is the problem of axotomy obviated, but in addition, these cultured neurons may represent a discrete gustatory subset of ganglion cells. Thus, in axolotls, we have a unique opportunity to explore the mechanisms by which putative gustatory neurons come to innervate their target receptor organs.

6.4.1.2 Cultures of Gustatory Ganglia with Target Tissue

Developing sensory ganglia have been paired *in vitro* with target explants using rat embryos (Rochlin and Farbman, 1998; Rochlin et al., 2000). In particular, both the trigeminal and geniculate ganglia have been raised in culture with explants of the mandibular portion of the first branchial arch, the second branchial arch, and various regions of the developing tongue. In the case of early ganglion explants, geniculate axons develop and grow into the matrix, despite the absence of exogenously supplied neurotrophins. The target tissue therefore is providing an as-of-yet unidentified factor that promotes neurite outgrowth. However, when cultured ganglia paired with target explants are supplemented with neurotrophins, these targets were found to be repellant to early axons. Surprisingly, the target is promoting growth, although not in a directed way, and is simultaneously, directionally repellant. The dominant chemorepellant factor at the time of these experiments appears to be a semaphorin, Sema3A. The transcript encoding this protein is present in the target regions at the appropriate time; and when a function-blocking antibody against the Sema3A receptor, neuropilin-1, is applied to the cultures, the repellant effect of the target is abolished. These studies reveal a complex mixture of growth-promoting and chemorepellant effects that vary with developmental stage. Clearly, this type of co-culture system will help reveal the molecular entities involved in the development of the sensory innervation of the taste periphery.

Another type of co-culture paradigm has been employed to study placodal neurogenesis in chickens (Begbie et al., 1999). Here, early epibranchial placodes were removed from chick embryos at a time when the placodes are first forming. Using a series of molecular markers, these researchers demonstrated that placodes cultured in unsupplemented collagen gel generate a small number of neurons. However, when placodal ectoderm was paired with pharyngeal endoderm at a distance, the number of neurons generated increased substantially. The molecular nature of the endodermally derived signal was identified as BMP7, a secreted TGF-β family member. BMP-7 was found expressed in the pharyngeal endoderm at the proper time, and application of BMP-7 to placodes cultured alone produced the same increase in neuronal number as did co-culture with the pharyngeal endoderm. Interestingly, pharyngeal endoderm possesses taste buds, and thus ultimately is the target tissue, in part, of the placodal neurons it has now been shown to induce (Begbie et al., 1999; Graham and Begbie, 2000).

We have also begun co-culture experiments using axolotl embryos (Gross and Barlow, 2000; 2001). In these studies, placodal ectoderm is paired with various targets, and the degree to which growth is directed, or not, is assessed using circular distribution statistics. In general, outgrowth from placodal explants of axolotls is much sparser than that observed for cultured rat ganglia. Thus, all fibers can be counted and each fiber assigned to a radial bin on a polar plot. A statistical analysis can then be performed to determine if the axon array differs from a random distribution. When placodal explants alone are assessed, the pattern of outgrowth is random. However, when placodes are raised with oropharyngeal endoderm, the presumed target of many of these axons, growth is now directed toward the target. Surprisingly, when placodes are paired with flank ectoderm, a tissue these fibers do not encounter *in vivo*, directed growth is also observed. However, other embryonic tissues, such as notochord (which is not innervated) and gut endoderm (which is not encountered by placodal neurons *in vivo*), do not trigger directed growth when artificially paired with placodal neurons *in vitro* (Gross and Barlow, 2001). Thus, pharyngeal endoderm and flank ectoderm likely share some conserved factor that promotes directed growth of placodally derived neurons. We can now begin to assess the molecular nature of this phenomenon, using function blocking antibodies to known neurotrophins and chemical cues, much as has been the trend in other experimental systems of axonal growth (Rochlin and Farbman, 1998; Rochlin et al., 2000).

6.5 SUMMARY: A CURRENT HYPOTHESIS OF TASTE BUD DEVELOPMENT

Our view of embryogenesis of taste buds has changed substantially during the past decade. The advent of molecular biology and improved organ culture approaches has permitted reevaluation of the older hypotheses based predominantly on morphological analyses. No longer can it be assumed that gustatory nerve fibers endow an indifferent, albeit competent, epithelium with special prowess to generate taste buds. Rather, the primary events in gustatory genesis lie in the epithelium itself, long before innervation occurs.

Based on the past decade of work in this field, the contemporary view of the origins of taste buds is that taste-bud bearing epithelial cells are specified positionally early in development, i.e., just after gastrulation. The specified competent epithelium then undergoes reorganization — probably through the agency of intraepithelial signaling — to establish taste-bud progenitor populations spaced within the competent field. The taste-bud progenitors in mice differ molecularly (Hall et al., 1999a; Jung et al., 1999) from the surrounding epithelium and shortly thereafter differ histologically as well (Farbman and Mbiene, 1991). At this early preneural stage of taste-bud development, the taste-bud progenitors begin to produce and secrete growth factors, including BDNF. The local production of BDNF facilitates invasion of the tissue by the BDNF-responsive gustatory nerves. In mammals, the approach or contact of the nerve fibers may then enhance cytodifferentiation and proliferation of progenitors to produce a young taste bud replete with multiple elongate cells exhibiting a variety of phenotypic characters. In amphibians, contact with the nerve is not necessary for elaboration of a histologically mature taste bud. Thus, our view of gustatory genesis has changed from one centered on the nerve fibers as the primary inducers to one focused on the epithelium as the primary determinant of taste-bud progenitors.

REFERENCES

Altman, J. and S. Bayer. 1982. Development of the cranial nerve ganglia and related nuclei in the rat, *Adv. Anat. Embryol. Cell Biol.,* 74, 1-90.

Arumae, U., U. Pirvola, J. Palgi, T.R. Kiema, K. Palm, M. Moshnyakov, J. Ylikoski, and M. Saarma. 1993. Neurotrophins and their receptors in rat peripheral trigeminal system during maxillary nerve growth, *J. Cell Biol.*, 122, 1053-1065.

Barlow, L.A. 2000a. Gustatory Development, in *Neurobiology of Taste and Smell,* T.E. Finger, W. Silver, and D. Restrepo, Eds. Wiley & Sons, New York, 393-422.

Barlow, L.A. 2000b. Taste buds in ectoderm are induced by endoderm: implications for mechanisms governing taste bud development, in *Regulatory Processes in Development: The Legacy of Sven Hörstadius. Proceedings of the Wenner-Gren International Symposium,* L. Olsson and C.-O. Jacobson, Eds. Portland Press, 185-190.

Barlow, L.A., C.-B. Chien, and R.G. Northcutt. 1996. Embryonic taste buds develop in the absence of innervation, *Development*, 122, 1103-1111.

Barlow, L.A. and R.G. Northcutt. 1994. Analysis of the embryonic lineage of vertebrate taste buds, *Chem. Senses*, 19, 715-724.

Barlow, L.A. and R.G. Northcutt. 1995. Embryonic origin of amphibian taste buds, *Dev. Biol.,* 169, 273-285.

Barlow, L.A. and R.G. Northcutt. 1997. Taste buds develop autonomously from endoderm without induction by cephalic neural crest or paraxial mesoderm, *Development*, 124, 949-957.

Begbie, J., J.F. Brunet, J.L. Rubenstein, and A. Graham. 1999. Induction of the epibranchial placodes, *Development*, 126, 895-902.

Bitgood, M.J. and A.P. McMahon. 1995. Hedgehog and Bmp genes are coexpressed at many diverse sites of cell-cell interaction in the mouse embryo, *Dev. Biol.,* 172, 126-138.

Chalfie, M. 1995. Green fluorescent protein, *Photochem. Photobiol.*, 62, 651-656.

Chalfie, M., Y. Tu, G. Euskirchen, W.W. Ward, and D.C. Prasher. 1994. Green fluorescent protein as a marker for gene expression, *Science*, 263, 802-805.

Chalmers, A.D. and J.M.W. Slack. 2000. The *Xenopus* tadpole gut: fate maps and morphogenetic movements, *Development*, 127, 381-392.

Chien, C.-B., D.E. Rosenthal, W.A. Harris, and C.E. Holt. 1993. Navigational errors made by growth cones without filopodia in the embryonic Xenopus brain, *Neuron*, 11, 237-251.

Cleaver, O. and P.A. Kreigstein. 1999. Expression from DNA injected into Xenopus embryos, in *Molecular Methods in Developmental Biology,* Xenopus and Zebrafish. Vol. 127. M. Guille, Ed. Humana, Totowa, NJ. 133-153.

Cleine, J.H. and J.M.W. Slack. 1985. Normal fates and states of specification of different regions in the axolotl gastrula, *J. Embryol. Exp. Morph.,* 86, 247-269.

Collazo, A., M. Bronner-Fraser, and S.E. Fraser. 1993. Vital dye labelling of *Xenopus laevis* trunk neural crest reveals multipotency and novel pathways of migration, *Development*, 118, 363-376.

Conn, P.M. 1999. Green fluorescent protein, in *Methods in Enzymology,* Vol. 302. J.N. Abelson and M.I. Simon, Eds. Academic Press, San Diego, 490.

Conover, J.C. and G.D. Yancopoulos. 1997. Neurotrophin regulation of the developing nervous system: analyses of knockout mice, *Rev. Neurosci.,* 8, 13-27.

Cook, M.H. and H.V. Neal. 1921. Are the taste buds of elasmobranchs endodermal in origin?, *J. Comp. Neurol.,* 33, 45-63.

Cooper, D. and B. Oakley. 1998. Functional redundancy and gustatory development in *bdnf* null mutant mice, *Dev. Brain Res.,* 105, 79-84.

D'Amico-Martel, A. and D.M. Noden. 1983. Contributions of placodal and neural crest cells to avian cranial peripheral ganglia, *Am. J. Anat.,* 166, 445-468.

Davies, A.M. 1988. The trigeminal system: an advantageous experimental model for studying neuronal development, *Development,* 103 (Suppl.), 175-183.

Davies, A.M. and R.M. Lindsay. 1985. The cranial sensory ganglia in culture: differences in the response of placode-derived and neural crest-derived neurons to nerve growth factor, *Dev. Biol.,* 111, 62-72.

Delarue, M., S. Sanches, K.E. Johnson, T. Darribere, and J.-C. Boucat. 1992. A fate map of the superficial and deep circumblastoporal cells in the early gastrula of *Pleurodeles waltl., Development*, 114, 135-146.

Eagleson, G., B. Ferreiro, and W.A. Harris. 1995. Fate of the anterior neural ridge and the morphogenesis of the *Xenopus* forebrain, *J. Neurobiol.,* 28, 146-158.

Eagleson, G.W. and W.A. Harris. 1989. Mapping of the presumptive brain regions in the neural plate of *Xenopus laevis*, *J. Neurobiol.,* 21, 427-440.

Enokido, Y., S. Wyatt, and A.M. Davies. 1999. Developmental changes in the response of trigeminal neurons to neurotrophins: influence of birthdate and the ganglion environment, *Development*, 126, 4365-4373.

Farbman, A.I. 1965. Electron microscope study of the developing taste bud in rat fungiform papilla, *Dev. Biol.,* 11, 110-135.

Farbman, A.I. 1968. Differentiation of fetal rat tongue grafts in the anterior chamber of the eye, *Anat. Rec.,* 160, 347.

Farbman, A.I. 1971. Differentiation of foetal rat tongue homografts in the anterior chamber of the eye, *Arch. Oral Biol.,* 16, 51-57.

Farbman, A.I. 1972a. Differentiation of taste buds in organ culture, *J. Cell Biol.,* 52, 489-493.
Farbman, A.I. 1972b. The taste bud: a model system for developmental studies, in *Developmental Aspects of Oral Biology,* H.C. Slavkin and L.A. Bavetta, Eds. Academic Press, New York, 109-123.
Farbman, A.I. and J.-P. Mbiene. 1991. Early development and innervation of taste bud-bearing papillae on the rat tongue, *J. Comp. Neurol.,* 304, 172-186.
Fritzsch, B., P.A. Sarai, M. Barbacid, and I. Silos-Santiago. 1997. Mice lacking the neurotrophin receptor trkB lose their specific afferent innervation but do develop taste buds, *Int. J. Dev. Neurosci.,* 15, 563-576.
Ganchrow, D., J.R. Ganchrow, R. Gross-Isseroff, and J.C. Kinnamon. 1995. Taste bud cell generation in the perihatching chick, *Chem. Senses*, 20, 19-28.
Gans, C. and R.G. Northcutt. 1983. Neural crest and the origin of vertebrates: a new head, *Science*, 220, 268-274.
Gilbert, S.F. 1997. *Developmental Biology,* Sinauer, Sunderland, MA.
Gimlich, R.L. and J.C. Gerhart. 1984. Early cellular interactions promote embryonic axis formation in *Xenopus laevis*, *Dev. Biol.,* 104, 117-130.
Graham, A. and J. Begbie. 2000. Neurogenic placodes: a common front, *Trends Neurosci.*, 23, 313-316.
Graveson, A.C., M.M. Smith, and B.K. Hall. 1997. Neural crest potential for tooth development in a urodele amphibian: developmental and evolutionary significance, *Dev. Biol.*, 188, 34-42.
Gross, J.B. and L.A. Barlow. 2000. Development of epibranchial placodes in *Ambystoma mexicanum*, *Dev. Biol.,* 222, 255.
Gross, J.B. and L.A. Barlow. 2001. Placodally derived sensory fibers grow preferentially, but not exclusively, toward oropharyngeal endoderm, *Chem. Senses,* in press.
Gurdon, J.B., H. Standley, S. Dyson, K. Butler, T. Langon, K. Ryan, F. Stennard, K. Shimizu, and A. Zorn. 1999. Single cells can sense their position in a morphogen gradient, *Development*, 126, 5309-5317.
Hall, J.M., T.E. Finger, D.K. MacCallum, and C.M. Mistretta. 1999a. Sonic hedgehog signaling in rodent tongue cultures, *Chem. Senses*, 24, 572.
Hall, J.M., J.E. Hooper, and T.E. Finger. 1999b. Expression of *Sonic hedgehog*, *Patched* and *Gli1* in developing taste papillae of the mouse, *J. Comp. Neurol.,* 406, 143-155.
Henry, G.L., I.H. Brivanlou, D.S. Kessler, A. Hemmati-Brivanlou, and D.A. Melton. 1996. TGF-β signals and a prepattern in *Xenopus laevis* endodermal development, *Development*, 122, 1007-1015.
Herrup, K. 1986. Cell lineage relationships in the development of the mammalian CNS: role of cell lineage in control of cerebellar Purkinje cell number, *Dev. Biol.*, 115, 148-154.
Herrup, K. and R.J. Mullen. 1979. Intercellular transfer of beta-glucuronidase in chimeric mice, *J. Cell Sci.*, 40, 21-31.
Holtfreter, J. 1996. Differentiation potencies of isolated parts of the anuran gastrula, by J. Holtfreter. Translated by Viktor Hamburger, *Dev. Dynamics*, 205, 217-222.
Hosley, M.A., S.E. Hughes, L.L. Morton, and B. Oakley. 1987a. A sensitive period for the neural induction of taste buds, *J. Neurosci.,* 7, 2075-2080.
Hosley, M.A., S.E. Hughes, and B. Oakley. 1987b. Neural induction of taste buds, *J. Comp. Neurol.,* 260, 224-232.
Imai, H., N. Osumi-Yamashita, Y. Ninomiya, and K. Eto. 1996. Contribution of early-emigrating midbrain crest cells to the dental mesenchyme of mandibular molar teeth in rat embryos, *Dev Biol.*, 176, 151-165.
Jacobson, M. and G. Hirose. 1978. Origin of the retina from both sides of the embryonic brain: a contribution to the problem of crossing at the optic chiasma, *Science*, 202, 637-639.

Jeffery, W.R. 2001. Cavefish as a model system in evolutionary developmental biology, *Dev. Biol.*, 231, 1-12.

Johnston, J.B. 1910. The limit between the ectoderm and entoderm in the mouth, and the origin of taste buds. I. Amphibians, *Am. J. Anat.,* 10, 41-67.

Jones, K.R., I. Farinas, C. Backus, and L.F. Reichardt. 1994. Targeted disruption of the BDNF gene perturbs brain and sensory neuron development but not motor neuron development, *Cell*, 76, 989-999.

Jung, H.S., V. Oropeza, and I. Thesleff. 1999. Shh, Bmp-2, Bmp-4 and Fgf-8 are associated with initiation and patterning of mouse tongue papillae, *Mech. Dev.*, 81, 179-182.

Kinnamon, J.C. 1987. Organization and innervation of taste buds, in *Neurobiology of Taste and Smell,* T.E. Finger and W.L. Silver, Eds. Wiley, New York, 277-297.

Kinnman, I. and H. Aldskogius. 1988. Collateral reinnervation of taste buds after chronic sensory denervation: a morphological study, *J. Comp. Neurol.,* 270, 569-574.

Landacre, F.L. 1907. On the place of origin and method of distribution of taste buds in *Ameiurus melas, J. Comp. Neurol.,* 17, 1-66.

Landacre, F.L. 1910. The origin of the cranial ganglia in *Ameiurus*, *J. Comp. Neurol.,* 20, 309-411.

Le Douarin, N.M., and C. Kalcheim. 1999. *The Neural Crest,* Cambridge University Press, Cambridge.

Liebl, D.J., J.P. Mbiene, and L.F. Parada. 1999. NT4/5 mutant mice have deficiency in gustatory papillae and taste bud formation, *Dev. Biol.*, 213, 378-389.

Lo, C.W. 1986. Localization of low abundance DNA sequences in tissue sections by *in situ* hybridization, *J. Cell. Sci.,* 81, 143-162.

Lumsden, A.G. and A.M. Davies. 1983. Earliest sensory nerve fibres are guided to peripheral targets by attractants other than nerve growth factor, *Nature*, 306, 786-788.

Mbiene, J.-P., D.K. MacCallum, and C.M. Mistretta. 1997. Organ cultures of embryonic rat tongue support tongue and gustatory papilla morphogenesis *in vitro* without intact sensory ganglia, *J. Comp. Neurol.,* 377, 324-340.

Miller, I.J. and D.V. Smith. 1988. Proliferation of taste buds in the foliate and vallate papillae of postnatal hamsters, *Growth, Dev. Aging*, 52, 123-131.

Misteli, T. and D.L. Spector. 1997. Applications of the green fluorescent protein in cell biology and biotechnology, *Nat. Biotechnol.*, 15, 961-964.

Mistretta, C.M. 1972. Topographical and histological study of the developing rat tongue, palate and taste buds, in *Third Symposium on Oral Sensation and Perception. The Mouth of the Infant,* J.F. Bosma, Ed. Charles C Thomas, Springfield, IL, 163-187.

Mistretta, C.M. 1998. The role of innervation in induction and differentiation of taste organs: introduction and background, *Ann. N.Y. Acad. Sci.,* 855, 1-13.

Mistretta, C.M., K.A. Goosens, I. Farinas, and L.F. Reichardt. 1999. Alterations in size, number, and morphology of gustatory papillae and taste buds in BDNF null mutant mice demonstrate neural dependence of developing taste organs, *J. Comp. Neurol.*, 409, 13-24.

Mistretta, C.M. and L.F. Haus. 1996. Temporal and spatial patterns of tenascin and laminin immunoreactivity suggest roles for extracellular matrix in development of gustatory papillae and taste buds, *J. Comp. Neurol.*, 364, 535-555.

Morris-Wiman, J., L. Brinkley, and R. Sego. 2000. An *in vitro* model for the study of taste papillae morphogenesis using branchial arch explants, *Brain Res. Brain Res. Protoc.*, 5, 172-181.

Narayanan, C.H. and Y. Narayanan. 1980. Neural crest and placodal contributions in the development of the glossopharyngeal-vagal complex in the chick, *Anat. Rec.,* 196, 71-82.

Northcutt, R.G., L.A. Barlow, C.B. Braun, and K.C. Catania. 2000. Distribution and innnervation of taste buds in the axolotl, *Brain Behav. Evol.,* 56, 123-145.

Northcutt, R.G. and K. Brändle. 1995. Development of branchiomeric and lateral line nerves in the axolotl, *J. Comp. Neurol.,* 355, 427-454.

Northcutt, R.G., K. Brändle, and B. Fritzsch. 1995. Electroreceptors and mechanosensory lateral line organs arise from single placodes in axolotls, *Dev. Biol.,* 168, 358-373.

Nosrat, C.A. 1997. Neurotrophins in the Development of the Gustatory System and Teeth. Karlolinska Institutet, Ph.D. thesis.

Nosrat, C.A., J. Blomlöf, W.M. ElShamy, P. Ernfors, and L. Olson. 1997. Lingual deficits in BDNF and NT3 mutant mice leading to gustatory and somatosensory disturbances, respectively, *Development*, 124, 1333-1342.

Nosrat, C.A., T. Ebendal, and L. Olson. 1996. Differential expression of brain-derived neurotrophic factor and neurotrophin 3 mRNA in lingual papillae and taste buds indicates roles in gustatory and somatosensory innervation, *J. Comp. Neurol.,* 376, 587-602.

Nosrat, C.A., D.K. MacCallum, and C.M. Mistretta. 1998. Early molecular events in gustatory papilla development are independent of nerve fibers, *Chem. Senses*, 23, 603-604.

Nosrat, C.A. and L. Olson. 1995. Brain-derived neurotrophic factor mRNA is expressed in the developing taste bud-bearing tongue papillae of rat, *J. Comp. Neurol.,* 360, 698-704.

O'Rourke, N.A. and S.E. Fraser. 1986. Dynamic aspects of retinotectal map formation revealed by a vital-dye fiber-tracing technique, *Dev. Biol.*, 114, 265-276.

Oakley, B. 1993a. Control mechanisms in taste bud development, in *Mechanisms of Taste Transduction,* S.A. Simon and S.D. Roper, Eds. CRC Press, Boca Raton, FL, 105-125.

Oakley, B. 1993b. The gustatory competence of the lingual epithelium requires neonatal innervation, *Dev. Brain Res.,* 72, 259-264.

Oakley, B., A. Brandemihl, D. Cooper, D. Lau, A. Lawton, and C. Zhang. 1998. The morphogenesis of mouse vallate gustatory epithelium and taste buds requires BDNF-dependent taste neurons, *Dev. Brain Res.,* 105, 85-96.

Oakley, B., D.E. LaBelle, R.A. Riley, K. Wilson, and L.-H. Wu. 1991. The rate and locus of development of rat vallate taste buds, *Dev. Brain Res.,* 58, 215-221.

Osumi-Yamashita, N., Y. Ninomiya, H. Doi, and K. Eto. 1994. The contribution of both forebrain and midbrain crest cells to the mesenchyme in the frontonasal mass of mouse embryos, *Dev. Biol.,* 164, 409-419.

Pasteels, J. 1942. New observations concerning the maps of presumptive areas of the young amphibian gastrula, (*Amblystoma* and *Discoglossus*), *J. Exp. Zool.,* 89, 255-281.

Pirvola, U., U. Arumae, M. Moshnyakov, J. Palgi, M. Saarma, and J. Ylikoski. 1994. Coordinated expression and function of neurotrophins and their receptors in the rat inner ear during target innervation, *Hear Res.*, 75, 131-144.

Reutter, K. 1978. Taste organ in the bullhead (*Teleostei*), *Adv. Anat. Embryol. Cell Biol.,* 55, 1-98.

Rice, F.L., K.M. Albers, B.M. Davis, I. Silos-Santiago, G.A. Wilkinson, A.M. LeMaster, P. Ernfors, R.J. Smeyne, H. Aldskogius, H.S. Phillips, M. Barbacid, T.M. DeChiara, G.D. Yancopoulos, C.E. Dunne, and B.T. Fundin. 1998. Differential dependency of unmyelinated and A delta epidermal and upper dermal innervation on neurotrophins, trk receptors, and p75LNGFR, *Dev. Biol.*, 198, 57-81.

Ringstedt, T., C.F. Ibanez, and C.A. Nosrat. 1999. Role of brain-derived neurotrophic factor in target invasion in the gustatory system, *J. Neurosci.,* 19, 3507-3518.

Rochlin, M.W. and A.I. Farbman. 1998. Trigeminal ganglion axons are repelled by their presumptive targets, *J. Neurosci.,* 18, 6840-6852.

Rochlin, M.W., R. O'Connor, R.J. Giger, J. Verhaagen, and A.I. Farbman. 2000. Comparison of neurotrophin and repellent sensitivities of early embryonic geniculate and trigeminal axons, *J. Comp. Neurol.*, 422, 579-593.

Ruiz i Altaba, A., R.M. Warga, and C.D. Stern. 1993. Fate map and cell lineage analysis, in *Essential Developmental Biology: A Practical Approach,* C.D. Stern and P.W.H. Holland, Eds. Oxford University Press, Oxford. 81-95.

Spemann, H. 1921. Die Erzeugung tierischer Chimären durch heteroplastische embryonale Transplantation zwischen *Triton cristatus* und *taeniatus*, *Roux' Arch. Entw. Mech.,* 48, 448-555.

Spemann, H. and H. Mangold. 1924. Über Induktion von Embryonen anlagen durch Implatation artfremder Organisatoren, *Arch. Microscop. Anat. Entwicklungsmechanik*, 100, 599-638.

Stone, L.M., T.E. Finger, P.P.L. Tam, and S.-S. Tan. 1995. Taste receptor cells arise from local epithelium, not neurogenic ectoderm, *Proc. Natl. Acad. Sci. U.S.A.,* 92, 1916-1920.

Stone, L.M., T.E. Finger, P.P.L. Tam, and S.-S. Tan. 1996. Serotonin-immunoreactive taste cells are related by lineage, but may derive from multiple progenitors, *Chem. Senses*, 21, 677.

Stone, L.S. 1922. Experiments on the development of the cranial ganglia and the lateral line sense organs in *Amblystoma punctatum*, *J. Exp. Zool.,* 35, 421-496.

Stone, L.S. 1932. Independence of taste organs with respect to their nerve fibers demonstrated in living salamanders, *Proc. Soc. Exp. Biol. Med.,* 30, 1256-1257.

Stone, L.S. 1940. The origin and development of taste organs salamanders observed in the living condition, *J. Exp. Zool.,* 83, 481-506.

Tan, S.-S. and S. Breen. 1993. Radial mosaicism and tangential cell dispersion both contribute to mouse neocortical development, *Nature*, 362, 638-640.

Torrey, T.W. 1940. The influence of nerve fibers upon taste buds during embryonic development, *Proc. Natl. Acad. Sci. U.S.A.,* 26, 627-634.

Trainor, P.A. and P.P.L. Tam. 1995. Cranial paraxial mesoderm and neural crest cells of the mouse embryo: co-distribution in the craniofacial mesenchyme but distinct segregation in branchial arches, *Development*, 121, 2569-2582.

Trainor, P.A., S.-S. Tan, and P.P.L. Tam. 1994. Cranial paraxial mesoderm: regionalisation of cell fate and impact on craniofacial development in mouse embryos, *Development*, 120, 2397-2408.

Vogel, K.S. and A.M. Davies. 1993. Heterotopic transplantation of presumptive placodal ectoderm changes the fate of sensory neuron precursors, *Development*, 119, 263-276.

Vogt, W. 1929. Gestaltungsanalyse am Amphibienkeim mit örtlicher Vitalfärbung. II. Gastrulation und Mesodermbildung bei Urodelen und Anuren, *Wilhelm Roux' Arch. Entwicklungmech. Org.,* 120, 385-706.

Whitehead, M.C. and D.L. Kachele. 1994. Development of fungiform papillae, taste buds, and their innervation in the hamster, *J. Comp. Neurol.,* 340, 515-530.

Willier, B.H., P.A. Weiss, and V. Hamburger. 1955. *Analysis of Development,* Hafner Publishing Co., New York.

Yang, H., I.B. Wanner, S.D. Roper, and N. Chaudhari. 1999. An optimized method for *in situ* hybridization with signal amplification that allows the detection of rare mRNAs, *J. Histochem. Cytochem.,* 47, 431-436.

Yee, C.L., K.R. Jones, and T.E. Finger. 2000. Brain-derived neurotrophic factor (BDNF) is present in diverse taste cell types of adult mice, *Chem. Senses*, 25, 626-627.

Yntema, C.L. 1937. An experimental study of the origin of the cells which constitute the VIIth and VIIIth cranial ganglia and nerves in the embryo of *Amblystoma punctatum*, *J. Exp. Zool.,* 75, 75-101.

Yntema, C.L. 1943. An experimental study on the origin of the sensory neurones and sheath cells of the IXth and Xth cranial nerves in *Amblystoma punctatum*, *J. Emb. Exp. Morph.,* 92, 93-118.

Yntema, C.L. 1944. Experiments on the origin of the sensory ganglia of the facial nerve in the chick, *J. Comp. Neurol.,* 81, 147-164.

Zalewski, A.A. 1976. The neural induction of taste buds in the salivary ducts of the lingual gland of Von Ebner, *Exp. Neurol.,* 52, 565-580.

Zernicka-Goetz, M., J. Pines, K. Ryan, K.R. Siemering, J. Haseloff, M.J. Evans, and J.B. Gurdon. 1996. An indelible lineage marker for *Xenopus* using a mutated green fluorescent protein, *Development*, 122, 3719-3724.

7 Mammalian Taste Receptors

Mark A. Hoon and Nicholas J. P. Ryba

CONTENTS

7.1 INTRODUCTION

Receptors play a critical role in all forms of sensory signaling. In the chemical senses taste and smell, detection of and discrimination between structurally diverse molecules are determined by interaction between these stimuli and their receptors. Taste and smell are encoded through these interactions and events downstream can only be involved in processing this information. Therefore, chemical stimuli that can be distinguished by animals must result in distinct patterns of receptor activation. Consequently, elucidating the nature, distribution, sensitivity, and specificity of taste receptors is fundamental to understanding the sense of taste itself. In this chapter, the methods that have been used to try to identify taste receptors for bitter, sweet, and umami tastes, and current knowledge about these receptors as well as problems that remain to be solved are reviewed.

0-8493-2329-0/02/$0.00+$1.50

Taste is a rather simple sense. It provides an animal with basic but essential information about whether or not it should ingest food. For humans, much of the pleasure derived from eating stems from the combination of this information with olfactory, somatosensory, and visual cues, as well as the general ambiance. Although large numbers of structurally diverse molecules elicit taste responses, just a few distinct modalities are distinguished by animals as diverse as insects and mammals. For example, sweet tastants are generally nutritious, include sugars and many amino acids, and are attractive behavioral cues for animals. Saltiness primarily conveys the presence of sodium and is of particular importance for animals with low-sodium diets. Sourness warns of acidity and bitterness implies toxicity. A number of mammals are also attracted to tastants that evoke a distinct sensation in humans, termed umami and typified by the taste of glutamate. The fact that diverse and distinct taste stimuli can be applied at will, coupled with the relative simplicity of taste, mean that deciphering the details involved in coding and decoding this sensory information might be accomplished relatively easily. The principles gleaned from these studies may be of significance in understanding how the brain interprets sensory input and indeed reveals general strategies of information processing.

What pathways are likely to be involved in mediating responses to these different modalities? It is generally believed that ion-fluxes resulting from changes in the composition of the external medium are responsible for salty and sour tastes. In the case of sour compounds, taste cell depolarization is hypothesized to result from H^+ blockage of K^+ channels[1] or activation of pH-sensitive channels;[2] salt transduction may be partly mediated by the entry of Na^+ via amiloride-sensitive Na^+ channels.[3] However, most of the molecular components of the sour and salty pathways have not been identified. Many different signaling pathways have also been proposed as mediating sweet and bitter taste, and most of these involve activation of G-proteins. Indeed, in different studies, a number of the second messenger pathways downstream of G-proteins have been reported as playing a role either in sweet or bitter taste signaling.[4-6] For example, it has been suggested that cAMP is a key messenger in sweet transduction.[7,8] However, recent work suggests that sweet receptor activation also involves cGMP,[9,10] and there is some evidence of calcium mobilization[11] in response to sweet tastants (perhaps through phospholipase C). For bitter transduction, there is also controversy as to whether phospholipase C or a cGMP-phosphodiesterase is the effector enzyme.[12-14] Indeed, it is possible that both pathways exist, but whether they do so in the same cell remains unknown.

It remains ambiguous as to how all these findings fit together, and even some of the most fundamental questions about taste remain unanswered. For example, it is known that taste receptor cells are grouped together in taste buds, but the specific fractions of these cells that respond to particular stimuli or even that synapse with sensory neurons have not been determined. There is evidence that more than one neuron can innervate a single taste bud and, indeed, that neurons can innervate multiple taste buds.[15] However, the details of these connections have yet to be worked out and, in their absence, it is still unclear whether taste buds or the individual taste receptor cells are the primary coding units. Extensive physiological studies demonstrate that there are differences in the sensitivity and selectivity of different regions of the oral cavity in most animals.[16-19] However, the significance of topographical differences in taste is unknown and there is little indication of what might be responsible for variations in sensitivity. Moreover, many different compounds can be classed psychophysically as tasting (e.g., bitter), but it is still unclear whether this similarity in response results simply from the stimulation of a common pool of receptor cells or is a product of signal processing.

One approach to answer some of these basic questions is described here and involves the identification of the molecular components that mediate signaling ideally the taste receptors with known specificity. These molecules can then be used to mark specific subsets of cells, to investigate their distribution, their responses to stimuli, and ultimately to trace their neural connections. However, conventional molecular studies have been hindered both by the small number of taste receptor cells and because taste buds are scattered throughout the oral cavity. In total, a rat has only about a thousand taste buds containing approximately 50,000 receptor cells.[20] The region of the tongue with the highest density of taste buds that can be isolated *en masse,* the circumvallate papilla, contains about half of these cells but is estimated to contain 10 times as many non-sensory cells. Both the small number of taste cells and their purity are problems because conventional biochemical and cloning approaches are most useful when applied to large quantities of homogeneous material (a minimum of 10^7 cells). Therefore, progress in the molecular biology of taste has also lagged behind that of the other senses — at least until recently.

7.2 A ROLE FOR G-PROTEIN COUPLED RECEPTORS IN TASTE

The low sensitivity of many taste responses is quite contrary to the high affinity seen for many classical membrane receptors including most G-protein coupled receptors (GPCRs). This and difficulties in identifying taste receptors have led to the suggestion that at least some taste responses may be mediated by other mechanisms, for example, general effects on the cell membrane or even by uptake of tastants into cells.[21] It is debatable if such mechanisms would allow for the discrimination of distinct taste modalities because they appear to be relatively nonselective. There is also long-standing evidence that receptors play a role in the detection of many tastants (although there is less strong evidence that all these receptors are cell-surface molecules). Enantiomers of a number of chemicals have distinct taste properties;[22] this is particularly notable for the well-known sweetener aspartame, L-Asp-L-Phe-OMe, for which only one of the four chiral forms is intensely sweet and the other forms are bitter.[23] Indeed, this sweet/bitter difference in taste is true for some amino acids, for example, Phe and Pro, where the D-form is sweet and the L-form bitter. Moreover, a number of proteins and polypeptides have been identified that taste intensely sweet (e.g., monellin, thaumatin, and curculin) or that modify the taste properties of other compounds (e.g., guamarin, miraculin, and curculin). These are unlikely to be taken up by cells within the time frame that could account for their responses.

Genetic characterization of both sweet and bitter responses in mice suggests that a number of loci alter the animal's ability to sense several distinct compounds.[24-31] While this in itself is not direct evidence for cell-surface receptors, the fact that changes in behavior appear to be paralleled by changes in nerve responses suggests that these genes are involved in peripheral processes.[32] Moreover, the clustering of a number of these bitter loci at the distal end of mouse chromosome 6 implied that distinct but tightly linked genes might encode a family of bitter taste receptors and this conclusion appears to be borne out by recent data (see Section 7.4).

In the past 10 years, the involvement of GPCRs in taste has received strong support from the identification and cloning of a taste-specific G-protein α-subunit, gustducin, by Margolskee and co-workers.[33] Although it now appears that gustducin is not exclusively expressed in taste tissue,[34-36] the high level of expression of this G-protein in a large subset of taste receptor cells strongly suggests that it is involved in mediating specific aspects of taste sensation. Moreover, mice in which this gene has been deleted show marked behavioral differences in their responses to a number of bitter compounds and less-pronounced changes in response to sweet compounds.[37] These mice respond normally to salty and sour tastants, indicating that such stimuli trigger other signal transduction pathways and that defects in sweet and bitter taste are likely to be downstream of specific gustducin-linked receptors rather than to result from a generalized loss of taste. One interesting aspect of the knockout phenotype is that the deficits in sweet and bitter responses are shifted in sensitivity rather than lost completely. This suggests that either there are multiple pathways for all the responses that have been measured or that (an)other G-protein(s) can substitute for gustducin in taste signaling.

As mentioned in the introduction, there are many studies describing events that may be downstream of gustducin or other G-proteins in taste. However, there is no clear consensus about the relative importance of several different possible pathways either for sweet or bitter stimuli. In part, this may be because all these pathways are important. However, measurements have often been made in homogenates of taste cell-containing tissue and may not always reflect physiologically significant events. Sweet tastants have been reported to result in an increase in cAMP as a result of G-protein stimulation of adenylate cyclase.[7] This in turn may activate protein kinase A, leading to phosphorylation of potassium channels and a depolarization of taste receptor cells.[38] In support of this, it has also been reported that cAMP and cGMP can mimic the effect of artificial sweeteners by blockade of a potassium conductance.[39] But other data suggest that sweet tastants stimulate a G-protein-dependent cGMP phosphodiesterase,[12] perhaps resulting in a decrease in cGMP with the opening of a cyclic nucleotide suppressible conductance.[8] The activation of this channel has been hypothesized to result in an increase in the level of intracellular calcium. Yet another model for sweet taste transduction involves a different second messenger, inositol triphosphate (IP_3), presumably generated by the G-protein-dependent activation of a phospholipase-C.[11,40] In this scenario, sweet-tastants are believed to stimulate IP_3 production, that in turn leads to release of calcium from intracellular stores. Recent molecular evidence may support the last of these models as phospholipase-Cβ_2 is expressed at a high level in a subset of taste receptor cells.[41] However, it remains unclear if these phospholipase-Cβ_2-expressing cells are responsive to sweet tastants *in vivo,* and there is as yet no physiological or behavioral evidence showing that phospholipase-C is necessary for sweet taste responses.

Similarly, there is evidence that G-protein-mediated signaling pathways function in the detection of bitter stimuli. For example, in reconstitution experiments, gustducin and transducin appear to be activated by taste receptor membranes in the presence of a number of different bitter compounds.[42] This might well be expected to activate a gustducin-stimulated cGMP phosphodiesterase that has been reported to be present in taste tissue,[12] and open the cyclic-nucleotide-suppressible channel[8] if these components are co-expressed in bitter responsive cells. More recently, it has been shown that the bitter compound denatonium can stimulate phospholipase-Cβ_2 *in vitro,* possibly through activation of gustducin and release of G$\beta_1\gamma_{13}$, G$\beta_3\gamma_{13}$ or G$\beta_3\gamma_3$ heterodimers.[14,41,43] Animal studies showing which behavioral and physiological responses to a range of bitter compounds depend on specific effector enzymes may be very useful in resolving these apparently disparate findings. Thus, behavioral and electrophysiological recordings from mice with knockouts of potentially important signaling molecules are likely to be important in investigating downstream signaling pathways.

During the past 10 years, there have also been a number of putative taste receptors that have been described and characterized to various degrees. These include putative sweet and bitter taste receptors that are generally thought to be GPCRs, potential components of salty and sour signaling pathways and umami receptors. It is unlikely that all of these play a role in taste signaling, and as discussed in Section 7.3, the evidence for several of these being even candidate taste receptors is very weak. Some of these receptors may represent the same molecular species characterized in different ways and to different extents, but it is only very recently that the first molecularly defined, functional, and genetically tested taste receptors have been reported.

7.3 METHODS USED TO IDENTIFY CANDIDATE TASTE RECEPTORS

Much of the rapid progress in understanding olfaction in eukaryotes has been made possible by molecular biological techniques. Therefore, it is not surprising that despite inherent difficulties of applying such methods to taste, most candidate taste receptors have been identified using a variety of molecular cloning approaches. This chapter section attempts to divide the methods systematically and to describe the candidate receptors that have been reported using particular strategies. We have set out to explain why these molecules are possible taste receptors but have also identified the questions that remain and how they might be answered. In addition, this chapter section highlights modifications of approaches that might be useful in discovering additional taste receptors.

7.3.1 Direct Purification and Characterization

The classical approach of protein purification might in itself allow functional characterization of receptors. More frequently, partially pure protein can be used to obtain sequence or generate antibodies for further molecular studies. Such methods have been used to great effect in other systems, for example, in characterizing molecular components of vertebrate and cephalopod vision where large quantities of receptor membranes can be easily purified. The application of these methods to taste is orders of magnitude more complex because it is only possible to obtain very small quantities of, at best, partially pure receptor cell membranes. A second major limitation to this type of study stems from the low apparent affinity of tastants. Even the most avid binding is expected to have a K_D in the 100 nM range, and thus to be too weak for most types of binding study necessary to follow purification. Three examples of this approach are detailed below, demonstrating the results of these studies and some of the difficulties in interpreting this type of work.

Cagan and Morris[44] developed an assay that measured specific binding of the amino-acid L-alanine to membranes derived from catfish taste receptor cells. In parallel experiments, monoclonal antibodies were raised using these purified membranes in the hope that taste receptors would be sufficiently highly expressed and immunogenic to generate antibodies that could then be characterized.[45] They then used inhibition of the alanine binding activity to identify a monoclonal antibody G-10, possibly directed against the alanine receptor. Using Western analysis, it was demonstrated that this antibody specifically recognized a protein of 113 kDa that was selectively expressed in taste tissue. Moreover, immunohistochemical analysis demonstrated that G-10 labeled the taste-pore region of a subset of taste-receptor cells.[46] Indeed, cells expressing this epitope generally did not have binding sites for lectin Pha-E, which has been shown to inhibit arginine but not alanine binding to catfish membranes. Thus, it is possible that the catfish alanine taste receptor contains the 113-kDa protein recognized by the monoclonal antibody G-10 and there may be little overlap between cells expressing this receptor and the arginine receptor. However, further identification of these receptors has not been possible and evidence that the 113-kDa protein is a component of the alanine receptor remains circumstantial. For example, binding of G-10 to proteins, either associated with the alanine receptor or simply highly expressed in the taste pore region, might nonspecifically inhibit alanine binding and therefore even the pattern of immunostaining might be misleading.

Another approach that has been attempted is to cross-link tastant to its receptor. For example, Nakao and colleagues[47] cross-linked the sweet-tasting protein thaumatin to taste papillae of monkeys. Analysis suggested that a protein (this time of ~50 kDa) could be functioning as a taste receptor for thaumatin. However, again no progress has been made in determining the molecular nature of the cross-linked protein and, in its absence, it is impossible to seriously conclude that the 50-kDa protein is a receptor for thaumatin.

More recently, Margolskee and co-workers[12,42,48] have used the activation of gustducin (or its close functional homolog transducin) in response to denatonium (and other bitter compounds) as a sensitive assay of taste receptor function. Membranes derived from bovine circumvallate papillae appeared to contain receptors that could be solubilized in detergent, reconstituted with transducin and activate the G-protein in response to denatonium. However, no further steps of purification were attempted and thus the molecular nature of this receptor remains unclear. Indeed, solubilization alone appeared to significantly reduce the activity of the receptor, suggesting that the task of purification might not be possible even if sufficient starting material were to be isolated.

7.3.2 Degenerate PCR Approaches

More than 10 years ago, data suggesting a role for G-proteins in sweet and bitter taste pathways and the discovery of specialized G-proteins in other sensory systems suggested that novel GPCRs and G-proteins might also have a role in taste. At about the same time, the power of PCR in identifying and cloning genes with at least limited identity to other known genes was beginning to be realized. Indeed it was using this type of degenerate PCR approach that Margolskee and co-workers[33] first identified gustducin. Similarly, Buck and Axel[49] used degenerate PCR to great effect in the identification and cloning of a multigene family of GPCRs encoding odorant receptors. Therefore, it is not surprising that this strategy was also widely used to look for taste receptors.

There have been a few reports of apparent success in using this type of technique to identify taste receptors for sweet or bitter compounds.[50,51] Alas, important controls as to whether these really are taste receptors are missing. For example, although putative GPCRs have been amplified from taste tissue using degenerate primers, there is no indication that these GPCRs are expressed in taste receptor cells. Indeed, these receptors bear striking similarity to odorant receptors or other known GPCRs; and because the coding sequence of many GPCRs lack introns, it is probable that they were amplified from contaminating genomic DNA. In analogous experiments, it was reported[52] that bees also express GPCRs closely related to mammalian olfactory receptors but there is no evidence for the existence of this class of receptor in the invertebrate genome and quite distinct GPCRs appear to mediate chemosensation in insects.[53-55] Therefore, the great strength of PCR — exponential amplification of small quantities of starting material — can also be a weakness and produce misleading results. The problems are exacerbated for receptors such as GPCRs because there are many different GPCRs involved in normal cellular signaling. Consequently, PCR amplification may produce many products and a strategy is likely to be required to identify potential taste receptors among these. Finally, none of the strong candidate taste receptors (see Section 7.4) is a typical GPCR (i.e., closely related to opsins or olfactory receptors). Consequently, degenerate PCR-primers designed using the sequence of these known sensory receptors may not be useful in identifying any taste receptors.

A more directed approach was used to search for potential umami receptors based on the hypothesis that these receptors might be either a known glutamate receptor that is also expressed outside taste, or a closely related homologue.[56] The results show that certain ionotropic glutamate receptors (specifically, NMDAR1, NMDAR2d, KA2, and δ-1) and the metabotropic glutamate receptor mGluR4 are present in lingual tissues that contain taste buds. Moreover, a low-level expression of mGluR4 in taste buds was suggested by *in situ* hybridization using radiolabeled antisense probes and a sensitive nonradioactive method.[56,57] Therefore, it is possible that this receptor, normally localized to the presynaptic region of glutaminergic neurons, is also involved in mediating umami taste (see Section 7.4.1).

7.3.3 Subtractive Screening

The circumvallate papilla contains the highest concentration of taste receptor cells that can be isolated without microdissection. Moreover, several hundred papillae produce sufficient RNA for making high-quality cDNA libraries. However, only about 5% of the mRNA isolated from punched circumvallate papillae is likely to come from taste buds. One would anticipate that any particular taste receptor would only be found in a subset of this 5% of the cells. Therefore, even if receptors are expressed at very high level within individual sensory cells, their fractional representation within such a library is likely to be very low. In our hands, homology-based screening using known GPCRs failed to identify candidate taste receptors, although several known receptors of this class were readily identified with appropriate probes. Thus, we and others have resorted to subtractive hybridization approaches in an attempt to enrich taste-specific sequences by selectively depleting cDNAs that are also present in lingual epithelium from regions containing no taste buds.[58,59] There are now several methods for performing DNA-subtraction to enrich in cell-specific transcripts. Several attractive approaches use PCR,[60-63] some tag the driver (nonspecific) DNA (e.g., with biotin)[64,65] and others make use of different biophysical properties of single- and double-stranded DNA or DNA/RNA hybrids.[58] The methodology determines the utility of the process: it must achieve significant enrichment of sufficiently complex starting material to adequately sample mRNAs expressed only in perhaps 1% of the cells. In practice, this is likely difficult to achieve except using PCR, and the subtractive library is unlikely to contain a full representation of taste-specific clones. PCR methods may also provide significant bias to a library, regardless of the care used to minimize differences in the length of fragments analyzed. Moreover, this fragmentation of DNA can itself produce problems because an interesting clone may no longer have sufficient sequence to identify it as a homolog of, for example, a GPCR. Nonetheless, if such clones are expressed at ~0.1% of the mRNA (approximately 10 times lower than the expression of odorant receptors in olfactory neurons) in 1% of cells, methods that enrich taste-specific sequences 100-fold will generate libraries from which only a few thousand clones need to be analyzed.

Subtraction alone is not likely to be sufficient to identify taste receptors but the rationale is that there will be receptors within the subtracted library. Most subtraction methods do not completely remove sequences that are expressed outside taste; therefore, additional enrichment may be achieved by directly screening clones at low density with a driver probe. Particularly useful methods of homing in on potential receptors include sequencing the ends of clones and performing *in situ* hybridization, both of which can be carried out on sufficient numbers of clones to be used as direct screening techniques. With the dawn of the post-genomic age, sequencing the ends of clones has gained enormous power because a direct match or ortholog can often be identified from the databases. This can both eliminate irrelevant sequences and highlight interesting clones. However, for a clone to have a role in taste, it must be expressed within the taste receptor cells. *In situ* hybridization of cRNA probes to sections containing taste buds provides a relatively rapid, sensitive screen that directly shows whether this is the case. Moreover, because both sense (control) and antisense cRNA can usually be generated directly from the library clone, a relatively large number of such screens can be performed in parallel.

While sequencing clones from this type of library, a clone (T1R-1) has been identified[66] that has significant homology to the class of G-protein-linked receptors that includes the calcium sensing receptor mGluRs and putative pheromone receptors (V2Rs).[67-71] *In situ* hybridization demonstrated that this GPCR was expressed in a small subset of taste receptor cells in the circumvallate papillae, as would be expected for a taste sensory receptor. Degenerate PCR strategies using regions conserved between T1R-1 and other members of this class of GPCRs revealed the existence of a second related receptor T1R-2 that is also expressed in a subset of taste receptor cells (see Section 7.4.2).

7.3.4 Single-Cell Approaches

Catherine Dulac (in Richard Axel's laboratory) pioneered the use of single cell techniques to identify and clone sensory receptors by adapting Brady and Iscove's single-cell methodology to identify putative pheromone receptors from the vomeronasal organ.[72] In principle, the methodology made use of the fact that while any particular pheromone receptor might be present at a very low level in whole tissue preparation, it is likely to be concentrated in individual cells. Moreover, other genes are much more likely to be equally expressed between cells. Therefore, the pheromone receptors were found by screening for abundant transcripts from one cell that were absent from a second. The only assumption made was that a receptor is expressed at a sufficiently high level to be detected by this type of screen. The heterogeneous nature of the taste bud and the low abundance of receptor cells make this technique particularly attractive. Taste cells may be comprised of significantly more distinct subtypes than found among vomeronasal organ neurons, implying that comparing cells at random will likely reveal signaling molecules in addition to taste receptors. Thus, as was the case for subtractive approaches, receptor identification is likely to rely on sequence data and *in situ* evidence. However, comparisons between, for example, two gustducin-containing cells might specifically identify receptors, if there are indeed different receptors present in these cells.

This type of approach to taste has been successfully applied in identifying taste-specific components (as judged by *in situ* hybridization), but has thus far failed to identify additional GPCRs as candidate taste receptors. Signal transduction components expressed in taste have been identified in this way.[14] In addition, we found novel transmembrane proteins selectively expressed in distinct subsets of taste receptor cells. These may turn out to encode receptors or co-receptors involved in taste signaling, but their predicted amino acid sequences provide no clues of function.

There are a number of problems that might account, in part, for the failure to find GPCRs. A technical concern is that whereas vomeronasal neurons could be recognized by their distinct morphology, the harsher dissociation conditions required to liberate taste receptor cells tend to make identification more difficult. One solution is to label the cells *in vivo,*[14,73] and an alternative is to define the cells using molecular markers after picking cells and preparing cDNA. However, the major assumption underlying the single-cell approach is that receptors are expressed at high levels.

Indeed, our experience is that only transcripts that are present at more than 1/5000 in the probe are capable of generating hybridization signal in conventional library screens. Problems may be exacerbated by the PCR process itself, which although optimized to retain representation by restricting the size of transcripts to <1000 bp, is still likely to introduce significant bias. Moreover, because the short transcripts are produced by limiting the original reverse transcription reaction, clones represent mainly 3′ nontranslated regions. This reduces the power of DNA sequencing and homology searches as tools for identification of interesting clones. In addition, the sensitivity of *in situ* hybridization is often lower when using shorter probes. Finally, the assumption that receptors are differentially expressed between the cells chosen for comparison may not be true and, in fact, many gustducin-containing cells appear to express relatively low levels of many members of a family of bitter taste receptors (see Section 7.4.3).

7.3.5 Expression Cloning

A recurring difficulty in chemosensory biology has been to establish the functional specificity of the receptors that have been cloned by homology-based or differential screening techniques. Thus, to date, many different candidate odorant, pheromone, and taste receptors have been identified but we have little idea of the response profiles of individual receptors. Indeed, *in vitro* activation has only been demonstrated for a handful of odorant and taste receptors.[56,74-78] Therefore, expression cloning of receptors is, in principle, an extremely attractive alternative. This type of approach has been used to clone many different GPCRs but has not yet been successfully applied to identify taste receptors. The utility of this type of approach is well-demonstrated by a recent study where responses of transfected culture cells to the basic amino acids lysine and arginine were used to clone a fish olfactory receptor that is related to T1R-1, T1R-2, and V2Rs.[77] A second example is the identification of the capscaisin (or vanilloid) receptor, which turns out to encode an ion channel that probably mediates the hot taste typically associated with chili peppers.[79]

There are several potential problems in an expression cloning approach. First, in general, expression cloning requires that the receptor not be composed of distinct subunits. Second, and for chemosensory receptors perhaps the major problem, may be getting robust surface expression of these proteins in culture cell lines. This appears to have been one of the major obstacles in determining the function of most of the putative odorant and taste receptors that have been identified to date. Thus, it is possible that a specific factor or chaperone is necessary to promote surface expression and/or appropriate folding of taste receptors, as is the case for several other sensory GPCRs. For example, *Drosophila* rhodopsin requires nina A, two color opsins need RanB and, in *caenorhabditis elegans,* odr4 is required for expression of some chemosensory receptors.[80-82] If this proves to be the case, it may be difficult to use expression cloning to identify taste receptors until such factors are identified. However, with the characterization of taste receptors of known functional specificity (see Section 7.4.3) and development of cell-based assay systems, it is possible that this approach will yet prove useful.

7.3.6 Genetic Approaches and Bioinformatics

For the past 30 years, it has been appreciated that the ability of mice and humans to detect certain sweet and bitter compounds is genetically determined. The approximate positions of several of the controlling genes have been mapped in mice by behavioral taste-choice assays.[24-31] Moreover, nerve recordings suggest that these differences are often peripheral[32] and, at least in the case of bitter taste, several tightly linked genes have been implicated in controlling responses to structurally distinct compounds. Therefore, it is likely that some of these loci encode taste receptors. Classically, positional cloning of genes has often used repeated backcrosses to reduce the length of donor DNA (mediating the phenotype) in the background of the other strain. By coupling this approach with determination of the donor interval, initially by use of genetic markers, it has been possible to identify candidate genes. Recently, the wealth of DNA sequence information derived from the human genome project means that a combined genetic-bioinformatics approach may be effective in rapidly identifying candidate genes. Furthermore, as the genomes of model organisms have been completed, it has turned out that even in the absence of genetic markers, bioinformatic approaches can be very powerful and directly identify potential chemosensory receptors.[53-55]

The classical genetic approach is beautifully illustrated by the cloning of the diacetyl receptor from *C. elegans*.[83] A behavioral assay was used to isolate worms that were capable of responding to attractive odorants but were unable to sense diacetyl. Genetic markers were then used to determine the approximate site of the defect in the mutant animals. The search for the gene was narrowed by complementation using a genomic cosmid and then a small sub-clone from this cosmid. In contrast, more recently, a pure informatics approach was used to identify a large family of putative taste receptors in *Drosophila*.[55] In this case, the assumption made was that the genes for these receptors were likely to encode a family of divergent seven-transmembrane helix receptors. Using an algorithm that searched genomic data for putative membrane spanning helices, Clyne et al., were able to identify 40 to 80 novel receptors of this type that may well include taste receptors for the fly.

A combined approach was used to identify a family of bitter taste receptors in humans and rodents. Genetic linkage studies in humans identified a region of chromosome 5 that influences perception of the bitter substance 6-*n*-propyl-2-thiouracil (PROP).[84] A bitter taste receptor might be responsible for these genetic differences. Therefore, DNA sequence databases were searched and a novel GPCR (T2R-1) at 5p15.2 was identified.[36] Computer searches using T2R-1, and reiterated with T2R-1-related sequences, revealed 19 additional human receptors (12 full-length and 7 pseudogenes). These new receptors, referred to as T2Rs, define a novel family of GPCRs distantly related to V1R vomeronasal receptors[72] and opsins. Additional studies demonstrated that T2Rs are strong candidate taste receptors and that members of this gene family are activated by bitter-tasting compounds (see Section 7.4.3). Independently and also using bioinformatics and genetics, but with a slightly different starting point based on the mapping of a number of bitter loci in mice, Matsunami et al.[85] found the same receptors.

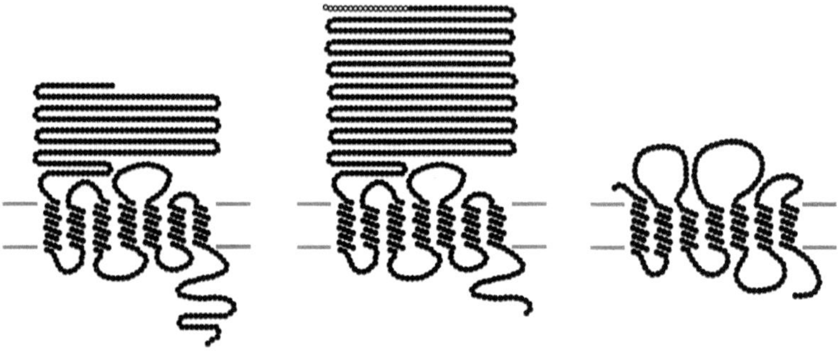

FIGURE 7.1 Proposed membrane topology of candidate taste receptors. These three classes of candidate taste receptors are predicted to span the membrane seven times. Like other GPCRs, it is predicted that the N-terminus is outside the cell and the C-terminus inside. In the case of the T1Rs, a predicted N-terminal signal peptide is present (schematically represented as open circles). Ligands are believed to bind to the long N-terminal extracellular domain of the mGluR4 and T1Rs but within the helical domain of T2Rs.

7.4 CANDIDATE TASTE RECEPTORS

As detailed above in Section 7.3, recent results point to strong candidates for some of the elusive taste receptors. This chapter section describes the characteristics, similarities, and differences of the putative umami receptors, T1Rs and T2Rs in some detail, indicating what is known as well as the problems that remain to be answered. Intriguingly, although all these receptors are believed to be GPCRs, they have marked differences in their overall structure (Figure 7.1). The T2Rs most closely resemble the classical seven-transmembrane helical fold with a putative binding site within the transmembrane domain typified by opsins, the β-adrenergic receptor, and odorant receptors. The T1Rs are of a different class of GPCR related to the extracellular calcium-sensing receptor and the second family of putative mammalian pheromone receptors (V2Rs). In this class of GPCRs, the long N-terminal extracellular domain is believed to encode the binding site.[86,87] Finally, the variant taste mGluR4 that has been proposed to function as the umami receptor also has a long N-terminus but lacks key features usually associated with this class of GPCR. For example, unlike mGluR4 from brain, the umami receptor has no signal peptide and a large portion of the extracellular domain, believed to be critical for the binding of glutamate, is missing.

7.4.1 Taste mGluR4: A Putative Umami Receptor

As explained in Section 7.3.2, the mGluR4 receptor was initially identified using a PCR-based search to examine the expression pattern of known glutamate receptors.[56] This approach revealed expression of mGluR4 in taste tissue, and tantalizing support for this receptor as a mediator of umami taste is that a relatively specific mGluR4 agonist (i.e., AP-4) also elicits umami taste.[88] However, one major problem was that this receptor is sensitive to glutamate at 100 to 1000 times lower concentration than is sensed by taste. A possible resolution could be that a less-sensitive form of mGluR4 acts as the taste glutamate receptor. PCR data and RNase protection analyses suggest that just such a receptor may be present in the tongue (taste mGluR4), and heterologous expression indicates that this form of mGluR4 responds to higher concentrations of glutamate and AP-4. These are important observations, but the data are complex. For example, RNase protection shows that there are approximately equal quantities of the taste and normal mGluR4 transcripts in the circumvallate papillae of juvenile mice and significantly more of the normal transcript in adult mice, yet adult mice are less sensitive to glutamate than juveniles. Unfortunately, RNase protection assays provide no information about the cellular expression pattern or subcellular localization of these receptors. Therefore, it may be that both forms are expressed in taste, although not necessarily within the same cells or the same subcellular compartments.

There is evidence from *in situ* hybridization studies that mRNA for the mGluR4 is found in a small subset of taste receptor cells in the circumvallate and foliate papillae, but as yet it is not clear if these are the cells that mediate umami taste.[57] In fact, electrophysiological studies indicate that larger subsets of taste receptor cells from various regions of the tongue respond either with a depolarization or hyperpolarization when exposed to physiological concentrations of glutamate.[89,90] It is not clear whether the cells expressing mGluR4 might correspond to either of these populations because the *in situ* results indicate low levels of expression in a few taste receptor cells but detailed quantitation is missing. Moreover, the *in situ* probe does not distinguish between the normal mGluR4 and the taste form. Indeed, because the entire protein sequence of the taste mGluR4 is within that of the normal receptor, immunological techniques are also powerless to resolve which cells express the taste mGluR4 or demonstrate its subcellular location. However, Western blots of taste tissue (where the two forms would be resolved because of their distinct sizes) might provide important evidence that the taste form is expressed as a protein. This is important because the transcript encoding this form of the receptor is not a normal, alternatively spliced product, but rather appears to be transcribed from a start site within an intron.

Very surprisingly, mGluR4 knockout mice in which the normal mGluR4 is eliminated respond with slightly increased sensitivity to umami stimuli.[91] Therefore, it may be that taste mGluR4 does not mediate umami taste. However, it is also possible that this knockout does not cause loss of the taste form of the receptor, in which case immunohistochemical analysis of taste tissue from these mice may be very valuable in establishing a role for taste mGluR4. These experiments remain to be done; and even if it turns out that the knockout mice do express taste mGluR4, it might be argued that another knockout specifically deleting this form would also need to be studied before one would know that this receptor was responsible for mediating umami taste.

The heterologous expression results suggest that taste-mGluR4 has lower affinity but otherwise similar function to the full-length receptor.[88] However, it is not clear why deletion of the first three exons (309 amino acids) should have this effect. First of all, whereas the full-length receptor has a signal peptide that exports the hydrophilic extracellular domain across the endoplasmic recticulum,[86] the truncated receptor does not. Moreover, the extracellular domain of a number of mGluRs, including mGluR4, function as glutamate binding proteins when expressed as soluble proteins *in vitro*.[86,87] Recently, the crystal structure of one of these was reported, and it can be readily discerned that the deleted residues make up critical elements of the bi-lobed glutamate binding domain.[92] These deletions include several of the tightly packed secondary structural elements of both lobes of the binding domain. Therefore, if taste-mGluR4 does indeed function as the umami receptor, very differently folded versions of the mGluR4 gene product must bind the same ligands. This might be of great interest to structural biologists and those interested in protein structural evolution.

7.4.2 T1Rs: Putative Receptors for Sweet and/or Bitter Compounds

The first of these receptors (T1R-1) was isolated using a subtractive screen and, like the mGluR4 receptor, is a member of the class of GPCRs related to the CaSR.[66] Unfortunately, thus far, it has been impossible to ascribe function to T1Rs and the evidence that these are sensory receptors is strong but purely circumstantial. This chapter section summarizes the data that suggest such a role for this small family of GPCRs.

Using a PCR approach, a second T1R, T1R-2 with ~40% amino acid identity to T1R-1, was isolated. As expected, T1R-1 and T1R-2 contain all the hallmarks of this superfamily of membrane receptors, including a predicted N-terminal signal peptide, seven putative transmembrane segments, a series of conserved cysteine residues in the extracellular domain, and several short sequence motifs scattered throughout the molecule. *In situ* hybridizations to a variety of non-taste tissues did not detect any T1R expression. However, as expected, T1Rs are abundantly expressed in subsets of taste receptor cells. T1R-1 is rare in the taste buds of circumvallate papillae (less than 25% of circumvallate taste buds contain a T1R-1-positive cell) but is expressed in all fungiform taste buds. In contrast, T1R-2 is almost undetectable in fungiform papillae but is expressed in all circumvallate taste buds (see Figure 7.2). This differential expression profile is also found in the palate and to a lesser extent in foliate taste buds. However, even in taste buds that express both receptors, there are only very few cells that co-express T1R-1 and T1R-2. The sequence data imply that T1R-1 and T1R-2 might well be activated by distinct ligands because the extracellular binding domains are less than 40 % identical. Moreover, because these receptors are mainly expressed in non-overlapping populations of taste receptor cells, their activation could be distinguished by the brain, and thus T1R-1 and T1R-2 potentially encode distinct modalities. Intriguingly, there is no correlation between T1R and gustducin expression. For example, in the foliate papillae, only ~20% of T1R-1-positive and ~10% of T1R-2-positive cells also express gustducin, and only ~15% of all gustducin-positive cells express either T1R. Thus, it is unlikely that gustducin functions as the Gα subunit that couples T1Rs to downstream effectors.

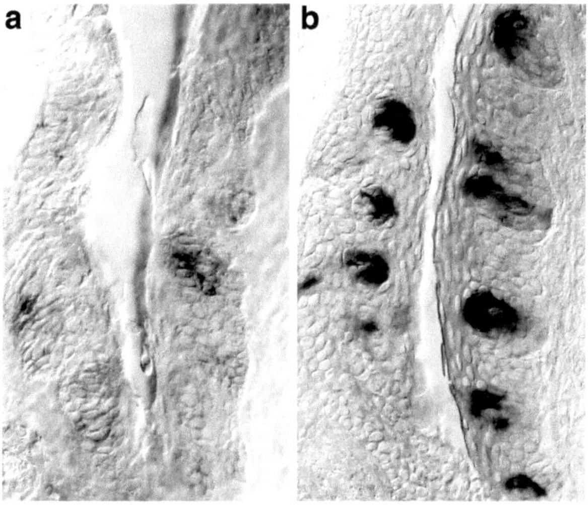

FIGURE 7.2 T1Rs exhibit marked differences in distribution. *In situ* hybridization of cRNA antisense probes of the T1Rs to sections of mouse circumvallate papillae reveals that these receptors are selectively expressed in a subset of taste receptor cells. A marked difference in the number of cells expressing mT1R-1 (a) and mT1R-2 (b) is observed. T1R-2 is expressed in all taste buds and is found in 25 to 30 % of all taste receptor cells. In contrast, only very few T1R-1 expressing cells are observed. In other regions of the oral cavity very different patterns of expression of these two receptors is detected (For further information see text and Hoon, M.A. et al., *Cell,* 96, 541, 1999.)

Mammalian taste receptor cells contain apical microvilli that are exposed to the oral cavity through a narrow opening known as the taste pore, which is hypothesized to be the site of tastant recognition by membrane receptors. Immunohistochemical detection of T1Rs demonstrated that they specifically localize to the apical surface of taste receptor cells, with antibodies to T1R-1 and T1R-2 selectively labeling the taste pore region. These results strongly implicate T1Rs as sensory receptors but do not indicate what sort of ligands they respond to.

What types of experiment might be used to determine the function of T1Rs? Convincing demonstration of function can be obtained using heterologous expression, but thus far, neither we nor others have succeeded in this. It is possible that genetic approaches can also shed light on the role of taste receptors: both T1R-1 and T1R-2 have been mapped to the distal end of mouse chromosome 4 close to *Sac*.[93] The *Sac*-locus appears to control responses to a large number of different sweet compounds.[29] It has been reported that T1R-1 and T1R-2 map proximal to *Sac*; nonetheless, the close proximity of these genes to the *Sac*-locus is tantalizing. Other approaches might include studying the behavioral and electrophysiological responses of mice in which the genes for T1Rs have been deleted. However, the close proximity of these genes to the *Sac*-locus means that great care has to be taken in interpreting the results. Less direct, but also attractive would be to mark the cells expressing T1Rs and to study the functional responses of these cells. However, any correlation between responses and T1R expression would need confirmation, perhaps by demonstrating loss of responses in cells where the T1R had been knocked out. Therefore, although T1Rs are attractive candidate receptors for sweet and/or bitter stimuli, much work will be required to determine their exact role in taste and it is conceivable that their function is not directly related to sweet or bitter taste discrimination.

7.4.3 T2Rs: Receptors for Bitter Compounds

As described in Section 7.3.6, genetic studies of taste coupled with bioinformatic approaches were used to identify this family of receptors. The T2Rs are separated from other candidate taste receptors by convincing demonstrations of function, localization to specific taste receptor cells, genetic implication of function in taste, and the fact that this is a large family.[36,78] Psychophysical studies of a large group of subjects indicated that a region of human chromosome 5p15 controls sensitivity to the bitter compound PROP.[84] Similarly, behavioral studies of mice have suggested a cluster of genes controlling aversion to a variety of bitter tasting compounds is located on distal chromosome 6 (see below). Using this information as a starting point, we searched for candidate GPCRs on 5p15 and found T2R-1.[36] Independently, Matsunami et al.[85] searched a region of human chromosome 12, homologous to distal mouse chromosome 6, and identified the same family of receptors. Like us, Matsunami et al. identified a family of genes from the sequence of the human genome. These genes are clustered in just a few locations: T2R-1 is found at 5p15, others are found on both arms of chromosome 7, and a large group is located at 12p13. Now that 80% of the sequence of the human genome is available (at least in draft form), we can identify 23 apparently functional genes and several additional pseudogenes.

The genetics of taste has been extensively studied in mice, where a number of loci influencing responses to sweet and bitter tastants have been mapped by behavioral taste-choice assays. The distal end of mouse chromosome 6 contains a cluster of genes that includes *Soa* (for sucrose octaacetate),[30] *Rua* (raffinose undecaacetate),[27] *Cyx* (cycloheximide)[94] and *Qui* (quinine).[26] Recombination studies indicated that these four loci are closely linked to each other, and to *Prp* (salivary proline rich protein).[95] We isolated more than 30 mouse T2R genes and examined their genomic organization and physical and genetic map locations. Mapping studies showed that these mouse genes, like their human counterparts, are clustered at only a few genomic locations. Each mouse genomic interval containing T2Rs is homologous to one containing the closest human counterparts. Two of these intervals are locations implicated in human bitter perception.[84,96] Another contains at least 25 mT2Rs, spans more than 400 kb, and maps to distal chromosome 6. Because *Prp* also maps to the distal end of mouse chromosome 6, we examined whether it is contained in this array of T2Rs. We isolated the mouse *Prp* gene and found that it lies within the large chromosome 6 T2R cluster. These results establish T2Rs as genes that might control perception of a number of bitter compounds in mice and humans and indicate that these GPCRs might function as bitter taste receptors.

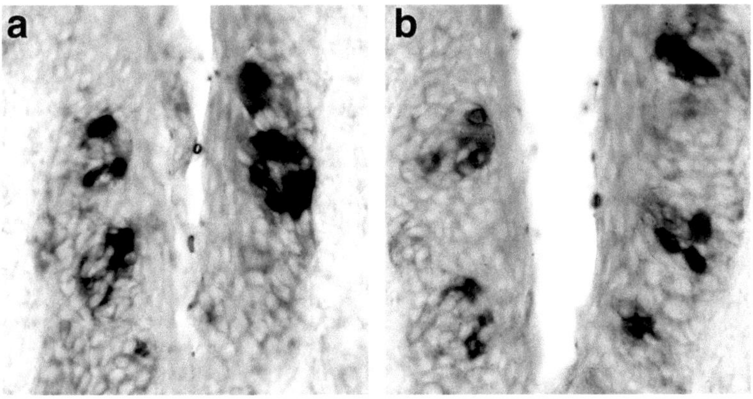

FIGURE 7.3 T2Rs are coexpressed with one another in a subset of taste receptor cells. *In situ* hybridization of cRNA antisense probes demonstrates that these receptors are selectively expressed in a subset of taste receptor cells. Hybridization of a single mT2R (a), or a mix of 20 mT2Rs (b) to sections of mouse circumvallate papillae detects a similar number of positive taste receptor cells. This indicates that many T2Rs are co-expressed in the same subset of taste receptor cells. Similar results were obtained when sections containing taste buds from other regions of the oral cavity were examined (For further information see text and Adler, E. et al., *Cell,* 100, 693, 2000.)

As expected, T2Rs are expressed in subsets of taste receptor cells[36,85] (and see Figure 7.3). We determined that each receptor is found in about 15% of the cells in a circumvallate taste bud and that all of the taste buds of the circumvallate papilla contain cells expressing T2Rs. Similar studies in foliate, geschmackstreifen, and epiglottis taste buds demonstrated that receptor probes label ~15% of the cells in every taste bud. In contrast, T2Rs are rarely expressed in fungiform papillae. We found that less than 10% of all fungiform papillae contain T2R-expressing cells. Interestingly, the few fungiform taste buds that do express T2Rs regularly contain multiple positive cells and generally appear clustered. This unexpected finding may provide important information related to taste coding: single fibers of the chorda tympani nerve innervate multiple cells in a fungiform taste bud, with the same fiber often projecting to neighboring papillae.[15,97] Thus, the non-random distribution of T2R-positive taste receptor cells and taste buds in fungiform papillae may reflect integration of responses of similarly tuned receptor cells at the level of innervation.

For functional assays, we used a cell-based reporter system to allow the responses of many T2Rs to a variety of sweet and bitter ligands to be tested.[78] T2Rs were co-expressed with Gα15, a G-protein α-subunit that has been shown to couple a wide range of receptors to phospholipase Cβ[75,98] resulting in increased intracellular calcium [Ca^{2+}]i, which can be monitored at the single-cell level using the FURA-2 calcium-indicator dye.[99] An initial problem that we encountered was that T2Rs were not efficiently targeted to the plasma membrane. However, tagging T2Rs with the first 39 amino acids of bovine rhodopsin (akin to the approach used by Krautwurst et al.[75] to express olfactory receptors) increased membrane expression.

We expected functional coupling to meet four criteria: tastant selectivity, temporal specificity, and receptor and G-protein dependence. Initially, 11 rhodopsin-tagged human T2R receptors were tested against bitter and sweet tastants. The intensely bitter tastant denatonium produced dose-dependent responses in cell transfected with hT2R-4 that met the above criteria.[78] However, the threshold of ~100 μ*M* for activation was over two orders of magnitude higher than the human psychophysical threshold for denatonium.[100] This could be due to poor functioning of this receptor in the heterologous expression system, or perhaps humans express another higher-affinity denatonium receptor. The mouse receptor mT2R-8 is a likely ortholog of hT2R-4: they share ~70% identity, while the next closest receptor is only 40% identical; these two genes are contained in homologous genomic intervals.[36] Mouse T2R-8 was also activated by denatonium; interestingly, hT2R-4 and mT2R-8 both displayed a limited range of promiscuity because they are also activated by high concentrations of PROP. No other tastants elicited significant responses from cells expressing hT2R-4 or mT2R-8.

Cells expressing another mouse receptor mT2R-5 specifically responded to cycloheximide with a threshold response at ~0.2 μ*M*.[78] Several different inbred strains of mice have been defined either as tasters or non-tasters for cycloheximide. Taster strains of mice have threshold responses to cycloheximide at ~0.25 μ*M*, while non-taster strains are indifferent to this toxin at concentrations below 1 μ*M*. A single dominant locus *Cyx* appears responsible for this difference and has been mapped to distal chromosome 6.[94] The mT2R-5 gene also maps to distal chromosome 6 and might correspond to the *Cyx*-locus. If this is indeed the case, either differences in expression of mT2R-5 between strains or in its sensitivity to cycloheximide might be expected. *In situ* hybridizations to tissue sections demonstrated that the expression profile of mT2R-5 is indistinguishable between taster and non-taster strains. However, mT2R-5 from non-taster strains has five amino-acid mutations when compared with the gene from taster mice. Moreover, mT2R-5 from the non-taster strains displayed shifted cycloheximide sensitivity with a threshold very close to that measured for non-taster strains to this bitter tastant. This strongly suggests that mT2R-5 corresponds to the *Cyx*-locus. Nonetheless, proof that mT2R-5 is *Cyx* will require complementation of the phenotype with appropriate transgenes or, better still, knockout of the gene in taster strains.

Double-label *in situ* hybridization demonstrated that T2Rs are expressed only in gustducin-containing cells, suggesting that T2Rs may activate this G-protein in response to bitter tastants.[36] Indeed, when rhodopsin-tagged mT2R-5 was reconstituted with gustducin, specific cycloheximide-dependent activation could be directly demonstrated.[78] In contrast, no coupling was seen with Gαs, Gαi, Gαq, or Gαo. These results show that some T2Rs are likely to be gustducin-linked bitter receptors and certainly suggest that most, if not all, T2Rs may play equivalent roles as gustducin-linked receptors for other bitter-tasting compounds but do not directly explain why so many structurally diverse molecules all evoke a single taste: bitter.

In olfaction, the ability of mammals to distinguish multiple odors has been suggested to result from the expression pattern of odorant receptors. Individual olfactory neurons express only 1 of about 1000 different odorant receptors, with the neurons that express a common receptor projecting to the same glomeruli (reviewed by Mombaerts et al.[101]). It is believed that mammals interpret patterns of glomerular activation to respond to a wide diversity of odorants with considerable discriminatory power.[102] Thus, to understand the logic of bitter signaling, it is critical to determine the expression pattern of T2Rs. In their study, Matsunami et al.[85] interpreted limited *in situ* hybridization data as suggesting little co-expression of these receptors in taste receptor cells. However, in studies using 12 rat T2Rs and 20 mouse receptors, we showed that a taste receptor cell expressing one T2R also contains most, if not all other T2Rs (Reference 36 and Figure 7.3). The pattern of *in situ* hybridization observed when sections were probed for expression of a single T2R or a mix of several or many receptors was quantitatively indistinguishable, implying co-expression of receptors in the same cell. This result was also confirmed using double-label *in situ* methods. Therefore, a logical explanation for the uniform bitter taste evoked by many structurally diverse toxins is that although they signal through distinct T2Rs, they activate similar or identical patterns of cells.

There are still a number of missing pieces in the puzzle. First, there appear to be missing receptors: while all T2R-expressing cells contain gustducin, the reverse is not true; and especially in fungiform taste buds, many cells express gustducin but no known T2Rs. Could these cells express a family of sweet taste receptors? Or do they express additional bitter responsive receptors? Additionally, while it is attractive to consider all T2Rs to be bitter receptors, functional data has only been obtained for a few. This is not because only a few different receptors or potential ligands have been tested: many of the human and mouse genes have been tested in the cellular assay using a large number of sweet and bitter compounds. There are likely to be no more than 30 human T2Rs; if these represent the only bitter receptors, it might be argued that about 1/30 of all compound/receptor pairs should result in activation. But clearly, this has not proved to be the case. It may simply be that the majority of these receptors do not couple efficiently to Gα15 under the conditions used and that we therefore do not measure activity. Given that T2Rs are likely to activate gustducin *in vivo*, one solution might be to measure activity by reconstitution of T2Rs with this G-protein. Another concern is that in order to drive surface expression of T2Rs, we have expressed rhodopsin-T2R chimeras. It may be that this reduces the functional coupling of many T2Rs and therefore it will be important to test other methods of obtaining cell-surface expression. These could include searching for specific chaperones or the use of alternative tagging strategies. However, there are other questions. For example, a number of distinct loci have been implicated as important in controlling the human response to PTC and PROP.[84,96] While it is reassuring that T2R-clusters map to several of these loci, it is difficult to understand how mutations in several different genes could simply account for the relatively similar PTC-tasting phenotypes that have been mapped to different loci. This suggests there might be a more complex interaction between the T2Rs, perhaps at the level of gene expression. Alternatively, it is conceivable that co-expression of receptors results in receptor complexes that respond sensitively to a wider range of ligands than stimulate the individual T2Rs.

7.5 CONCLUDING REMARKS

There has been great progress in the past few years in describing candidate taste receptors. In particular, the functional and expression data reported for the T2Rs point to logic in the coding of modalities at the level of the tongue that had not previously been observed. It will be fascinating to see how this field develops and, in particular, uncovering the answers to some of the questions posed in Section 7.4 will be important to understanding the sense of taste. Eventually, it is hoped that description of receptor function and their expression pattern on the tongue will illuminate the problem that the brain needs to solve in decoding taste. For example, one can already state that T1Rs are not found in the cells containing T2Rs; the T1Rs may turn out to be sweet receptors. If so, then it might be concluded that the pattern of activation of distinct subsets of cells determines whether a stimulus is sweet or bitter. However, it may be that there is greater overlap in the response characteristics of particular cells in taste, as hinted by electrophysiological recordings. Moreover, it should be noted that most natural foods are not simply sweet, sour, salty, or bitter but a combination of these flavors. Thus, it is likely that the brain makes decisions critical for the survival of the animal based on more than a simple analysis of which cells are activated in the tongue.

REFERENCES

1. Kinnamon, S.C., Dionne, V.E., and Beam, K.G., Apical localization of K^+ channels in taste cells provides the basis for sour taste transduction, *Proc. Natl. Acad. Sci. U.S.A.,* 85, 7023, 1988.
2. Gilbertson, T.A., Avenet, P., Kinnamon, S.C., and Roper, S.D., Proton currents through amiloride-sensitive Na channels in hamster taste cells. Role in acid transduction, *J. Gen. Physiol.,* 100, 803, 1992.
3. Heck, G.L., Mierson, S., and DeSimone, J.A., Salt taste transduction occurs through an amiloride-sensitive sodium transport pathway, *Science,* 223, 403, 1984.
4. Kinnamon, S.C. and Cummings, T.A., Chemosensory transduction mechanisms in taste, *Annu. Rev. Physiol.,* 54, 715, 1992.
5. Margolskee, R.F., The molecular biology of taste transduction, *Bioessays,* 15, 645, 1993.
6. Margolskee, R.F., The biochemistry and molecular biology of taste transduction, *Curr. Opin. Neurobiol.,* 3, 526, 1993.
7. Striem, B.J., Pace, U., Zehavi, U., Naim, M., and Lancet, D., Sweet tastants stimulate adenylate cyclase coupled to GTP-binding protein in rat tongue membranes, *Biochem. J.,* 260, 121, 1989.
8. Kolesnikov, S.S. and Margolskee, R.F., A cyclic-nucleotide-suppressible conductance activated by transducin in taste cells, *Nature,* 376, 85, 1995.
9. Miwa, K., Kanemura, F., and Tonosaki, K., Tastes activate different second messengers in taste cells, *J. Vet. Med. Sci.,* 59, 81, 1997.
10. Krizhanovsky, V., Agamy, O., and Naim, M., Sucrose-stimulated subsecond transient increase in cGMP level in rat intact circumvallate taste bud cells, *Am. J. Physiol. Cell. Physiol.,* 279, C120, 2000.

11. Bernhardt, S.J., Naim, M., Zehavi, U., and Lindemann, B., Changes in IP3 and cytosolic Ca^{2+} in response to sugars and non-sugar sweeteners in transduction of sweet taste in the rat, *J. Physiol. (London),* 490, 325, 1996.
12. Ruiz-Avila, L., McLaughlin, S.K., Wildman, D., McKinnon, P.J., Robichon, A., Spickofsky, N., and Margolskee, R.F., Coupling of bitter receptor to phosphodiesterase through transducin in taste receptor cells, *Nature,* 376, 80, 1995.
13. Rosenzweig, S., Yan, W., Dasso, M., and Spielman, A.I., Possible novel mechanism for bitter taste mediated through cGMP, *J. Neurophysiol.,* 81, 1661, 1999.
14. Huang, L., Shanker, Y.G., Dubauskaite, J., Zheng, J.Z., Yan, W., Rosenzweig, S., Spielman, A.I., Max, M., and Margolskee, R.F., Ggamma13 colocalizes with gustducin in taste receptor cells and mediates IP3 responses to bitter denatonium, *Nat. Neurosci.,* 2, 1055, 1999.
15. Whitehead, M.C., Ganchrow, J.R., Ganchrow, D., and Yao, B., Organization of geniculate and trigeminal ganglion cells innervating single fungiform taste papillae: a study with tetramethylrhodamine dextran amine labeling, *Neuroscience,* 93, 931, 1999.
16. Frank, M.E., Contreras, R.J., and Hettinger, T.P., Nerve fibers sensitive to ionic taste stimuli in chorda tympani of the rat, *J. Neurophysiol.,* 50, 941, 1983.
17. Nejad, M., The neural activities of the greater superficial petrosal nerve of the rat in response to chemical stimulation of the palate, *Chem. Senses,* 11, 283, 1986.
18. Frank, M.E., Taste-responsive neurons of the glossopharyngeal nerve of the rat, *J. Neurophysiol.,* 65, 1452, 1991.
19. Cummings, T.A., Daniels, C., and Kinnamon, S.C., Sweet taste transduction in hamster: sweeteners and cyclic nucleotides depolarize taste cells by reducing a K^+ current, *J. Neurophysiol.,* 75, 1256, 1996.
20. Bradley, R.M., Stedman, H.M., and Mistretta, C.M., Age does not affect numbers of taste buds and papillae in adult rhesus monkeys, *Anat. Rec.,* 212, 246, 1985.
21. Peri, I., Mamrud-Brains, H., Rodin, S., Krizhanovsky, V., Shai, Y., Nir, S., and Naim, M., Rapid entry of bitter and sweet tastants into liposomes and taste cells: implications for signal transduction, *Am. J. Physiol. Cell. Physiol.,* 278, C17, 2000.
22. Shallenberger, R.S., Acree, T.E., and Lee, C.Y., Sweet taste of D and L-sugars and amino-acids and the steric nature of their chemo-receptor site, *Nature,* 221, 555, 1969.
23. Mazur, R.H., Schlatter, J.M., and Goldkamp, A.H., Structure-taste relationships of some dipeptides, *J. Am. Chem. Soc.,* 91, 2684, 1969.
24. Warren, R.P. and Lewis, R.C., Taste polymorphism in mice involving a bitter sugar derivative, *Nature,* 227, 77, 1970.
25. Lush, I.E., The genetics of tasting in mice. I. Sucrose octaacetate, *Genet. Res.,* 38, 93, 1981.
26. Lush, I.E., The genetics of tasting in mice. III. Quinine, *Genet. Res.,* 44, 151, 1984.
27. Lush, I.E., The genetics of tasting in mice. IV. The acetates of raffinose, galactose and beta-lactose, *Genet. Res.,* 47, 117, 1986.
28. Whitney, G. and Harder, D.B., Single-locus control of sucrose octaacetate tasting among mice, *Behav. Genet.,* 16, 559, 1986.
29. Lush, I.E., The genetics of tasting in mice. VI. Saccharin, acesulfame, dulcin and sucrose, *Genet. Res.,* 53, 95, 1989.
30. Capeless, C.G., Whitney, G., and Azen, E.A., Chromosome mapping of Soa, a gene influencing gustatory sensitivity to sucrose octaacetate in mice, *Behav. Genet.,* 22, 655, 1992.

31. Boughter, J.D., Jr. and Whitney, G., Behavioral specificity of the bitter taste gene Soa, *Physiol. Behav.,* 63, 101, 1997.
32. Bachmanov, A.A., Reed, D.R., Ninomiya, Y., Inoue, M., Tordoff, M.G., Price, R.A., and Beauchamp, G.K., Sucrose consumption in mice: major influence of two genetic loci affecting peripheral sensory responses, *Mamm. Genome,* 8, 545, 1997.
33. McLaughlin, S.K., McKinnon, P.J., and Margolskee, R.F., Gustducin is a taste-cell-specific G protein closely related to the transducins, *Nature,* 357, 563, 1992.
34. Hofer, D., Puschel, B., and Drenckhahn, D., Taste receptor-like cells in the rat gut identified by expression of alpha-gustducin, *Proc. Natl. Acad. Sci. U.S.A.,* 93, 6631, 1996.
35. Hofer, D. and Drenckhahn, D., Identification of the taste cell G-protein, alpha-gustducin, in brush cells of the rat pancreatic duct system, *Histochem. Cell. Biol.,* 110, 303, 1998.
36. Adler, E., Hoon, M.A., Mueller, K.L., Chandrashekar, J., Ryba, N.J.P., and Zuker, C.S., A novel family of mammalian taste receptors, *Cell,* 100, 693, 2000.
37. Wong, G.T., Gannon, K.S., and Margolskee, R.F., Transduction of bitter and sweet taste by gustducin, *Nature,* 381, 796, 1996.
38. Avenet, P., Hofmann, F., and Lindemann, B., Signaling in taste receptor cells: cAMP-dependent protein kinase causes depolarization by closure of 44 pS K-channels, *Comp. Biochem. Physiol. A.,* 90, 681, 1988.
39. Tonosaki, K. and Funakoshi, M., Cyclic nucleotides may mediate taste transduction, *Nature,* 331, 354, 1988.
40. Spielman, A.I., Nagai, H., Sunavala, G., Dasso, M., Breer, H., Boekhoff, I., Huque, T., Whitney, G., and Brand, J.G., Rapid kinetics of second messenger production in bitter taste, *Am. J. Physiol.,* 270, C926, 1996.
41. Rossler, P., Kroner, C., Freitag, J., Noe, J., and Breer, H., Identification of a phospholipase C beta subtype in rat taste cells, *Eur. J. Cell. Biol.,* 77, 253, 1998.
42. Ming, D., Ruiz-Avila, L., and Margolskee, R.F., Characterization and solubilization of bitter-responsive receptors that couple to gustducin, *Proc. Natl. Acad. Sci. U.S.A.,* 95, 8933, 1998.
43. Rossler, P., Boekhoff, I., Tareilus, E., Beck, S., Breer, H., and Freitag, J., G protein betagamma complexes in circumvallate taste cells involved in bitter transduction, *Chem. Senses,* 25, 413, 2000.
44. Cagan, R.H. and Morris, R.W., Biochemical studies of taste sensation: binding to taste tissue of ^{3}H- labeled monellin, a sweet-tasting protein, *Proc. Natl. Acad. Sci. U.S.A.,* 76, 1692, 1979.
45. Goldstein, N.I. and Cagan, R.H., Biochemical studies of taste sensation: monoclonal antibody against L-alanine binding activity of catfish taste epithelium, *Proc. Natl. Acad. Sci. U.S.A.,* 79, 7595, 1982.
46. Finger, T.E., Bryant, B.P., Kalinoski, D.L., Teeter, J.H., Bottger, B., Grosvenor, W., Cagan, R.H., and Brand, J.G., Differential localization of putative amino acid receptors in taste buds of the channel catfish, *Ictalurus punctatus, J. Comp. Neurol.,* 373, 129, 1996.
47. Shimazaki, K., Sato, M., and Nakao, M., Photoaffinity labeling of thaumatin-binding protein in monkey circumvallate papillae, *Biochim. Biophys. Acta,* 884, 291, 1986.
48. Ming, D., Ninomiya, Y., and Margolskee, R.F., Blocking taste receptor activation of gustducin inhibits gustatory responses to bitter compounds, *Proc. Natl. Acad. Sci. U.S.A.,* 96, 9903, 1999.
49. Buck, L. and Axel, R., A novel multigene family may encode odorant receptors: a molecular basis for odor recognition, *Cell,* 65, 175, 1991.

50. Matsuoka, I., Mori, T., Aoki, J., Sato, T., and Kurihara, K., Identification of novel members of G-protein coupled receptor superfamily expressed in bovine taste tissue, *Biochem. Biophys. Res. Commun.,* 194, 504, 1993.
51. Abe, K., Kusakabe, Y., Tanemura, K., Emori, Y., and Arai, S., Multiple genes for G protein-coupled receptors and their expression in lingual epithelia, *FEBS Lett.,* 316, 253, 1993.
52. Danty, E., Cornuet, J.M., and Masson, C., Honeybees have putative olfactory receptor proteins similar to those of vertebrates, *C. R. Acad. Sci. III,* 317, 1073, 1994.
53. Clyne, P.J., Warr, C.G., Freeman, M.R., Lessing, D., Kim, J., and Carlson, J.R., A novel family of divergent seven-transmembrane proteins: candidate odorant receptors in Drosophila, *Neuron,* 22, 327, 1999.
54. Vosshall, L.B., Amrein, H., Morozov, P.S., Rzhetsky, A., and Axel, R., A spatial map of olfactory receptor expression in the *Drosophila* antenna, *Cell,* 96, 725, 1999.
55. Clyne, P.J., Warr, C.G., and Carlson, J.R., Candidate taste receptors in Drosophila, *Science,* 287, 1830, 2000.
56. Chaudhari, N., Yang, H., Lamp, C., Delay, E., Cartford, C., Than, T., and Roper, S.D., The taste of monosodium glutamate: membrane receptors in taste buds, *J. Neurosci.,* 16, 3817, 1996.
57. Yang, H., Wanner, I.B., Roper, S.D., and Chaudhari, N., An optimized method for *in situ* hybridization with signal amplification that allows the detection of rare mRNAs, *J. Histochem Cytochem.,* 47, 431, 1999.
58. Hoon, M.A. and Ryba, N.J.P., Analysis and comparison of partial sequences of clones from a taste-bud- enriched cDNA library, *J. Dent. Res.,* 76, 831, 1997.
59. Asano-Miyoshi, M., Kusakabe, Y., Abe, K., and Emori, Y., Identification of taste-tissue-specific cDNA clones from a subtraction cDNA library of rat circumvallate and foliate papillae, *J. Biochem. (Tokyo),* 124, 927, 1998.
60. Timblin, C., Battey, J.F., and Kuehl, W.M., Application for PCR technology to subtractive cDNA cloning: identification of genes expressed specifically in murine plasmacytoma cells, *Nucleic Acids Res.,* 18, 1587, 1990.
61. Hara, E., Kato, T., Nakada, S., Sekiya, S., and Oda, K., Subtractive cDNA cloning using oligo(dT)30-latex and PCR: isolation of cDNA clones specific to undifferentiated human embryonal carcinoma cells, *Nucleic Acids Res.,* 19, 7097, 1991.
62. Sharma, P., Lonneborg, A., and Stougaard, P., PCR-based construction of subtractive cDNA library using magnetic beads, *Biotechniques,* 15, 610, 612, 1993.
63. Diatchenko, L., Lau, Y.F., Campbell, A.P., Chenchik, A., Moqadam, F., Huang, B., Lukyanov, S., Lukyanov, K., Gurskaya, N., Sverdlov, E.D., and Siebert, P.D., Suppression subtractive hybridization: a method for generating differentially regulated or tissue-specific cDNA probes and libraries, *Proc. Natl. Acad. Sci. U.S.A.,* 93, 6025, 1996.
64. Davis, M.M., Cohen, D.I., Nielsen, E.A., Steinmetz, M., Paul, W.E., and Hood, L., Cell-type-specific cDNA probes and the murine I region: the localization and orientation of Ad alpha, *Proc. Natl. Acad. Sci. U.S.A.,* 81, 2194, 1984.
65. Welcher, A.A., Torres, A.R., and Ward, D.C., Selective enrichment of specific DNA, cDNA and RNA sequences using biotinylated probes, avidin and copper-chelate agarose, *Nucleic Acids Res.,* 14, 10027, 1986.
66. Hoon, M.A., Adler, E., Lindemeier, J., Battey, J.F., Ryba, N.J.P., and Zuker, C.S., Putative mammalian taste receptors: a class of taste-specific GPCRs with distinct topographic selectivity, *Cell,* 96, 541, 1999.
67. Brown, E.M., Gamba, G., Riccardi, D., Lombardi, M., Butters, R., Kifor, O., Sun, A., Hediger, M.A., Lytton, J., and Hebert, S.C., Cloning and characterization of an extracellular Ca^{2+}-sensing receptor from bovine parathyroid, *Nature,* 366, 575, 1993.

68. Nakanishi, S., Molecular diversity of glutamate receptors and implications for brain function, *Science,* 258, 597, 1992.
69. Herrada, G. and Dulac, C., A novel family of putative pheromone receptors in mammals with a topographically organized and sexually dimorphic distribution, *Cell,* 90, 763, 1997.
70. Matsunami, H. and Buck, L.B., A multigene family encoding a diverse array of putative pheromone receptors in mammals, *Cell,* 90, 775, 1997.
71. Ryba, N.J.P. and Tirindelli, R., A new multigene family of putative pheromone receptors, *Neuron,* 19, 371, 1997.
72. Dulac, C. and Axel, R., A novel family of genes encoding putative pheromone receptors in mammals, *Cell,* 83, 195, 1995.
73. Okabe, M., Ikawa, M., Kominami, K., Nakanishi, T., and Nishimune, Y., 'Green mice' as a source of ubiquitous green cells, *FEBS Lett.,* 407, 313, 1997.
74. Zhang, Y., Chou, J.H., Bradley, J., Bargmann, C.I., and Zinn, K., The Caenorhabditis elegans seven-transmembrane protein ODR-10 functions as an odorant receptor in mammalian cells, *Proc. Natl. Acad. Sci. U.S.A.,* 94, 12162, 1997.
75. Krautwurst, D., Yau, K.W., and Reed, R.R., Identification of ligands for olfactory receptors by functional expression of a receptor library, *Cell,* 95, 917, 1998.
76. Zhao, H., Ivic, L., Otaki, J.M., Hashimoto, M., Mikoshiba, K., and Firestein, S., Functional expression of a mammalian odorant receptor, *Science,* 279, 237, 1998.
77. Speca, D.J., Lin, D.M., Sorensen, P.W., Isacoff, E.Y., Ngai, J., and Dittman, A.H., Functional identification of a goldfish odorant receptor, *Neuron,* 23, 487, 1999.
78. Chandrashekar, J., Mueller, K.L., Hoon, M.A., Adler, E., Feng, L., Guo, W., Zuker, C.S., and Ryba, N.J.P., T2Rs function as bitter taste receptors, *Cell,* 100, 703, 2000.
79. Caterina, M.J., Schumacher, M.A., Tominaga, M., Rosen, T.A., Levine, J.D., and Julius, D., The capsaicin receptor: a heat-activated ion channel in the pain pathway, *Nature,* 389, 816, 1997.
80. Baker, E.K., Colley, N.J., and Zuker, C.S., The cyclophilin homolog NinaA functions as a chaperone, forming a stable complex *in vivo* with its protein target rhodopsin, *Embo. J.,* 13, 4886, 1994.
81. Ferreira, P.A., Nakayama, T.A., Pak, W.L., and Travis, G.H., Cyclophilin-related protein RanBP2 acts as chaperone for red/green opsin, *Nature,* 383, 637, 1996.
82. Dwyer, N.D., Troemel, E.R., Sengupta, P. and Bargmann, C.I., Odorant receptor localization to olfactory cilia is mediated by ODR-4, a novel membrane-associated protein, *Cell,* 93, 455, 1998.
83. Sengupta, P., Chou, J.H., and Bargmann, C.I., ODR-10 encodes a seven transmembrane domain olfactory receptor required for responses to the odorant diacetyl, *Cell,* 84, 899, 1996.
84. Reed, D.R., Nanthakumar, E., North, M., Bell, C., Bartoshuk, L.M., and Price, R.A., Localization of a gene for bitter-taste perception to human chromosome 5p15, *Am. J. Hum. Genet.,* 64, 1478, 1999.
85. Matsunami, H., Montmayeur, J.P., and Buck, L.B., A family of candidate taste receptors in human and mouse, *Nature,* 404, 601, 2000.
86. Han, G. and Hampson, D.R., Ligand binding to the amino-terminal domain of the mGluR4 subtype of metabotropic glutamate receptor, *J. Biol. Chem.,* 274, 10008, 1999.
87. Tsuji, Y., Shimada, Y., Takeshita, T., Kajimura, N., Nomura, S., Sekiyama, N., Otomo, J., Usukura, J., Nakanishi, S., and Jingami, H., Cryptic dimer interface and domain organization of the extracellular region of metabotropic glutamate receptor subtype 1, *J. Biol. Chem.,* 275, 28144, 2000.

88. Chaudhari, N., Landin, A.M., and Roper, S.D., A metabotropic glutamate receptor variant functions as a taste receptor, *Nat. Neurosci.,* 3, 113, 2000.
89. Lin, W. and Kinnamon, S.C., Responses to monosodium glutamate and guanosine 5′-monophosphate in rat fungiform taste cells, *Ann. N.Y. Acad. Sci.,* 855, 407, 1998.
90. Lin, W. and Kinnamon, S.C., Physiological evidence for ionotropic and metabotropic glutamate receptors in rat taste cells, *J. Neurophysiol.,* 82, 2061, 1999.
91. Chaudhari, N. and Roper, S.D., Molecular and physiological evidence for glutamate (umami) taste transduction via a G protein-coupled receptor, *Ann. N.Y. Acad. Sci.,* 855, 398, 1998.
92. Kunishima, N., Shimada, Y., Tsuji, Y., Sato, T., Yamamoto, M., Kumasaka, T., Nakanishi, S., Jingami, H., and Morikawa, K., Structural basis of glutamate recognition by a dimeric metabotropic glutamate receptor, *Nature,* 407, 971, 2000.
93. Fuller, J.L., Single-locus control of saccharin preference in mice, *J. Hered.,* 65, 33, 1974.
94. Lush, I.E. and Holland, G., The genetics of tasting in mice. V. Glycine and cycloheximide, *Genet. Res.,* 52, 207, 1988.
95. Azen, E.A., Lush, I.E., and Taylor, B.T., Close linkage of mouse genes for salivary proline-rich proteins (PRPs) and taste, *Trends Genet.,* 2, 199, 1986.
96. Conneally, P.M., Dumont-Driscoll, M., Huntzinger, R.S., Nance, W.E., and Jackson, C.E., Linkage relations of the loci for Kell and phenylthiocarbamide taste sensitivity, *Hum. Hered.,* 26, 267, 1976.
97. Miller, I.J., Jr., Branched chorda tympani neurons and interactions among taste receptors, *J. Comp. Neurol.,* 158, 155, 1974.
98. Offermanns, S. and Simon, M.I., G alpha 15 and G alpha 16 couple a wide variety of receptors to phospholipase C, *J. Biol. Chem.,* 270, 15175, 1995.
99. Tsien, R.Y., Rink, T.J., and Poenie, M., Measurement of cytosolic free Ca^{2+} in individual small cells using fluorescence microscopy with dual excitation wavelengths, *Cell. Calcium,* 6, 145, 1985.
100. Saroli, A., Structure-activity relationship of a bitter compound: denatonium chloride, *Naturwissenschaften,* 71, 428, 1984.
101. Mombaerts, P., Wang, F., Dulac, C., Vassar, R., Chao, S.K., Nemes, A., Mendelsohn, M., Edmondson, J., and Axel, R., The molecular biology of olfactory perception, *Cold Spring Harb. Symp. Quant. Biol.,* 61, 135, 1996.
102. Rubin, B.D. and Katz, L.C., Optical imaging of odorant representations in the mammalian olfactory bulb, *Neuron,* 23, 499, 1999.

8 Researching Isolated Taste Receptor Cells: Deciphering Transduction Cascades with Patch-Clamp and Calcium-Imaging Techniques

M. Scott Herness

CONTENTS

0-8493-2329-0/02/$0.00+$1.50

8.1 INTRODUCTION

The taste receptor cell is an ideal example of biology's complicated solution to an essentially technological problem. The task of this cell is to serve as a detector; taste receptor cells act to relay the presence of chemicals in the oral cavity to the central nervous system (CNS). To accomplish its mission, the taste receptor cell has evolved a myriad array of signaling mechanisms; seven-transmembrane receptors, G-proteins, second messengers, and ion channels have been choreographed in still ill-understood manners to signal the identity and intensity of gustatory stimuli. At its core, the manner of this orchestration, which on the surface appears to so seamlessly initiate perceptions of sweet, sour, salty, and bitter, is unknown to us. Its scientific study has thus far merely underscored its complexity and those familiar with the techniques outlined in this chapter do not underestimate its difficulty.

As with any complex scientific problem, the collective results of several methodologies will be required to elucidate its fundamental workings. This chapter serves as an apposite testament to the multidimensional analyses being applied to the chemical senses. Cross-comparison of data obtained with these analyses — behavioral, electrophysiological, and molecular approaches — will be essential in extricating its mechanisms. Inherent in most of these individual approaches is some level of reductionism. A portion of the chemical senses — a subnucleus of solitary tract, a single fiber from the chorda tympani nerve — has been isolated for study. This reductionism is particularly evident in approaches applied to the study of taste receptor cells, where isolated currents from individual cells are studied in hopes of resynthesizing this information to meaningful mechanisms. Although our understanding of taste receptor cells has profited immensely from this approach, it is not without jeopardy. Reducing gustatory science to a single cell is a dangerous premise when the population of cells is as heterogeneous as taste receptor cells.

The study of taste receptor cells revolves around the central problem of how these cells detect the presence of environmental stimuli and relay this information to the CNS. Although much of our understanding of these processes is incomplete, tremendous progress has been made in the past decade. One major advance in our understanding of taste receptor cell function has come from the application of the patch-clamping recording technique to taste receptor cells. The author's laboratory has taken the strategy (like most research, partly forced and partly planned) of studying isolated taste receptor cells. At the time, isolating taste receptor cells was an inescapable consequence of the patch-clamp method. Patch-clamp recording could analyze electrical properties of small cells in a manner never before possible, yet its prerequisite was an exposed piece of clean membrane. Procedures for isolating taste cells soon appeared for both amphibians and mammals. The scope of this chapter is to discuss some of the technical aspects of how isolated taste receptor cells are studied with whole-cell patch-clamp recording and with calcium imaging toward understanding transduction and electrical excitability.

8.2 PREPARING ISOLATED TASTE CELLS FOR STUDY

8.2.1 THE NEED FOR ISOLATING TASTE RECEPTOR CELLS AROSE FROM PATCH CLAMPING

Science often takes significant advances when new experimental methods allow exploration of previously recalcitrant areas of research. The inaugural application of the patch-clamp technique to taste receptor cells, in the late 1980s, produced such a quantal step in our understanding of peripheral gustation. For technical reasons (subsequently outlined in more detail), a thorough understanding of the types, properties, and modulation of ion channels in these cells was previously not possible.

The patch-clamp technique arose fortuitously from the observation that, under proper conditions, a clean glass pipette in contact with a clean cell membrane could form a very tight seal that possessed high mechanical strength and high electrical resistance. This seal produced resistances on the order of gigaohms ($10^9\ \Omega$) and quickly became universally referred to as the gigaseal. It was the advantages in the electrical and mechanical properties of this seal that made patch-clamping possible. The very high resistance allowed one to record very small currents — currents, in fact, never recorded before. (Simply put, since $V = IR$, then for the same voltage V, as R increases I will decrease. Thus, gigaohm resistances at physiological voltages allowed recording of picoamperes of currents. This corresponds to the current through a single ion channel.) As with any new technique, borders between superstition and science were sometimes blurred and some magic was required for forming gigaseals on new cell types. One such superstition was the absolute requirement of dissociated cells. Initially, to apply this technique to taste receptor cells, one first needed to devise a method for isolating these cells. Subsequently, further development showed that gigaseals could be formed with intact cells, such as in slice preparations. Techniques first arose with amphibian cells, which are heartier than mammalian cells and sustain the insults and injuries of the dissociation procedure with more rebound. However, the taste cells of rats and mice were soon to follow suit.

Patch-clamp studies have provided invaluable new information on transductive utensils of taste receptor cells. However, with every gain there is a loss. Work on isolated cells is necessarily accompanied by several companion caveats. The non-native state of a dissociated cell, the unknown degree of enzymatic alteration, the uncertain alterations of external, and internal ionic environments all bear witness to the fact that isolated cells provide no more than a ball-park figure to true physiology. Dissociated cells merely approximate a natural condition.

Although patch-clamping taste receptor cells had its origin in dissociated preparations, this method has evolved to include the study of an individual cell from a single isolated taste bud and, more recently, within an intact epithelium. The methods for patch-clamping are virtually identical for all three preparations and are described subsequent to describing each of these preparatory techniques.

8.2.2 Single Cell Isolation

As previously stated, the historical starting point for the patch-clamp analysis of cells was the isolated cell preparation. From 1987 to 1989, when patch-clamping was emerging as a scientific endeavor, techniques for dissociating taste cells appeared, chronologically, in frog,[2] mudpuppy,[26] mouse,[43] and rat.[17] Although slightly variant strategies were required for the particular anatomies of these species, all involved isolating small pieces of tissue containing taste buds, subjecting them to enzymatic treatment, and searching among the vast number of dissociated cells for isolated taste cells under the microscope.

Dissociation involves breaking down the extracellular matrix with the use of enzymes and calcium-free media. Enzymes of choice include collagenase, which breaks down the extracellular collagen matrix, and various proteases, such as papain. Because only a few cells are needed for an afternoon's experiments, these techniques are more than adequate for patch analysis. It should be emphasized that a counter-intuitive principle applies to cell yield for these techniques — improving yield is not necessarily progress…fewer cells treated less harshly may be vastly superior for physiological recording. The final arbiter is cell health, not cell number.

8.2.2.1 Amphibians: Frogs and Mudpuppies

Amphibian taste receptor cells, i.e., frogs (usually *Rana catesbiana* or *Rana pipiens*) or mudpuppies (*Necturus maculosus*), were the first to be patch-clamped, largely because as poikilothermic and hence with reduced metabolic requirements, these cells are hardier than mammalian and better withstand the rigor of enzymatic insult. Additionally, both had histories of study in gustation and therefore were good candidates for further, more detailed explorations. Although cells are harvested in different manners, both techniques utilize the unique morphology of taste receptor cells as a final identifier.

In frog tongues, taste receptor cells are located in taste discs, similar to but distinct from fungiform papilla. These discs can be easily harvested by cutting their stems with fine scissors. This reduces the number of non-gustatory cells as opposed to a whole tongue approach. Excised discs are incubated in collagenase (0.5 to 1.0 mg/ml) containing Ringer's (110 m*M* NaCl, 1 m*M* $CaCl_2$, 0.5 m*M* $MgCl_2$, 3.5 m*M* K_2HPO_4, pH 7.2) and stirred for about 60 min, followed by incubation in Ca^{2+}-free/EGTA (1 m*M*) Ringer's solution for another 60 min. Cells are then rinsed in normal Ringer's and triturated (in the recording dish) in normal Ringer's to dissociate the cells. Isolated cells are allowed to settle to the bottom of the dish. Taste receptor cells are identified by their spindle flask shape compared to the associated cells (wing cells or mucous cells).[2]

A different procedure[26] for harvesting cells is utilized in the mudpuppy. A piece of lingual epithelium containing taste bud-bearing structures and its underlying musculature is excised from the animal and rinsed in 100 m*M* ammonium chloride (buffered to pH 7.2 with 5 m*M* HEPES) to remove mucus. This tissue block is incubated in collagenase (1.5 mg/ml) in amphibian Ringer's (112 m*M* NaCl, 2 m*M* KCl, 8 m*M* $CaCl_2$, and 5 m*M* HEPES, pH 7.2) containing 1 mg/ml albumin for about 30 min or until the lingual epithelium can be peeled from the underlying musculature. The more adherent taste buds remain on the underlying tissue on prominent connective tissue papillae. This tissue piece is rinsed in calcium-free amphibian Ringer's with EGTA (0.5 m*M*) to loosen the taste cells for about 15 minutes. Taste cells are drawn into fire-polished pipettes and plated in recording chamber.

8.2.2.2 Mammals: Rats and Mice

Techniques for mammalian taste receptor cell isolation work best for structures containing numerous taste buds, such as circumvallate and foliate papillae, but can be applied to other structures, such as fungiform or naso-incisor ducts.

In general, proteases, in combination with divalent-free solutions, work best for dissociation of mammalian tissue. In particular, papain, a proteolytic enzyme isolated from the latex of the Carica papaya fruit, is very successful for isolation of viable cells. It has wide specificity, acting to hydrolyze large proteins into smaller peptides and amino acids. Native crystalline papain is unreactive until it has been acted upon by mild reducing agents such as cysteine. Papain is also an esterase.

Circumvallate or foliate tissue is first excised from the tongue and blocked into small pieces. Papain solution is prepared at 14 U/ml in a divalent-free bicarbonate buffered solution (26 m*M* $NaHCO_3$, 2.5 m*M* NaH_2PO_4, 20 m*M* glucose, 65 m*M* NaCl, 20 m*M* KCl, 1 m*M* EDTA) with 1mg/ml cysteine. The solution is cloudy and becomes acidic when cysteine is added, but then becomes clear when equilibrated in the CO_2 incubator. It is convenient to include the pH-sensitive dye Phenol Red (2 μl/ml) to monitor the change in acidity. Tissue blocks are incubated in papain/divalent-free solution for about 2 hr in a 5% CO_2 incubator (for example, at 32°C). A tissue block is then transferred to the recording chamber in ECF (see Table 8.1) and dissociated. Typically, the epithelium falls off from the underlying muscle layers and single taste cells are released from the epithelium with mild agitation. Examples of dissociated taste receptor cells are presented in Figure 8.1.

TABLE 8.1
Some Sample Compositions of Recording Solutions for Isolating Varying Currents in Voltage-Clamp Mode

	Standard		Sodium Currents		Potassium Currents		Calcium Currents	
	ICF	ECF	ICF	ECF	ICF	ECF	ICF	ECF
NaCl	0	126	0	126	0	126	0	91
KCl	140	5	140	5	140	5	0	5
$CaCl_2$	1	2	1	2	1	2	1	0
$MgCl_2$	2	2	2	2	2	2	2	2
NaHEPES	10	5	8	5	10	5	10	26
Glucose		10		10		10		10
NaH_2PO_4		1.25		1.25		1.25		1.25
4-AP			10				10	
TEA				20				20
TTX						0.001		0.001
$BaCl_2$								25
$CsCl_2$							140	

All concentrations are millimolar.

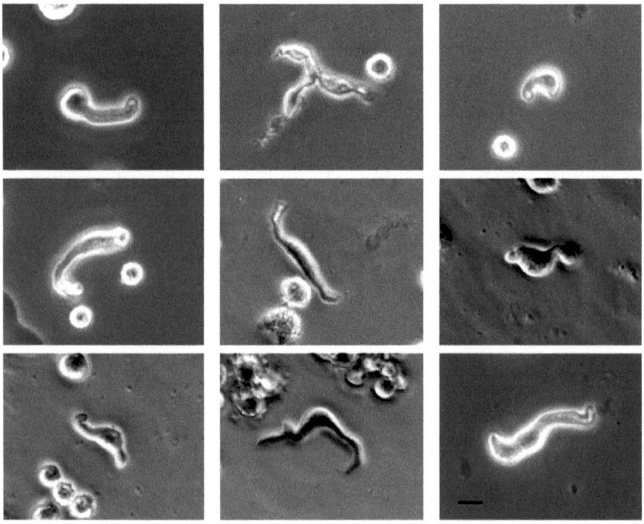

FIGURE 8.1 Dissociated taste receptor cells from rat circumvallate and foliate papillae. Examples are presented to illustrate the varying morphologies of the taste receptor cells. Nuclei (noted by the somatal swelling) may be located in either the basolateral region of the cell or its mid-region. Cells may have one or two processes, which may be thin or thick. The characteristic elongated morphology of taste receptor cells allows them to be easily distinguished from the smaller spherical epithelial cells. The scale bar represents ten microns.

A similar technique has been reported[43] using pronase as the dissociative enzyme in mice. Elastase, a neutral nonspecific protease, is also included to help eliminate a thin transparent glycocalix left by pronase treatment.

8.2.3 Single Taste Buds

An alternative to the single-cell approach is the isolation of single intact taste buds. This approach has been described for rats[4] and hamsters.[15] An advantage of this technique is that cell-to-cell contacts are maintained and cell-to-cell communication within the bud may be less damaged. Obviously, these communication avenues are eliminated in the isolated cell preparation. However, accessibility and sampling are a problem with this technique because only cells at the periphery are accessible for patch-clamp analysis. Cells within the center of the bud, which may be of different age and/or sensitivities, are not sampled.

In overview, this technique involves first removing the lingual epithelium, which contains the taste buds, and then extracting the taste buds, one by one, from the epithelium. This procedure has been described for the anterior tongue, where individual taste buds are accessible in fungiform papillae. An enzyme cocktail is injected into the muscle layer of the tongue immediately ventral to the dorsal epithelium. This cocktail includes 0.5 mg/ml collagenase, 2.5 U/ml dispase, and 1.0 mg/ml trypsin inhibitor in a mammalian Ringer's solution (140 m*M* NaCl, 5 m*M* KCl, 10 m*M* HEPES, 10 m*M* glucose, 10 m*M* Na pyruvate, 1 m*M* $CaCl_2$, 1 m*M* $MgCl_2$, pH 7.4). After a variable time period (e.g., 30 min), the epithelium is removed, pinned ventral surface up in a dissecting chamber, and incubated in calcium-free Ringer's (containing 2 m*M* BAPTA in place of $CaCl_2$ and $MgCl_2$). Individual taste buds can be harvested using a suction electrode.

8.2.4 Isolated Epithelium

Among the disadvantages of the isolated cell preparation is that one cannot apply hyper- or hypo-osmotic solutions in dissociated cell preparations. The best situation would involve the closest mimic to the *in situ* situation. Additionally, paracellular pathways may influence taste transduction mechanisms. More recently, techniques have been developed for *in situ* recording in an epithelial slice. Such preparations, although difficult, have the tremendous advantage of maintaining the normal polarity of the cells. Hence, apical (mucosal) and basolateral (serosal) divisions remain separate in the recording chamber and solutions can be changed independently. Importantly, such a setup restricts stimulation to the apical division and much more closely mimics the *in situ* recording situation.

A forerunner to this preparation was reported by Béhé et al.[4] and has subsequently been followed by two independent reports[13,31] utilizing refinements to this technique. All methods employ the same basic strategy: the lingual epithelium (with taste buds intact) is removed from the underlying musculature (as previously described) and placed in a recording chamber that maintains the polarity of the serosal and mucosal surfaces. Taste receptor cells are accessible to micropipettes for patch-clamping by a ventral approach. It is the placement of the epithelium in the recording chamber that keeps the apical and basolateral surfaces separate.

The basic procedure for isolating anterior lingual epithelium does not differ significantly from that first reported by Mistretta.[33] The lingual musculature is injected with 0.5 ml of 0.4% collagenase (type 1A) or an enzyme cocktail mixture (collagenase/dispase, trypsin inhibitor as previously described) dissolved in physiological Ringer's. Lingual epithelium is peeled off after incubation; times from 3 to 30 min after injection have been reported. Success has been reported in both rat[4,31] and murine tissue.[13] The intricacies of this procedure lay more in the recording chamber than the isolation procedure. Stability is a key issue, and vibrations of the epithelium caused by changes of solution can produce detrimental results in the quality of the patch-clamp recording.

The future of this approach is extremely promising. A similar approach can also be applied to calcium imaging using confocal microscopy. It is clear that many highly significant results will be gleaned from these approaches.

8.3 PATCH-CLAMP RECORDINGS OF ISOLATED TASTE RECEPTOR CELLS

The technique of patch-clamp recording has revolutionized electrophysiology. Similar to most experimental revolutions, it was the result of an advanced technological feat and in its wake lies a still-ongoing experimental discourse. At its inception, the revolution of patch-clamp recording lay in its near-perfect voltage control of a small patch of membrane, allowing the real-time recording of ionic current through a single ion channel. Much to the delight of electrophysiologists, single-channel conductances, conductance substates, mechanisms of ion channel block quite suddenly became experimentally accessible factors. This was not, however, of great benefit to the chemical senses. The usefulness of the technique was quickly extended when it was discovered that, due to the mechanical strength of the gigaseal, one could not only record from the patch of membrane occluding the microelectrode tip, but that the converse was also possible. One could record from the cell by removing the patch. It is this configuration of the patch-clamp technique that transformed our understanding of taste receptor cells.

Prior to the advent of patch-clamp recording, little was understood regarding the electrical properties of taste receptor cells. Such analysis required the application of voltage-clamp protocols and these protocols were not possible on cells as small and slender as taste receptor cells due to technical limitations. Voltage-clamping was only possible on small cells with a single electrode (as opposed to the two-electrode voltage-clamp technique). The high resistance of microelectrodes required to impale small cells with the least amount of damage (up to 200 MΩ) produced unacceptably high series resistance. Acceptable voltage-clamping of the membrane across such high series resistance was simply not possible. Patch-clamping, on the other hand, was able to dramatically reduce the resistance of the recording pipette, consequently reducing the series resistance as well, while concomitantly reducing the amount of injury to the cell. Thus, for the first time, voltage-clamp recording of taste receptor cells was possible. The whole-cell configuration of the patch-clamp technique allowed the isolation and characterization of individual ionic currents in taste cells — all currents not previously recorded. Thus, the field was finally in the position to begin the process of relating individual currents to transduction cascades. Meaningful hypotheses relating activation of membrane receptors to impulse generation was no longer merely a creative narrative.

Now, over a decade after the inception of patch-clamp recordings, many excellent reviews exist on the whole-cell configuration of the patch clamping technique.[6,14,22,29,34,42]

8.3.1 Equipment for Patch-Clamping

The patch-clamp station has the familiar appearance of most electrophysiology rigs. True to the stereotype, its most arcane features are the most obvious, whereas its most essential features are the most concealed. What one sees — a cage made of copper shielding sitting on an oversized, heavily weighted table, patch boards dangling with cables, plastic tubing leading from reservoir bottles to the microscope stage — are accessories to the single most important item of the recording rig, the feedback resistor across the headstage (located inside the probe). The signal is recorded across this resistor with a combination of collateral equipment — an inverted microscope to view the cells, micromanipulators to position the pipette to the cell, the amplifier to record the signal, and a computer to run the experiment. Other essential pieces of equipment include a pipette puller to manufacture the pipettes and an incubator to dissociate the cells.

Amplifiers designed for sharp microelectrode voltage-clamping are generally not adequate for patch-clamp applications. The difference between a patch-clamp amplifier and a regular amplifier lies in the headstage. Current flow through the electrode is recorded across a resistor of high impedance that causes a voltage drop proportional to the measured pipette current. An operational amplifier is used to automatically adjust the voltage source to maintain a constant pipette potential at the desired reference potential. Conventional intracellular amplifiers have too low a headstage resistance for patch-clamp applications.

Inverted microscopes are usually used because of easy access to the stage for pipette and changes of solution. Unless imaging is to be performed on the same setup, optics are not of primary importance. However, the use of phase contrast objectives is of great benefit in working with dissociated cells. The microscope rests on a vibration isolation table that acts to reduce vibrations at the pipette tip. Such vibrations, although only occurring over a distance of a micron, are hazardous to the seal. Because taste receptor cells are not plated to the bottom of the recording dish, cells are generally held in the solution by the pipette. This also acts to reduce vibration, by removing the cell from the chamber, and mitigates deleterious effects of drift by the pipette micromanipulator. In this experimenter's experience, the drift associated with hydraulic microdrives is unacceptable and piezoelectric devices are superior in this respect. Attached to the table is the Faraday cage, a hallmark of the electrophysiological rig. While not completely essential, the shielding it provides from extraneous electrical signals is sometimes indispensable. Removing 60-cycle interference by shielding is infinitely preferable to off-line filtering. A final acquisition cost will be the purchase of commercially available software to run the experiment and analyze data.

8.3.2 The Patch Pipette Electrode

Patch-clamp recording derives its name from the small patch of membrane that occludes the tip of the pipette electrode. The pipette tip isolates this "patch" of membrane that in one case (single-channel recording) is recorded from to the exclusion of the cell, or in the other case (whole-cell recording) is destroyed to record from the cell to the exclusion of the patch. Obviously, the pipette that isolates this patch of membrane from the remainder of the cell is of critical importance. In general, one wants to find a model pipette design that provides consistency in its ability to establish the whole-cell configuration and then remain invariant. Salient design features include tip diameter, shank length, and to a lesser degree glass composition. Excellent discussions can be found in the literature.[9,7,38,39] Essentially, one should design a low-resistance pipette (e.g., 3 to 4 MΩ), keep it as clean as possible, and, with luck, begin data acquisition.

Patch pipettes require blunter tips than microelectrodes designed for sharp intracellular recording. Such short-shanked pipettes are produced on multi-stage pullers, which pull the pipette in two to five stages. Pipette pullers pass large currents through a metal (platinum or tungsten) filament that then heats to high temperature. Tension is applied to the glass capillary so that the glass begins to stretch as it softens. As the pull reaches a predetermined velocity, the current is switched off and simultaneously a puff of nitrogen gas quickly cools both the filament and the glass capillary. This procedure can be performed several times (three to five times) until, finally, the glass capillary pulls into two pipettes. The advantage of such a multi-stage procedure for fabricating pipettes is largely in reducing the length of the shank of the pipette. Some of the pipette resistance occurs in the shank, as well as the tip, so that pipettes with shorter shanks have lower series resistance. Series resistance is the single most important source of error in whole-cell recording and is subsequently discussed in more detail.

Prior to recording, some investigators prefer some post-pulling processing of their pipettes. This commonly includes fire polishing and/or slyguard coating. Fire-polishing involves smoothing the pipette tip under microscopic control by manipulating it close to a heat source. This smoothing of the pipette tip will increase the resistance to a measurable degree and hence one should start with a lower resistance pipette than that finally desired. Although this smoothing of the tip is thought to improve gigaseal formation (see Section 8.3.3), many investigators find fire polishing unnecessary. Slyguard coating is an attempt to reduce pipette capacitance by coating most of the shank with an insulating material. Many find this additional step optional as well.

Pipettes are filled by drawing the recording solution, a psuedo-intracellular fluid (ICF; see Table 8.1 for some recipes), into the tip, backfilling the shank with a thin needle (e.g., 30 gauge), and tapping the bubbles out at the interface. Pipettes should not be overfilled, in order to minimize stray capacitance and background noise. The shank need only be filled to make contact with the silver/silver chloride electrode of the pipette holder. Remember that in entering whole-cell configuration, one is effectively increasing the cell volume by several thousand-fold.

Finally, although often forgotten, it is worth emphasizing that the patch pipette is not an electrode. The electrode is a solid electric conductor through which an electric current enters or leaves an electrolytic cell or other medium. The pipette is actually a salt bridge, a conduit that connects the electrode — the silver/silver chloride wire — to the cell. The actual recording is not made at the pipette tip; it takes place at the electrolytic surface of the two silver/silver chloride wires. Neglect of these electrodes has large unfavorable consequences on the recording. Although silver/silver chloride electrodes are relatively resilient to insult and injury (e.g., scratching the wire by constant removal and insertion of patch pipettes has little daily consequence if the electrodes are maintained) they should be regularly chloridized and are often stored shorted together in a electrolytic solution to minimize them as a source of voltage offsets in the recordings.

8.3.3 The Gigaseal and the Voltage Clamp

With cells isolated and a pipette designed and fabricated, the next step in recording currents is obtaining the gigaseal. The fortuitous observation that the high-resistance seal between the glass pipette and the cell membrane possessed great mechanical strength allowed for the development of the whole-cell configuration of the patch-clamp technique. After the formation of the coveted gigaseal, it was discovered that one could apply negative pressure to the inside of the pipette and actually rupture the patch of membrane occluding the tip of the pipette without damaging the seal. With the high-resistance seal intact, one now has the analogous configuration of an intracellular recording with a very significant difference: one electrode confluent with the cell's interior without significant damage. With a seal resistance much higher than the cell's membrane resistance (a situation rarely encountered with the sharp microelectrode technique), current through the pipette is forced to travel through the membrane rather than around the pipette. Thus, good voltage control of the membrane is possible and, for the first time, taste receptor cells could be adequately voltage-clamped.

Gigaseals are obtained on a hit-or-miss basis. One can hope to optimize conditions to facilitate the success rate, but uncontrollable factors, such as the condition of the cell membrane or the glass pipette, make such achievement an anticipated but non-guaranteed event. The pipette is lowered into the bath solution, clamped with a zero millivolt potential, and the junction potential is zeroed electronically. (At best, this is an imperfect procedure because liquid junction potentials change by an unpredictable amount when the whole-cell configuration is established and the pipette, originally in contact with the ionic composition of the bathing solution becomes in contact with the ionic composition of the cytoplasm.) After nulling the pipette potential, the resistance is monitored. Most patch-clamp amplifiers have built-in stimulators for measuring the pipette resistance. This resistance is measured continuously and the hoped-for event of gigaseal formation often becomes manifest to the experimenter as the gradual disappearance of the square-wave pulses on the oscilloscope screen into the now less-noisy baseline.

To obtain a gigaseal, the pipette is positioned under visual guidance to the isolated cell and pressed against its side. With taste receptor cells, this site is almost always on the somatal swelling, the cell's largest target. One can alternatively apply a small amount of suction and cause the cell to hop to the pipette tip. The pipette resistance will increase when the tip encounters the membrane, but will not approach gigaohms. Although gigaseals occasionally form spontaneously, for the most part, they do not. Instead, a second step is required. While monitoring seal resistance with a series of square-wave pulses, negative pressure is applied to the pipette interior through the side port of the pipette holder. Such negative pressure can be produced by mouth suction or with the use of a syringe, as dictated by personal preference. As the resistance increases, the size of the square-wave pulses will diminish. As a true gigaseal forms, the pulses diminish into the noise with only brief capacitive transients remaining. Using a negative holding potential, such as –70 or –80 mV, often assists gigaseal formation. Additionally, if the cell accidentally enters whole-cell configuration during the gigaseal process, it will be immediately held at a healthy resting potential. The formation of the gigaseal, although equally exciting or frustrating, is only half the battle. After gigaseal formation, one must enter whole-cell configuration. Negative pressure is now reapplied to the pipette interior but toward a different goal. Again, monitoring the seal resistance with continuous square-wave pulses, the formation of whole-cell configuration is apparent by the sudden appearance of capacitive transients associated with the cell membrane capacitance. The steady-state magnitude of the pulse, measuring resistance, will only increase slightly. If the gigaseal is lost during this process, square waves will appear without any additional capacitance added to the circuit and, generally, the resistance will drop precipitously.

The formation of whole-cell configuration is exciting but one should not linger too long in its grandeur. Within 10 to 30 min, the recording conditions may deteriorate to an unacceptable quality. Thus, one should be prepared with the protocol (voltage changes, stimuli, etc.) for the experiment of interest. To properly perform voltage-clamp experiments, a modicum of understanding of its circuit is required. Historically, voltage-clamp experiments have been performed in a variety of ways. In its five decades, the evolution of the voltage-clamp technique has included the original axial wire design, sucrose gap, two-electrode voltage-clamp, continuous single electrode, and discontinuous single electrode. Many of these early forms were, not surprisingly, results of limitations of cellular preparation (e.g., that axial wire suits the squid giant axon well, but hardly any other cell type). Nevertheless, its use successfully elucidated the ionic basis of the action potential. The whole-cell recording configuration of the patch-clamp technique performs voltage-clamping in the form of a continuous single electrode clamp. This is important to bear in mind because the hellion of continuous single electrode clamp — series resistance — always needs to be considered.

In the ideal voltage-clamp, the amplifier imposes a voltage on the membrane such that the membrane potential (Vm) is equal to the clamped potential (Vclamp). When Vm deviates from Vclamp, current is injected to maintain Vm = Vclamp. This is illustrated graphically in Figure 8.2. The single most important element of the amplifier is the current-to-voltage converter, which is contained in the headstage. Current is recorded across a feedback resistor within the headstage of the amplifier

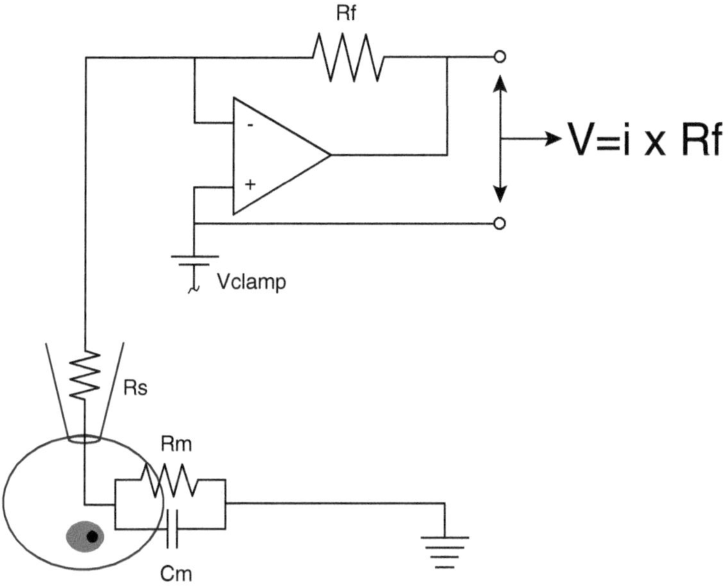

FIGURE 8.2 A schematic diagram of the fundamental circuit for the voltage-clamp recording. The most essential component is the current-to-voltage converter of the headstage. Current flow measured across a resistor of high impedance (Rf) causes a voltage drop that is proportional to the measured current through the pipette. An operational amplifier automatically adjusts the voltage to maintain a constant pipette potential at a desired reference potential (Vclamp). Hence, the membrane potential is held constant (i.e., voltage-clamped) although its resistance and current flow may change. Because this is a single-electrode voltage-clamp technique, the pipette introduces a resistance that is in series (Rs) with the cell membrane resistance (Rm).

(Rf). This feedback resistor is often switchable on patch-clamp amplifiers; for example, a 500 MΩ feedback resistor allows measurement of currents up to 20 nA and 50-GΩ feedback resistor is used for single-channel currents. In the current follower circuit, a single electrode is used to both pass current and record the voltage simultaneously. Therefore, the voltage-clamp is not across the membrane alone, but also across the series resistance (i.e., the pipette acts as a voltage divider). For small cells, Rm $\gg$ Rs; but for large cells, Rm is close to Rs. As will be described, Rs limits the speed and steady-state accuracy of the voltage-clamp.

8.3.4 Compensating for Capacity and Series Resistance

After achieving whole-cell recording conditions, adjustments should be made to optimize the voltage-clamp of the membrane. These include both adjustments to improve the accuracy and speed of the voltage-clamp, as well as cosmetic changes. The two most significant adjustments include compensating for circuit capacitance and for the series resistance.

8.3.4.1 Capacity Compensation

With every change of voltage, the circuit capacitance will change, either charging or discharging depending on the polarity of the voltage step. This capacitance arises from several sources, most notably from the pipette and the cell membrane. Current flow through capacitors is unavoidable and no compensation protocol will eliminate this current; at best, it can be cosmetically hidden. However, compensation can serve more than a cosmetic function. It is possible for this current to overwhelm the circuitry of the patch-clamp amplifier. Feedback current through the headstage of the amplifier is limited (depending on the size of the feedback resistor) and electronic compensation of capacitive current can reduce the current flowing through the headstage and prevent saturation from occurring. It is important to emphasize that while capacity compensation might eliminate the current following through the headstage of the amplifier, it does not eliminate the current flowing through the membrane. Therefore, there is no improvement in the performance of the clamp, either its speed or accuracy (both of which are limited by the series resistance).

Capacity compensation is generally offered on the amplifier in two stages: (1) electrode capacitance compensation, performed prior to formation of the gigaseal; and (2) whole-cell capacitance, performed after entering the whole-cell mode. These capacitive currents are easily distinguished based on their time-courses, the former with a much faster time constant than the latter. Pipette capacitance compensation is performed just after the pipette has been placed in the bath. It is obvious at the make and break of a series of square-wave pulses, applied to monitor the pipette resistance. Pipette capacitance is not in series with any resistance because it represents a stray capacitance to ground. However, uncompensated pipette capacitance can have a detrimental effect on the stability of the series-resistance correction circuitry and hence should be reduced. Many amplifiers supply fast magnitude (mag) and time constant (τ) controls as well as slow magnitude and time constant controls. The mag controls compensate for the net charge and the τ control adjusts for the time constant. By properly setting the fast and slow magnitude and fast and slow τ controls, a current can be induced in a capacitor connected to the headstage input to exactly equal the pipette current. In this case, no current needs to be supplied by Rf and there is no transient on the output. The fast controls compensate for a major part of the pipette capacitance. However, a small amount of the total capacitance can only be represented as a capacitor in series with a series resistance component. This takes longer to charge to its final value, due to the series resistance, and is compensated by slow controls. These controls are iteratively adjusted. Again, these controls do not eliminate capacitance in the bath, but do reduce the capacitative current flowing through the headstage. The actual capacitance can only be reduced by minimizing the length of the pipette inserted into the bath solution or by thickening the wall of the glass pipette, effectively eliminating its capacitance, such as by slyguard coating.

The next step in capacitive compensation occurs when the membrane capacitance has been added to the circuit after breaking into whole-cell mode. Cell membrane capacitance is larger and slower than electrode capacitance. Its RC time constant is determined by the product of the membrane capacitance and the series resistance*. It is impossible to record meaningful currents during this transient because the membrane potential has not settled. Obviously, one wants it to be as brief as possible, thus minimizing this dead time. Because the cell membrane capacitance is relatively invariant, it can only be reduced by minimizing the series resistance. And because the membrane capacitance is in series with the series resistance, it is iteratively compensated by use of both the series resistance compensation and the whole-cell capacitance compensation controls. This transient is commonly suppressed in patch-clamp recording, although this does not improve the meaningfulness of this time period. Its compensation reduces current flow through the headstage and is cosmetic to the output. Additionally, it supplies a numeric measure of the membrane capacitance, which in itself is a very useful number.

8.3.4.2 Series Resistance Compensation

The single most important consideration used to assess the quality and accuracy of voltage-clamp when recording in the whole-cell configuration is unquestionably the error introduced by the series resistance. This error is an inherent problem of the continuous single-electrode voltage-clamp used in whole-cell recording. Series resistance error arises because the voltage-clamp occurs over two major resistances — the pipette resistance and the cell membrane resistance — that, in terms of an equivalent electric circuit, are in series with one another. Unavoidably, any voltage drop that is occurring between the two silver/silver chloride electrodes will do so across both of these resistances. This means that the clamped potential is never the same as the membrane potential because some voltage drop also occurs across the pipette series resistance. It is worthwhile to emphasize that a series resistance error is always present in this technique (because one cannot eliminate the electrode from the circuit, one cannot eliminate the series resistance error). It is the task of the experimenter to determine the extent and significance of this error.

The extent of this error varies from preparation to preparation because the magnitudes of these resistances will change from electrode to electrode and from cell to cell. The series resistance error will introduce characteristic artifacts into the recorded currents that affect both the magnitude and the time-course of the voltage-clamped current. These artifacts occur in opposite directions for inward and outward currents. Inward currents have enhanced magnitudes, left-shifted current-voltage relationships, and unusually fast activation. Outward currents are the opposite. Recognizing these artifacts is fundamental to producing data of acceptable quality.

Technically, the series resistance includes all resistances between the two silver/silver chloride electrodes that are in series with the membrane. These include the resistance of the bath solution (and salt bridge, if any), the resistance of the cytoplasm, the resistance of the electrode, and the resistance associated with the

* The true time constant is $\tau = [(R_S R_m)/(R_S + R_m)]C_m$ but because in general $R_m \gg R_S$, the equation is roughly equivalent to $\tau = R_S C_m$.

pipette solution. Of these, only the pipette resistance is of real consequence. However, the series resistance is not equal to the pipette resistance. In fact, it is commonly up to twice as high. There may be various reasons for this discrepancy, including partial occlusion of the pipette tip in whole-cell mode. Lower-resistance pipettes tend to have less dramatic increases in the series resistance.

It is worthwhile to consider in some detail the effects that the series resistance error will have on both inward and outward currents. In addition to affecting the ionic currents, series resistance also affects the capacitive transient because the capacitance charges with a time constant of RsCm. In cases of high series resistance, the transient will appear as a slower decaying trace; this feature may be useful in diagnosing preparations with high series resistance. This error is negative when the current is outward (such as a potassium current) and positive when the current is inward (such as a sodium current). To understand the basis for these errors, it is important to emphasize that the membrane potential is not equal to the clamped potential and voltage-sensitive ion channels respond to the membrane potential, not the clamped potential.

In the case of outward currents, such as potassium currents, for a suprathreshold depolarizing step potential (e.g., –70 mV to +40 mV), the membrane potential Vm is smaller than clamped potential Vclamp (the true Vm is less than 40 mV) and therefore fewer channels open than expected. Additionally, when these potassium channels open, their driving force is reduced by the series resistance error (the driving force is Vm – Vk, a result of the membrane potential and the potassium reversal potential, Vk). Both these factors act to make the potassium current artifactually small. Hence, the current is smaller than expected because the channels are actually responding to a lower voltage step than expected. These deviations become larger as the currents become larger. The current-voltage relationship is shifted to the right, i.e., more voltage is needed to evoke an equivalent amount of current.

Series resistance errors are more serious when dealing with inward currents. An analogous situation occurs for a depolarizing membrane potential intended to evoke inward currents, such a sodium current. For a typical step potential to evoke sodium currents, such as –70 to –30 mV, Vm is again not equal to Vclamp because some of the voltage drop occurs across the series resistance. Hence, Vm < –30 mV (i.e., more toward 0 mV). For sodium channels, such a step is more effective at opening channels, and more channels open than expected. Additionally, because in effect a larger command potential is felt by the membrane with more sodium channels opening, the I-V relationship is distorted and shifted to the left, and less voltage is required to evoke an equivalent amount of current. Unlike potassium currents, whose I-V plot appears merely attenuated by series resistance error, and hence not easily recognized, there are characteristic deformations of the I-V plot for sodium currents that unmistakably bellow series resistance error to the educated electrophysiologist. The abrupt drop in the current-voltage plot from below threshold to maximum inward current within a 10-mV increment of command potential is a clear sign of high series resistance error.

Series resistance compensation employs positive feedback to attempt to compensate (often called correction) for the voltage drop across the series resistance.

Essentially, the desired membrane potential is achieved by supplying a proportionately larger clamp potential. This form of compensation is very helpful but it is limited. Because this is positive feedback circuitry, as one approaches 100% compensation, the amount of supplied current approaches infinity and is hence unachievable. Additionally, the feedback is positive so that the stability of the circuitry cannot be maintained; as compensation approaches 100%, the circuitry becomes an oscillator. Compensation of 80% is common.

8.3.5 Acquiring Data

8.3.5.1 Filtering Signals

The electric signal that is recorded is the product of a variety of sources. These include not only electrical properties of the cell membrane and the recording electrode, but also thermal noise of the electric components of the amplifier, high-frequency noise picked up by the probe from neighboring instruments, and even radio waves. Amplifiers pick up and amplify all signals, not just the signals of interest. Filtering allows one to pull a desired signal from the noise. Because filtering is inexorable, a basic knowledge of its mechanism is required for acquiring meaningful data.

In point of fact, data are filtered in controlled and uncontrolled ways. In practice, the purpose of filtering is essentially to reduce the amount of high-frequency noise, and so a low-pass filter is employed. (A low-pass filter attenuates high frequencies, allowing low frequencies to pass unaltered; a high-pass filter does the opposite.) By eliminating high-frequency signals, low-pass filters help to reduce background noise, thus increasing the signal-to-noise ratio. It is desirable to create a window of frequencies that are of interest, filtering signals that are higher or lower. (In practice, for whole-cell recording, the lower limit of this window is simply DC current and only high frequencies are filtered.) Such a window is referred to as the bandwidth. The sharpness of the borders of this window, called the cutoff frequencies, varies with the type of filter. The cutoff frequency is defined as the frequency at which the power of the output is reduced by one half. Colloquially, it is referred to as the –3 db frequency because:

$$dB = 10\log_{10}\frac{P_{out}}{P_{in}} \tag{8.1}$$

And when $P_{out}/P_{in} = 0.5$, db = –3.01. Thus, attenuation of the amplitude by –3db corresponds to about half the input magnitude. Above the cutoff frequency, there is increasing attenuation with increasing frequency. Cutoff frequencies will vary with application; for example, single-channel data, with events much faster than whole-cell currents, require less filtering at the high end but more at the low end to prevent shifts in the baseline.

Filters can be either analog or digital. In the vast majority of cases, data are run through an analog low-pass filter built into the patch-clamp amplifier prior to digitization. Off–line filtering can be performed, although it is time-consuming, requires increased data storage, and provides little useful information. Typical amplifiers provide low-pass filters with –3db settings in the range of 0.5 to 20 kHz. Intrinsic filtering also occurs by the preparation itself, over which the experimenter has very little control. This intrinsic filtering occurs due to the RC components of the pipette and cell. The series resistance and the membrane capacitance actually form a single-stage RC filter that significantly affects the true time-course resolution of a recording. For a taste receptor cell, with a membrane capacitance of 6 pF and a 20-MΩ series resistance, these RC components, considered in isolation, have a –3-db cutoff frequency; that is:

$$F = \frac{1}{2\pi \cdot R_s \cdot C_m} \tag{8.2}$$

of 1300 Hz. Fortunately, this is much higher than biologically interesting signals. Typically, more reduction of the bandwidth is necessary with a low-pass filter that is built into the amplifier. Analog filtering of the amplifier filters the entire bandwidth, including all signals that might be external to the preparation. Several types of filters might be appropriate for this extrinsic filtering. Butterworth or Tschebycheff filters have the steepest attenuation slope and provide the best reduction in high-frequency noise; and although they are superior in the steepness of their cutoff frequencies, they may distort a step pulse by producing an overshoot ring or a damped oscillation. The most common choice is a Bessel filter (4 or 8 pole) with the slope of the attenuation proportional to the number of poles in the filter. Most common are four poles; eight poles are more expensive but more efficient. Cutoff frequencies of 1 to 10 kHz are typical for whole-cell recording.

Because analog filters are most often applied, data are digitized after filtering. The sampling rate refers to the frequency of the digitization. For example, a sampling rate of 10 Hz would produce 10 data points to account for 1 sec of data. Obviously, this might be satisfactory to describe events on a time scale of minutes, but could not reproduce an action potential. The sampling theorem — known as the Nyquist theorem — states that data should be sampled at a frequency twice the bandwidth of the signal in order to prevent an artifactual increase in the noise by a phenonmenon known as aliasing. In practice, data are sampled at much faster rates, 5 to 20 times the –3 db frequency, sometimes called oversampling. Sampling rates will vary, depending on the duration over which the current is measured. For example, the brief time-course of sodium currents, measured over a 10-msec period, would be sampled at a faster rate than potassium currents measured over a 500-msec period. Overly fast sampling rates for the current of interest simply produce unnecessarily large data files without adding any additional meaningful information.

8.3.5.2 Leak Subtraction Protocols

Inevitably, whole-cell recording occurs with an imperfect seal. Some current flow occurs through the seal resistance in addition to the current flow through the membrane. Unlike membrane current, there is nothing selective about current flow through the seal. It is linear, i.e., ohmic and nonrectifying in both directions. Because of these features, it is possible, if a region of membrane voltage can be found that is devoid of any active currents, to measure this current and subtract it from the recorded currents. Indeed, most software programs contain subtraction protocol routines to enable the investigator to more purely isolate active, voltage-gated currents from passive linear currents. To optimize these protocols to a particular cell preparation, these protocols employ user-defined variables. A series of leak subtraction voltage pulses is evoked and summated prior to the command step potential that is intended to evoke the active currents. To minimize leak subtraction pulses evoking any active currents, it is common to divide the leak subtraction pulse into four sub-pulses, each of one quarter amplitude. The current evoked by these four pulses is summated and electronically subtracted from its corresponding command potential. Because the leak current is ohmic and recorded over a region of the I-V plot that does not evoke any active currents, the leak subtraction pulses may be of either polarity. Commonly, they are of opposite polarity to the command potential; however, for cells with large inwardly rectifying currents (such as K_{IR}), the pulses may be of the same polarity as the test pulse. In general, a settling time is also allowed, for example if the cell has a long time constant. Leak subtraction pulses are often executed before the waveform. Their total approximates the leakage component for that command voltage. This procedure does not improve the quality of the recorded signal; it merely provides cosmetic enhancement.

8.3.6 Whole-Cell Recording and Isolating Currents

The voltage-clamp technique was invented not only to be able to directly measure the ionic current that flows across the membrane, but also to measure isolated individual components of that total current. The ability to isolate currents is one of the great strengths of the voltage-clamp technique. Currents are isolated through various manipulations that include voltage, pharmacology, ion substitution, and kinetics. Most currents can be isolated in a pure enough form to allow their biophysical characterization; some, however, are more recalcitrant. To isolate a current of interest, currents arising from channels other than the channel type of interest must be eliminated from the current record. Leakage and capacitative currents can be removed as previously described, although these are merely cosmetic changes. Currents arising from non-desired ion channels are removed at the source; their permeant ion may be removed, they may be blocked by pharmacology, or inactivated through manipulation of the membrane voltage. See Box 8.1 for an example of how to isolate voltage-dependent sodium currents and Box 8.2 for an example of how to study modulation of voltage-dependent potassium currents by cyclic nucleotides.

Taste receptor cells of the rat express a variety of different types of ionic currents. These include sodium currents,[18] calcium currents,[3] several types of potassium currents,[8] chloride currents,[19] inwardly rectifying potassium currents,[45] and possibly cyclic nucleotide gated currents.[30] To record only one of these currents, one must isolate it to the exclusion of all other conductances within the cell. To do so, varying strategies, often used in amalgamation, are employed. A combination of ion dependence, kinetics, pharmacology, and membrane voltage can be used to best isolate a particular type of voltage-dependent current.

BOX 8.1
Isolating Voltage-Dependent Ionic Currents
(From Herness, M. et al., *Amer. J. Physiol. Cell Physiol.*, 272, C2005, 1997. With permission.)

Although it is always of interest to record whole cell currents from a cell, these currents represent the algebraic summation of all the active conductances in the cell. For a variety of reasons, it is useful to isolate a particular conductance from all the active conductances in the cell. This not only helps to decipher the underlying basis of the electrical excitability of the cell, but it also allows study of the basis of modulation of this excitability, since individual classes of ion channels may be modulated (*e.g.*, by phosphorylation) within the cell. This example illustrates how voltage-dependent sodium currents are isolated from the whole cell currents.

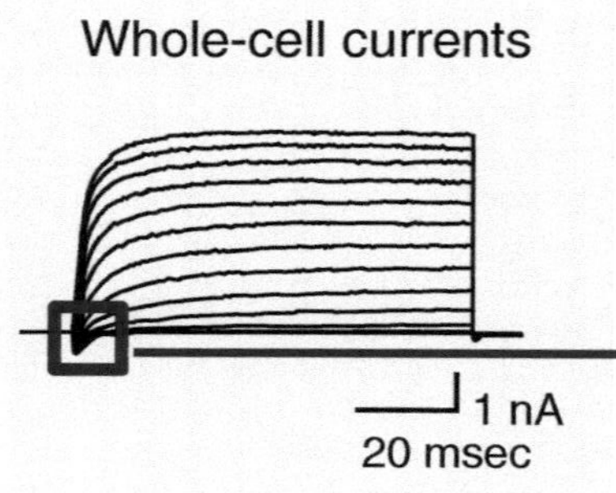

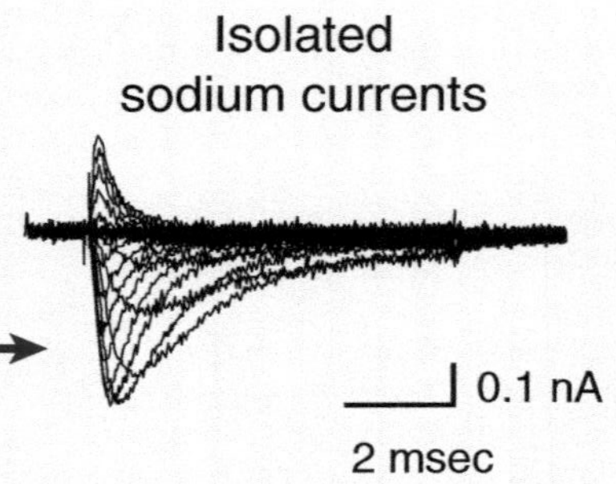

By using different recording solutions for both the bath and the pipette (see TABLE 1), the potassium conductances may be blocked, leaving the sodium current intact. The voltage protocol is then combined with a leak subtraction protocol resulting in nicely isolated sodium current. Note the difference in time course and current magnitude between the two current records. As well, the difference in sampling rate between the two figures more faithfully reproduces the true time course of the sodium currents. The outward portions of the sodium currents where previously obscured by the outward potassium currents. Both the kinetics of the sodium currents, its current voltage relationship, and its reversal potential could not have been accurately determined from the whole-cell records. (For a more detailed description see Herness and Sun, 1995).

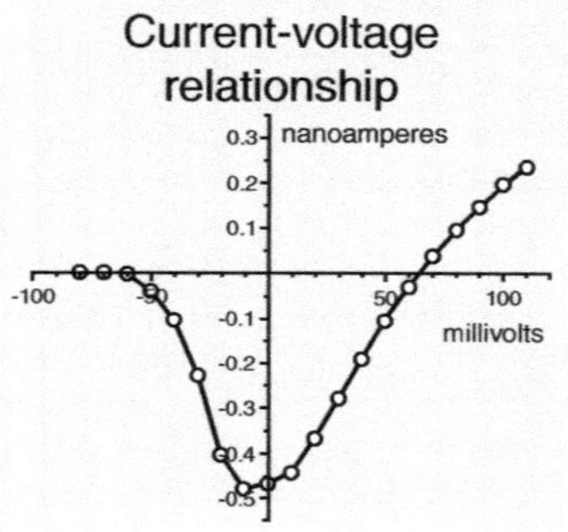

The current-voltage relationship of this current for this cell accurately identifies its voltage dependent. Typical features for sodium currents from taste receptor cells include activation around -50 mV, peak currents occurring at -10 mV, and a reversal potential between +60 and +70 mV. These data may be used to calculate the conductance of the sodium current (g), represented at right (solid circles).

$$g = \frac{I}{V - V_r}$$

A Boltzman relationship is used to describe these data (solid line). A decline in conductance at very positive potentials from its asymptotic maximum of 8.3 nS is common in most cells. It may be due to the inability of single ion channels to obey the constant field equation at high potentials.

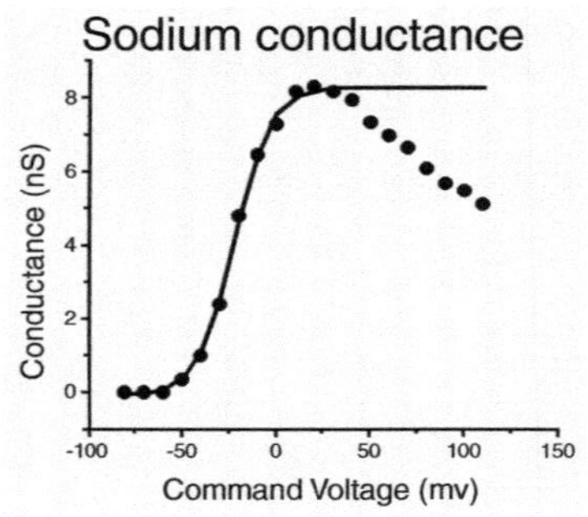

BOX 8.2
Studying Second Messengers with Patch-Clamp Recording
(From Herness, M. et al., *Amer. J. Physiol. Cell Physiol.*, 272, C2005, 1997. With permission.)

cAMP Inhibition of K^+ Currents is Dependent Upon PKA Phosphorylation

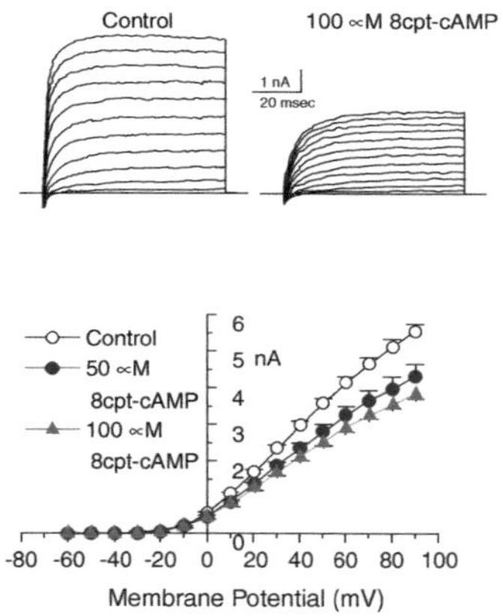

cAMP inhibits K^+ currents

The whole cell configuration of the patch clamp technique is useful for studying the effects of second messenger systems. This example illustrates how the effect of cAMP on potassium channels is investigated. In the panel at left, outward potassium currents are recorded prior to and during the application of the membrane permeant analog of cyclic AMP, 8-cpt cAMP. Application of this drug causes inhibition of the outward potassium current.

This raises the question of how cAMP acts to inhibit potassium channels. In the panel at right, the specific inhibitor of protein kinase A (PKA), PKA inhibitor (PKI), is applied to the intracellular compartment by adding it to the pipette solution (ICF) and allowing it to diffuse into the cell after forming whole cell configuration. Data in the

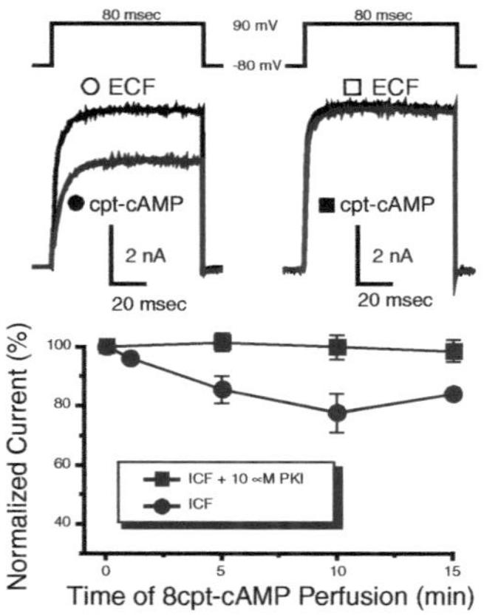

cAMP requires PKA to inhibit K Currents

panel at the left show that the 8cpt-potassium current (solids circles) when normal ICF but is ineffective in (solid squares) when the ICF contains concluded that PKA is necessary for the current. This proposed mechanism is panel at right. See Herness *et al* 1997

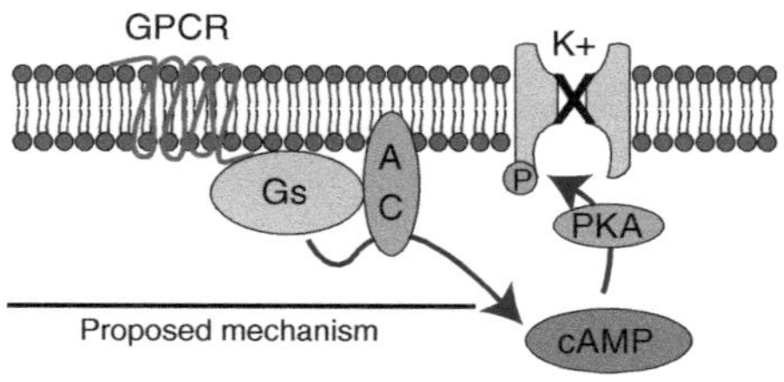

Proposed mechanism

cAMP effectively inhibits the pipette solution contains inhibiting the potassium current PKI. From these data it is actions of cAMP on potassium illustrated schematically in the for more details.

A simple example of kinetic isolation involves measuring sodium and potassium currents at different times after the start of a command potential. For example, sodium current is completely inactivated within milliseconds of the start of the command pulse, prior to the development of potassium current. This simple difference in kinetics provides an easy first-pass manipulation for examining these major currents. However, one should caution that for detailed characterization, these currents must be more properly isolated.

Ionic substitution was one of the first methods used to isolate currents. For example, if a channel were permeable to sodium, shifts in the external sodium concentration, a relatively easy maneuver during a recording session, should predictably alter the magnitude and, perhaps more importantly, the reversal potential of the recorded current. Sodium could be replaced by a large impermeant cation, such as choline or N-methyl-D-glucamine, to eliminate the sodium current. Specific pharmacological agents provide powerful tools. It may be possible to block all the currents except the one of interest, or a blocker for the specific current may be used and pre- and post-records subtracted to obtain the current of interest. For example, the TTX-sensitive current could be used to extract sodium current from total ionic current. Similarly, subtraction of charybdotoxin-sensitive current provides a record of the small-conductance, calcium-activated potassium current. See Table 1 of Standen et al.[44] and Table 1 in Sontheimer[42] for commonly used blockers of ion channels. These observations lead to the preparation of custom ionic solutions that favors a particular current of interest. Dialysis of the intracellular milieu with the patch intracellular solution is an advantage for isolating currents. Combining specific ionic compositions of the ECF and ICF with additional pharmacological agents is the usual strategy for isolating a given current. Some sample recipes are provided in Table 8.1.

Isolating subsets of channels may be more problematic. For example, outward potassium currents may consist of delayed rectifier channels, transient potassium current, and calcium-activated potassium current. The particular voltage dependence of certain classes of voltage-sensitive channels can be exploited toward the purpose of current isolation. The voltage dependency of calcium-activated potassium current, for example, is different from that of transient potassium current. Calcium-activated potassium current may be enriched if evoked from a holding potential of –50 rather than –80 mV because the more depolarized holding potential causes inactivation of transient (A-type) current. Low- or high-threshold calcium currents can similarly be activated separately by different holding potentials.

8.3.7 Perforated Patch

Soon after the development of the patch-clamping technique, it became apparent that some currents quickly diminished in magnitude or disappeared after the establishment of the whole-cell configuration. This phenomenon was termed run-down. It is thought that some electrophysiological responses, such as many calcium currents, depend on the integrity of cytoplasmic constituents involved with cellular metabolism (e.g., those dependent on second messenger systems). In this sense, exchange between the pipette contents and the cytoplasm is a two-edged sword that cuts on the fore and aft. On the one hand, it greatly facilitates the control of the ionic composition of the cytoplasm and the introduction of pharmacological agents to the cell's interior, both a great advantage for many recording situations. At the same time, it similarly leaches endogenous agents from the cell, a great problem for some currents.

A modification of the whole-cell recording configuration was developed to dramatically delay the onset of run-down. This technique greatly reduces the exchange between the cytoplasm and pipette, and hence keeps the cytoplasmic constituents intact. It does so by entering whole-cell recording without rupturing the portion of the cell membrane that occludes the pipette tip. Rather, it selectively permeabilizes this patch of membrane with the insertion of antibiotic channels — either nystatin or amphotericin — thus allowing electrical connectivity without the exchange of larger molecules. Several excellent reviews of this technique are available.[21,27,38,49] In overview, after establishing a membrane seal, no attempt is made to rupture the occluding patch of membrane. Rather, the highly lipophilic ionophore channels, present in the pipette solution, insert themselves into the occluding patch by diffusion. This establishes electrical contact without disruption of the membrane. In our experience, amphotericin is superior to nystatin for taste receptor cells.

To be successful for whole-cell recording, these channels need to fulfill several criteria. The pores need to create an electrical conduit between the pipette and the cell interior without allowing exchange of larger molecules, which would defeat its purpose. These ionophores are largely nonselective monovalent cation channels that are impermeable to molecules larger than the size of glucose. There is little selectivity for sodium over potassium; however, the cation selective pore is impermeable to divalent ions such as Ca^{2+} and Mg^{2+}. Thus, second messenger processes dependent on metabolic constituents such as ATP or GTP or calcium are not disrupted. Importantly, these ionophores show little voltage dependence and will therefore not add any rectifying properties to the recording current, which would otherwise render them unsuitable. Another important criterion for the success of this technique is that there be no lateral diffusion of the antibiotic outside of the patch membrane. The technique would be inoperable if there were. If the channels were able to diffuse outside of the patch, not only would voltage-clamp be impossible with the large leak conductance added to the membrane, but the cell would ultimately be killed (which is, after all, their evolved purpose). Nystatin does not cross the membrane, and therefore does not enter the cytoplasm. Although these channels are monovalent cation channels, they do possess a finite permeability to monovalent anions as well. This permeability can create a Donnan potential based on the chloride gradient across the membrane and the nonpermeable anions contained inside the cell. To minimize such artifacts, the chloride concentration of the pipette solution is reduced and replaced with an impermeant anion such as SO_4^{2-}. Typically, the pipette solution contains about 71 m*M* chloride and 75 m*M* SO_4^{2-}.

In practice, the presence of these ionophores in the pipette solution causes the formation of gigaseals to be problematic. It is thought that diffusion of the ionophores out of the pipette prior to the formation of a gigaseal destabilizes the membrane. The usual strategy to overcome this problem is to fill the tip of the pipette with standard intracellular solution and backfill the shank with ionophore solution. A gigaseal may be formed as usual and, within several minutes, diffusion of the ionophore to the tip of the pipette allows channel insertion to occur. This channel insertion is observed by continuously monitoring the seal resistance with a series of square-wave pulses and observing the gradual addition of cell membrane capacitance to the circuit. The values of the series resistance and cell capacitance will plateau, after which the recording session can begin.

Because these ionophores are so highly lipophilic, the preparation of stock and working solutions is a bit tedious. Additionally, absorption of the ionophore from solution occurs so that these solutions are typically prepared fresh every few hours. Detailed protocols are available.[21,27,38,49] Because the high series resistance is always a concern with the perforated patch technique, the use of large tipped pipettes (and hence lowest resistance) is encouraged. Tips that are too large can result in complete disruption of the membrane patch and need to be empirically determined. In our experience, tips with resistances of 5 to 7 MΩ work well.

The perforated patch technique is severely limited by the inherently high series resistance. As previously outlined, such high series resistance has deleterious effects on the biophysical characteristics of currents, particularly inward currents. For this reason, it is unsuitable for characterization of sodium currents. The series resistance problem is variable, particularly because the insertion of channels is not a controlled event — a plateau of the series resistance can occur quickly or continue to change throughout the recording session. A great advantage of this technique is that recordings can occur for greatly protracted periods when compared to standard whole-cell recording. Whereas the latter can typically last for 20 to 30 min, perforated patch typically last for 2 hr. In itself, this makes perforated patch a highly desirable method for the repeated application and washout of stimuli or pharmacological agents to the cell.

8.3.8 Running an Experiment: Appropriate Controls

Given the many technical aspects of patch-clamping that must be considered for the fruition of a successful experiment, it is not surprising that aspects of experimental design are sometimes lacking in discussions of the technique. Because every experiment is unique, replete with its own feats and foibles, universal statements regarding the proper design of a patch-clamp experiment are difficult to make. However, it cannot be overstated that each experiment is only as good as its best control. Thus, a brief discussion regarding control experiments in patch-clamp design should help in the goal of compelling data collection.

One of the most serious variables to control is time. Inevitability, the quality of the recording will deteriorate during the recording session. The experimenter must ensure that data collection occurs during a period of acceptable quality. A simple and essential experiment to perform is to simply monitor current magnitude over time. With whole-cell recording of dissociated taste receptor cells, it should be possible, under good conditions, to record a stable magnitude for periods of 30 to 60 min. The experimenter should routinely perform this control experiment until convinced of the quality of the recordings and familiar with signs of deterioration. After that, a current magnitude can be monitored as an internal control during an experiment. In practice, cells often deteriorate rapidly, changes in the input resistance can be quite sudden (e.g., when losing the seal), and current magnitudes are swiftly distorted.

In addition to time, isolated currents should be tested for their purity. Ion substitution experiments should have predictable changes in the reversal potential and current magnitude. Isolated currents should demonstrate appropriate pharmacology. Finally, for drug experiments, the experimenter should search for inactive enantiomers as well as test for repeatability across cells.

8.4 CALCIUM IMAGING

It is not surprising that a cell whose evolved function essentially rests entirely within the domain of signaling would utilize, among other second messenger systems, such a ubiquitous signaling system as calcium. Such is the case for the taste receptor cell. It is becoming apparent that taste receptor cells employ dynamic changes in the intracellular calcium concentration in both signaling and regulatory mechanisms. Stimuli such as sweet, bitter, neuropeptides, and neurotransmitters have been demonstrated to produce large and transient increases in intracellular calcium.[1,5,12,28,35,36,50] Taste receptor cells appear to exert rigid homeostatic control over calcium and it is very probable that the usual regulator mechanisms — calcium extrusion systems (Ca-ATPase, Na/Ca exchange), calcium influx mechanisms (voltage and ligand channels), release and uptake by intracellular organelles (endoplasmic reticulum and mitochondria), and buffering by calcium-binding proteins — are viable in taste receptor cells. The study of such calcium regulation in taste cells, however, is at a nascent stage. Unlike the patch-clamp analysis, the application of calcium imaging to taste receptor cell function (although no newer a technique) is relatively recent. A dilatory factor contributing to delayed exploration of this complex and essential signaling mechanism may lie in the acquisition cost of an imaging system.

As expected, because calcium is so essential in the signaling and regulatory mechanisms of the cell, its intracellular concentration is tightly regulated. In a typical cell, through mechanisms still under very active study, the intracellular concentration of calcium is held very low (20 to 200 n*M*) with respect to the calcium concentration of the extracellular fluid (0.2 to 1 m*M*). Intracellular calcium levels increase rapidly (within milliseconds) via two basic mechanisms of entry: from extracellular sources through plasma membrane calcium channels or from intracellular stores (such as in the endoplasmic reticulum) via IP_3-gated calcium channels. In either case, the driving force for calcium entry is very high and calcium can be elevated rapidly. Additionally, calcium-binding proteins in the cytoplasm restrict the free diffusion of calcium, thus allowing local alterations of cytoplasmic concentrations of calcium to produce local actions within the cell. Calcium regulation may therefore occur along both spatial and temporal dimensions in taste receptor cells.

The strategy for the optical measurements of such temporal and spatial fluctuations of intracellular calcium concentrations involves introducing a fluorescent dye into the cell's interior whose spectral properties are a function of calcium binding. A variety of such compounds exist (fura-2, fluo-3, calcium green, indo-1, to name a few). The parent structure of these compounds is a calcium chelator to which fluorescent substituent groups have been added. Screening of many such compounds produced some whose change in fluorescent properties as a result of calcium binding made them useful for dynamic monitoring of intracellular calcium fluxes. After several generations of indicators, the use of fura-2, because of its unique spectral properties, has become widespread. The following focuses on ratiometric analysis of taste receptor cells with fura-2.

8.4.1 Equipment

At the heart of experiments designed to monitor dynamic changes in intracellular calcium levels is the acquisition of a fluorescent image. To acquire such an image, a fluorescent microscopy station that allows excitation at dual wavelengths, perfusion of a recording chamber, and dynamic capture of an image under low light conditions is required. This necessarily involves a microscope with fluorescent objectives, an excitation component, an emission cube, a camera, and a computer. In general, such stations utilize an inverted microscope, due to its ease of use. The inverted design greatly facilitates physiological manipulations, such as stimulation of cells by exchanging bathing solution. A recording chamber can be easily adapted to the inverted stage (homemade or commercial designs) and should have an interchangeable bottom that uses a coverslip. The coverslip will provide the best optics for the acquisition of the image, providing both sharp focus and minimal filtering of UV excitation, and it can be coated (e.g., with poly-L-lysine or Cell-tak) to facilitate adherence of dissociated cells, a requirement for changes of the bathing solution while imaging. (The frustration of viewing the imaged cell's disappearance as it is whisked away by the change of solution lies in every imager's future.) Corrosion of the substage parts and mechanisms due to the occasional leaks of saline or culture medium is all but expected.

One of the most important considerations of the microscope, because optics are fundamental, is the objective. Standard objectives have low ultraviolet (UV) transmission and hence are not suitable for imaging. A fluorescent objective of high numerical aperture is required to ensure high efficiency in its light collecting (which varies with the square of the numerical aperture). Typically, this means having to deal with oil, but the optical advantages far outweigh its tedious if not messy character. As an example, a digital camera with a 1280 × 1024 photoarray combined with a 40× 1.3 numeral aperture yields a theoretical resolution of 0.15 micron per pixel, which lies at the limit set by light diffraction. A higher power objective (e.g., 100×), would therefore not improve spatial resolution of light-gathering power and would actually yield fewer photons per pixel; additionally, it would provide a reduced field of view, a disadvantage for the simultaneous measurement of several cells.

There are two procedures for producing UV excitation wavelengths: interference filters or grating monochromator. These methods differ with respect to both efficiency and cost. Both operate to limit the total emitted spectrum from a UV source to a desired wavelength. For calcium imaging, xenon arc lamps are superior to mercury as an excitation source because mercury lamps have prominent emission lines, especially at 360 nm, within the range of fura-2 excitation. (If a desired excitation wavelength coincides with an emission line, it will be more intense than expected.) Xenon, on the other hand, gives a relatively uniform spectral output. Mercury lamps do put out more total UV energy than xenon but not necessarily at the best wavelengths for calcium imaging. Moreover, a 75- or 150-watt xenon puts out more than enough energy required for fura-2 imaging. Interference filters or monochromators act to limit this total UV output to a desired wavelength. They operate by different mechanisms and therein lie their strengths and weaknesses. Interference filters are assembled in filter wheels and are alternated by spinning. Each filter allows only a certain bandwidth of the total UV to pass. Because the filter wheel must spin between two positions (e.g., one filter for 340 nm and another for 380 nm), wavelengths cannot be changed as rapidly as with a monochromator. Fewer wavelengths are available because each desired wavelength requires a different filter. Additionally, spinning of the filter wheel produces mechanical artifacts that must be mitigated. Newer models attach to the microscope via fiber-optic cables capable of passing UV wavelengths and hence the filter wheel (and its associated vibration) can be physically removed from the microscope. Interference filters are the less expensive of the two options, about 25% of the acquisition price of a monochromator. A monochromator, on the other hand, can produce a near-continuous spectrum of wavelengths. As its name implies, a monochromatic light is produced from a polychromatic source. Monochromators utilize a set of geometrically fixed slits at its entrance port (the chopper) and an internal dispersive element (the diffraction grating) that is turned. The angle at which it is turned determines the wavelength of the output light. The result is a versatile, selectable, monochromatic source of light that is freely adjustable in wavelength and bandpass. It has the added advantage of narrowing the rejection of stray light, which requires insertion of neutral density filters for filter wheels. Additionally, monochromators allow spectral scanning. This feature is highly desirable. Because all optics are imperfect, every setup possesses unique optical characteristics. Scanning the dye over a broad range of wavelengths allows the empirical determination of the actual peaks for an individual setup. It is a great advantage to know how the dye operates in a particular system because the experimental excitation wavelengths can be adjusted to compensate for imperfections in the optics. Such compensation is easy with a grating monochromator because it is equally facile at producing any desired wavelength, whereas new filters would be required for a filter wheel. The disadvantage of monochromators is their acquisition price.

The fluorescent light produced by excitation of a calcium-sensitive dye is dim and requires a low light level camera for its detection. These cameras can detect light at lower levels than the human eye. Whereas earlier calcium imaging literature commonly referred to the use of SIT (silicon intensified target) cameras, these have now effectively been replaced by CCD (charge coupled device) cameras. SIT cameras are analog cameras with low noise and good spatial resolution but with slow response to changes in image intensities. The CCD camera, on the other hand, is a digital camera that utilizes solid-state, charge-coupled light-sensing detectors to convert collected photons to collected electrons. For the most part, CCD cameras are not sensitive enough for low-light applications and one of two possible modifications is required to extend the sensitivity to low-light levels: a CCD camera preceded by an intensifier (ICCD) or a cooled CCD (CCCD). Intensifiers operate by boosting the incident light to detectable levels for a standard CCD camera. Intensifying the signal is not equivalent to increasing the signal because intensifiers increase the noise level as well (more specifically, they increase the signal shot noise but not the camera noise). CCCDs operate by cooling the array of photodetectors; this greatly reduces the inherent noise of the photodetectors (i.e., the dark current) and extends the signal-to-noise ratio to low-light levels.

The frame rate imposed by the readout of the CCD camera is one of the limiting factors for the speed of data acquisition, which can vary from second to millisecond time scales. CCDs operate with a two-dimensional photoarray of independent potential wells, with each well corresponding to a pixel. When a CCD imager is exposed to light, charge accumulates in the potential wells, proportional to the light intensity and the exposure time (called integration). A programmed sequence, known as readout, causes all the charge packets in the photoarray to be shifted toward a serial register, then shifted out of the serial register toward the output amplifier, where accumulated electrons are read. This process requires time. The fastest frame rate for a CCD is calculated from the total number of pixels and the readout speed. To facilitate acquisition, CCD cameras have varying architectures, either full frame, frame transfer, or interline transfer. These architectures transfer the array either to a second parallel array for readout while the original array is integrating, or from a single line to a second parallel line. Frame transfer or interline transfer thus have portions of the photoarray that are protected from light, resulting in a reduction of the total pixel size and the resolution, with an increase in the frame readout speed. Another variable manner of improving frame transfer is by binning a pixel with adjacent pixels. The total signal can be read out at one time, and the signal increases with increases in readout noise. This produces higher sensitivity (a larger collecting area for a single pixel) at the expense of spatial resolution. For a cooled CCD with 1280 × 1024 resolution, frame rates of 9 to 56 Hz are possible using normal (i.e., zero binning) to 8 × 8 binning. Some cameras allow subarray scans, in which only a portion of the array is read from the CCD. This maintains the original resolution, reduces data size, and is useful for applications that require a higher frame rate. A final comment on cameras regards the quantum efficiency, which varies with wavelength (a phenomenon known as spectral response). Since fura-2 emits around 510 nm, it is essential to ensure high quantum efficiency at this wavelength.

8.4.2 Choice of Dye

Fluorescent dyes used for calcium imaging experiments fall into two broad categories: single or dual wavelength. The choice of dye is strongly dependent on the type of experiment to be performed. Each has spectral characteristics that can be exploited for different situations. The structure of these dyes is based on the tetracarboxylic acid chelator BAPTA with added side chains that produce fluorescence (Figure 8.3). The electron-withdrawing effect of a bound calcium ion significantly affects this fluorescence.

In response to changes in calcium concentration, single-wavelength dyes, such as fluo-3 or calcium green, produce changes in the intensity of the emitted fluorescence (fluo-3 is essentially nonfluorescent unless bound to Ca^{2+}), whereas its spectral

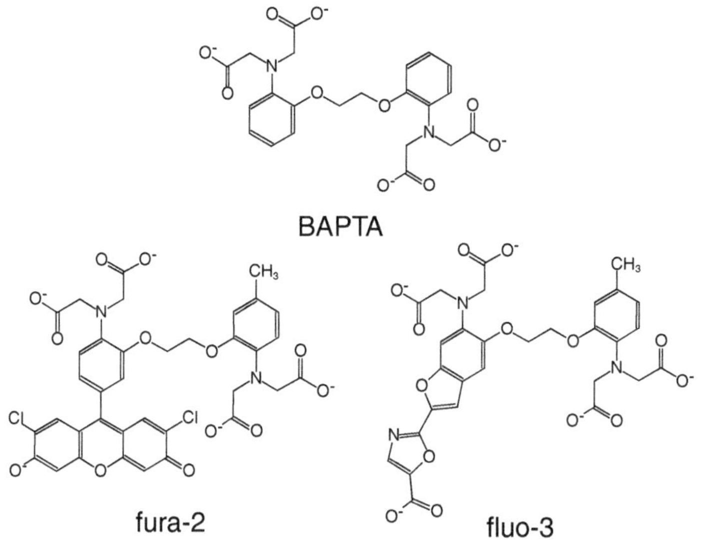

FIGURE 8.3 Structures of the fluorescent calcium indicator dyes fura-2 and fluo-3. These dyes are based on the parent structure of the calcium chealtor BAPTA to which fluorescent side groups have been added. The success of these compounds as calcium indicators is due to the electron-withdrawing power of a bound calcium ion that affects the fluorescent properties of the side chains. These compounds are not lipid soluble and hence do not readily enter the cell. The charged carboxylate groups are masked with ester protection groups, increasing the lipophilicity of these compounds and allowing easier cell penetration. The masked ester groups are cleaved by intracellular nonspecific esterases, trapping the dye intracellularly at concentrations much higher than the incubation solution.

maxima (i.e., the most effective wavelength for excitation) remains essentially unchanged. For example, fluo-3 is excited maximally at 506 nm but emits light at 525 nm several times brighter in the presence of calcium (Figure 8.4). Fluo-3 lacks significant shifts in emission or excitation wavelength upon binding to Ca^{2+}. Dual-wavelength, also called ratiometric, dyes are more complex and their advantages lie within this complexity. Ratiometric indicators such as fura-2 exhibit not only changes in intensity with changes in calcium binding, but also changes in their excitation spectra (Figure 8.4). Unbound fura-2 dye is maximally excited at 362 nm, whereas when bound to calcium, the excitation peak shifts to 335 nm. Moreover, in the nanomolar to micromolar range, the intensity at 335 nm significantly increases with increasing calcium concentration. Thus, for fura-2, significant shifts are observed in the excitation spectra but not in the emission spectra. Taking the ratio of intensity measurements at two wavelengths (generally chosen as 340 and 380 nm) becomes a good indicator of the amount of calcium bound, with an added advantage that the measurement becomes independent of many artifactual considerations which cancel in the ratio process.

When intensity is monitored only at a single wavelength, such as with fluo-3, factors unrelated to changes in calcium (e.g., changes in cell thickness, loss of

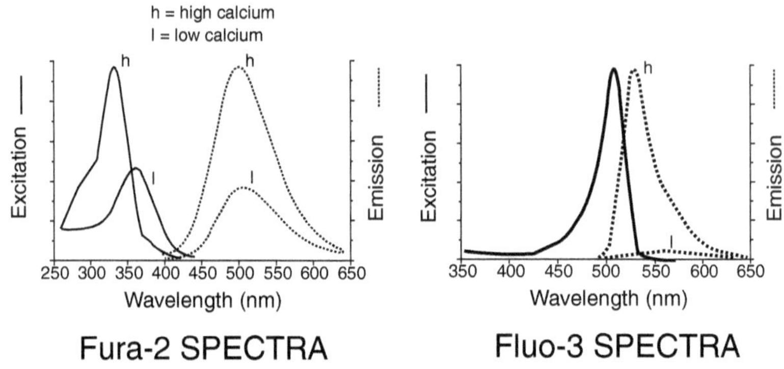

FIGURE 8.4 Schematic diagrams of the excitation and emission spectra for the dual-wavelength calcium indicator fura-2, and the single-wavelength calcium indicator fluo-3. Not only does the emission intensity increase for fura-2 with increasing intracellular calcium, but the most effective excitation wavelength shifts to a lower most effective wavelength as the calcium concentration increases. This shift in the most effective excitation wavelength forms the physical basis of the ratiometric procedure. Fluo-3, on the other hand, has the same excitation maximum at both low and high calcium. However, the intensity of the emitted fluorescence increases with increasing calcium concentration.

indicator from a cell by leakage) can confound interpretation of the data. For example, decreases in the amount of dye will decrease the intensity of emission independent of calcium concentration. A ratio measurement, using a dual-wavelength dye, minimizes the effect of many artifacts that are unrelated to changes in calcium. The calcium measurement becomes independent of dye concentration, path length, thickness of the specimen, and overall absolute level of illumination and sensitivity of the detector. Temporal measurements are obviously much slower with ratio measurements than with single-wavelength measurements because two measurements — each at a different point in time — are required. Thus, very dynamic changes in calcium are lost within the "bin" of the ratio process. As well, heavy divalent cations without unpaired electrons, such as Zn, Cd, and Pb, also bind fura-2 quite avidly. For example, the Kd values are in the nanomolar range for Zn and picomolar range for Cd.

8.4.3 Making the Fura-2 Solution and Loading Cells

8.4.3.1 Preparing Fura-2 Stock Solutions and Loading Cells

The process of introducing a fluorescent intracellular calcium indicator to the cytoplasm is referred to as loading. For loading to be successful, a water-soluble dye must cross the highly lipophilic barrier of the cell membrane. To accomplish this, the highly polar features of the dye, especially the negative charges of the carboxylates, are masked with temporary protecting groups that increase its lipophilicity. Once inside the cell, these protecting groups, which are ester substituents, are removed by intracellular esterases, returning the dye to its polar form and trapping it within the confines of the plasma membrane. Because the dye is trapped intracellularly, the final intracellular concentration is much higher than the incubation concentration. The protecting groups are mixed esters of ethylene glycol [$(CH_2)_2(OH)_2$], called acetomethoxy esters, and are abbreviated as AM. Because these membrane-permeant ester forms (fura-2 AM) are large uncharged organic molecules, they are of limited aqueous solubility. The solubility of fura-2 AM ester in pure water is 100 μ*M*,[25] and would be lower in the ionic strength of physiological saline. Fine particles of ester often adhere well to the outer surface of cells or to the extracellular matrix and can contribute to a calcium-insensitive fluorescent signal. To counteract this problem, a mild, nonionic, low-toxicity surfactant, Pluronic F-127, is commonly added to the AM mixture. Pluronic F-127 acts as a dispersing agent and is presumed to sequester AM ester in micellar form.

In practice, fura-2 and Pluronic F-127 stock solutions are prepared separately. Immediately prior to use, the two stocks are mixed and diluted with an ECF solution that contains either serum or bovine serum albumin (BSA). Serum proteins have a salutary effect on cells and can improve loading efficiency. The stock of the AM ester, due to its limited aqueous solubility, is dissolved in DMSO at a final concentration of 2 m*M* (but can range from 0.1 to 10 m*M*). This stock is aliquoted into convenient volumes (e.g., 25 μl aliquots will produce 10 ml final working solution) and can be stored in the freezer for months. The Pluronic F-127 stock is similarly prepared in DMSO at 20% concentration and can be stored at room temperature. Care should be taken to ensure that the DMSO is dry; when exposed to air, DMSO slowly absorbs moisture that can result in premature cleavage of the AM esters or precipitation of Pluronic F-127. (A solution of DMSO melts at 18°C, a solution that remains liquid at 0 to 4°C probably contains significant water.) Fura-2 AM and Pluronic F-127 stocks are mixed 1:1 (e.g., to produce 50 μl) and this mixture is diluted in 10 ml ECF that contains BSA at 10 mg/ml (or fetal calf serum). Sometimes, sulfinpyrazone or probenicid can be added to inhibit organic anion transporter activity, to minimize organic anion transporters that might cause internalization or extrusion of the dye.

Dissociated taste receptor cells can be incubated in loading solution for 15 min to 1 hr at room temperature. A more physiological temperature, such as 37°C, may speed up the loading process but will also increase compartmentalization (see Section 8.4.3.2). After incubation, cells are washed with normal ECF. Because the enzymatic processing of AM esters consists of sequential hydrolysis of several ester groups, and it is only when all AM groups are hydrolyzed that the molecule becomes properly Ca # sensitive, extra processing time should be allowed after rinsing. The inclusion of serum or bovine serum albumin (3%) helps remove unhydrolyzed AM dye bound to the cell surface or to the coverslip.

8.4.3.2 Assessing Compartmentalization

Ideally, fluorescent imaging is designed to measure dynamic changes in the cytosolic calcium levels. However, because fura-2 enters the cell in a membrane-permeant form, the dye is free to enter membrane-bound organelles as well. Some portion of the dye may be then be trapped within these organelles. These compartments (e.g., endoplasmic reticulum and mitochondria) have high calcium levels that saturate the dye and contribute a high fluorescence signal that is essentially independent of varying cytoplasmic levels. Internal endocytotic mechanisms may also contribute to compartmentalizing the dye to intracellular organelles.

The degree of internal compartmentalization can vary, depending on loading conditions and cell viability and therefore should be addressed in every preparation. A simple procedure that involves the sequential use of two agents that permeabilize the membrane — digitonin and Triton X-100 — can be employed. Digitonin primarily tends to permeabilize the plasma membrane, whereas the detergent Triton X-100 more readily enters the cell and can permeabilize the membrane of intracellular organelles. Assessing compartmentalization helps determine the amount of the fluorescent signal derived from sources independent of the cytoplasmic calcium levels. A simple experiment to determine compartmentalization involves measuring total fluorescence from a cell, followed by permeabilization with digitonin, and subsequently by Triton X-100. As the plasma membrane is permeabilized, the total cellular fluorescence decreases and a reticular pattern of staining, indicative of the compartmentalized dye, is observed. Further treatment with Triton X-100, which permeabilizes the cell membrane and membranes of the cellular organelles, should eliminate the remaining fluorescence. This test is performed in highly buffered, low extracellular calcium designed to mimic the resting intracellular calcium concentration (generally about 1 μ*M*). In this manner, membrane permeabilization does not result in a calcium influx that would cause the remaining fluorescent signal to be meaningless.[23]

Compartmentalization can be minimized by reducing temperature; however, pharmacological approaches are also sometimes used. Because endocytotic processes are temperature dependent, reducing the loading temperature tends to reduce compartmentalization. However, the enzymatic cleavage of the AM ester is also a temperature-dependent process. Extrusion mechanisms for organic anions can be blocked by inhibitors such as probenecid (millimolar range) and sulfinpyrazone (micromolar range). Both are hydrophobic organic acids, are practically insoluble in water, and must be neutralized with a stoiochiometric amount of base before an aqueous stock solution can be prepared. These drugs not only stop extrusion, but also can cause cellular stress such as blebbing. In general, slowing of compartmentalization by dropping the temperature is preferable. Additionally, blockers of intracellular compartmentalization, such as probenecid, may secondarily affect cellular function, such as calcium waves.

8.4.4 Illumination System Running an Experiment

8.4.4.1 Experimental Design and Controls

Many features of the execution of a typical imaging experiment are illustrated in Figure 8.5. In this example, a pair of dissociated taste receptor cells loaded with fura-2 are imaged prior to, during, and after application of the bitter tastant caffeine (10 m*M*). The change in color, measured with a ratiometric procedure, indicates that these cells respond to caffeine with an increase in cytoplasmic calcium levels. The top panel of images represents an overlay of the fluorescent intensities recorded with 380-nm excitation wavelength (presented in green) or 340-nm excitation wavelength

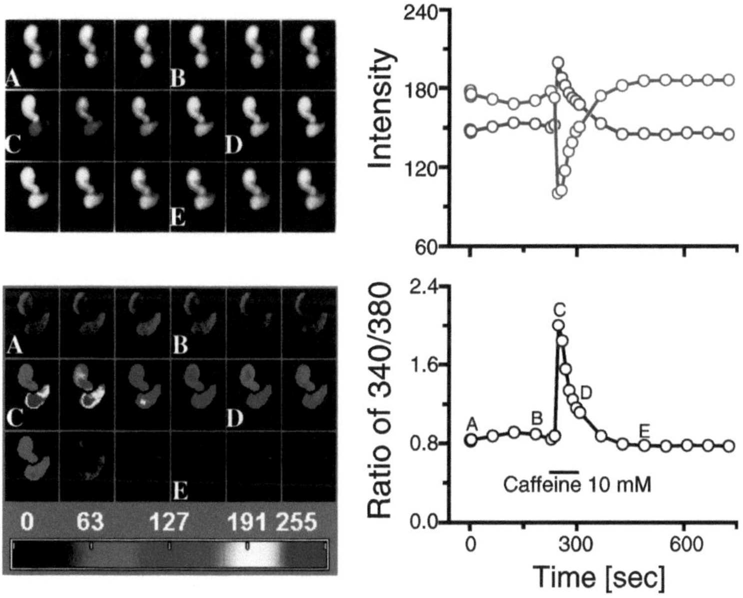

FIGURE 8.5 (see color insert following page 238) Example of a pair of dissociated taste receptor cells responding to the application of the bitter tastant caffeine with an increase of cytoplasmic calcium levels using ratiometric imaging with the calcium indicator fura-2. The upper panel of images represents an overlay of the fluorescent intensities recorded using 380-nm excitation (green) or 340-nm excitation (red). Notice that when the cell responds to the caffeine stimulus (marked "C"), the emission from the 340-nm excitation greatly increases while that of the 380-nm excitation decreases. The intensities of the individual wavelengths for the lower of the two cells are presented in the upper-right-corner graph. Note that when the stimulus is applied, the two wavelengths cross over. This causes the 340/380 ratio to increase above unity. The calculated ratios are presented in the graph in the lower right. Note how the cross-over of the 340/380 intensities and the increase in the 340/380 ratios coincide when caffeine is presented. The spatial localization of these changes in the ratio is presented in the montage of the cellular images presented in the lower left. Here, ratios have been calculated from the raw data and presented in a psuedocolored format as indicated in the calibration bar.

(presented in red). When the cells respond to the caffeine stimulus (marked "C"), the emission from the 340-nm excitation greatly increases, while that of the 380-nm excitation decreases. The intensities of the individual wavelengths for the lower of the two cells are presented in the upper-right corner graph. Note that when the stimulus is applied, the two wavelengths cross over. The individual exposures of these wavelengths had been purposefully set prior to the experiment so that their intensities were close enough to one another to allow this cross to occur during a response. This causes the 340/380 ratio to increase above unity. The calculated ratios for this cell are presented in the graph in the lower right of Figure 8.5. As expected, the crossover of the 340/380 intensities and the increase in the 340/380 ratios coincide when caffeine is presented. The spatial localization of these changes in the ratio is presented in the montage of the cellular images presented in the lower left. Here, ratios have been calculated from the raw data and presented in a psuedocolored format as indicated in the calibration bar. Using this psuedocolor format, blue represents low ratios and red represents high ratios. Many different arrangements of psuedocolor presentation are possible.

Monitoring a few salient variables during the course of an experiment helps to ensure the ultimate acceptable quality of the data. For example, it is important to monitor the individual wavelengths in real-time, not just the ratio. This will provide a useful monitor of bleaching of the photodye. The amount of bleaching is monitored by the downward tendency in the intensities at 340 and 380 nm. The ratio procedure cancels this photobleaching effect. As well, the capture speed should be varied during the experiment. Baseline measurements need not be obtained at fast acquisition rates because this will facilitate photobleaching. Adequate baseline should be recorded before any experimental manipulation, but images need only be captured on a minute, rather than second, time scale. During the manipulation, capture speed can be increased as determined by experimental need and limitations of the acquisition system.

Many of the experimental controls are inherent in the ratio procedure. For example, such variables as changes in the dye concentration, path length, cell shape, and bleaching of dye are controlled by the ratio procedure. Changes in any of these variables are roughly equivalent at the two wavelengths and hence cancel in the ratio procedure. However, it is also good to remember that although the ratio procedure cancels many variables, the two wavelength images are captured sequentially and thus do not reflect the status of the dye at the same time.

Procedures for background subtraction are software specific. In general, software should include the ability to select a region of interest that includes only background, subtract the portion of the fluorescence at each respective wavelength that originates from background, and produce a corrected ratio. Background is usually best performed off-line to conserve computer resources. Background subtraction helps to control for endogenous fluorescence of some stimuli (e.g., quinine).

Pharmacological controls relating to the drug of interest should also be employed. Inactive enantiomers or similar drugs with different targets are used to ensure that the effect is specific. For application of taste stimuli to isolated taste receptor cells, the same problems as with patch-clamping are involved (e.g., osmotic effects due to high concentrations) and need to be considered.

8.4.4.2 Estimating Intracellular Calcium Concentrations

Although the dynamic changes in ratio provide extremely valuable qualitative information about both the magnitude and temporal properties of responses of cells to stimuli, at times it is useful to relate ratio information in a more semi-quantitative fashion to intracellular calcium concentration. Discussions of this topic can be found in some recent reviews.[24,41,46]

Intracellular calibration of fura-2 can be carried out *in situ* according to the technique of Grynkiewicz.[16] Empirical measurements are made at saturating concentrations of calcium, yielding a maximum 340/380 ratio (R_{max}) and at minimum calcium concentrations to yield a minimum ratio (R_{min}). Intermediate ratios (R) can be related to calcium concentration by

$$[\text{Ca}\,\#]_i = K_d \frac{R - R_{\text{min}}}{R_{\text{max}} - R} \frac{S_{f2}}{S_{b2}} \tag{8.3}$$

where K_d is the effective dissociation constant* (224 n*M* at 37°C), S_{f2} is the free dye signal at 380 nm, and S_{b2} is the signal of Ca #-bound dye at 380 nm. Experimentally, S_{f2}/S_{b2} is simply the ratio of the measured fluorescence intensity when all the indicator is calcium-free to the intensity measured when all the indicator is Ca # bound; both measurements are taken at 380 nm. To obtain these measurements, the calcium permeability of the plasma membrane is experimentally altered by the addition of the ionophore ionomycin. Cells are incubated in zero calcium ECF containing 2 m*M* EGTA to produce R_{min} measurements, followed by ECF containing 2 m*M* calcium to produce measures with the dye saturated.

Some experimental problems are encountered with these procedures. In practice, most cells, including dissociated taste receptor cells, do not tolerate "zero" calcium environments well, although this is more problematic for patch-clamp analysis than calcium imaging. Supplementing zero calcium ECF with extrernal magnesium (e.g., 5 to 20 m*M*) is helpful. Because inomycin transports calcium poorly in low-calcium environments, measurements of R_{min} are made up to 90 min after introduction of ionomycin. A final step includes permeabilizing the cell with digitonin to release cytosolic dye so that the background signal can be measured.

8.5 CONCLUDING REMARKS

Researching isolated taste receptor cells with the technique of patch-clamp recording has been a profitable avenue for understanding the basic mechanisms of how these cells are able to transduce the presence of chemical stimuli in the oral cavity into an electrical event that is decipherable by the nervous system. Although it was long thought that these transduction events involved cascades of receptors, G-proteins, second messengers, and ion channels, it was not until the application of patch-clamp recordings that these utensils became experimentally accessible. Unlike many other cell types, currents through ion channels in taste receptor cells had not been previously recorded. Due to technical problems associated with sharp microelectrode techniques, such as the high input resistance of the recording electrode and the low input resistance of the cell due to injury, previously it was essentially impossible to voltage-clamp these cells. Without the ability to voltage-clamp these cells, it was not possible to isolate individual ionic currents.

* The Kd is determined *in vitro* and is not the same as the *in situ* Kd, which will vary with the cytoplasmic makeup, and hence from cell type to cell type. In practice, the *in situ* Kd is not an easily experimentally determined variable.

The importance of attaining a fundamental understanding of the ionic currents inherent within this population of receptor cells cannot be overstated. Ultimately, all information processing within the cell must converge on the ion channel because its lexis with peripheral afferents is based on action potentials. For example, many second messenger systems likely converge within the cell to influence action potential firing. Along this line, the application of calcium imaging tools will yield significant new information in the study of gustatory transduction mechanisms.

Finally, the caveats of reductionism must be applied to the study of isolated taste receptor cells. Taste receptor cells are an extremely heterogeneous lot and experimenters should refrain from generalizing data for the sake of simplicity. It is expected that mechanistic observations will not be repeatable across all cells. Rather, when considering transduction mechanisms, many significantly overlapping subpopulations of taste receptor cells likely exist. It is within this complexity that the true means of gustatory coding lie.

REFERENCES

1. Akabas, M.H., Dodd, J., and Al-Awqati, Q. A bitter substance induces a rise in intracellular calcium in a subpopulation of taste cells, *Science,* 242, 1047, 1988.
2. Avenet, P. and Lindeman, B. Patch-clamp study of isolated taste receptor cells of the frog, *J. Membrane Biol.,* 97, 223, 1987.
3. Béhé, P., DeSimone, J.A., Avenet, P., and Lindemann, B. Membrane currents in taste cells of the rat fungiform papilla. Evidence for two types of Ca currents and inhibition of K currents by saccharin, *J. Gen. Physiol.,* 96, 1061, 1990.
4. Béhé, P., DeSimone, J.A., Avenet, P., and Lindemann, B. Patch-clamp recording from taste buds of maintained epithelial polarity: a novel approach, in *Proceedings of the Tenth International Symposium on Olfaction and Taste,* Døving, K.B., Ed., Univ. of Oslo, Telemark, 1989, 271.
5. Bernhardt, S.J., Naim, M., Zehavi, U., and Lindemann, B. Changes in IP_3 and cytosolic Ca^{2+} in response to sugars and non-sugar sweeteners in transduction in sweet taste in the rat, *J. Physiol.,* 490, 325, 1996.
6. Cahalan, M. and Neher, E. Patch clamp techniques: an overview, *Meth. Enzymol.,* 207, 3, 1992.
7. Cavalié, A., Grantyn, R., and Lux, H.-D. Fabrication of patch clamp pipettes, in *Practical Electrophysiological Methods*, Kettenmann, H. and Grantyn, R., Eds., Wiley-Liss, New York, 1992, 235–240.
8. Chen, Y., Sun, X.-D., and Herness, M.S. Characteristics of the action potentials and their underlying outward currents in rat taste cells, *J. Neurophysiol.*, 75, 820, 1996.
9. Corey, D.P. and Stevens, C.F. Science and technology of patch-recording electrodes, in *Single-Channel Recording*, Sakmann, B. and Neher, E., Eds., Plenum Press, New York, 1983, 53–68.
10. Dunn, K.W., Major, S., Meyers, J.N., and Maxfield, F.R. Applications of ratio fluorescence microscopy in the study of cell physiology, *FASEB J.,* 8, 573, 1994.
11. Dunn, K.W. and Maxfield, F.R. Ratio imaging instrumentation, *Methods Cell Biol.,* 56, 217, 1998.

12. Fujiyama, R., Miyazaki, T., Miyamoto, T., Okada, Y., Mizuno, A., and Inokuchi, T. Intracellular free calcium concentration in human taste bud cells increases in response to taste stimuli, *FEBS Lett.,* 434, 47, 1998.
13. Furue, H. and Yoshii, K. A method for in-situ tight-seal recordings from single taste bud cells of mice, *J. Neurosci. Methods,* 84, 109, 1998.
14. Gilbertson, T. Patch clamping taste cells in hamster and rat, in *Experimental Cell Biology of Taste and Olfaction Current Techniques and Protocols,* Spielman, A.I. and Brand, J.G., Eds., CRC Press, Boca Raton, FL, 1995, 317–328.
15. Gilbertson, T., Roper, S.D., and Kinnamon, S.C. Proton currents through amiloride-sensitive Na^+ channels in isolated hamster taste cells: enhancement by vasopressin and cAMP, *Neuron,* 10, 931, 1993.
16. Grynkiewicz, G.M., Poenie, M., and Tsien, R.Y. A new generation of Ca^{2+} indicators with greatly improved fluorescence properties, *J. Biol. Chem.,* 260, 3440, 1985.
17. Herness, M.S. A dissociation procedure for mammalian taste cells, *Neuroscience Lett.,* 106, 60, 1989.
18. Herness, M.S. and Sun, X.-D. Voltage-dependent sodium currents recorded from dissociated rat taste cells, *J. Membr. Biol.,* 146, 73, 1995.
19. Herness, M.S. and Sun, X.-D. Characterization of chloride currents and their noradrenergic modulation in rat taste receptor cells, *J. Neurophysiol.,* 82, 260, 1999.
20. Herness, M.S., Sun, X.-D., and Chen, Y. cAMP and forskolin inhibit potassium currents in rat taste receptor cells by different mechanisms, *Am. J. Physiol. (Cell Physiol.),* 272, C2005, 1997.
21. Horn, R. and Korn, S.J. Prevention of rundown in electrophysiological recording, *Methods Enzym.,* 207, 149, 1992.
22. Jones, S.W., Whole-cell and microelectrode voltage clamp, in *Neuromethods,* Vol. 14, Neurophysiological Techniques: Basic Methods and Concepts, Boulton, A.A., Baker, G.B., and Vanderwolf, C.H., Eds., Humana Press, Clifton, NJ, 1990, 143–192.
23. Kao, J.P.Y., Harootunian, A.T., and Tsien, R.Y., Photochemically generated cystolic calcium pulses and their detection by fluo-3, *J. Biol. Chem.,* 264, 8179, 1989.
24. Kao, J.P.Y. Practical aspects of measuring [Ca^{2+}] with fluorescent indicators, *Methods in Cell Biol.,* 40, 155, 1994.
25. Kao, J.P.Y., Alderton, J.M., Tsien, R.Y., and Steinhardt, R.A., Active involvement of C_2^{2+} in mitotic progression of Swiss 3T3 fibroblasts, *J. Cell. Biol.,* 111, 183, 1990.
26. Kinnamon, S., Cummings, T.A., and Roper, S.D. Isolation of single taste cells from lingual epithelium, *Chem. Senses,* 13, 355, 1988.
27. Korn, S.J., Marty, A., Connor, J.A., and Horn, R. Perforated patch recording, in *Methods in Neurosciences,* Vol. 4, Conn, P.M., Ed., Academic Press, San Diego, 1991, 364–373.
28. Lu, S.-G., Kaya, N., and Herness, M.S. Cholecystokinin increases intracellular calcium levels in rat posterior taste receptor cells, *Chem. Senses* (abstr.), 25, 685, 2000.
29. Marty, A. and Neher, E. Tight-seal whole-cell recording in *Single-Channels Recording,* 2nd ed., Sakmann, B. and Neher, E., Eds., Plenum Press, New York, 1995, 31–53.
30. Misaka, T., Kusakabe, Y., Emori, Y., Gonoi, T., Arai, S., and Abe, K. Taste buds have a cyclic nucleotide-activated channel, CNGgust, *J. Biol. Chem.,* 272, 22623, 1997.
31. Monroe, W.T., Smith, D.V., and Gilbertson, T.A. The MU chamber: a new method to record electrophysiological responses of taste receptor cells to gustatory stimuli, *Chem. Senses,* 21, 644, 1996.
32. Moreton, R.B. Optical techniques and Ca^{2+} imaging, in *Cellular Neurobiology. A Practical Approach*, Chad, J. and Wheal, H., Eds., IRL Press at Oxford University Press Oxford, 1991, chap. 11, 205–222.

33. Mistretta, C.M. Permeability of tongue epithelium and its relation to taste, *Am. J. Physiol.,* 220, 1162, 1971.
34. Ogden, D. and Stanfield, P. Patch clamp techniques for single channel and whole-cell recording, in *Microelectrode Techniques. The Plymouth Workshop Handbook*, Ogden, D., Ed., The Company of Biologists Limited, Cambridge, 1994, chap. 4, 53–78.
35. Ogura, T. and Kinnamon, S.C. IP_3-independent release of Ca^{2+} from intracellular stores: a novel mechanism for transduction of bitter stimuli, *J. Neurophysiol.,* 82, 2657, 1999.
36. Ogura, T., Mackay-Sim, A., and Kinnamon, S.C. Bitter taste transduction of denatonium in the mudpuppy *Necturus maculosus*, *J. Neurosci.,* 17, 3580, 1997.
37. Rae, J. Cooper, K., Gates, P., and Watsky, M. Low access resistance perforated patch recordings using amphotericin B, *J. Neurosci. Meth.,* 37, 15, 1991.
38. Rae, J.L. and Levis, R.A. Glass technology for patch clamp electrodes, *Methods Enzymol.,* 207, 66, 1992.
39. Sakmann, B. and Neher, E. Geometric parameters of pipettes and membrane patches, in *Single Channel Recording 2nd ed.,* Sakmann, B., and Neher, E., Eds., Plenum Press, New York, 1995, chap. 21, 637–650.
40. Scheenen, J.J. W., Hofer, A.M., and Pozzan T. Intracellular measurement of calcium using fluorescent probes. in *Cell Biology: A Laboratory Handbook.* 2nd ed., Celis, J.E., Ed., Academic Press, San Diego, 1998, Vol. 3, 363–374.
41. Silver, R.B. Ratio imaging: practical considerations for measuring intracellular calcium in pH in living tissue, *Methods in Cell Biology,* 56, 237, 1998.
42. Sontheimer, H. Whole cell patch-clamp recordings, in *Neuromethods,* Vol. 26, Patch-Clamp Applications and Protocols, Boulton, A.A., Baker, G.B., and Walz, W., Eds., Humana Press, Totowa, NJ, 1995, 37–74.
43. Spielman, A.I., Mody, I., Brand, J.G., Whitney, G., MacDonald, J.F., and Salter, M.W. A method for isolating and patch-clamping single mammalian taste receptor cells, *Brain Res.,* 503, 326, 1989.
44. Standen, N.B., Davies, N.W., and Langton, P.D. Separation and analysis of macroscopic currents, in *Microelectrode Techniques: The Plymouth Workshop Handbook,* 2nd ed., Ogden, D., Ed., The Company of Biologists Limited, Cambridge, 1994, chap. 3, 37- 52.
45. Sun, X.-D. and Herness, S. Characterization of inwardly-rectifying potassium currents from dissociated rat taste receptor cells, *Am. J. Physiol: Cell Physiol.,* 271, C1221, 1996.
46. Takahashi A., Camacho, P., Lechleiter J.D., and Herman, B. Measurement of intracellular calcium, *Physiological Rev.,* 79, 1089, 1999.
47. Tsien, R.Y. and Harootunian A.T. Practical design criteria for a dynamic ratio imaging system, *Cell Calcium,* 11, 93, 1990.
48. Tsien, R.Y. Monitoring cell calcium, in *Calcium as a Cellular Regulator,* Carafoli, E. and Klee, C., Eds., Oxford University Press, New York, 1999, chap. 2, 28–54.
49. Walz, W. Perforated patch-clamp technique, in *Neuromethods,* Vol. 26: Patch-Clamp Applications and Protocols, Boulton, A., Baker, G., and Walz, W., Eds., Humana Press, Totowa, NJ, 1995, 155–171.
50. Zviman, M.M., Restrepo, D., and Teeter, J.H. Single taste stimuli elicit either increases or decreases in intracellular calcium in isolated catfish taste cells, *J. Membr. Biol.,* 149, 81, 1996.

9 Rapid Kinetic Measurements in Chemosensory Systems

Gulshan Sunavala, Wentao Yan, Joseph G. Brand, and Andrew I. Spielman

CONTENTS

0-8493-2329-0/02/$0.00+$1.50

9.1 INTRODUCTION

Many cellular processes are rapid and transitory. Consequently, their monitoring requires extremely fast methods such as electrophysiology or imaging. Application of rapid reaction techniques to biochemical studies have generally followed the paradigms inherent in rapid-mixing systems. Rapid kinetics systems have been used in a wide variety of biochemical studies. These include the study of enzyme kinetics (Barman and Gutfreund, 1965; Walinder and Joshi, 1974; Fersht and Jakes, 1975; Langowski et al., 1984; Anderson et al., 1991), protein folding (Blond-Elguindi et al., 1988; Radford et al., 1992; Jennings and Wright, 1993; Jennings et al., 1993; Mullins et al., 1993; Varley et al., 1993), formation of second messengers in signal transduction (Breer et al., 1990; Boekhoff et al., 1990; Boekhoff and Breer, 1992;

Boekhoff et al., 1994; Spielman et al., 1996; Rosenzweig et al., 1999; Huang et al., 1999 Yan et al., 2001), kinetics of calcium release in cells (Inesi, 1988; Calviello and Ciesi, 1989; Champeil et al., 1989; Orlowski and Champeil, 1991), and at intrasynaptosomal junctions (Tareilus and Breer, 1992), receptor-mediated ion fluxes (Cash et al., 1991; Ly and Michaelis, 1991; Mihic et al., 1991), the rate of dephosphorylation of phospholamban (Antipenko et al., 1997a, 1997b, 1999), and fast time measurements of drug transport and their effects in cells (Cano-Gauci et al., 1990; Bilgin et al., 1992).

Rapid reactions can be essentially studied in two ways. The first method, such as in flash photolysis (DeVault, 1964; Greenwood, 1964) and in relaxation methods (Eigen and DeMaeyer, 1964), utilizes perturbation of a biochemical system from its equilibrium with the simultaneous physical measurement of the extent of the biochemical reaction.

The second method, the one that is the focus of this chapter, utilizes direct mixing of reactants with subsequent quenching, or rapid withdrawal of the samples from the reacting system for subsequent measurement of the extent of the reaction. The progress of a chemical reaction initiated by rapid mixing of two reagents may be followed either by the continuous-flow or the stopped-flow methods. In the former, and historically the older of these, the mixed reagents are observed at a point downstream from the mixing chamber, where because of the continuous flow, the composition of the reaction mixture remains invariant with time. With the stopped-flow technique, on the other hand, a stationary portion of the mixture of reactants is examined and the reaction is followed in the abruptly stopped fluid.

Flow methods generally derive time resolution by rapid movement of a mixture from one point to another in a closed tube and are generally based on the rapid generation of a state of non-equilibrium in biochemical systems. All flow methods for the measurement of reaction rates are variations based either on the continuous- or stopped-flow methods.

With the emergence of rapid kinetic techniques, many investigators have been encouraged to probe into certain reaction mechanisms that have eluded them for years. Currently, some of the rapid kinetic systems being used include the stopped-flow, the quench-flow, the rapid filtration, and the rapid mix-freeze (mixicon) systems. Combinations of some of these systems have also been used to enhance the accuracy of measurements under certain experimental conditions. This chapter describes the background, principles of operation, setup, and utilization of a rapid kinetic system — the quench flow module — with applications to the chemosensory field.

9.2 BACKGROUND

The pioneering work in the development of rapid-mixing systems was performed in the early 1920s and 1930s (Hartridge and Roughton, 1923, Roughton and Millikan, 1936). These studies laid the foundation and elucidated the basic points in the understanding of the mechanics of these apparati. Many of the flow apparati described in the scientific literature were designed with a specific objective in view and, therefore, incorporated features that made them valuable for the study of that particular problem. Further, in an effort to extend the range of application of flow methods down to milliseconds, designers have constantly made changes to increase the accuracy and sensitivity of the system.

9.2.1 Continuous Flow Apparatus

Almost eight decades have elapsed since Hartridge and Roughton (1923) first developed a procedure for following the course of chemical reactions having half-lives of less than 10 seconds. They were the pioneers of the classical continuous-flow apparatus. They studied thermal and optical changes by utilizing fluid-flow at steady state and taking readings at various points along the flow tube to measure the extent of the reaction (Roughton and Chance, 1963). The main disadvantage with this method was the expenditure of large volumes of solutions to obtain meaningful data (Gutfreund, 1969). However, during those early years, due to the limitations of available detection equipment, this was the only method accessible for securing time resolution.

It was important, therefore, to design instrumentation that could use small amounts of reactants without compromising time resolution. With the introduction of the modified continuous flow apparatus, Millikan (1936) tried to balance fluid economy against time resolution. Instead of the conventional compressed gas-driven apparatus, Millikan delivered the reactants through syringes that were simultaneously activated by a single motor. This was not only more convenient and economical, but also provided simplicity, precision, and reproducibility of operation. In recent years, modern instruments have individual stepping motors for each syringe, which facilitates accurate adjustment of speeds and gear ratios and allows for variations in flow rate.

For the study of faster reactions, the continuous-flow apparatus can be adapted for rapid quenching experiments. With the continuous quench-flow method (Lymn and Taylor, 1970), a continuous stream of mixed reactants is quenched after flowing through a tube. Varying the distance between mixing and quenching enhances the versatility of present-day instruments.

9.2.2 Pulsed-Flow Apparatus

The pulsed-flow system is a continuous-flow method in which the duration of the mixture delivery is extremely short; hence, the name pulsed flow. Here, the reactants pass through the mixer, fill the reaction chamber and, after an incubation time, are expelled into the observation tube. The earlier systems used high-pressure pneumatic drives (Chance, 1964), with the fastest possible reaction time being 100 msec. Now, with the new motor-driven syringes and better design of reaction chambers, the reaction time has been shortened to 12 msec (Langowski et al., 1984).

A pulsed quench-flow apparatus can be adapted with readily exchangeable units, either to permit observation by a variety of physical techniques or for quenching of the mixture.

9.2.3 Stopped-Flow Apparatus

One of the early developments in the stopped-flow method was the apparatus of Chance (1963). In this apparatus, observations started soon after mixing and were made on a stationary reacting system. The important feature of this appliance was the savings in the amount of reactants when compared to the conventional continuous-flow systems of that time. Most of the early stopped-flow apparati were based on Chance's design. Later, Gibson and Milnes (1964) introduced an effective stopped-flow apparatus in which the syringes were driven by hydraulic pressure. Revisions in the design of this apparatus have been done to make it more suitable for particular applications (Sturtevant, 1964), such as for use in nonaqueous systems (Allen et al., 1960) where the use of gas-driven syringes was preferred. Further, with the emergence of faster recording systems, time resolution of the apparatus has been considerably shortened.

Fast chemical and biochemical reactions were often studied by rapid-mixing techniques in which a physical quantity that changed during the course of the reaction was monitored. In the widely used stopped-flow system, the progress of the reaction was most often studied by recording the change in absorbance after the flow of the reactants was stopped (Gibson and Milnes, 1964; Sturtevant, 1964). Other methods of observation included fluorescence (Chen et al., 1969; Gibson, 1969), calorimetry (Balko et al., 1969; Berger and Chick, 1968; Pennington et al., 1969), electrode measurement (Rossi-Bernardi and Berger, 1968), polarographic recording (Chance, 1950), and dual-wavelength recording (Gibson, 1969). Lately, technologically advanced Modular Optical Systems (MOS 200, MOS 400/CD and MOS 400/DW) have been linked to the stopped-flow apparati to enhance the sensitivity of measurements (Doyle et al., 1991; Speros et al., 1991, Jennings and Wright, 1993, Jennings et al., 1993, Mann and Matthews, 1993; Shearer, et al., 1993). These systems use a UV polarizer with a photoelastic modulator for either circular dichroism (MOS 400/CD) or rapid-shutter dual-wavelength measurements (MOS 400/DW).

Unfortunately, even with the advances in technology, some reactions are incapable of being studied by the stopped-flow method because the changes in parameters during the reaction cannot be observed rapidly enough. However, if these reactions can be stopped swiftly by some chemical reagent, they could be studied by quench-flow techniques. Hence, some of the newer designs of these apparati have been modified to permit their use in rapid quenching experiments.

9.2.4 Quenching and Chemical Sampling

Until a few years ago, the major contributions to the study of transients and intermediates in biochemical reactions used rapid reaction techniques coupled to the physical observation of some quantity, which was proportional to the progress of the reaction. The combination of rapid-flow techniques with rapid quenching and subsequent analysis of the reaction mixture has added considerably to the scope of fast reaction techniques. As mentioned, the continuous-flow, pulsed-flow, and stopped-flow apparati can be easily modified for rapid quenching experiments.

The quench flow method involves three major steps: (1) mixing of known volumes of two solutions in the mixing chamber; (2) passing the mixed components through the flow tube; and (3) quenching the mixture. Knowing the flow rate of the reaction mixture, the incubation time (time for which the mixture rests in the flow tube or the reaction chamber), and the length of the flow tube, the age of the mixture can be determined.

In the quenching process, it is important that the quenching agent be able to freeze the reaction mixture instantaneously without affecting the reaction characteristics under study. Any sluggishness in arresting the reaction would give rise to an artificial quenching time. In quenching experiments, the only limitations are given by the efficiency and the possible side effects of the quenching agent, and by the availability of methods of analyzing the quenched reaction mixture.

The first generation of instruments used mechanical or pneumatic methods to drive the syringes, while the subsequent second-generation instruments incorporated a single motor to drive all the syringes at the same time. Although the latter were far superior to the former, one still had to compromise on the small proportion of the mixture that could actually be collected for analysis.

The newly emerging third-generation apparati (e.g., Quench-Flow Module — QFM5 — Molecular Kinetics, Pullman, WA) have incorporated refinements that were lacking in their predecessors. Some important features include the elimination of mechanical valves after the last mixer, allowing a considerable reduction of the dead volumes; the arrangement of individual stepping motors for every syringe, providing variable ratio mixing; and the incorporation of machined aging lines, permitting their easy replacement. Moreover, these instruments are versatile and can be used in single or double mixing experiments. In addition, they can be operated in the continuous mode or the interrupted mode, with a number of mixing sequences.

9.3 PRINCIPLE OF OPERATION OF FLOW APPARATI

The essential feature of rate measurements of biochemical reactions is the ability to make observations of the extent of a reaction as a function of time; that is, at different ages of the reaction mixture. The age of the reaction mixture at the point of observation is given by the volume between the point of mixing and the point of observation divided by the flow rate [ml/(ml/sec) = sec]. The volume under consideration is a function of the diameter of the flow tube and the distance traveled.

The principle of operation of a flow apparatus lies in the establishment of a steady state, resulting in a reaction mixture of a defined age at any one point in the flow system. If one can sample only at one point, one has to vary the age of the reaction mixture by either changing the flow rate or the distance between mixing and observation points.

In a flow apparatus designed to measure rapid reactions, it is necessary to achieve a turbulent flow for mixing and for maintenance of a homogenous cross-section of the mixture. Accuracy of measurements largely depends on a thorough mixing of the reagents before the start of observation. The primary aim of mixing is to produce maximum turbulence in the flow while avoiding a violent spinning motion of the liquid that might favor the entrapment of air (cavitation) within it. To obtain reliable observations, the lowest flow rate permissible would depend on the achievement of a turbulent flow, whereas the fastest rate would depend on the force that can be applied without the occurrence of cavitation.

Cavitation takes place when the external pressure on the liquid is less than the vapor pressure. The phenomenon itself is described in physical terms as a drop in pressure across an orifice in a flow stream that reduces the local pressure to a point where the water vaporizes. Proper mixer designs and controlled backpressure can avoid cavitation. Mixer designs should be consistent with the requirement for avoiding flow past sharp edges and just enough angulation of the offset jets to avoid cavitation due to vortex flow. Another way to avoid cavitation is by employing backpressure. Optimum backpressure ensures that the pressure drop at any point in the stream never falls below the vaporization pressure of any component in the flow stream. This backpressure can be obtained either hydraulically or pneumatically. It is known that pressure rises at the point of constriction, while it drops at the cavitation point. Hence, appropriate constrictions in the design of the mixer would increase the hydraulic pressure and thus avoid cavitation. Alternatively, in the pneumatic way, backpressure would be exerted on the flow stream exiting from the collection port.

It should be borne in mind that the accuracy of measurements tend to be illusionary and doubtful if observations are made below the dead time of the instrument; that is, the minimum time required for the reaction mixture to flow from the mixing chamber to the point of observation. Thus, the apparent limitation of a rapid-flow system is the age of the reaction mixture, which depends on the distance between the point of mixing and the point of observation. Neither of these are true points.

In the early years, the study of transient intermediates in biochemical reactions was limited to rapid-reaction techniques coupled to absorbance or fluorescence monitoring devices (Gibson and Milnes, 1964; Chen et al., 1969; Gibson, 1969) that recorded the progress of the reaction. With the recent application of quenching techniques to rapid-flow systems, the study of transient intermediates has re-awakened the interest of many investigators. Although quenching techniques require subsequent analysis of many samples and are therefore more time-consuming than direct recording techniques, they are increasingly used because they permit the study of the kinetics of intermediates that cannot be observed by other methods.

The quench flow apparatus described here has been extensively applied during the past decade to a variety of biochemical systems (e.g., Breer et al., 1990; Radford et al., 1992; Varley et al., 1993; Jennings and Wright, 1993; Jennings et al., 1993; Spielman et al., 1996; Antipenko et al., 1997a; 1997b; 1999). This apparatus has proven to be convenient, reliable, and capable of providing physiologically relevant data.

9.4 DESCRIPTION OF THE QUENCH FLOW APPARATUS

One of the most useful quench flow apparati on the market is the Quench Flow Module-5 (QFM-5 BioLogic, Claix, France; Molecular Kinetics, Pullman, WA) (Figure 9.1). The description in this chapter details aspects not found together in either the Instruction Manual or in research publications. These details will be helpful to those interested in setting up the Quench Flow Module, and the principles underlying the setup can be used when working with most of the existing instruments on the market.

The QFM-5 consists of the mechanical subsystem and the electronic controller/power supply, both computer-controlled.

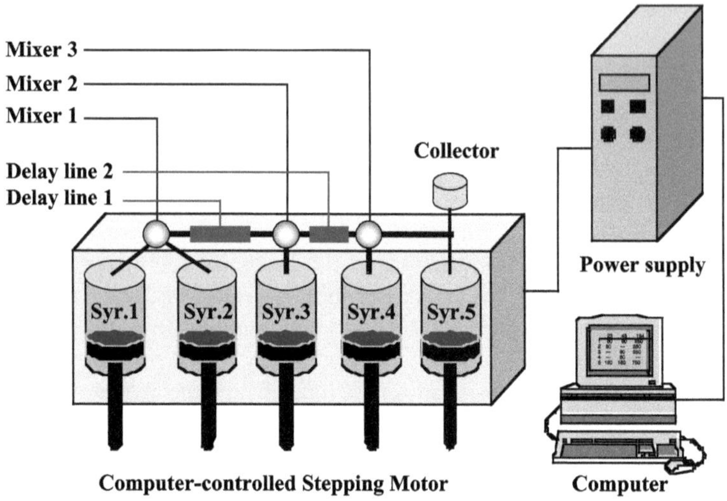

FIGURE 9.1 Design of a Quench-Flow Apparatus.

9.4.1 Mechanical Subsystem

The mechanical subsystem of the equipment has four stainless steel or plastic-coated (PEEK) driving syringes (Syr. 1 to 4) and a waste collection syringe (Syr. 5). During activation, the four driving syringes are programmed to move upward (ejecting) while the waste syringe (Syr. 5) moves downward (intaking) at the same time. The amount of intake for Syr. 5 is either the sum of the volumes ejecting from the other four syringes or is zero, in which case the mixture can go to the collector. The three mixers are responsible for proper mixing of the reactants (Figures 9.1 and 9.2).

In the design of this apparatus, Syr. 1 and 2 connect to Mixer 1, Syr. 3 to Mixer 2, and Syr. 4 to Mixer 3. The two aging loops (Delay Lines) are situated between Mixers 1 and 2, and between Mixers 2 and 3. This arrangement incorporates features for both single and dual mixing quench-flow experiments. Nonetheless, operation in the single mixing mode can also be achieved. In single mixing experiments (wherein only two reactants are mixed), the reaction is initiated by driving the reactants from the syringes (1 and 3 or 2 and 3) into the mixing chamber. The mixture is then stopped by the quenching solution, delivered from Syr. 4, prior to diverting it either to the waste syringe or to the collection port. However, in double mixing experiments, two reactants from Syr. 1 and 2 are initially combined in Mixer 1, before the third reactant is propelled from Syr. 3 into the initial mixture in Mixer 2. All reactants then pass through Delay Line 2, after which the reaction is quenched and the contents collected.

9.4.2 Power Supply

In reality the power supply contains five individual but synchronized power supplies, all independently driven by their separate microprocessor boards. The mother board synchronizes all five power supplies to deliver precise impulses to the stepping

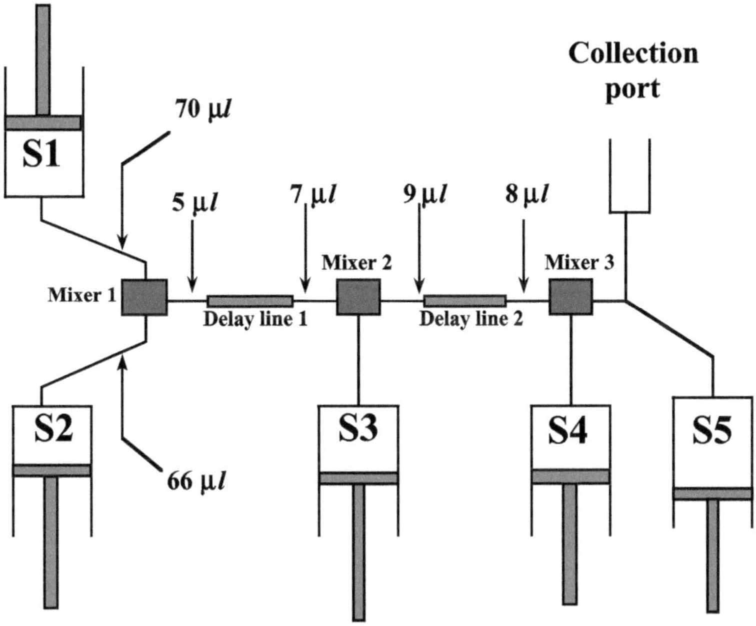

FIGURE 9.2 Line diagram showing the volumes of various flow lines.

motors (the minimum volume per mechanical step is 3.14 μl in the four driving syringes (Syr. 1 to 4) and 6.28 μl in the waste syringe (Syr. 5). The use of microprocessor-controlled stepping motors allows for high performance without the use of valves to divert the mixture. Microprocessor control allows accuracy to 0.098 μl (drive) and 0.196 μl (waste) per microstep due to sub-dividing each mechanical step into 32 microsteps.

9.4.3 Computer

This system is operated from a computer that permits automatic or manual movement of the syringes. The computer contains a set of driving sequences (designed by the user, see below) that control the sequence of the reaction and the precise time and volume delivered by each syringe. The parameters that can be adjusted are the number of driving sequences, the duration of the sequences, and the volume injected by each of the syringes during each phase. The independence of the four syringes allows for a higher versatility of injection sequences.

9.4.4 Aging Loops

An aging loop or delay line determines the time that the mixture travels between mixers. The length of the hollow loop can have varying measures, and, therefore, for adjustment of the aging of the reaction, several preset aging loops are provided with the system. The length and diameter of the hollow loop can have varying measures. These are fixed-volume delay lines of 7.1, 20, 50, 100, and 200 μl, respectively. The attribute of having interchangeable delay lines greatly enhances the ease of operation and the versatility of the instrument. The presence of four driving syringes and two variable delay lines permits the use of this instrument for dual and single mixing experiments with a variety of driving sequences.

9.4.5 Mixers

Obtaining maximum mixing efficiency in a flow system has long been a problem in the field of rapid kinetics. To increase fluid economy and achieve shorter reaction times, more rapid discharges at higher speeds, together with better mixers and improved detection systems were required. Through the years, a number of mixer designs were introduced and revised in an attempt to achieve adequate mixing in the shortest possible time. The current ball mixer introduced in the mid-1960s and extended to flow systems (Berger et al., 1968) pushed the resolution time for mixing to less than 200 μsec.

9.5 Initial Setup

The installation of the QFM-5 program in the computer allows the setting of various parameters and the selection of the port for communication with the power supply. This software is provided on a floppy disk by the distributors, but no specific driving sequences are contained in it.

The critical decisions in the use of the quench flow are the choice of delay lines, the flow rate, and the type of operational mode (continuous or interrupted) to be employed.

The continuous mode is generally used when short aging times are desired; that is, in the range of 10 to 100 msec. In this mode, solutions injected in the sequential phases are pushed through the delay lines as a continuous fluid stream. The aging of the mixture will therefore depend only on the flow rate and the distance traveled (length of delay line 2) by the mixture after merger.

When long aging times (over 100 msec to several seconds) are desired, the interrupted mode is used. After the initial washing phase, the actual solutions are injected and the mixture is allowed to rest in the delay line for a programmed period of time before it is pushed out, quenched, and collected. This interruption (aging) can be easily extended to get the desired aging time, up to 10 s.

The use of a particular mode is not a rigid norm, for the same aging time can theoretically be achieved using either mode. A variety of sequences for any age-time can be designed using different modes, different delay lines, or a combination of the two.

Some criteria to keep in mind when designing sequences include:

- Flow rate/syringe should be about 1 ml/sec.
- Flow rate through delay line 2 (i.e., the flow rate through the mixer) should be between 1 and 2 ml/sec, although the instrument is capable of faster flow rates.

- Flow rate through delay line 2 (i.e., volume S2 + S3/time duration in the phase) should be maintained the same in all phases (this does not apply to the washing phase).

9.5.1 Components of a Driving Sequence for Single Mixing Experiments

The age of the reaction mixture is determined by the volume and the flow rate, and can be derived using the formula:

$$\text{Time of aging}\left(T_{\text{age}}\right)=\frac{\text{Volume of delay line 2}\,(V)}{\text{Flow rate in delay line 2}\,(F)} \tag{9.1}$$

Example: In Table 9.1, phase 4:

V = Volume of delay line 2 would be the inter-mixer volume (in this case, 9 + 100 + 8 = 117 μl)
T_{age} = Desired time at which one wants to collect the sample (e.g., 75 msec).
In this formula the T_{age} and the V are given
F must be calculated ($F = T_{age}/V$; 75 msec/117 μl = 1.56 ml/sec).

The designing of a sequence rests primarily on one's understanding of the fluid-column concept. When liquid is pushed into a tube, it travels as a continuous column of fluid and the propulsion of additional liquid merely pushes this entire column forward. Thus, after the initial phase, the system is filled with a column of fluid, and any additional injection of fluid will merely push the column forward by exactly the same volume as that of the incoming amount.

To obtain meaningful data, it is important that one sample (shot) does not contaminate the subsequent sample. Therefore, the system needs to be washed, purged and shots have to be discarded to avoid cross-contamination. This requires understanding of the following terms: washing phase, concentration-purge phase, discarding-purge phase and the collect phase.

9.5.1.1 The Washing Phase

The washing phase (phase 2 in Table 9.1) is the initial phase of a driving sequence, where a relatively large volume of buffer is pushed through the lines. The washing solution not only evacuates the mixture standing in delay line 2 from the earlier shot, but also flushes the lines free of air bubbles. To reduce cross-contamination, a washing shot is highly recommended in all cases, especially when there is a long time lag between shots. Because a large volume of fluid is pushed through the lines during this phase, the flow rate should be relatively low to avoid back pressure and the consequent dilution of the concentration purge that follows. A flow rate of 1.5 ml/sec/phase is considered adequate to ensure uniform washing of the lines without any build up of back pressure. As an added precaution, one may wish to introduce a time pause (phase 3, 500 msec) after the washing phase in the driving sequence, as illustrated in Table 9.1.

TABLE 9.1
Example of a Driving Sequence

	Quench-Flow Parameters									
Phase	**1**	**2**	**3**	**4**	**5**	**6**	**7**	**8**	**9**	
Time (msec)	**0**	**200**	**500**	**301**	**0**	**104**	**0**	**0**	**0**	**Syr. Vol.**
S1: Not used	0	0	0	0	0	0	0	0	0	S1------μl
S2: Stimulant	0	300	0	176	0	60	0	0	0	S2------μl
S3: Tissue	0	0	0	176	0	60	0	0	0	S3------μl
S4: Quencher	0	0	0	176	0	60	0	0	0	S4------μl
S5:	0	300	0	528	0	C	0	0	0	S5------μl
Drive Sequence 1(1:9)										

Note: Delay line 1: 7.1 μl/Delay line 2: 100 μl

The fact that the flow rate in the washing phase may not be exactly the same as in the following phases is of no importance because it does not influence the aging time.

In Table 9.1, phase 2 is the washing phase where 300 μl of the buffer is pushed through in 200 msec. In this phase, no other driving syringe is activated and this buffer passes toward the waste (S5). This is followed by a 500-msec pause (phase 3). In the subsequent phase 4 (concentration-purge phase), during the 301 msec all three syringes (2 to 4) are activated to deliver 176 μl each of the reactants. Those from S2 and 3 are mixed first, passed through delay line 2, and quenched by the incoming 176 μl perchloric acid from S4. This reaction mixture also travels toward the waste (S5). In the final collect phase (phase 6), a total of 120 μl (S2+S3) is pushed forward in 104 msec, and quenched by the solution (60 μl) from S4. This entire 180 μl is then diverted to the collection port (C, phase 6). Omission of phase 5 in this case is irrelevant. It is used in the interrupted mode.

9.5.1.2 The Concentration Purge Phase

The concentration-purge phase (phase 4 in Table 9.1) follows the washing phase. Of the solution mixture that is forced through delay line 2 in the concentration purge phase, the leading edge and a fraction of the mixture enters mixer 3, while the trailing portion is retained in the delay line itself.

In the continuous mode, the concentration purge, in principle, allows the purging of delay line 2 a few times by the real solution mixture (phase 4). The amount of purge necessary to achieve a required output concentration at the end of the delay line can be estimated using a theoretically calculated concentration profile graph. This plot reveals that if the exact volume of the line (100 μl) is pushed through the delay line, only 50% of the solution in the delay line is exchanged. At least three times the volume (300 μl) of mixture is required to flow through the delay line to achieve 80% exchange. And to reach 95 to 100% replacement, ten times the volume of the delay line needs to be pushed through! In experiments in which tissue is scarce and difficult to obtain, such a volume is prohibitive. It is evident that the volume of the concentration purge increases with the length of delay line 2. Above a certain aging time, the size of the delay line must increase to maintain an optimum flow rate of 1 to 2 ml/sec for correct mixing. This, in turn, implies the need for an exceedingly large volume of purge and it is in such situations that a switch to the interrupted mode becomes necessary.

In our experiments (Rosenzweig et al., 1999), three times (3×) the volume of delay line 2 was selected as a fair compromise between the percentage of recovery and the volume of purge required to equilibrate delay line 2. Table 9.1, shows 176 + 176 = 352 μl (volume of the concentration purge); actual volume (nominal + dead volumes) of delay line 2 = 117 μl; thus an approximate threefold increase of the purge volume over the delay line 2 volume was achieved.

During the concentration purge phase, it is necessary to activate all the syringes to avoid concentration errors that are likely to occur in the later phases. Furthermore, pushing the quenching solution during this phase helps to wash mixer 3 prior to the final collect phase (Table 9.4, phase 6).

9.5.1.3 The Discarding Purge Phase

This phase is introduced only in the interrupted mode and immediately follows the incubation or the waiting period. In the waiting period, when the mixture in delay line 2 is allowed to age, there occurs a slow contamination of the mixture at both the ends of the delay line. Hence, it is important to avoid the leading and trailing edges of the mixture to prevent any discrepancies in the results.

The discarding purge helps in eliminating the leading edge prior to the collect phase, where only the uncontaminated middle fraction of the mixture is pushed forward, quenched, and collected. The volume of the discarding purge should be sufficient to eliminate the leading edge but not large enough to incorporate the trailing edge (Tables 9.4 and 9.5, phase 6). Hence, it is necessary to discard the initial 20 to 30 μl and the last 30 to 40 μl of the aged mixture from delay line 2.

9.5.1.4 The Collect Phase

The collect phase is the final phase in the driving sequence (Table 9.3, phase 6; and Table 9.5, phase 8). The main difference in this phase (from previous steps) is that the solution is diverted to the collection port instead of being pulled into syringe 5 (S5). The volumes pushed by the syringes in this phase will determine the total volume at the collection port.

After mixing, the mixture flows through delay line 2, ejecting the trailing edge of the earlier phase. Therefore, in the continuous mode, the solution that enters mixer 3 is actually the latter part of the concentration purge phase and the initial part of the collect phase. The trailing end of the mixture pushed in this phase will now be retained in delay line 2 and consequently the correct volume will be recovered at the collection port. However, in the interrupted mode, the fresh mixture of reactants pushes only a part of the aged solution forward (that was resting in delay line 2) to avoid the inclusion of the contaminated trailing edge.

Note:

1. The flow rate of the concentration purge and collect phases, in the continuous and the interrupted mode, should be the same to ensure uniform aging of the entire sample to be recovered.
2. A flow rate of 1 to 2 ml/sec through each mixer is necessary for efficient mixing.

9.5.2 Designing Driving Sequences for Single Mixing Experiments

9.5.2.1 Using the Continuous Mode

During the creation of driving sequences in the continuous mode, the time of each phase has to be calculated in order to record the required aging of the mixture.

In Table 9.2, phase 2 is the washing phase. The solution pushed through in this phase moves toward syringe 5 (S5) as it washes the delay lines (1 and 2). In this phase, although S5 moves downward by exactly the same volume (V_w) as the amount pushed by the washing syringe (V_w), only the leading edge of this fluid stream will actually enter S5. A large part of this solution essentially fills up the flow lines.

The concentration purge (phase 4) immediately follows the washing phase. The actual solutions that are injected and mixed equilibrate in delay line 2. All the syringes (i.e., S4 also) need to be activated in this phase, so as to prevent subsequent concentration errors. Finally, the collect phase (phase 6) is activated. Aging occurs as the mixture flows through the entire length of delay line 2. At the input of the delay line, the age of the mixture is zero, whereas at its output, the age will depend on the distance traveled and the flow rate of the mixture. The flow rates in phases 4 and 6 must be the same in order to achieve uniform aging.

The aging time desired is substituted into the following formula to calculate the duration time of the phase, T_p:

$$T_{\text{age in phase}} = \frac{dl2 \times T_p}{2 \times (V_p)} \tag{9.2}$$

where

$dl2$ = Volume of delay line 2

T_p = Duration (in msec) of the phase

V_p = Volume injected by a particular syringe in that phase

Thus,

$$T_{\text{age in phase 4}} = \frac{dl2 \times T_{p4}}{2 \times (V_{p4})} \qquad T_{\text{age in phase 6}} = \frac{dl2 \times T_{p6}}{2 \times (V_{p6})} \tag{9.3}$$

where

TABLE 9.2
Basic Design of Driving Sequences Using the Continuous Mode

	Quench-Flow Parameters									
Phase	**1**	**2**	**3**	**4**	**5**	**6**	**7**	**8**	**9**	
Time (msec)	**0**	T_w	**0**	T_{p4}	**0**	T_{p6}	**0**	**0**	**0**	**Syr. Vol.**
S1: Not used	0	0	0	0	0	0	0	0	0	S1------µl
S2: Stimulant	0	V_w	0	V_{p4}	0	V_{p6}	0	0	0	S2------µl
S3: Tissue	0	0	0	V_{p4}	0	V_{p6}	0	0	0	S3------µl
S4: Quencher	0	0	0	V_{p4}	0	V_{p6}	0	0	0	S4------µl
S5: Waste	0	V_w	0	$3V_{p4}$	0	C	0	0	0	S5------µl
Drive Sequence 1(1:9)										

Note: Delay line 1: 7 µl; Delay line 2: Variable, 1:1 mixing ratio; T_w = Duration (in msec) of the washing phase; V_w = Volume injected by the particular syringe in the washing phase; T_{p4} = Duration of phase 4; V_{p4} = Volume injected by a particular syringe in phase 4; T_{p6} = Duration of phase 6; V_{p6} = Volume injected by a particular syringe in phase 6.

T_{p4} = Duration of phase 4
V_{p4} = Volume injected by a particular syringe in phase 4
T_{p6} = Duration of phase 6
V_{p6} = Volume injected by a particular syringe in phase 6

During calculations, the actual total volume — and not just the nominal volume of delay line 2 — should be taken into account (i.e., delay line 2 volume + 17 µl of the connecting flow lines, Figure 9.2). The volume in the denominator is multiplied by a factor of 2 (in a 1:1 mixing ratio sequence) to give the total volume flowing through delay line 2. An example is given in Table 9.3. Thus, Equation 9.3 gives:

$$T_{\text{age in phase 4}} = \frac{dl2 \times T_{p4}}{2 \times \left(V_{p4}\right)} \qquad T_{\text{age in phase 6}} = \frac{dl2 \times T_{p6}}{2 \times \left(V_{p6}\right)}$$

$$50 = \frac{67 \times T_{p4}}{2 \times (101)} \qquad 50 = \frac{67 \times T_{p6}}{2 \times (60)}$$

$$T_{p4} = 150.7 = \sim 151 \qquad T_{p6} = 89.5 = \sim 90$$

TABLE 9.3
A Driving Sequence to Achieve 50-msec Aging of the Mixture Using a 1:1 Mixing Ratio

	Quench-Flow Parameters									
Phase	**1**	**2**	**3**	**4**	**5**	**6**	**7**	**8**	**9**	
Time (msec)	**0**	**133**	**0**	**151**	**0**	**90**	**0**	**0**	**0**	**Syr. Vol.**
S1: Not used	0	0	0	0	0	0	0	0	0	S1------μl
S2: Stimulant	0	150	0	101	0	60	0	0	0	S2------μl
S3: Tissue	0	0	0	101	0	60	0	0	0	S3------μl
S4: Quencher	0	0	0	101	0	60	0	0	0	S4------μl
S5: Waste	0	150	0	303	0	C	0	0	0	S5------μl
Drive Sequence 1(1:9)										

Note: Delay line 1: 7 μl/Delay line2: 50 μl (Total delay line 2 Vol. = 67 μl), 1:1 mixing ratio.

Immediately after the washing shot (phase 2), the concentration purge is initiated. Here, a total of 202 μl (volumes of S2 and 3) is flushed through delay line 2. The actual volume of delay line 2 (50-μl delay line) is 67 μl and hence a minimum of three times this volume ($3 \times 67 = 201$) is needed to achieve an 80% output concentration (concentration we selected for our experiments). The collect phase (phase 6) is finally activated and the aged solution from delay line 2 is quenched and collected. The maintenance of the same flow rate in phase 4 (Flow rate = Volume in dl2/Time duration of the phase, i.e., 202/151 = 1.3 ml/sec) and phase 6 (Flow rate = 120/90 = 1.3 ml/sec) is necessary for achieving equal aging of the entire sample that is eventually recovered.

9.5.2.2 Using the Interrupted Mode

The interrupted mode is used when long reaction times are required. In this mode, *only* the 200 μl delay line is used between mixers 2 and 3. Table 9.4 provides the basic design for driving sequences using the interrupted mode.

Aging time is achieved in the interrupted mode by calculating the duration time of the phase from the formula:

$$T_{\text{age of the phase}} = T_i + \frac{dl2 \times T_p}{2 \times V_p} \tag{9.4}$$

T_i = Time of incubation of the mixture in *dl*2
T_p = Duration of the phase
V_p = Volume injected by a syringe in that phase
$dl2$ = Volume of delay line 2

TABLE 9.4
Basic Design for Driving Sequences Using the Interrupted Mode

	Quench-Flow Parameters									
Phase	**1**	**2**	**3**	**4**	**5**	**6**	**7**	**8**	**9**	
Time (msec)	**0**	**T_w**	**0**	**T_{p4}**	**T_i**	**T_{p6}**	**0**	**T_{p8}**	**0**	**Syr. Vol.**
S1: Not used	0	0	0	0	0	0	0	0	0	S1------µl
S2: Solution1	0	V_w	0	V_{p4}	0	V_{p6}	0	$2(V_{p8})$	0	S2------µl
S3: Solution2	0	0	0	V_{p4}	0	V_{p6}	0		0	S3------µl
S4: Quencher	0	0	0	V_{p4}	0	V_{p6}	0	V_{p8}	0	S4------µl
S5: Waste	0	V_w	0	$3V_{p4}$	0	$3V_{p6}$	0	C	0	S5------µl
Drive Sequence 1(1:9)										

Note: Delay line 1: 7 µl/Delay line 2: 200 µl (Total delay line 2 Vol. = 217 µl), 1:1 mixing ratio; T_w = Duration of washing phase; V_w = Volume injected by a particular syringe in the washing phase; T_{p4} = Duration of phase 4; V_{p4} = Volume injected by a particular syringe in phase 4; T_i = Time of incubation of the mixture in *dl2* (used to age the solution beyond the time frame possible with the continuous mode); T_{p6} = Duration of phase 6; V_{p6} = Volume injected by a particular syringe in phase 6; T_{p8} = Duration of phase 8; V_{p8} = Volume injected by a syringe in phase 8.

$$T_{\text{age phase 4}} = T_i + \frac{dl2 \times T_{p4}}{2 \times V_{p4}} \quad T_{\text{age phase 6}} = T_i + \frac{dl2 \times T_{p6}}{2 \times V_{p6}} \quad T_{\text{age phase 8}} = T_i + \frac{dl2 \times T_{p8}}{2 \times V_{p8}} \tag{9.5}$$

The delay line 2 is washed with a large amount of buffer in the washing phase (phase 2). Phase 4 is the concentration purge phase where the solutions are injected into mixer 3 and the mixture then flows to purge delay line 2. When driving syringe 4 (S4) in this phase, it is essential to activate the syringe and also to wash mixer 3. The mixture one wants to collect now ages in delay line 2 for the calculated amount of time (T_i). After this waiting period (phase 5), a short discarding phase (phase 6) eliminates a probably contaminated leading edge. The collect phase (phase 8) then follows and the reaction mixture is collected from the collection port. The volume one wants to recover will be determined by the total volume pushed by the syringes in the collect phase. Because, in the last phase, the solution injected through the syringes (1 or 2, and 3) merely pushes out the aged solution from delay line 2, one can push twice the volume of buffer instead of wasting a precious reactant. The discarding-purge and the collect phases must have the same flow rate as the concentration purge to maintain uniform aging of the entire mixture collected. The driving sequence for 200-msec aging time using the interrupted mode is presented in Table 9.5.

TABLE 9.5
Driving Sequence for 200-msec Aging Time, Using the Interrupted Mode

	Quench-Flow Parameters									
Phase	**1**	**2**	**3**	**4**	**5**	**6**	**7**	**8**	**9**	
Time (msec)	**0**	**200**	**0**	**149**	**119**	**15**	**0**	**45**	**0**	**Syr Vol**
S1: Not used	0	0	0	0	0	0	0	0	0	S1------μl
S2: Stimulant	0	300	0	200	0	20	0	120	0	S2------μl
S3: Tissue	0	0	0	200	0	20	0	0	0	S3------μl
S4: Quencher	0	0	0	200	0	20	0	60	0	S4------μl
S5: Waste	0	300	0	600	0	60	0	C	0	S5------μl
Drive Sequence 1(1:9)										

Note: Delay line 1: 7 μl/Delay line2: 200 μl (Total delay line 2 Vol. = 217 μl), 1:1 mixing ratio; *C* = collector.

$$T_{\text{age phase 4}} = T_i + \frac{dl2 \times T_{p4}}{2 \times V_{p4}} \quad T_{\text{age phase 6}} = T_i + \frac{dl2 \times T_{p6}}{2 \times V_{p6}} \quad T_{\text{age phase 8}} = T_i + \frac{dl2 \times T_{p8}}{2 \times V_{p8}}$$

$$200 = 119 + \frac{217 \times T_{p4}}{2 \times 200} \quad 200 = 119 + \frac{217 \times T_{p6}}{2 \times 20} \quad 200 = 119 + \frac{217 \times T_{p8}}{120} \tag{9.6}$$

$$T_{p4} = 149.3 = \sim 149 \text{ msec} \quad T_{p6} = 14.9 = \sim 15 \text{ msec} \quad T_{p8} = 44.7 = \sim 45 \text{ msec}$$

The concentration purge (phase 4) follows the washing phase (phase 2), where the real mixture that is to be aged is pushed through the delay line 2. At the beginning of the waiting phase (phase 5), the mixture at the input of the delay line will have a zero age, whereas at the end it will have an age of 81 msec:

$$T_{\text{age in phase 4}} = \frac{217 \times 149}{2 \times 200} = 80.8 = \sim 81 \text{msec} \tag{9.7}$$

At the end of the waiting phase, the mixture at the input of the delay line would have aged 119 msec, while at the end of the delay line the age of the mixture will be 119 + 81 msec (= 200 msec). The syringes are then activated again (phase 6) in an attempt to discard the contaminated leading edge of the aged solution. Finally, the solution pushed in phase 8 expels the correctly aged solution into mixer 3 to be quenched by perchloric acid. Note that the flow rates (2.6 ml/sec) in the discarding purge (phase 6) and the collect (phase 8) phases were the same as in the concentration purge (phase 4) in order to maintain uniform aging of the mixture. The solution that reaches the last mixer will therefore have 200-msec aging of the leading edge and 119 + 81 msec aging for the trailing edge. Thus, a constant age of the mixture flowing through the last mixer is achieved. The aged volume is then collected in the collection port.

Using the above principles and formulae, one can design any aging time sequence between 10 msec and several seconds. Maintaining an optimum flow rate can be achieved using the right combination of delay lines and volumes. Currently in mammalian taste and olfactory research, the only limitation is the amount of tissue available. Hence, the design of each sequence was done with an aim to conserve tissue without compromising the results. Although our experiments use a mixing ratio of 1:1:1 (tissue:stimulant:quench), other ratios are possible due to the individual control of the QFM-5 syringes by individual stepping motors.

9.5.3 Calibration of the QFM-5 System

9.5.3.1 Recovery of Bovine Serum Albumin (BSA)

One can estimate the volume delivered by the syringes during different flow times either by weighing the recovered solution or by analysis of some marker substance dissolved in the solution, such as BSA. To test the efficiency of recovery after mixing equal amounts of solutions among syringes 2, 3, and 4, experiments were performed using BSA. Syringes 2 and 4 are filled with water, and syringe 3 with BSA. The absorbance of BSA is measured before and after mixing. It is expected that the absorbance of mixed BSA sample (at 1:1:1 mixing ratio) would drop to one third the initial absorbance. Only 80 to 85% of the 1/3 absorbance should be expected because for all sequences in the continuous mode, we use only 3 times the volume of delay line 2 in the concentration purge phase. If, for example, the absorbance of BSA (before mixing) is 1.2 AU, after a 1:1:1 mixing the absorbance of BSA in the collection port should be 0.400. At the estimated 80% output concentration, this should be only 0.320. The experimental values are considered acceptable if the variation is not greater than ±5% of this theoretical value. To test the recovery for each individual syringe, one could vary the placement of BSA in any of the four syringes.

Various driving sequences were tested using the QFM-5 and the absorbance of the solution recovered at the collection port was recorded. Absorbance was measured at 280 nm. The absorbance of water (as minimum value) and unmixed BSA (as maximum value) were taken as the baseline values.

The data reveal that the desired output concentration of ~80% was achieved for the particular time points using sequences with the delay line mentioned in Table 9.6.

9.5.3.2 Test of Mixing and Time Resolution

A number of methods have been introduced to measure the extent to which fast reactions have progressed at the point of quenching (Gutfreund, 1969). An empirical approach used by Barman and Gutfreund (1966) to test the accuracy of the instrument was the hydrolysis of 2,4-dinitrophenylacetate (DNPA) by sodium hydroxide

TABLE 9.6
List of Delay Lines Whose Sequences Gave ~80% Output Concentration Using BSA

Time (msec)	Delay line 2 (μl)
010	7.1/20
025	20/50
050	50/100
075	100
100	100/200
200	200

(NaOH). This methodology was chosen to validate our driving sequences and test the QFM-5. Examples of other calibration reactions described in the literature include:

- The reduction of 2,6-dichlorophenolindophenol by ascorbic acid (Tonomura et al., 1978)
- The association of Mg^{2+} ions with 8-hydroxyquinoline (Brissette et al., 1989)

Studying the rates of quenching of enzyme reactions can also add useful information about the efficiency of the instrument. The three enzyme reactions frequently used are:

- Trypsin-catalyzed hydrolysis of N-benzoyl-L-arginine ethyl ester (BAEE; Barman and Gutfreund, 1966)
- Chymotrypsin-catalysed hydrolysis of N-β-(2-furyl) acryloyl-L-tyrosine ethyl ester (Barman and Gutfreund, 1966)
- Alkaline decomposition of phosphoryl-alkaline phosphatase (Aldridge et al., 1964)

To test the quench-flow apparatus, we preferred to use the alkaline hydrolysis of 2,4-dinitrophenyl acetate because of its simplicity and reliability. A detailed description of this experiment is found in the User Manual of the manufacturer of the QFM system.

9.6 MAINTENANCE

Regular maintenance and the upkeep of the quench flow is necessary for efficient and smooth running of the machine. After taking apart the mechanical subsystem, its individual components can be cleaned. It is recommended that the equipment be cleaned at least once a month. However, if solutions used for the experiments have a high salt content or are highly viscous, more frequent cleaning may be required.

9.6.1 Washing of the System

Before, after, and in between each experiment, washing the machine with copious amounts of water is mandatory. Distilled water added into the reservoirs is pumped up and down in the syringes via manual control. Each syringe is washed separately and the wash discarded. This operation is repeated at least two or three times to remove all traces of the solution. Flushing of the mixers and the lines is done by pushing water through the reservoirs using a 5-ml disposable plastic syringe. The valve handles are turned to connect the reservoir to the collection port and a suction tip is attached to the collection port to facilitate rapid washing of the system. Alternatively, after filling the syringes, the valve can be turned to connect the syringe to the collection port, and the syringes emptied by the upward activation of their respective plungers. The delay lines are removed and washed with a jet of water, from a wash bottle, forced through the loop, dried, and replaced.

Note:

- It is best to wash the system with plain water without the use of any organic solvent. Organic solvents tend to dissolve the connections within the system, causing leakage and, therefore, inaccuracy in mixing.
- Further, soaking the delay lines in strong basic or acidic solutions should be avoided because the delay lines undergo deformation, and experimental reproducibility and efficiency of the system are hampered.

9.6.2 Care of the Valves

Continued use of the machine increases the stickiness in the movement of the valve handles. If difficulty is encountered when turning the valves, they need to be removed and cleaned. The removal and cleaning of the valve with a soft dampened tissue will reestablish the smooth movement of the handle.

9.6.3 Care of the Plunger Tip

During normal use, minimal leakage from the driving syringes is inevitable. This leak increases when great demands are placed on the seal at the syringe tips. In experiments that use viscous solutions and high flow rates, an increase in leakage occurs.

Proper care of the apparatus often alleviates excessive operational leaks. At the end of the day, it is advisable to fully withdraw the plungers from their syringes. This allows for the drying of the syringe barrels and for the white teflon tips on the plunger heads to expand between uses, thereby creating a better seal with the sides of the syringe barrel during experiments. Frequent cleaning of these heads with an ethanol-dampened wipe should also be done. To prevent dust accumulation, cover the system when not in use.

Prolonged use of the apparatus causes wear of the teflon tips and occasionally they should be inspected for scratches, defects, and deformities.

9.6.4 Care of the Syringe Barrels

The cleaning of the syringe barrels can be done after disengaging the top part of the subsystem. Once the plungers are fully withdrawn, the top plate is removed. The top unit is separated from its lower motor component by lifting it up (approximately 4.5 kg). On up-turning this upper unit, the insides of the barrels are visualized and cleaned with wet cotton-tipped applicators.

9.7 Troubleshooting Errors in Quench-Flow Techniques

Problems are likely to arise during the running of this equipment. Listed below are some of the problems we encountered and remedied.

9.7.1 Leakage

If a delay line does not fit snugly into the channel, a leak is likely to occur. This is noticeable by the reduction in the collected volume. Although minimal, this leak may affect the results where the volume collected is small. To minimize this problem, the conical head of the delay line can be coated with a very thin film of high-vacuum grease, taking care to avoid occluding the orifices.

9.7.2 Loss of Recovery Material at the Collection Port

Formation of backpressure at the collection port can cause a reduction in the volume of the recovery. The reason for avoiding backpressure at the collection port can be appreciated when the entire driving sequence is analyzed. During the washing phase, when the solution is pushed through the lines, only the leading front enters syringe 5 (waste syringe) and the remainder fills the line itself. However, at this point, syringe 5 has been computerized to move downward by exactly the same volume that was pushed. This gives rise to a discrepancy in the amount of solution pushed and the amount entering syringe 5. Hence, backpressure at the collection port during the collect phase is likely to permit some volume (i.e., to be collected) to travel toward syringe 5, leading to a decrease in the recovered volume. To alleviate this problem, it is recommended that a 1-ml plastic syringe with the plunger withdrawn to maximum be placed in the collection port. The syringe, below the withdrawn plunger, should be pierced with holes for release of pressure. The volume collected in the syringe should be below the holes to avoid losing any sample.

9.7.3 Back Pressure Due to Precipitate Formation

In certain experiments, where high concentrations of solutions are used, formation of a precipitate may be seen in the collected mixture. Formation of the precipitate is most likely to occur when the pH of the mixture suddenly changes. This occurs in mixer 3, where the quenching solution hits the aged mixture. Precipitation may partially occlude the line leading to syringe 5, causing excessive backpressure and therefore contamination of the recovered sample. Forcing water through the reservoir of syringe 5 while applying a vacuum at the port enables flushing of this line between individual shots.

9.7.4 Adherence of Small Molecules and Proteins

Adherence of proteins and stimulant molecules to connecting and delay lines may pose a problem in obtaining consistent results. This could be alleviated by the inclusion in buffers of a non-denaturing zwitterionic detergent (such as CHAPS or NP-40 at 0.05%).

9.7.5 Air Bubbles in the Flow Lines

Small air bubbles trapped within the lines cause serious problems in the proper flow of the solutions. Hence, it is absolutely essential to have the apparatus completely free of air bubbles. First, one can use degassed buffers. Second, the flow lines should be purged of air bubbles by pushing large amounts of liquid from the syringes. The washing shots in our designed sequences help to accomplish this, but as a precautionary measure, before and during an experiment it is wise to discard the first shot after placement of each new delay line.

9.7.6 Collection of Unquenched Sample

During the collect phase, 120 μl of the aged mixture is pushed into mixer 3 where it is quenched and diverted to the collection port. While retrieving this sample, care should be taken to avoid aspirating any fluid from the connecting line (connecting mixer 3 to the base of the port), because there lies an increased risk of sucking out the unquenched mixture from delay line 2. Hence, during the collection of the sample, it would be safe to allow the tip of the pipette to rest in the corner of the base of the collection port (and remove only 180 μl or less), thereby ensuring that no unquenched mixture is unknowingly pipetted out.

9.7.7 Diffusion from One Working Syringe to Another

If a long interval is allowed to elapse between successive ejections of solutions from the working syringes, some diffusion from one side of the flow line to the other occurs, and this diffusion may reach proportions sufficient to influence the subsequent shot. Problems of this nature are likely to be encountered in cases where concentration of the reactants show wide disparity. The only remedy to this problem is to make shots in close succession, thereby reducing the amount of diffusion of the reactants. Furthermore, the inclusion of an extra washing shot that is discarded and an adequate concentration purge minimize this error.

9.8 THE USE OF QFM FOR MONITORING SECOND MESSENGERS IN CHEMOSENSORY SYSTEMS

9.8.1 Experimental Design for the Study of Second Messengers

9.8.1.1 Solutions

- *MOPS buffer used for both taste and olfactory experiments:* consists of 50 m*M* MOPS [3-(N-morpholino)propanesulfonic acid], 10 m*M* EGTA, 2.5 m*M* $MgCl_2$, 1 m*M* DTT, 81 μ*M* $CaCl_2$ (calculated free Ca^{2+} = 10 n*M*), and 100 m*M* NaCl, pH 6.9. Each stimulant is dissolved in the MOPS buffer + 0.1% Na-cholate, 1 m*M* ATP, μ*M* GTP. It is important to check and adjust the pH of the solutions particularly if GTP or ATP are added), because even a difference of 0.1 pH units will affect the free calcium concentration. Mg^{2+}- and Ca^{2+} ions present in the tissue are not included in the calculation. In both tissue and stimulant buffers, the amount of $CaCl_2$ is calculated using the Chelator program (Schoenmakers et al., 1992) to give a free calcium concentration of 10 n*M* in the reaction mixture.
- *Ringer Solution:* consists of 1.6 m*M* K_2HPO_4, 25 m*M* $NaHCO_3$, 120 m*M* NaCl, 5 m*M* KCl, and 7.5 m*M* glucose, pH 7.4.
- *TME-buffer:* consists of 10 m*M* Tris/HCl, 2 m*M* EGTA, and 3 m*M* $MgCl_2$, pH 7.4.

9.8.1.2 Taste and Odorant Solutions

For each experiment, taste compounds are freshly dissolved at 4 times the final concentration and mixed with 2× MOPS buffer before applying into the QFM. At a 1:1 ratio when tissue and stimulants are mixed, a 2× stimulant is diluted to the desired concentration. Odorants insoluble in water are prepared as a 100 m*M* stock solution in dimethylsulfoxide (DMSO) and mixed into buffer in an ultrasonic bath prior to use. Odorants are aliquoted in small quantities and stored in liquid nitrogen to avoid oxidation. The highest DMSO concentration is included in all control experiments (Tareilus et al., 1995).

9.8.1.3 Stop Solution

Formation of the second messengers is stopped by injecting ice-cold 9% perchloric acid. This causes rapid denaturation of tissue proteins and instant freezing of the generated second messengers.

9.8.1.4 Isolation of Gustatory Tissue

Taste papillae are collected based on a procedure described earlier (Spielman et al., 1989; Spielman and Brand, 1995). Tissue taken from approximately 25 mice tongues is used for a given experiment. Taste-bud rich epithelium and nongustatory control are gathered by the methods of Striem et al. (1991) and Hayashi et al. (1996). Briefly, the epithelium, including taste and nongustatory control tissue, is peeled off from rodent tongues after injection of 4 mg/ml collagenase. One vallate and two foliate papillae and appropriate nongustatory control tissue are then removed from each peeled epithelium in the presence of protease inhibitor cocktail and enzyme inhibitors (1 mg/ml, specific for serine, cysteine, aspartic, and metalloproteinases, [Sigma Chemical Co., St Louis, MO]) (Yan et al., 2001). The collected tissue samples contain primarily taste tissue and minimal epithelial contamination. Control tissue is removed from the dorsal eminence of the peeled epithelium, devoid of any taste papillae. Tissue collected is washed several times and homogenized in MOPS buffer pH 6.9 in a tightly fitted glass/glass tissue homogenizer and centrifuged at 1000 *g* for 20 min at 4°C. This supernatant (homogenate) at a concentration of about 50 μg/ml is used for rapid kinetic measurements of second messengers.

9.8.1.5 Isolation of Olfactory Cilia

Studies on the molecular mechanisms that underlie transduction events in the chemosensory membranes of olfactory receptor cells have been greatly facilitated by isolating olfactory cilia. These organelles are supposed to contain the molecular machinery necessary to mediate olfactory signaling and can be easily subjected to biochemical studies.

Ciliary preparations from rat olfactory epithelia are isolated using the calcium-shock procedure as described by Anholt et al. (1986), Chen et al. (1986), and Tareilus et al., (1995). The olfactory epithelium is dissected from rat nasal septum and ethmoid turbinates and collected in Ringer solution at 4°C. Cilia are detached from the epithelium by raising the calcium concentration in the Ringer solution to 10 m*M*. After incubation for 20 min on an end-over-end shaker, the deciliated epithelia are separated by centrifugation for 10 min at 7700 g. The supernatant is collected and the pelleted epithelia are subjected to the same procedure. The supernatants are combined and centrifuged at 27,000 *g* for 15 min. The pelleted cilia are resuspended in TME medium and immediately stored at –70°C (Tareilus et al., 1995). The protein concentration of the cilia preparation is measured using the method of Bradford (1976, see further).

The amount of tissue required for each quench-flow experiment is dependent on the experimental design: time-course or a single time-point. To help calculate the amount (volume) of tissue, Table 9.7 shows, for each time point, the minimum volume that is required. For example, for a full time-course including 0, 25, 50, 75, 100, and 200 msec, one would need a minimum of 4470 μl of tissue for basal and 4789 μl for a single concentration of stimulant. The tissue should be at least 50 μg/ml. Therefore, the total tissue needed for basal and stimulant for a single time course is 9259 μl (for each, gustatory and nongustatory set). This includes 500 μl for each time-course to compensate for the dead volume (350 μl) in the QFM when

TABLE 9.7
Calculation of the Amount of Tissue Required for a Time Course of Basal and Stimulated Sets Using the Previously Shown Driving Sequences, for Triplicate (basal) and Quadruplicate (stimulated) Values

Time Point (msec)	Delay Line 2	Tissue/Shot	Number of Shots		Total Tissue (in μl)	
			Basal	Stimulated	Basal	Stimulated
0	Done manually	60 μl tissue + 60 μl PCA + 60 μl Buff. or Stimul.	3	3	180	180
25	20 μl	116 μl (56+60)	4	5	464	580
50	50 μl	161 μl (101+60)	4	5	644	805
75	100 μl	236 μl (176+60)	4	5	944	1180
100	100 μl	236 μl (176+60)	3[a]	4[a]	708	944
200	200 μl	220 μl (200+20)	4	5	880	1100
		Compensate for dead volume			500	500
		For protein determination			150	—
Total			22	27	4470	4,789

[a] If the same delay line is used in subsequent steps there is no need to discard an extra shot.

the tissue is first loaded and fills up the connecting lines. An additional 100 to 150 μl is also required for protein concentration determination.

The experimental design shown in the time-course experiment (Table 9.7) includes the retrieval of triplicates (for basal) and quadruplicates (for stimulated) at each time point. Each time a new delay line is placed, one shot is discarded. Therefore, an extra shot is included. Note the exception at the 100-msec time point, where the 100-μl delay line is the same as that used in the previous step with the delay line already filled.

9.8.1.6 Protein Determination

The amount of protein in the tissue homogenate is estimated using the Bradford dye-binding procedure (Bradford, 1976). Bovine serum gamma globulin (0.3 g/ml) is used as the standard and the absorbance is measured at 595 nm. A standard of 2, 4, 6, 10, 15, 20, and 30 μg gamma-globulin is prepared. Typically, 48 μl of tissue sample is assayed. This amount represents 80% of 60 μl, which is the amount of tissue injected in the collect phase of each driving sequence (for example, Table 9.3, phase 6). Only about 80% of tissue is in fact recovered at the end of delay line 2.

9.8.1.7 Tissue Incubation with Stimulants, Inhibitors, or Antibodies

Tissue is always prepared on the day of the experiment, 1 hr before the experiment to avoid protein degradation and kept at 4°C. The protein concentration of the homogenate or membrane preparation in the QFM is about 50 μg/ml such that each assay tube receives 1.5 to 5 μg protein. Tissue is loaded into the QFM in small batches just seconds prior to injection, at 22°C. Before injection and to avoid protein sedimentation, each batch of tissue is mixed by rapidly moving the syringe plunger up and down three times, followed by the injection of the tissue. For experiments in which tissue needs to be incubated with antibodies, small batches are aliquoted and antibody added to the tissue. The sample is left on ice to equilibrate for 30 to 90 min.

9.8.1.8 Instruments

Some commercial suppliers of quench-flow and stopped-flow machines:

- BioLogic (Claix, France) (http://www.bio-logic.fr/rapid-kinetic/qfm.html) and Molecular Kinetics (Pullman, WA) (http://www.molecularkinetics.com) (SFM-3/4 — Stopped-Flow, QFM-5 — Quench-Flow)
- KinTek Corporation, Austin, TX (RQF3 — Quench-Flow and SF2001 — Stopped Flow) (http://www.kintek-corp.com/index.html)
- Applied Photophysics, Surrey, U.K. (RX 1000) (http://www.apltd.co.uk/)
- Hi-Tech Scientific, Wiltshire, U.K., (http://www.hi-techsci.co.uk/scientific/noframes/rapidquench.html) SF-61, Single Mixing Micro Stopped-Flow System; Hi-Tech KinetAsyst™ (SF-61DXZ) Double Mixing Stopped-Flow System; HPSF-56 High Pressure Stopped-Flow System

9.8.1.9 Extraction of Second Messengers: Sample Preparation for All Assays

After experiments, each sample is stored at –20°C until assayed. Extraction of water-soluble second messengers is done according to previously described methods (Palmer et al., 1989; Palmer and Wakelam, 1990). Each tube is adjusted individually. The mix is centrifuged at 1200 *g* for 5 min at 4°C and the supernatant is assayed for second messengers (Spielman et al., 1996; Rosenzweig et al., 1999).

9.8.1.10 IP_3 Binding Assay

IP_3 is monitored using a competitive binding assay (Palmer et al., 1989). The assay system (purchased from Amersham) as a kit, or components of it) includes IP3 standard, assay buffer, tracer, a [^{3}H] IP_3 with high specific activity, and a binding protein, prepared from bovine adrenal gland. This assay has been extensively described elsewhere (Spielman et al., 1996). One can separately buy all the components of the kit and prepare the bovine adrenocortical IP_3-binding protein (Palmer and Wakelam, 1990) and save 60 to 70% of the cost of a kit.

9.8.1.11 cAMP and cGMP Binding Assays

We use the [^{125}I]cAMP-based assay with good sensitivity and reproducibility over the range of 0.5 to 14 pmoles. The cAMP assay is a double-antibody radioimmunoassay that uses a pre-reacted antibody-antigen complex. The operating principle of the method is the competition between a radioactive and nonradioactive antigen for a fixed number of antibody binding sites (NEN catalog: NEK 033). The commercially available assay kit gives pico and femtomolar sensitivity to cAMP (NEN-PerkinElmer). The assay buffer is 50 m*M* MOPS, pH 7.5, 4 m*M* EDTA, and 2% bovine serum albumin as a stabilizer. Each experiment includes tissue blanks to monitor endogenous cyclase activity. Data are expressed as concentration (pmoles) of cAMP per microgram of protein. The assay has a coefficient of variation of less than 7%. Similar to the cAMP, the [^{125}I]cGMP-based assay shows good reproducibility and little variation among triplicates (about 3 to 5%). The cGMP assay is a double-antibody radioimmunoassay that uses a pre-reacted antibody-antigen complex. The principle of the method is the competition between a radioactive and non-radioactive antigen for a fixed number of antibody binding sites (NEN Catalog: NEX-133). A detailed description of the principle is available in the manufacturer's catalog. While cAMP and cGMP assays are prone to fluctuations in sensitivity because of the labile nature of the radiolabel, these assays have been used in our lab for over 8 years with consistent results.

9.9 CONCLUSION

The quench-flow instrument requires technical expertise and a thorough understanding of the principles behind its design. Similar to other established methods such as patch-clamping or calcium imaging, it requires a lag-time to learn, design, maintain, and troubleshoot. During the past 10 years, starting with the pioneering work of Heinz Breer and his group at Stuttgart University, the quench- and stopped-flow instruments have yielded invaluable information about chemosensory signaling mechanisms that cannot be obtained by other means.

ACKNOWLEDGMENTS

The authors would like to thank the following individuals for their help at various stages of our Quench Flow experiments: Heinz Breer, Erwin Tareilus, Ingrid Boekhoff, Sophia Rosenzweig, Max Dasso, Hajime Nagai, Laura Wong, Doug Bayley, Taufiqul Huque, Paul Iskander, Robert Margolskee, Ya Yu Van, James Mueller, and Yves DuPont. Work in the authors' laboratory was supported by NIDCR (DC-10754 to AIS), Proctor and Gamble and U.S.-Israel Binational Agricultural Research and Development Fund (IS2518) (to AIS), and the Department of Veterans Affair (to JGB).

REFERENCES

Aldridge, W.N., Barman, T.E., and Gutfreund, H., The rate of formation and decomposition of phosphorylphosphatase (*Escherichia coli*), *Biochem. J.*, 92, 23C-25C, 1964.

Allen, C.R., Brook, A.J.W., and Caldin, E.F., Measurement of the rates of fast reactions at temperatures down to –120°C, *Trans. Faraday Soc.*, 56, 788-793, 1960.

Anderson, K.S., Miles, E.W., and Johnson, K.A., Serine modulators substrate channeling in tryptophan synthase. A novel intersubunit triggering mechanism, *J. Biol. Chem.,* 266, 8020-8033, 1991.

Anholt, R.R.H., Aebi, U., and Snyder, S.H. A partially purified preparation of isolated chemosensory cilia from the olfactory epithelium of the bullfrog, *Rana catesbeiana, J. Neurosci.,* 6, 1962-1969, 1986.

Antipenko, A.Y., Spielman, A.I., and Kirchberger, M.A. Comparison of the effects of phospholamban and jasmone on the calcium pump of cardiac sarcoplasmic reticulum, *J. Biol. Chem.* 272, 2852-2860, 1997a.

Antipenko, A.Y., Spielman, A.I., Sassaroli, M., and Kirchberger, M.A. A comparison of the kinetic effects of phospholamban phosphorylation and monoclonal phospholamban antibody on the calcium pump in purified cardiac sarcoplasmic reticulum membranes, *Biochemistry,* 36, 12903-12910, 1997b.

Antipenko, A.Y., Spielman, A.I., and Kirchberger, M.A. Contributory factors toward differences in the kinetics of the phospholamban-regulated calcium pump when studied in crude cardiac microsomes and light cardiac sarcoplasmic reticulum vesicles, *J. Membr. Biol.,* 167, 257-265, 1999.

Balko, B., Berger, R.L., and Friauf, W., Stopped-flow calorimetry for biochemical reactions, *Anal. Chem.,* 41, 1506-1511, 1969.

Barman, T.E. and Gutfreund, H., The application of the rapid-flow sampling technique to the chemical identification of intermediates in trypsin-catalyzed reactions, *Proc. Natl. Acad. Sci., U.S.A.,* 53, 1243-1245, 1965.

Barman, T.E. and Gutfreund, H., Optical and chemical identification of kinetic steps in trypsin- and chymotrypsin-catalyzed reactions, *Biochem. J.,* 101, 411-416, 1966.

Berger, R.L. and Chick, Y.B., Differential microcalorimeter for biochemical reaction studies, *Rev. Sci. Inst.,* 39, 362-368, 1968.

Berger, R.L., Balko, B., and Chapman, H.F., High resolution mixer for the study of the kinetics of rapid reactions in solution, *Rev. Sci. Inst.,* 39, 493-498, 1968.

Bilgin, N., Claesens, F., Pahverk, H., and Ehrenberg, M., Kinetic properties of *Escherichia coli* ribosomes with altered forms of S_{12}, *J. Mol. Biol.,* 244, 1011-1027, 1992.

Blond-Elguindi, S., Friguet, B., and Goldberg, M.E., Renaturation of guanidine-unfolded tryptophan synthetase by multi mixing stopped-flow dilution in D20, *FEBS Lett.,* 241, 251-256, 1988.

Boekhoff, I., Tareilus, E., Strotmann, J., and Breer, H., Rapid activation of alternative second messenger pathways in olfactory cilia from rats by different odorants, *EMBO J.*, 9, 2453-2458, 1990.

Boekhoff, I. and Breer, H., Termination of second messenger signaling in olfaction, *Proc. Natl. Acad. Sci., U.S.A.,* 89, 471-474, 1992.

Boekhoff, I., Michel, W.C., Breer, H., and Ache, B.W., Single odors differentially stimulate dual second messenger pathways in lobster olfactory receptor cells, *J. Neurosci.,* 14, 3304-3309, 1994.

Bradford, M.M., A rapid and sensitive method for the quantitation of microgram quantities of protein utilising the principle of protein-dye binding, *Anal. Biochem.,* 72, 248-254, 1976.

Breer. H., Boekhoff, I., and Tareilus, E., Rapid kinetics of second messenger formation in olfactory transduction, *Nature (London),* 344, 65-68, 1990.

Brissette, P., Ballou, D.P., and Massey, V., Determination of the dead time of a stopped-flow fluorometer, *Anal. Biochem.,* 181, 234-238, 1989.

Calviello, G. and Ciesi, M., Rapid kinetic analysis of the Ca^{+2}-release channels of skeletal muscle sarcoplasmic reticulum: the effect of inhibitors, *Biochemistry,* 28, 1301-1306, 1989.

Cano-Gauci, D.F., Busche, R., Tummler, B., and Riodan, J.R., Fast kinetic analysis of drug transport in multidrug resistant cells using a pulsed quench-flow apparatus, *Biochem. Biophys. Res. Commun.,* 167, 48-53, 1990.

Cash, D.J., Subbarao, K., Bradbury, J.R., and Mayes, G.M., Filter assay technique and quench-flow experiments: examples of receptor-mediated transmembrane ion-exchange measured with membrane vesicles, *J. Biochem. Biophys. Meth.,* 23, 151-161, 1991.

Chance, B., The reactions of catalase in the presence of the notatin system, *Biochem. J.,* 46, 387-402, 1950.

Chance, B., Accelerated and stopped flow methods using spectrophotometric measurements, in *Technique of Organic Chemistry,* 2 ed., Vol. VIII, Friess, S.L., Lewis, E.S., and Weissberger, A., Eds., Interscience Publishers, New York, 728-748, 1963.

Chance, B., The pulsed-flow apparatus, in *Rapid Mixing and Sampling Techniques in Biochemistry,* Chance, B., Eisenhardt, R.H., Gibson, Q.H., and Lonberg-Holm, K.K., Eds., Academic Press, New York, 125-130, 1964.

Champeil, P., Combettes, L., Berthon, B., Doucet, E., Orlowski, S., and Claret, M., Fast kinetics of calcium release induced by myo-inositol trisphosphate in permeabilized rat hepatocytes, *J. Biol. Chem.,* 264, 17665-17673, 1989.

Chen, R.F., Schechter, A.N., and Berger, R.L., Stopped-flow fluorometry with available instrumentation, *Anal. Biochem.,* 29, 68-75, 1969.

Chen, Z., Pace, U., Heldman, J., Shapira, A., and Lancet, D. Isolated frog cilia: a preparation of dendritic membranes from chemosensory neurons, *J. Neurosci.,* 6, 2146-2154, 1986.

DeVault, D., Photochemical activation apparatus with optical laser, in *Rapid Mixing and Sampling Techniques in Biochemistry,* Chance, B., Eisenhardt, R.H., Gibson, Q.H., and Lonberg-Holm, K.K., Eds., Academic Press, New York, 165-173, 1964.

Doyle, M.L., Speros, P., LiCata, V.J., Gingrich, D., Hoffman, B.M., and Ackers, G.K., Linkage between cooperative oxygenation and subunit assembly of cobaltous human hemoglobin, *Biochemistry,* 30, 7263-7271, 1991.

Eigen, M. and DeMaeyer, L., A combination of temperature-jump and flow techniques, in *Rapid Mixing and Sampling Techniques in Biochemistry,* Chance, B., Eisenhardt, R.H., Gibson, Q.H., Lonberg-Holm, K.K., Eds., Academic Press, New York, 175-179, 1964.

Fersht, A.R. and Jakes, R., Demonstration of two reaction pathways for the aminoacylation of tRNA. Application of the pulsed quenched-flow technique, *Biochemistry,* 14, 3350-3356, 1975.

Gibson, Q.H. and Milnes, L., Apparatus for rapid and sensitive spectrophotometry, *Biochem. J.,* 91, 161-171, 1964.

Gibson, Q.H., Rapid mixing: stopped-flow, in *Methods in Enzymology,* Vol. XVI, Kustin, K., Ed., Academic Press, New York, 187-228, 1969.

Greenwood C., Photochemical activation apparatus using flash tubes, in *Rapid Mixing and Sampling Techniques in Biochemistry,* EChance, B., Eisenhardt, R.H., Gibson, Q.H., Lonberg-Holm, K.K., ds., Academic Press, New York, 157-164, 1964.

Gutfreund, H., Rapid Mixing: continuous flow, in *Methods in Enzymology,* Vol. XVI, Kustin, K., Ed., Academic Press, New York, 229-249, 1969.

Hartridge, H. and Roughton, F.J.W., A method of measuring the velocity of very rapid chemical reactions, *Proc. Roy. Soc. Lond.,* A104, 376-394, 1923.

Hayashi, Y., Zviman, M.M., Brand, J.G., Teeter, J.H., and Restrepo, D., Measurement of membrane potential and $[Ca^{2+}]_i$ in cell ensembles: Application to the study of glutamate taste in mice, *Biophys. J.,* 71, 1057-1070, 1996.

Huang, L., Shanker, Y.G., Dubauskaite, J., Zheng, J.Z., Yan, W., Rosenzweig, S., Spielman, A.I., Max, M., and Margolskee, R.F., Gγ13 colocalizes with gustducin in taste receptor cells and mediates IP_3 responses to bitter denatonium, *Nature Neurosci.*, 2, 1055-1062, 1999.

Inesi, G., Rapid kinetic characterization of active transport and passive release of calcium in vesicular fragments of longitudinal and junctional sarcoplasmic reticulum, *Brazil. J. Med. Biol. Res.,* 21, 1241-1249, 1988.

Jennings, P.A. and Wright, P.E., Formation of a molten globule intermediate early in the kinetic folding pathway of apomyoglobin, *Science,* 262, 892-896, 1993.

Jennings, P., Finn, B., Jones, B., and Matthews, C.R., A re-examination of folding mechanism of DHFR from *E-coli* verification and refinement of a four channel model, *Biochemistry,* 32, 3783-3789, 1993.

Langowski, J., Urbanke, C., and Schuppe, E., Construction of microprocessor-controlled pulsed quench-flow apparatus for the study of fast chemical and biochemical reactions, *Anal. Biochem.,* 142, 91-97, 1984.

Ly, A.M. and Michaelis, E.K., Solubilization, partial purification and reconstitution of glutamate- and N-methyl-D-aspartate-activated cation channels from brain synaptic membranes, *Biochemistry,* 30, 4307-4316, 1991.

Lymn, R.W. and Taylor, E.W., Transient state phosphate production in the hydrolysis of nucleoside triphosphates by myosin, *Biochemistry,* 9, 2975-2983, 1970.

Mann, C. and Matthews, C.R., Structure and stability of an early folding intermediate of E-coli TRP aporepressor measured by far UV-CD and ANS binding, *Biochemistry,* 32, 5282-5290, 1993.

Mihic, S.J., Wu, P.H., and Kalant, H., Differences among effects of sedative-hypnotic drugs on GABA-mediated chloride flux: quench-flow studies, *Brain Res.,* 555, 259-264, 1991.

Millikan, G.A., Photoelectric methods of measuring the velocity of rapid reactions. III. A portable micro-apparatus applicable to an extended range of reactions, *Proc. Roy. Soc. Lond.,* A155, 277-292, 1936.

Mullins, L.S., Pace, C.N., and Raushel, F.M., Investigation of ribonuclease T_1 folding intermediates by hydrogen-deuterium amide exchange-two-dimensional NMR spectroscopy, *Biochemistry,* 32, 6152-6156, 1993.

Orlowski, S. and Champeil, P., Kinetics of calcium dissociation from its high-affinity transport sites on sarcoplasmic reticulum ATPase, *Biochemistry,* 30, 352-361, 1991.

Palmer, S., Hughes, K.T., Lee, D.Y., and Wakelam, M.J.O., Development of a novel, Ins (1,4,5) P_3-specific binding assay. Its use to determine the intracellular concentration of Ins (1,4,5) P_3 in unstimulated and vasopressin-stimulated cells, *Cell. Signal.,* 1, 147-156, 1989.

Palmer, S. and Wakelam, M.J.O., Mass measurement of inositol 1,4,5-trisphosphate using a specific binding assay, in *Methods in Inositide Research,* Irvine, R.F., Ed., Raven Press, New York, 127-134, 1990.

Pennington, S.A., Brown, H.D., Berger, R.L., and Evans, W.J., Analytical application of microcalorimetry: a stop-flow mixing device, *Anal. Biochem.,* 32, 251-257, 1969.

QFM-5 Quench-Flow Module User's Manual, Bio-Logic (Instruments de Laboratories), Claix, France; Molecular Kinetics, Inc., Pullman, WA.

Radford, S.E., Dobson, C.M., and Evans, P.A., The folding of hen lysozyme involves partially structured intermediates and multiple pathways, *Nature,* 358, 302-307, 1992.

Rosenzweig, S., Yan, W., Dasso, M., and Spielman, A.I., Possible novel mechanism for bitter taaste mediated through cGMP, *J. Neurophysiol.*, 81, 1661-1665, 1999.

Rossi-Bernardi, L., and Berger, R.L., The rapid measurement of pH by the glass electrode, *J. Biol. Chem.*, 243, 1297-12302, 1968.

Roughton, F.J.W. and Millikan, G.A., Photoelectric methods of measuring the velocity of rapid reactions, I. General principles and controls, *Proc. Roy. Soc. Lond.*, A155, 258-269, 1936.

Roughton, F.J.W. and Chance, B., Rapid reactions, in *Technique of Organic Chemistry*, 2 ed., Vol. VIII, Friess, S.L., Lewis, E.S., and Weissberger, A., Eds. Interscience Publishers, New York, chap. XIV, 704-727, 758-778, 1963.

Schoenmakers, T.J.M., Visser, G.J., Flik, G., and Theuvenet, P.R., Chelator: an improved method for computing metal ion concentrations in physiological solutions, *BioTechniques*, 12, 870-874, 1992.

Shearer, G.L., Kim, K., Lee, K.M., Wang, C.K., and Plapp, B.V., Alternate pathways and reactions of benzyl alcohol and benzaldehyde with horse liver alcohol dehydrogenase, *Biochemistry*, 32, 11186-11194, 1993.

Speros, P.C., LiCata, V.J., Yonetani, T., and Ackers, G.K., Experimental resolution of cooperative free energies for the ten ligation species of cobalt (II)/iron (II)-CO hemoglobin, *Biochemistry*, 30, 7254-7262, 1991.

Spielman, A.I., Brand, J.G., and Wysocki, L., A rapid method of collecting taste tissue from rats and mice, *Chem. Senses*, 14, 841-846, 1989.

Spielman, A.I., Nagai, H., Dasso, M., Breer, H., Boekhoff, I., Huque, T., Whitney, G., and Brand, J., Rapid kinetics of second messenger production in bitter taste, *Am. J. Physiol.*, 270, C926-931, 1996.

Spielman, A.I. and Brand, J.G., Collection of taste tissue from mammals, in *Experimental Cell Biology of Taste and Olfaction*, Spielman, A.I. and Brand, J.G., Eds., CRC Press, Boca Raton, FL, 25-32, 1995.

Striem, J.B., Naim, M., and Lindemann, B., Generation of cyclic AMP in taste buds of the rat circumvallate papilla in response to sucrose, *Cell Physiol. Biochem.*, 1, 46-54, 1991.

Sturtevant, J.M., A stopped-flow apparatus, in *Rapid Mixing and Sampling Techniques in Biochemistry*, Chance, B., Eisenhardt, R.H., Gibson, Q.H., and Lonberg-Holm, K.K., Eds., Academic Press, New York, 89-102, 1964.

Tareilus, E. and Breer, H., Rapid kinetics of depolarization-induced changes in intrasynaptosomal calcium concentrations, *Neurochem. Intl.*, 20, 275-279, 1992.

Tareilus, E., Boekhoff, I., Spielman, A.I., and Breer, H., Approaches for monitoring rapid kinetics of second messenger signaling, in *Experimental Cell Biology of Taste and Olfaction*, Spielman, A.I., and Brand, J.G., Eds., CRC Press, Boca Raton, FL, 193-202, 1995.

Tonomura, B.I., Nakatani, H., Ohnishi, M., Yamaguchi, I., and Hirami, K., Test reactions for a stopped-flow apparatus, *Anal. Biochem.*, 84, 370-383, 1978.

Varley, P., Gronenborn, A.M., Christensen, H., Wingfield, P.T., Pain, R.H., and Clore, G.M., Kinetics of folding of all-β sheet protein interleukin-1β, *Science*, 260, 1110-1113, 1993.

Walinder, O. and Joshi, J.G., Mechanism of action of rabbit muscle phosphoglucomutase. Rate of enzyme phosphate turnover studied with a rapid mixing technique, *J. Biol. Chem.*, 249, 3166-3169, 1974.

Yan, W. Sunavala, G., Rosenzweig, S., Dasso, M., Brand, J.G., Spielman, A.I. Bitter taste transduced by PLCβ2-dependent rise in IP3 and α-gustducin-dependent fall in cyclic nucleotides, *Am. J. Physiol.-Cell Physiol.*, 280, C742-C751, 2001.

METHODS IN CHEMOSENSORY RESEARCH

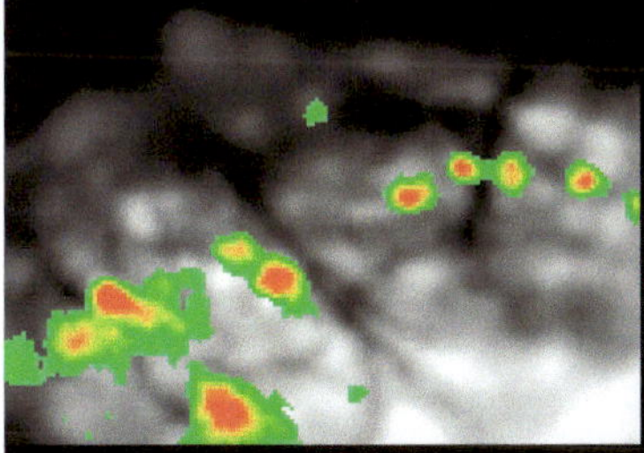

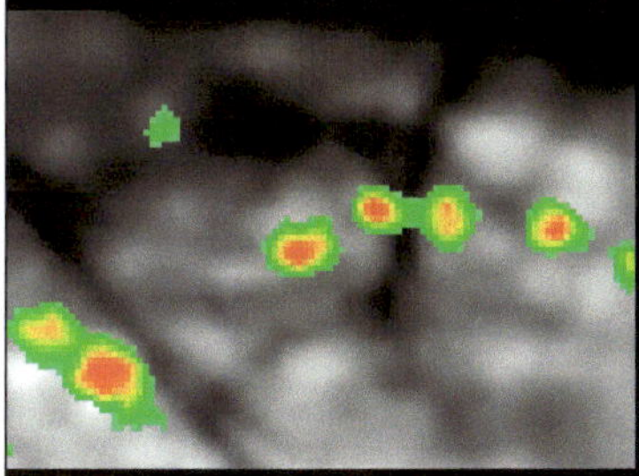

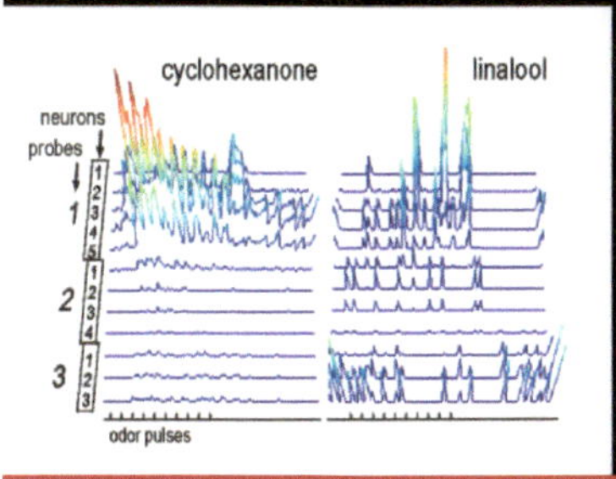

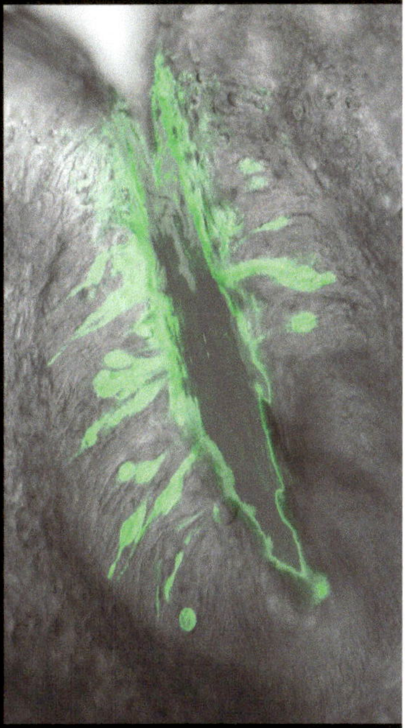

COLOR PAGES

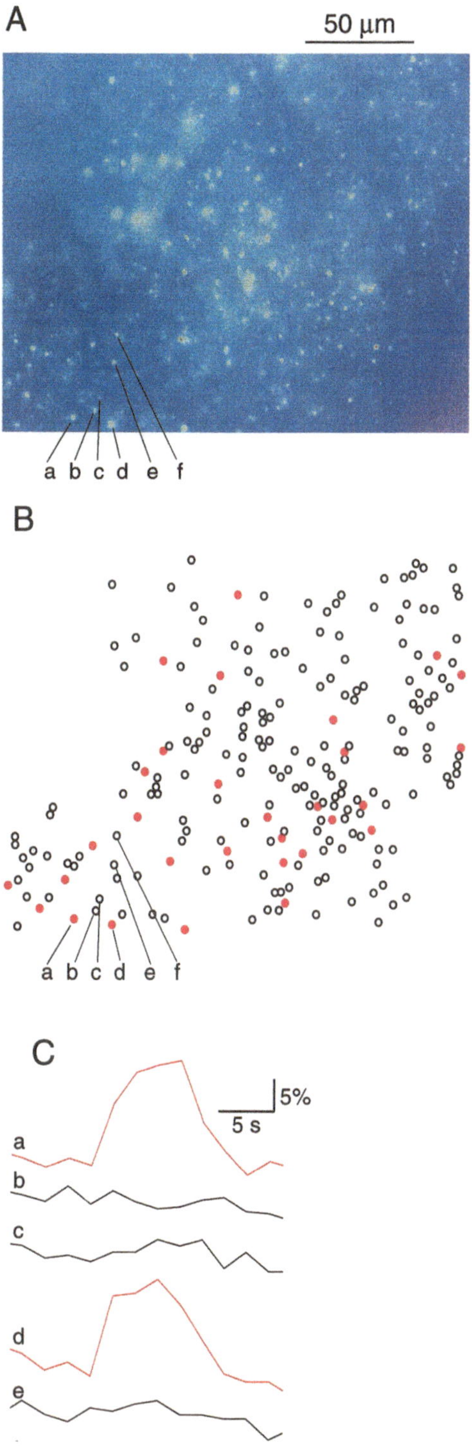

FIGURE 3.8 Mapping odor responses of mouse ORNs in the intact epithelial preparation by calcium imaging. A: Many dendritic knobs can be seen as bright spots under fluorescence illumination in an *en face* view, after a swatch of epithelium is loaded with 15 μ*M* calcium green-1 AM. B: 216 dendritic knobs are mapped for further analysis. A circle represents a single cell. Red-filled circles (30 out of 216) stand for cells that respond to a 4-sec odor stimulus, a mixture containing the following six odors at 10 μ*M*: acetophenone, anisaldehyde, n-amyl acetate, benzaldehyde, cineole, and 3-heptanone. C: The fluorescence intensity changes (ΔF/F) induced by an odor stimulus are measured and calculated for individual dendritic knobs; the traces corresponding to cells "a–f" are shown. (Modified from Ma, M. and Shepard, G.M., *Proc. Natl. Acad. Sci. U.S.A.,* 97, 12869, 2000.)

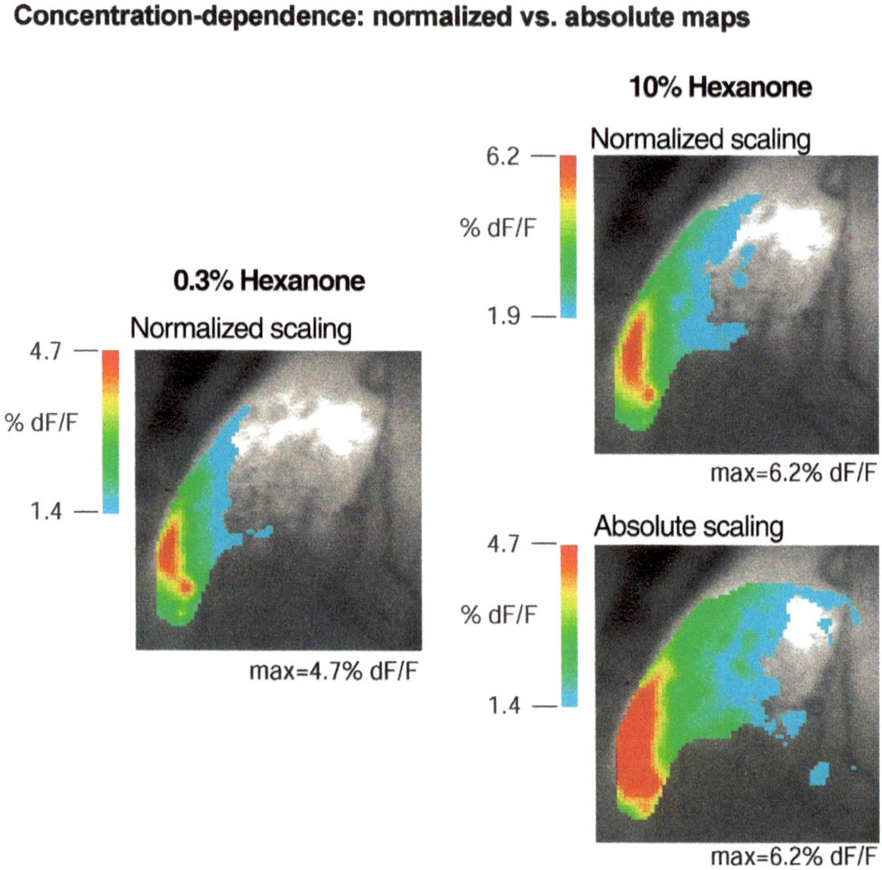

FIGURE 5.5 Normalized maps of receptor neuron input to the turtle olfactory bulb are concentration invariant. The left panel shows a pseudo-color map of the response to a 0.3% dilution of saturated vapor of 2-hexanone. The map is normalized to the maximum signal amplitude for this trial. The right panels show pseudo-color maps of a response to a 10% dilution of 2-hexanone. The map on the top is normalized to its maximum signal amplitude. The map on the bottom (absolute scaling) shows the same data using the same scaling as for the 0.3% trial. The figure shows a concentration-dependent increase in the number of glomeruli activated above a given absolute level, but shows a concentration invariance in the relative levels of input to all glomeruli activated by an odorant. (4× image magnification; the field of view is approximately 4 mm × 4 mm.)

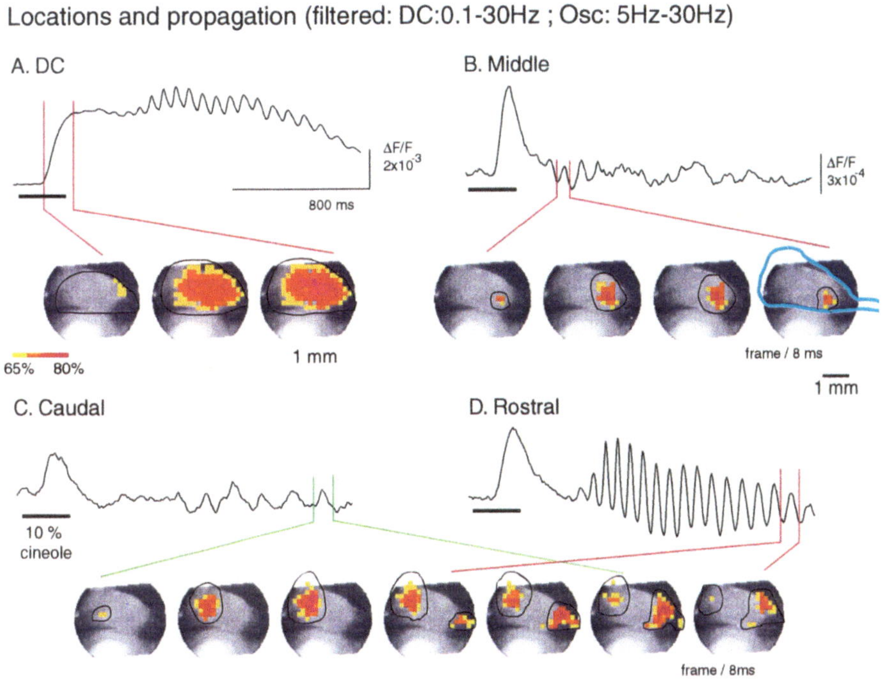

FIGURE 5.7 The locations and propagation of the four components from the trial shown in Figure 5.5. Multiframe pseudo-color displays of the signals are overlaid on the image of the olfactory bulb. The red color and the black contour lines label the areas where the signals are larger than 80% and 20% of the peak. The DC component in this animal covers almost the entire bulb. The other three panels show the position and propagation of one cycle (indicated by the red and green lines) of the three different oscillations. The black horizontal bars indicate the time of the odor command-pulse. The data are filtered by a high-pass digital RC (5 Hz) and low-pass Gaussian (30 Hz) filters. (Modified from Lam, Y. et al., *J. Neurosci.,* 20, 749-762, 1999.)

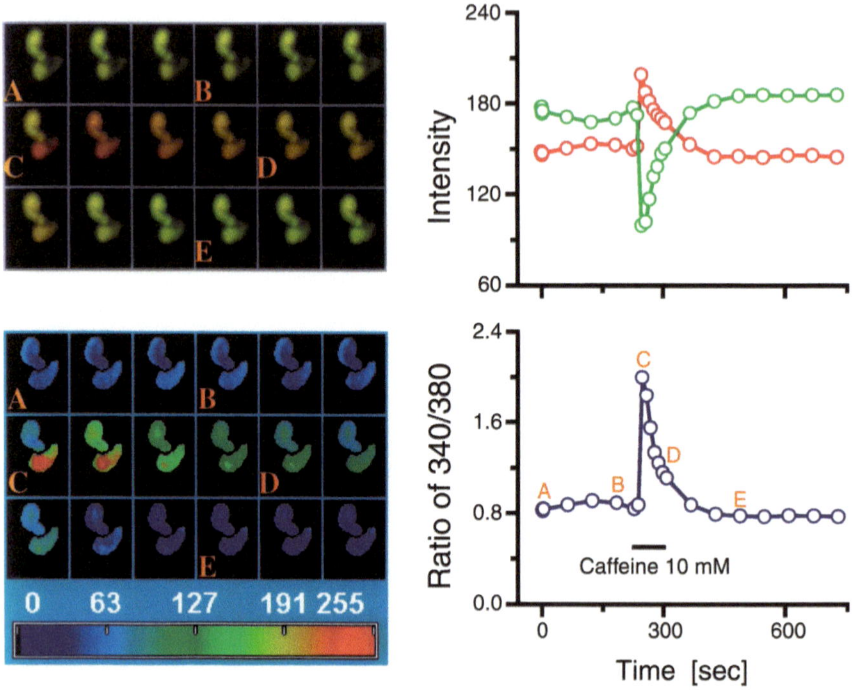

FIGURE 8.5 Example of a pair of dissociated taste receptor cells responding to the application of the bitter tastant caffeine with an increase of cytoplasmic calcium levels using ratiometric imaging with the calcium indicator fura-2. The top panel of images represents an overlay of the fluorescent intensities recorded using 380-nm excitation (green) or 340-nm excitation (red). Notice that when the cell responds to the caffeine stimulus (marked "C"), the emission from the 340-nm excitation greatly increases while that of the 380-nm excitation decreases. The intensities of the individual wavelengths for the lower of the two cells are presented in the upper-right-corner graph. Note that when the stimulus is applied, the two wavelengths cross-over. This causes the 340/380 ratio to increase above unity. The calculated ratios are presented in the graph in the lower right. Note how the cross-over of the 340/380 intensities and the increase in the 340/380 ratios coincide when caffeine is presented. The spatial localization of these changes in the ratio is presented in the montage of the cellular images presented in the lower left. Here, ratios have been calculated from the raw data and presented in a psuedocolored format as indicated in the calibration bar.

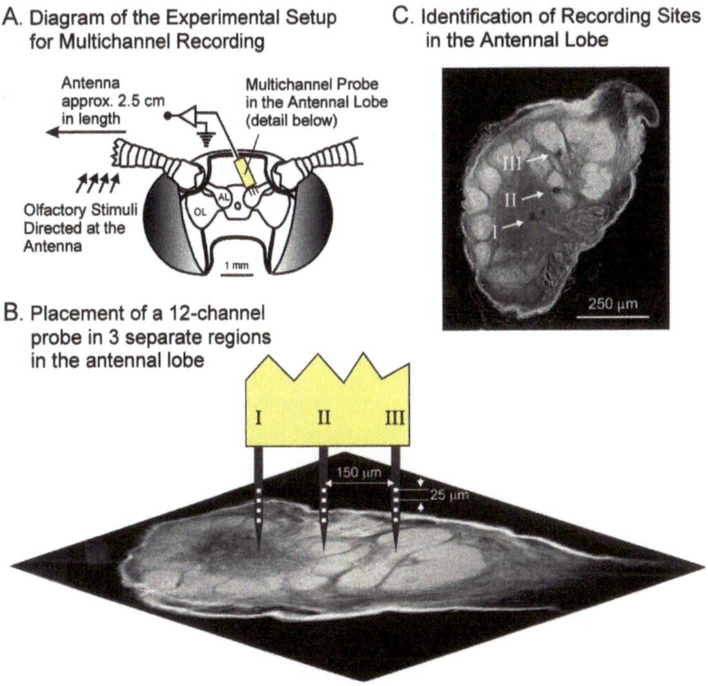

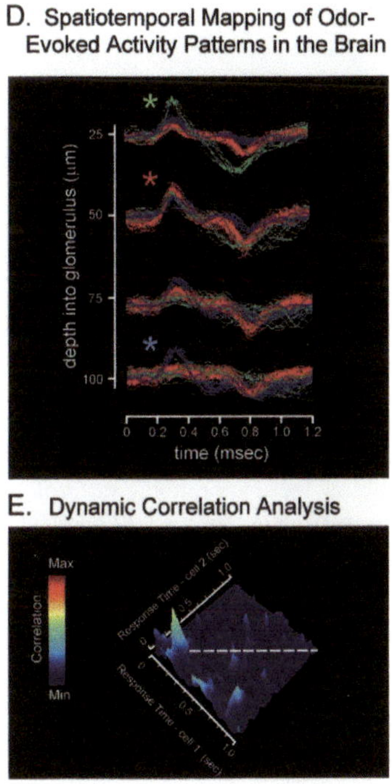

FIGURE 13.5 Method for obtaining simultaneous multineuronal recordings from the antennal (olfactory) lobe of the moth *Manduca sexta*. A: Diagram showing the placement of a silicon microelectrode array in the antennal lobe. B: A three-pronged probe configuration, with four recording sites per prong, is shown to scale with a plane projection of the antennal lobe (with glomeruli visible as the lighter structures). C: The exact locations of the three prongs in the glomerular neuropil can be identified because their penetration sites are visible in a confocal micrograph. D: This method is well-suited for studying the dynamic and distributed representations of odors across groups of olfactory glomeruli. For example, these records show the superimposed spike waveforms from three neurons at four depths through a glomerulus. The trace showing the maximum spike amplitude recorded for each neuron is indicated by an asterisk: the neuron represented by the green traces was closest to the top, while the blue neuron was localized to the base of the glomerulus. E: Dynamic cross-correlation analysis is then used to examine the temporal activity relationships between neurons in the ensemble. A 100-msec odor pulse was presented at time = 0. Note how these two neurons were synchronized only during a brief window of time following the stimulus presentation.

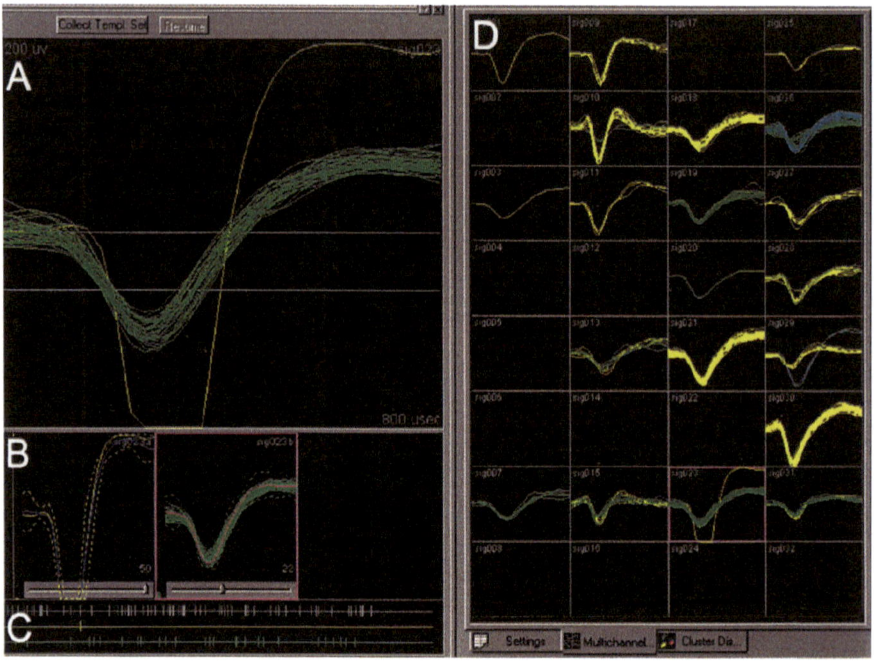

FIGURE 14.3 An example of the acquisition of IC units used in our laboratory (Plexon Inc., Dallas, TX). A: 800-μm segments of the channel 23 waveforms that have cleared the amplitude threshold. Note the two well-isolated single units (in green and yellow). B: Below the large window, the units are presented individually. C: Below these are running rasterplots of unit activity. D: The right half of the screen presents the waveforms from all 32 wires (16 in each IC). Several well isolated units can be seen, and others may be isolable using offline resorting.

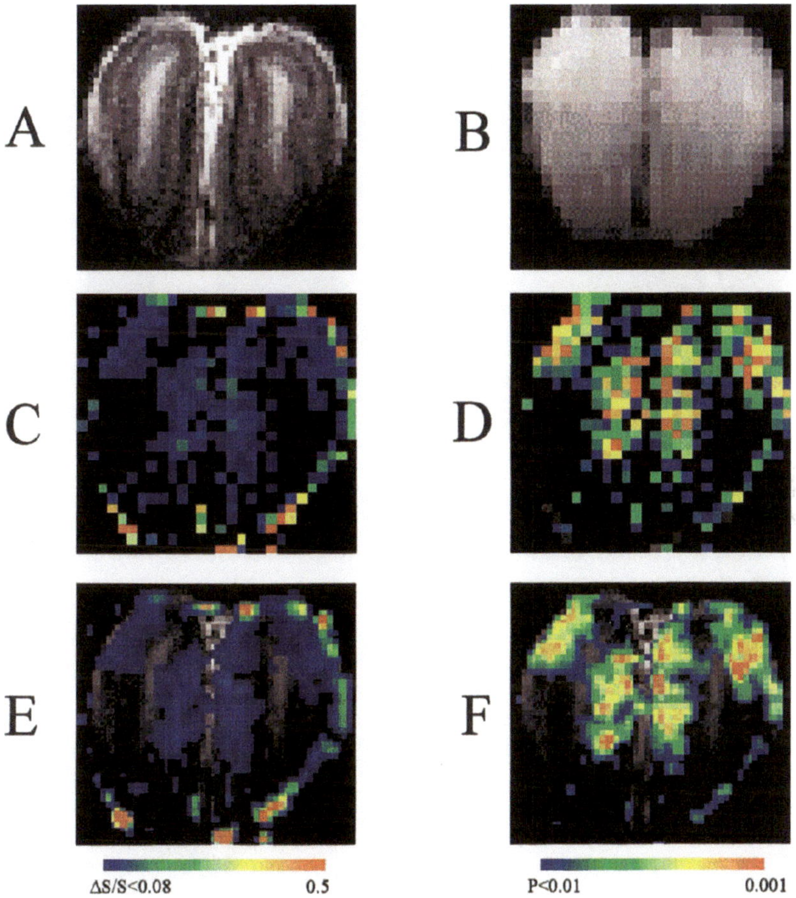

FIGURE 18.3 Data processing in which all images are from the same slice. A: T_1-weighted image of a slice located about 2/3 of the OB from anterior. B: T_2-weighted image. C: $\Delta S/S$ image, the rat is exposed to isoamylacetate for 2 min. D: *t* Image in the same exposure experiment as in C. E: Odor map produced by overlaying C on A. F: Odor map produced by overlaying D on A.

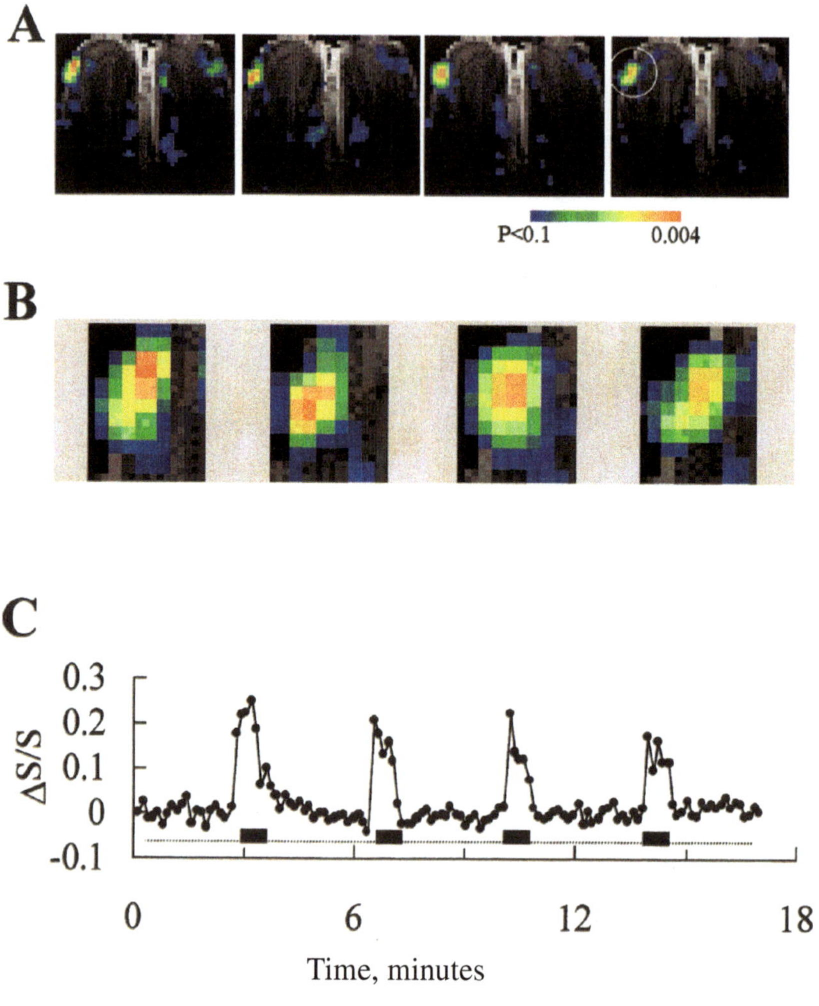

FIGURE 18.4 Odor maps of one slice generated by four repeated exposures (36 sec) of isoamylacetate. A: The activity patterns. B: The enlarged view of the highly activated focus circled in A. C: The corresponding time-course of B.

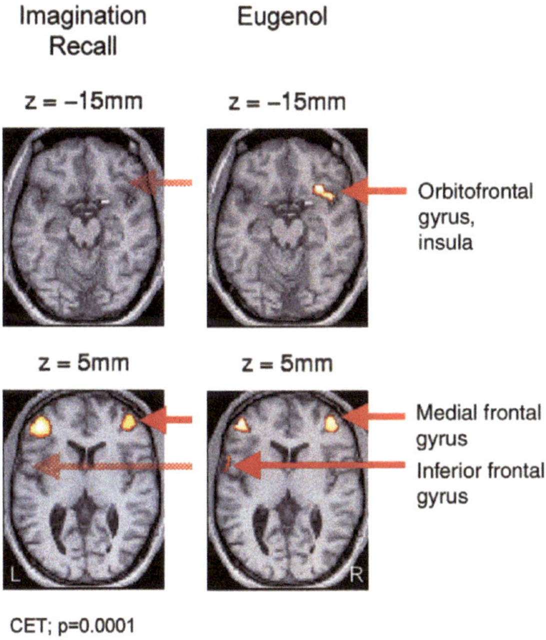

FIGURE 19.1 FMRI comparison between activations induced by the perception of eugenol vs. imagination of the smell of eugenol. Group analysis of 13 subjects using SPM'99. Activity differences between perception vs. non-perception and imagination vs. non-imagination. Two selected horizontal slices showing activities in the frontal and insular cortex. Shaded arrows indicate missing activities either in the perception or the imagination condition. The inferior and orbitofrontal activities are missing in the imagination condition. Not shown in this graph is that the study revealed activity in the posterior part of the medial frontal cortex only present in the imagination condition.

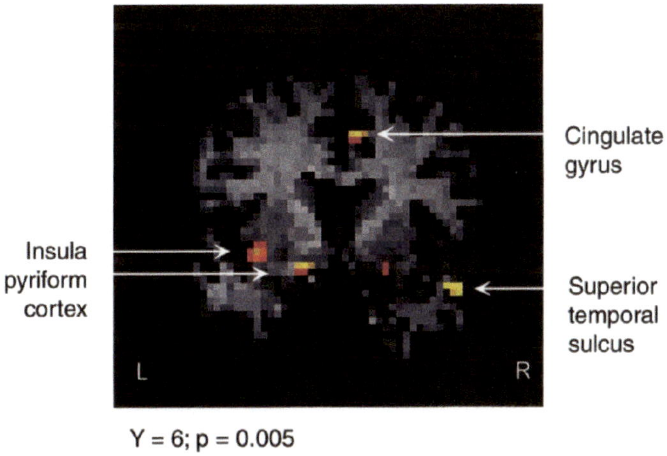

FIGURE 19.2 FMRI example of a single subject. Activities induced by the perception of the unpleasant hydrogen sulfide: pyriform cortex, insula, superior temporal sulcus, and also the cingulate cortex are significantly activated. The cingulate activity is more diffusely distributed in the posterior/anterior direction interindividually.

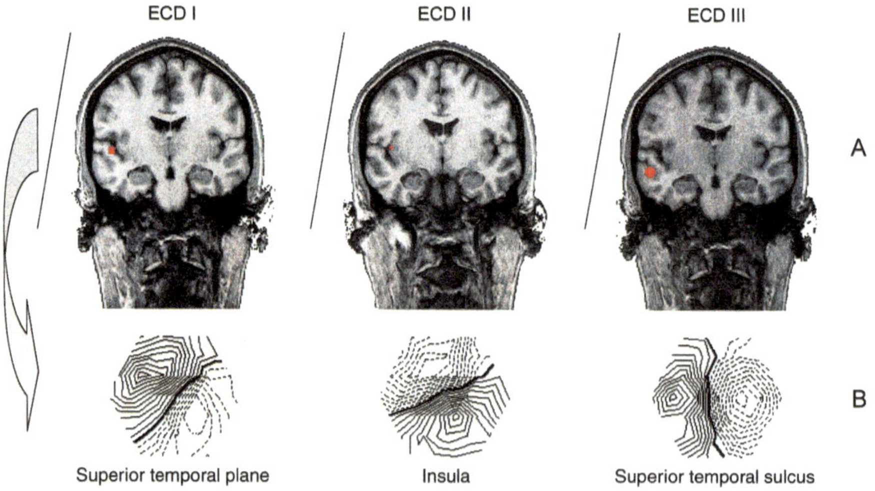

FIGURE 19.3 MSI activities after vanillin stimulation. Example from a single subject. A: MRI graphs including the localizations (size of spots indicates strength of dipole) of the three major equivalent current dipoles (ECDI, ECD II, ECD III) in the superior temporal plane the insular cortex, and the superior temporal sulcus. The activities are separated by approximately 100 msec starting with ECD I at 200 msec after stimulus onset. B: Isomagnetic field line maps (dashed and solid lines indicate the polarity of the magnetic field) recorded with a 36-channel sensor system from the surface of the left hemisphere.

10 Electrophysiological Recordings of Mammalian Taste Nerves

Vicktoria Danilova, Yuri Danilov, Thomas Roberts, Donald Elmer, and Gören Hellekant

CONTENTS

0-8493-2329-0/02/$0.00+$1.50

10.1 INTRODUCTION

Since the first recordings from sensory nerves (Adrian and Zotterman, 1926), the single and multifiber recording technique has been a consistent and frequently used method to collect information on the peripheral senses. In taste, as for the other senses, the experimental methods have been modified extensively. Most notable among these modifications are improved anesthetics and analgesics, refined surgery that allows recovery of the animal, improved control of stimulus parameters, low-noise amplifiers providing better signal-to-noise ratios, and application of computers and software to every aspect of the nerve recordings, from selection of the action potentials to the final publication.

This chapter describes methods to record from the chorda tympani proper (CT) and glossopharyngeal (NG) nerves in mammals ranging in size from the 20-g mouse to the 50-kg calf. The chapter summarizes the authors' knowledge of animal taste nerve recordings and data processing with the aim to facilitate research efforts in this area, and is organized into sections that reflect the general temporal sequence of typical experiments, recordings of the nerve signals, and final processing of the data. Problems, caveats, and pitfalls associated with each step are discussed within each chapter section. Similar suggestions or observations may occur in more than one section in order to stress their importance.

10.2 PREPARATIONS BEFORE SURGERY

The first risk encountered during initiation of anesthesia is aspiration of stomach contents, which can be diminished by food deprivation prior to anesthesia. The period of food deprivation depends on the species. Because the passage of food through the digestive system is fastest in carnivores, 4 to 8 h may be sufficient in the dog and cat. Primates should be food-deprived overnight. We have never food-deprived rodents and have not experienced any problems. Ruminants can be food-deprived over 24 to 48 h.

The other factor of concern in extended surgery is dehydration. In many species (e.g., rodents), fluid consumption is closely associated with food intake. Therefore, the need for food deprivation must be balanced against the risk of dehydration. The risk is highest in young animals such as piglets.

10.2.1 PREMEDICATION

Premedication is advised because it decreases the stress on the animal; it also lowers the needed dosage of anesthetics. Premedication is generally given by i.m. injection, which is the easiest and safest route of administration. The borderline between premedication and anesthesia is not obvious. In many species, we use short-lasting anesthetics as premedication extended into general anesthesia.

In lemur, macaque, and chimpanzee, we have found that intramuscular ketamine (10 mg/kg b.w.) has a fast onset and large safety margin. It will take effect within 5 min and will last for 30–45 min, which provides sufficient time to insert an intravenous cannula. Ketamine can be supplemented with acepromazine that suppresses muscle movements, a beneficial feature in nerve recordings

In pigs, we have found that intramuscular (6 mg/kg b.w.) Telazol® rapidly and painlessly induces a short-lasting anesthesia. It is a combination of tiletamine (a ketamine-like drug) and zolazepam (a diazepam-like drug). Telazol® has also been reported to be an excellent drug in chimpanzees. In cattle, we use i.m. injection of xylazine (0.2 mg/kg b.w.), generally available as Rompun®. It provides rapid sedation and muscle relaxation.

In cat, intramuscular (1 mg/kg b.w.) xylazine combined with intramuscular (15 mg/kg b.w.) ketamine given in a separate syringe causes sedation within 10 min. In dog, xylazine (2 mg/kg b.w.) combined with i.m. ketamine (5 mg/kg b.w.; given in a separate syringe) provides sedation within 10 min. In both species, sedation is followed by a short-lasting surgical anesthesia.

Atropine is generally included as a part of premedication although we do not routinely administer atropine because of its effect on heart rate. Without atropine, the heart rate can be used as a monitor of anesthetic depth.

After premedication, it is advisable to return the animal to its cage or an environment with which it is familiar with, because then the desired relaxation is achieved faster and with less dosage.

10.3 ANESTHESIA

Today, many new anesthetics are available and the reader is encouraged to consult veterinary anesthesiologists, other investigators, and the literature. The choice of anesthetic drug depends on whether the animal will be kept on parenteral or inhalation anesthesia.

10.3.1 PARENTERAL ANESTHESIA

For parenteral anesthesia, our first drug of choice has been a barbiturate because it functions in most species. There are several kinds available; differing mainly in the duration of their effects, which vary between ultrashort and long-lasting. The choice is a compromise between the ability to quickly adjust depth of anesthesia, which makes an ultrashort barbiturate the best choice, and the convenience of not having to inject almost continuously with the need to maintain a steady depth of anesthesia, which makes a long-lasting barbiturate the best choice.

We use sodium pentobarbital, a barbiturate with intermediate duration, in rodents, calves, lemurs, marmosets, and macaques. Its dosage cannot simply be calculated based on animal weight, especially when dealing with a species for which there are no surgical data available, although generally, 20 to 40 mg/kg b.w. can be used. Because of the rapid onset of anesthesia of barbiturates given intravenously, vital signs (respiration, heart rate, and color of mucous membranes and tongue) and reflexes (nociceptive, blink, muscle tonus, and tongue retrieval) can be used to determine how much should be given.

In small species, it is crucial to dilute the barbiturate; the smaller the species, the more dilute the barbiturate. In mice, it was diluted to 5 to 6 mg/ml and injected intramuscularly. Recently, in mice, we have switched to a mixture containing 2 ml of 10 mg/ml ketamine, 1 ml of 20 mg/ml xylazine, and 8.5 ml saline. It gives a good anesthesia for approximately 1 h. The mixture is injected intramuscularly at a dose of 0.05 ml/10 g b.w.

10.3.2 INHALATION ANESTHESIA

Over the years, we have found that it is generally better to use inhalation anesthesia. It allows one to better control the depth of the anesthesia and the animal, if desired, is awakened within a few minutes after turning off the anesthetic.

We use isoflurane or halothane through a vaporator in a system that vents the expired gas mixture into an exhaust system. The anesthetic is mixed with either oxygen or air. We have used halothane, for up to 20 h of anesthesia at a concentration between 0.8 and 1.0%, in rabbits, pigs, calves, and rhesus monkeys. In chimpanzee, we have used halothane in oxygen and nitrous oxide for up to 12 h without more than temporary discomfort after the anesthesia. In mice, isoflurane (0.4 to 0.6%) in oxygen will maintain the mouse for up to 8 h.

10.3.3 Depth of Anesthesia

Textbooks in surgery and anesthesia outline different levels and planes of anesthesia and the means to detect these. It is not necessary to know all of them, but it is useful to have some basic knowledge. We initially monitor the depth of anesthesia using pain reflexes from squeezing the paw, the eyelid-closing reflex, and the lingual reflex. The lingual reflex is very resilient to anesthesia. It is triggered when the tongue is pulled and the presence of the reflex manifests itself as a retraction of the tongue. All these reflexes should be abolished during general anesthesia. We use respiration as our monitor which, with the depth of anesthesia, changes from a shallow faster type to a slower more abdominal type. During a long experiment, the respiration should be a calm, mostly abdominal diaphragmatic pattern and not shallow.

Monitoring gas levels in inspired and expired air offers more control of the surgical status of the animal. We use an Ohmeda 5250 RGM system in most species. This system monitors several parameters, such as levels of CO_2 in inspired and expired air, respiratory rates, heart rate, and blood oxygen level. Monitoring CO_2 in the expired air can prevent acidosis or alkalosis. It is important to maintain an adequate flow rate of gases in the inhalation system to prevent back-breathing of expired CO_2. In rhesus monkey, we use a flow rate of at least 2 l/min.

It is sometimes difficult to abolish all motor reflexes with general anesthesia. In these situations, administration of a neuromuscular blocking agent can be useful in suppressing these motor reflexes. In a non-terminal experiment, it is necessary to use a compound with short-lasting effects such as succinyl choline. Then the animal must be put on a respirator.

10.3.4 Intubation

To diminish the risk of aspiration of foreign material into the lungs, especially in taste experiments when the oral cavity is flooded with fluids, the trachea should be intubated. This can be done by intubation through the oral cavity or by tracheotomy. Because of the trauma associated with tracheotomy, oral intubation is recommended in non-terminal experiments.

If the route of intubation through the oral cavity is chosen, it must be done before surgical depth of anesthesia has been reached so that the cough reflex can be used. The reflex opens up the entrance to the trachea so that the tube can be guided into the trachea. Frequently, the animal coughs at the surgeon when intubated; a face mask and eyeglasses will provide protection. If the anesthesia is too deep, this reflex, which is crucial for entering the trachea instead of esophagus, disappears.

We have found condensation inside the endotracheal tube, as it exits the mouth, in synchrony with respiration to be the best indication that the tube is in the trachea and not in the esophagus. The air expired from the lungs is at body temperature and saturated with water that will condense on the area where the walls of the tube are exposed to room temperature. It is possible to have air from the stomach coming out of the tube synchronous with the respiration, but it will not be saturated with water. Since acquiring the Ohmeda gas monitor, we use CO_2 fluctuations to determine whether or not the tube is inside the trachea.

A set of different-sized laryngoscopes is invaluable and saves time. Sometimes, they make the difference between failure and success. Because the trachea has no capacity to expand and a good fit of the endotracheal tube is necessary, one should have several sizes of endotracheal tubes available. We use commercial pediatric endotracheal tubes with cuffs in animals larger than 1 kg b.w. Tubes with inflatable cuffs prevent fluid from entering the lungs by sealing the space between the tube and the trachea. It also helps to secure the position of the tube. The cuff should be moderately inflated with a syringe filled with air; otherwise, excessive pressure will cause necrosis.

In animals less than 1 kg b.w., such as rats, commercial tubes are not available and other solutions must be found. For these species, we use a suitable size of semisoft laboratory tubing or make tubes of polyethylene plastic.

The other solution to keep the lungs free of foreign material during anesthesia is tracheotomy. The skin is incised with a longitudinal cut below the larynx, and the trachea exposed by blunt dissection to cause minimum trauma and bleeding. The trachea is then opened between two cartilages. This can be done with virtually no bleeding. The tube can be removed even after several hours of anesthesia, the tissue around the trachea realigned, and the wound — but not the trachea — sutured. Our experience is that suturing the trachea creates more problems than non-suturing. Making sure that there is no bleeding, realignment of the tissue around the trachea and merely suturing the wound in surrounding tissue and skin appears to suffice. We have successfully used tracheotomy in many non-terminal experiments.

Finally, it is important to ensure that the end of the endotracheal tube is somewhere between the larynx and the bifurcation of trachea so that obstruction of airflow to one or the other lung lobes does not occur.

In calves, we have found it necessary to also intubate the esophagus to drain stomach content. Microorganisms in the ruminant forestomachs continuously produce gases. Intubation will ease the intraruminal pressure and prevent stomach contents from entering the oral cavity and interfere with the taste stimuli.

10.3.5 Heart Rate Monitoring

Intact circulation is crucial for survival during surgery and the heart rate is a crucial parameter of the surgical status. The deeper the anesthesia, the slower the heart rate. Heart rate should not be allowed to drop significantly from the initial level. Pigs and rabbits appear to be particularly prone to cardiac arrest during surgery. We therefore monitor heart rate continuously throughout the experiment. There are several ways to do this. Because every electrophysiological laboratory has an oscilloscope available, it is easy to monitor the ECG using two intracutaneous needles. We use subdermal needle electrodes (Grass™) connected to a regular AC amplifier and then to an oscilloscope or a loudspeaker, or both. The trace on the oscilloscope will show the QRS complex of the animal's ECG, which can then be used to trigger a heart rate counter. There are other ways to monitor the heart (e.g., esophageal devices), but anything that involves the mouth creates a problem in taste recordings. There are also several commercial ECG devices. Many of these devices monitor, in addition to heart rate, several other parameters such as oxygen in capillary blood and CO_2 levels in air flow. We have found them extremely useful in larger animals, but of limited use in small animals.

Cardiac arrhythmia in rhesus monkeys is treated with 0.02 to 0.1 ml of 1% lidocaine intravenously, together with a similar dose of atropine. If the heart does not stabilize, the experiment is interrupted.

10.3.6 Respiratory Arrest

Short-lasting respiratory arrest occurs infrequently, sometimes at the initiation of anesthesia, and should not create a problem if the infusion of anesthesia is temporarily stopped. If needed, the thorax can be compressed a few times until spontaneous respiration is resumed.

It is highly advisable to have access not only to an animal ventilator, but to have one in the surgery room adapted for the species in question. Thus, the ventilator should be set to the approximate tidal volume of the animal before anesthesia, so that no time is lost if artificial respiration is needed.

10.3.7 Body Temperature

To maintain and control body temperature is vital. We use thermocouples to monitor the rectal temperature and show it on an LCD display. The smaller the animal, the easier it becomes either hypo- or hyperthermic. It does not take many minutes for a small anesthetized mouse to become hypothermic when kept at room temperature, while the room temperature very slowly affects a large animal. Hypothermia of about 1 to 2°C does not hurt and will deepen the anesthesia. However, it will also retard awakening after the experiment. Hyperthermia is never an advantage and increases the stress on the animal. This is especially important in pigs, which, due to their heavy layer of fat, more easily than in most other species can become hyperthermic.

10.3.8 Fluid Replacement

An anesthetized animal loses fluid continuously. To maintain the animal, the fluid should be replaced. This will also shorten postoperative recovery. The method of replacement is dependent on the presence or absence of an intravenous cannula. A regular intravenous device for pediatric or human use suffices in animals larger than 1 kg b.w. Animals below the size of a rat are usually not so easy to cannulate intravenously. Then fluid can be replaced with subcutaneous injections every hour or so. This works well also in very small animals such as the hamster or mouse.

The simplest replacement fluid is saline, Ringer's solution, or a similar electrolyte. A better replacement fluid is 5% dextrose and lactated Ringer's solution, which we use routinely at a dose of about 3 to 4 ml/kg b.w./h. If the animal is to survive it is important that the solution be sterile.

It is also important that not too much fluid be given because this will cause edema of both tongue and lungs. It is not necessary to elaborate on the possible effects of fluid in the lungs. Less well-known is that more extensive edema of the tongue will occlude the circulation to the taste buds so that no taste response can be recorded. Visual inspection will reveal a swollen tongue. If the edema has persisted for some time, it may be too late from the point of a successful recording to inject a diuretic, but it will facilitate recovery. We use furosemide, 0.5 to 2 mg/kg intramuscularly.

10.3.9 Position of the Head and Tongue

The dissection and the recordings are greatly facilitated if the head can be fixed. We use head holders of our design in animals of less than 5 kg b.w. In rodents, we have a number of different-sized head holders made of plastic. They consist of a plate that fits in the mouth of the animal and has a hole for the incisors. The plate is held against the hard palate or the molars, with a second plate that pushes against the dorsal surface of the head with a wing nut.

In marmosets of 300 to 400 g b.w., the head holder clamps the head between the occipital bones and the dorsal ridges of the orbits that enclose the eyeball and related structures. In rhesus monkeys weighing 2 to 7 kg b.w., we fix the head between a bar across the upper molars and two-finger like projections coated with semisoft material onto the ventral edge of the orbits. This clamps the head between these four points of contact: one or two upper molars on each side and the maxillary processes of the malar bones.

In animals weighing more than 5 to 7 kg, a head holder causes too much trauma at the points of contact, so we do not use one. On the other hand, because of the greater weight, their heads are more likely to stay in a given position. In larger animals (e.g., a calf or a pig), a number of small cushions filled with sand are useful to stabilize the head and can serve as replacements for the head holder.

10.3.10 Other Considerations

It is important that the surface of the tongue be kept moist during surgery, especially when the aim is to record from the CT (chorda tympani); otherwise, it will take some time before the dried-out fungiform taste buds will function properly. A gauze occasionally drenched in saline or Ringer's solution on the exposed surface will prevent drying.

To improve hygiene and collect waste, we keep the body of the larger animals in a clear plastic bag during surgery and recording. This also diminishes any drop in body temperature during the experiments. Diapers offer another possibility to improve hygiene.

10.4 SURGERY

There are two reasons for promoting good hygiene: to protect the animal and to protect the surgeon. It is important to prevent contamination in both directions. This is particularly important when using primates, which harbor infectious agents pathogenic to humans to a greater extent than non-primates. Aseptic procedures are well-described in surgical textbooks. Normal disinfection and procedures related to surgery are advised in survival experiments. The area should be scrubbed with alcohol as well as disinfectant. Hair at the incision site and surrounding area should be fully removed; otherwise it will interfere with dissection and healing.

On the other hand, there is a major difference between the infectious agents present in an experimental laboratory, where only healthy animals are handled, and the situation in an animal hospital where accumulation of pathogens from sick animals can occur. People with no personal experience in animal invasive experimentation do not appear to fully comprehend this difference. This does not negate the importance of hygiene and cleaniness.

During surgery, we also remove blood from instruments with 3% hydrogen peroxide, which is kept available at all times. We do not recommend any kind of heat sterilization of the instruments during surgery. An open flame dulls fine tips; heated glass beads coagulate proteins and also dull the tips.

In the following, the approaches used to dissect the CT and NG nerves are described. It may not be necessary to point out that this knowledge can only be acquired through hands-on experience and practical training; however, upon recognizing certain landmarks and anatomical structures, one should be able to reach and identify these nerves in any species.

All dissections are made under a binocular surgical microscope movable in all directions. There is an abundance of nerves, blood vessels, and other structures in the neck region. These structures are not fixed in position; therefore, they may not be found in exactly the same position in relation to other structures and may move when the animal is turned into a new position. Subcutaneous structures vary between species. It is therefore vital to keep the anatomical landmarks in mind.

We have used the same approach in all species. These include (1) laboratory species: rat, mouse, hamster, guinea pig, rabbit; (2) farm animals: pig and calf; (3) domestic species: cat, and dog; and (4) a number of non-human primates: tree shrew (*Tupaia glis*), slow loris *(Nycticebus coucang)*, bush baby (*Galago senegalensis*), mongoose lemur (*Lemur mongoz*), mouse lemur (*Microcebus murinus*), white ear-tufted marmoset (*Callithrix jacchus jacchus*), tamarin or yellow-handed marmoset (*Saguinus midas*), rhesus monkey (*Macaca mulatta*), grivet *(Cercopithecus aethiops)*, baboon *(Papio anubis)*, gibbon (*Hylobates lar*), and chimpanzee (*Pan troglodytes*). In all these species, we reach the nerve without cutting any bones through an incision along the lateral border of the mandible, ventral to the jaw. This leaves the mandible and the mandibular joint, as well as the vessels going through the mandible, intact.

10.4.1 Dissection of the Chorda Tympani Nerve in Rodents

The animal is positioned so that its head is tilted backward. After cutting through the skin, we use mostly blunt dissection along the medial side of the mandibular bone. To better expose the area, a retractor is applied on the mandibular angle and pulled toward the nose. We continue to dissect deeper along the flat medial pterygoid muscle. The muscle connects the medial side of the mandible with a process on the ventral side of the skull, the pterygoid process of the sphenoid and palatine bones. We dissect until the posterior trunk of the mandibular nerve appears. The trunk consists of four nerves. The medial one is the lingual nerve. The CT branches from the lingual nerve and is much thinner than the 4 nerves and runs oblique across the mandibular trunk along the lateral side of the muscle. There are large blood sinuses around or next to the CT. It is easy to cause bleeding because the wall of the sinus is very thin. We have found Greene's (1968) *Anatomy of the Rat* very useful.

10.4.2 Dissection of the Chorda Tympani Nerve in Pigs and Calves

The heads are long and the major problem is not in finding the nerve, but the size of the animal and the related problem of positioning the head so that the CT can be dissected under a surgical microscope. The CT is especially long and a large part can be seen before it is covered by the masseter muscle. After freeing the nerve, absorbable gelatin foam is placed underneath it to make it more accessible for recording.

10.4.3 Dissection of the Chorda Tympani Nerve in Cats and Dogs

The ventral approach can be used in both species and presents no particular difficulties.

10.4.4 Dissection of the Chorda Tympani Nerve in Primates (Rhesus Monkey, *Macaca mulatta*)

Although the same muscles, nerves, and blood vessels can be identified, the relative distance between the bulla tympani and the tongue is shorter in primates than in non-primates. The proportion between the vertical and longitudinal dimension of the wound is also different. Because the animal should survive with as little trauma and risk of infection as possible, the methods should be adjusted accordingly. We use the procedure in the rhesus monkey as our model description.

After incision through the skin and superficial muscle layers, the lobes from the parotid gland are the first structures that obstruct the view. To diminish the trauma, these lobes are not cut through, but loosened at their anterior border so they can be retracted caudally. This will expose the mandibular bone, which then is approached just caudal to its caudal edge. As in the rat, it is advised to dissect through the tissue that adheres to the posterior edge of the mandible as close to the mandible as possible. Thus, the dorsal border of the wound will be the parotid gland, while the ventral border is at the level of the angle of the mandible.

The continuing approach is now slightly in the anterior direction, so as to reveal the caudal and medial face of the medial pterygoid muscle. The medial side of the muscle is then followed closely along the caudal edge in the direction of the skull base.

We primarily use blunt dissection for two reasons: less bleeding and less trauma. However, in many cases, it may pay to mobilize about 1 cm of the caudal part of the pterygoid muscle, because it is attached along the mandible. This will expose

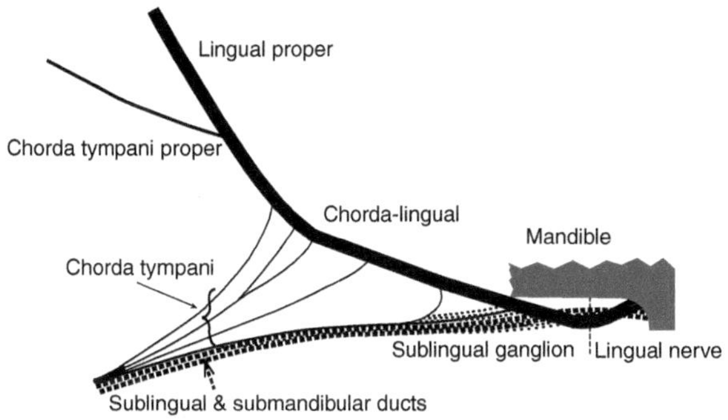

FIGURE 10.1 A generalized diagram of the anatomy of the nerve supply to the mandibular and sublingual salivary glands. (The nomenclature is adopted from Langley, J.N., Textbook of Physiology, E.A. Schaefer, Ed., Young J. Pentland, Edinburgh, 1898).

the lingual nerve at its junction with the CT. There are several other nerves with which the CT can be confused, but this junction is distinct.

Further dissection of the CT nerve is difficult because the CT is more or less completely surrounded with a network of vascular sinuses that anastomose and have very thin walls. We know of no way other than to ligate them because they cross over or surround the nerve. The use of cautery is too dangerous because it might damage the nerve through heat and the hemostasis is only temporary. Diffuse capillary bleeding can be stopped with absorbable gelatin foam.

In our opinion, the rhesus monkey dissection is the most challenging of all the dissections we have performed.

10.4.5 Nomenclature Considerations with Regard to the Chorda Tympani Nerve

The term chorda tympani is being used by different scientists to describe two different anatomical structures: the chorda tympani nerve that contains taste fibers and the chorda tympani nerve that supplies the sublingual and mandibular salivary glands (see Figure 10.1). To avoid confusion, it is important to return to the usage described by Langley (1898), in which the part of the seventh nerve conveying taste fibers and passing through the middle ear be called chorda tympani proper. In the section entitled "The chorda tympani and the nerve cells with which it is connected," Langley wrote the following: "The submaxillary, the sublingual glands, and the glands of the tongue, receive secretory and vaso-dilator fibers from the chorda tympani. The chorda tympani arises from the seventh nerve; it leaves this in the fallopian canal, runs across the tympanum, and joins the lingual branch of the third division of the fifth nerve. The nerve thus formed may be called the *chorda-lingual;* it extends roughly up to the dorsal edge of the sublingual gland; here nearly all the fibers for the submaxillary gland, and about half of those for the sublingual gland, leave the lingual fibres, generally in four or five delicate strands, lying close together. These strands, with the tissue around them, are easily dissected out as a single bundle, and the bundle of nerve strands is called the chorda tympani, although it is a part only of the chorda tympani proper. The chorda tympani curves backwards towards the gland ducts, and accompanies them into the glands. Other fine filaments, coming from the chorda tympani proper, are given off from both sides of the lingual, as it runs forward over the sublingual gland; most of these end in this gland, but a few fibres, varying in number in different animals, run back and supply the submaxillary gland. Finally, a few fibers, from the chorda tympani proper, continue their course in the lingual, and supply the glands and blood vessels in the area of distribution of the lingual nerve in the tongue."

By introducing the term proper, Langley realized and avoided the bewilderment that is bound to occur when the same term is used for the two different anatomical structures: the nerve that has no taste fibers and supplies the sublingual and submaxillary glands, and the nerve that contains, not only all secretory fibers to the chorda tympani, but also taste and other fibers. Figure 10.1, which is a generalized diagram of the anatomy of this nerve supply, demonstrates the fundamental difference between the two chorda tympani nerves, both in regard to fiber types and area of innervation. This is important because the difference between the two nerves is not limited to taste fibers. The fact is that there are more differences between the chorda tympani and the chorda tympani proper than between most peripheral nerves.

In 1973, we reintroduced the term "chorda tympani proper" for the chorda tympani that contains taste and other fibers to distinguish it from the nerve supplying the mandibular and lingual salivary glands, also called the chorda tympani nerve by researchers working on salivation (Hellekant and Kasahara, 1973). We also reintroduced the terminology "lingual proper nerve" for the nerve it joined and "chorda-lingual" according to Langley. While the use of lingual proper and chorda-lingual seem to have been accepted by other investigators, for example, Izumi and Karita (1995), the use of the term "chorda tympani proper" still lags behind.

It can be claimed that omitting this label may be of little importance in communicating results within the same field, be it gustation or salivation, but it is bound to cause misunderstanding when researchers from other fields attempt to read the literature on gustation and salivation. A brief search in the literature brings up a number of references using the same term for different nerves (see, for example, Grant et al., 1989; Danilova et al., 1998). Furthermore, as our taste field of research expands and interacts with, for example, salivation, we will need to be more precise in our terminology. This confusion can be avoided by using the term "chorda tympani proper" for the nerve through the middle ear.

10.4.6 Dissection of the Glossopharyngeal Nerve in Rodents

The incision is done somewhat more ventral than for the CT. We do not use the same entrance as for the tracheotomy because it is more difficult to create a pool of the liquid used to prevent drying of the nerve. Once through the skin, we primarily use blunt dissection. The digastric muscle, a muscle that in contrast to all other muscles in the neck region has two bellies joined with a tendon in the middle, serves as guide. We then cut the digastric muscle through the middle tendon and remove the posterior part. The posterior part of the hyoid bone is cut, which frees the stylohyoid muscle. After removal of the muscle, two small nerves are visible. The dorsal one is the NG. We cut it near to its exit from the posterior lacerated foramen. Because the nerve runs between the external and internal carotid arteries, it is necessary to be cautious. The hypoglossal nerve is cut to prevent tongue movements. To better expose the vallate and foliate taste buds, we cut the angle of the mouth and the masseter muscle bilaterally, and then clamp the tip and stretch the tongue.

10.4.7 Dissection of the Glossopharyngeal Nerve in Pigs

The first structure encountered is a large complex of salivary and lymphoid tissue, which is removed to expose the digastric muscle. The digastric muscle can be moved ventrally with a retractor. This will expose the NG and hypoglossal nerves and arteries. The NG runs dorsal and medial to the much larger hypoglossal nerve, which is close to the branches of the carotid artery. NG is medial to the external maxillary artery whose medial side it crosses. It runs in a rostral direction and disappears under and medial to the hyoid bone. Centrally, the nerve merges with the ovoid gray petrosal ganglion. The nerve distinguishes itself with a very typical meandering course. The angle of the mouth is cut to better expose the taste buds and the tongue is stretched with one or two sutures through the tongue.

10.4.8 Dissection of the Glossopharyngeal Nerve in Calves

The incision is made sagittally, medial to the right mandibular angle. The superficial muscle layers are divided and the intermediate tendon of the digastric muscle located. Blunt dissection medial to the tendon and medial to the ventrocaudal edge of the stylohyoid bone is performed. The hypoglossal nerve is located at the ventrocaudal plane of the stylohyoid bone. By further dissection, medial to the dorsocranial part of the stylohyoid bone, the NG is exposed. It is easily distinguished from the hypoglossal nerve because it is much thinner. The NG is composed of numerous fascicles separated by connective tissue and blood vessels. It contains a relatively small proportion of taste fibres. Frequent anastomoses between the bundles makes them difficult to separate. Furthermore, there are large blood vessels within the nerve. The angle of the mouth is cut to better expose the taste buds.

10.4.9 Dissection of the Glossopharyngeal Nerve in Primates (Rhesus Monkey, *Macaca mulatta*)

The incision is made somewhat more ventral than for the CT. Because the nerve is surrounded by soft tissue, muscles, and other structures, it is more mobile than the CT. The nerve runs dorsal and medial (deeper than) to the much larger hypoglossal nerve, which is close to the branches of the carotid artery. It is medial to the external maxillary artery whose medial side it crosses. It runs in a rostral direction and disappears under (medial to) the hyoid bone. Centrally, the NG merges with the ovoid gray petrosal ganglion. The nerve distinguishes itself with a very typical meandering course. The vallate and foliate taste buds can be exposed with a retractor in the angle of the mouth.

10.5 POST-RECORDING PROCEDURES

In survival experiments, the wound should be inspected under the surgical microscope for any non-tissue residues and rinsed with sterile saline. We have found 3% hydrogen peroxide to be excellent for removing blood clots and to refresh the tissue after recording and before closing. We also mechanically remove any residues on the exposed tissue and its surroundings before suturing. We use nylon sutures for closing the skin, supplemented with absorbable suture if deeper tissue needs to be sutured. The thickness of the suture depends on the size of the animal. We use interrupted mattress sutures for the skin. Interrupted sutures cause less disturbance of circulation and if the wound needs to be drained, this can be done without risk of opening the whole wound. If there is a risk that the animal may tear out the suture, we use subcuticular sutures. It is a good idea to use an antibiotic as prophylaxis against infection. This could be given with surgical premedication.

When the surgical procedures are completed, the animal is returned to a heated cage to recover from anesthesia. Care should be taken so that the animal is not exposed to a high temperature, which can easily happen with an infrared light too close to the cage. We try to position the heat source so that the cage is not uniformly heated. This allows the animal to adjust its temperature according to its need. Post-procedural analgesia should be considered.

10.6 NEURAL RECORDINGS

10.6.1 Stimulation

Taste stimulation is done either with a tongue chamber or an open flow system over the tongue We are using an open flow, gravitationally driven system, the Taste-O-Matic, which we developed over more than 20 years ago. It was first described in 1980 (Hellekant and Roberts, 1995) but has been modified and expanded since then. However, the general outline still applies.

The stimulus flows are controlled through magnetic valves. A total of 31 stimuli can be prepared and are available during the same session without reloading the Taste-O-Matic. The researcher is fully integrated with the Taste-O-Matic system through software, a computer, and an electrostatic chart recorder capable of recording single nerve impulses. We have included a schematic illustration of our stimulation and recording system as Figure 10.2.

Increasingly, we are analyzing gustatory effects of compounds produced in microgram quantities. Derivatives of the bitter compound denatonium benzoate and the sweet protein brazzein are examples of this. These compounds can be dissolved and loaded into 1-ml syringes and dropped into an instrument we call the Micro Taste-O-Matic. This device holds the syringes in register, targeting a selected area of the tongue, and is activated by all system valve codes above 31. A large array of stimuli can be tested and in any order, including in alternation with tastants delivered by the conventional Taste-O-Matic system, simply by selecting the desired valve code. Valve code, nerve response, and other parameters are integrated with the system as with the conventional Taste-O-Matic system.

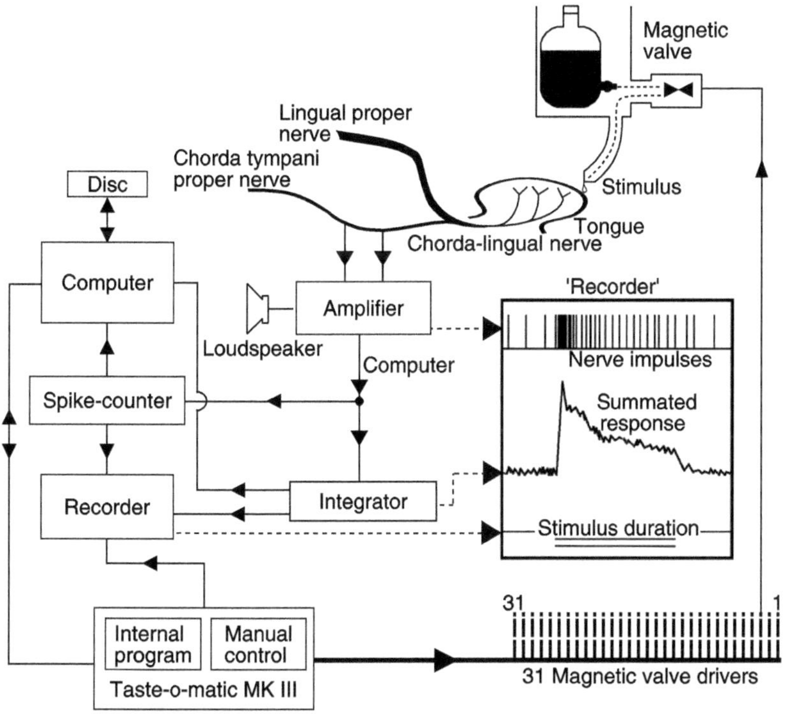

FIGURE 10.2 A block diagram of the recording and stimulation equipment used, together with a diagrammatic illustration of a summated and single-fiber recording.

We record the nerve activity before, during, and after stimulation. To obtain a relevant response to a stimulus, we subtract the activity before a stimulation from the activity recorded during stimulation. The appropriate duration of a stimulation depends on the area stimulated and compound used. The taste buds on fungiform papillae are more accessible than those in the vallate and foliate moats.Therefore, we usually use a shorter stimulation time for the CT recordings (5 to 10 sec), while the stimulation time for the foliate and vallate area has to be at least 10 sec to allow the stimulus to reach these taste buds. To facilitate stimulation, some investigators position a pipette into the trenches of the taste papillae to infuse the stimuli directly to the taste buds (Frank, 1991). This may shorten the time needed for stimulation. The foliate and vallate area can also be made more accessible with a cut in the angle of the mouth. This is particularly useful in species with elongated heads (rodents, pig, and calf).

Furthermore, stimulation time is influenced by differences in the taste profile (temporal pattern, temporal intensity) between compounds. Most stimuli elicit a response with an initial phasic maximum and a protracted, slowly fading static part. NaCl is a typical representative of this type of stimulus. However, some stimuli elicit no phasic response and the response grows slowly after a latency of a few seconds even from the fungiform taste buds. For example, the responses to sweet proteins have a long latency and develop over several seconds. Consequently, using only 5 sec of the responses may not give the appropriate results.

10.6.2 Removal of Stimuli

We think it is important to switch between rinse and stimulus as fast as possible to avoid dilution of stimulus. To avoid mechanical and thermal stimulation, we maintain constant flow and temperature. The taste buds not only respond to tastants, but also to temperature changes and mechanical stimulation. The mechanical artifact is best avoided by switching between stimulus and rinsing medium without any interruption in flow over the tongue.

With regard to thermal artifacts, any difference in temperature between stimulus and rinse will create an artifact. This is particularly important when the taste stimulus is colder than the rinsing medium. Therefore, the temperature difference between tastants and rinsing solution should be as small as possible. We keep it within 0.5°C. This is achieved by keeping tastants and rinsing medium in the same thermostatically controlled bath. These factors appear to be even more important when recording from the NG, which has a larger proportion of nongustatory sensory nerve fibers.

Many investigators use distilled water to rinse the tongue for removal of stimuli and as solvent for the taste stimuli. We have found over the years that distilled water is detrimental to a stable recording condition. Thus, during prolonged rinses with distilled water, the response to NaCl will keep growing, while the response to sweeteners will keep diminishing. For example, the response to aspartame, a good sweet stimulus in the rhesus monkey, is almost abolished after an hour of distilled water rinse of its tongue. Further, the baseline of the summated recording (i.e., the nerve activity between stimulations) will become more and more unstable and the reproducaibility of the responses will diminish. Tap water offers a solution but its composition varies from time to time and from location to location.

We have solved this problem by switching over to rinsing with a solution that is based on the ion contents found in "normal" saliva. This artificial saliva gives nerve recordings that are stable and taste responses that are reproducible for many hours. It is more important to include the major ions present in saliva at approximate normal proportions than try to achieve some ideal level because, as is well-known, the composition of natural saliva varies with the flow rate. In rats, pigs, calves, and primates, we use the recipe presented in Table 10.1. In mice, we use a 50% concentration of our standard recipe. In hamster, in which the relationship Na/K is less than 1, we use a composition based on a study by Rehnberg et al. (1992) (see Table 10.1).

10.6.3 Nerve Recordings

We use conventional electrophysiological equipment for extracellular recordings. Extracellular recordings imply that nerve signal voltages will be in the microvolt to 1 millivolt range. Apparatus should include a low-noise, AC-coupled differential

TABLE 10.1
Composition of Artificial Saliva in Different Species

	Standart Artificial Saliva	Mice Saliva	Hamster Saliva
NaCl	4 m*M*	2 m*M*	
KCl	10 m*M*	5 m*M*	15 m*M*
$NaHCO_3$	6 m*M*	3 m*M*	6.6 m*M*
$KHCO_3$	6 m*M*	3 m*M*	28 m*M*
HCl	3.6 m*M*	1.8 m*M*	
$CaCl_2$	0.5 m*M*	0.25 m*M*	
$MgCl_2$	0.5 m*M*	0.25 m*M*	
K_2HPO_4	0.24 m*M*	0.12 m*M*	
KH_2PO_4	0.24 m*M*	0.12 m*M*	
pH	7	7	8.5

preamplifier with a bandpass of 100 Hz to 10 kHz and an impedance greater than 50 mΩ. A dual-trace storage oscilloscope is preferable, and an audio amplifier and a speaker are essential. Our equipment includes a window comparator for selecting and counting single nerve impulses and an electrostatic recorder with a frequency response approaching 10 kHz, allowing it to record individual nerve impulses on paper with some accuracy.

During recording, the nerve is suspended, but not stretched, by a silver wire whose diameter is roughly that of the nerve itself. The indifferent electrode is a 5-mm diameter silver disc cemented to a glass foot of similar diameter formed at the end of a 20-cm piece of 5-mm outer diameter glass tubing through which the shielded cable passes. The silver foot is applied to tissue as close to the nerve as practical and away from the heart and other muscles. In small animals such as mice, we use a silver plate shaped like an arrow and insert it into the masseter muscle as the indifferent electrode. The animal is also grounded via a silver electrode, surgically implanted. Ordinary RG-174/U coax makes an excellent shielded electrode cable for extracellular recording. By taking full advantage of differential recording and carefully conducting all shielded and ground leads to a common ground point at the preamplifier, we have been successful in recording at several remote primate facilities without the benefit of a shielded room.

We use two methods to record the nerve activity: (1) a summated recording technique, showing the overall nerve activity; and (2) a single fiber recording method, showing the nerve impulse activity in one single taste fiber.

10.6.4 The Summated or Integrated Recording Technique

Summated activity shows the total activity as a trace whose distance from an arbitrary baseline is related to the total nerve activity. It can be used as a guide to choose relevant concentrations of stimulus for single fiber recordings and behavioral work.

There are several methods to measure the nerve activity. The simplest is to use the impulse frequency as a measure. This, however, does not take into account the differences in amplitude of the nerve impulses. This seems to be an important parameter for the information because when this characteristic is omitted, the taste response reflects less well the differences in taste stimuli. We use an instrument that takes both factors into account (Hellekant and Roberts, 1995).

The taste response to most stimuli consists of a strong initial burst of nerve activity, the phasic response, followed by a period of adaptation, the tonic response. The phasic response is the transient part, which occurs during the first few seconds of taste stimulation. The tonic response follows after the initial burst and is affected by the rate of adaptation. It may be important to point out that both phases are true features of the taste response. Both phases are evident in recordings from single fibers, for example, when the individual nerve impulses are recorded.

The equipment used for the summation or integration, the integrator, can have different time constants. Integrators with a long time constant suppress the phasic part. This gives a less faithful picture of the taste response than an integrator with a shorter time constant. Therefore, our integrator has a full-wave absolute value circuit with a time constant of 100 msec so that it displays the two phases. This is important to point out because some investigators suppress the phasic response by using integrators that are unable to show the phasic response. We strongly advise the use of a unit that can show both the phasic and the tonic phase of a taste response.

10.6.5 Maximum Amplitude vs. Surface Area

There is a direct relationship between the intensity of a taste sensation and the size of the summated taste nerve response. Most investigators use either maximum amplitude or surface/integrated area under the summated response as a measure. Because there is a good correlation between these parameters, in most cases it matters little which one is used.

When one records a slow, elongated response as for thaumatin, however, using maximum amplitude probably gives a less true picture of the intensity of the taste sensation than the use of surface area. Comparisons between NG and CT responses present a particular problem because of differences in the onset of response. We then stimulate the two areas over a longer stimulation period and use the integrated area as a measure of response.

10.6.6 The Temporal Profile

The taste of a stimulus can change during the time it is on the tongue, sometimes dramatically. The way this happens can be very different for different compounds. A comparison of the taste profiles of brazzein and sucrose illustrates this. During the first few seconds after brazzein application to the tongue, there is no sweetness at all, while the sweetness of sucrose is tasted within a second. This correlates well with the difference in onset and rise of the taste responses. Summated recordings from brazzein and sucrose show that the sweetness of sucrose reaches its maximum within a few seconds, while the sweetness of brazzein grows slowly. We call this change of taste over time the temporal profile of a tastant. The temporal profile can further be divided into several phases, which have been described elsewhere (Hellekant et al., 1991).

10.6.7 Single Fiber Recording in Large Animals

With this technique, the response in one or a few taste fibers is recorded. It gives information on the impulse activity in individual taste fibers.

Several methods for obtaining single-fiber recordings have been used. We and others have over the years tried micro-electrodes, plates with fine holes through which the fiber grows, and other methods. However, most single-fiber data in the literature have been obtained essentially with the same method as Zotterman and earlier workers used, i.e., teasing the nerves into fine filaments. We have successfully applied this method in species from the size of hamster and larger and found that single fibers from peripheral nerves are usually relatively easy to separate and record. We have in the past used needles on a smooth glass surface covering the indifferent electrode as a dissection stage. Presently, we use the finest possible forceps (Dumont #4) to tease the filaments apart.

Usually, a single fiber lasts for about an hour and there is not enough time to do multiple presentations of all 32 stimuli. While many stimuli are presented once, the four basic standards (NaCl, sucrose, quinine, and acid) are tested several times during stimulation as a control of the recording conditions.

10.6.8 Single-Fiber Recording in Small Animals

In very small animals such as mice, single-fiber recordings present major difficulties. The diameter of the mouse CT or glossopharyngeal nerve is in range of 80 to 120 μm, which makes it very difficult to isolate and maintain single fibers. To solve this problem, we use a neurophysiological technique to record multiunit activity, adapted from multineuron recordings from the CNS (for a review, see Lewicki, 1998). In multiunit recordings, the main problem is that the vast majority of spikes have similar amplitudes. Regular window discriminators are practically useless for spike separation. The solution is to sort the action potentials by shape. The procedure is divided into acquisition of signal and pattern recognition.

There has been significant progress in multichannel data acquisition hardware and software in recent years. Neuroscientists now have a wide selection of equipment, including commercial offerings from Bionics, CED, DataWave Technologies, Multichannel Systems, Plexon, RC Electronics, and others. These systems can acquire multielectrode and multineuron signals, as well as other non-spike signals commonly used in neurophysiological experiments (EEG, ECG, VEP, etc.).

We use one channel of DataWave Technologies' acquisition system "Experimental Workbench 32" for multiple spike separation. The basic data flow is presented in Figure 10.3. The spike train from the output of the main recording system and binary coded stimuli parameters with time marks can be directly connected to the acquisition system during the experiment (online), or can be recorded on tape recorder and analyzed later (offline).

10.7 MULTIFIBER RECORDINGS

In general, during the 2 to 4 h of the recording from whole nerve (mice) or small multifiber bundle (monkey), we can acquire 50,000 to 500,000 waveforms. On

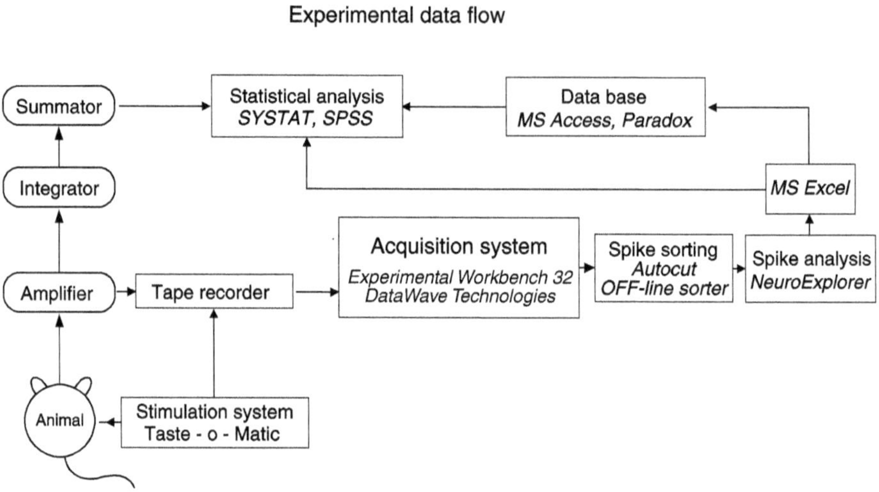

FIGURE 10.3 Experimental data flow for multifiber recordings.

average, we separate from 8 to 20 fibers. After all controls, 2 to 18 single-fiber spike trains remain for further analysis. Each single fiber is represented by 1000 to 25,000 impulses (waveforms).

10.7.1 Step 1: Transferring Analog Electrical Signals to Digitized Spike Waveforms

The acquisition system performs several basic operations: spike detection, waveform acquisition, recording of stimulus parameters, and time marks.

10.7.1.1 Spike Detection

First, we adjust the input signal, choose the optimal gain, and arrange the signal detection level. For detection of a spike, we use the peak detection approach, which means that as soon as the signal crosses the lower detection level, the system initiates the procedures for peak detection and waveform acquisition. The detection level can be chosen to acquire the largest spike only, or to the contrary, to include all spikes even barely separated from the background noise. In the latter case, computer memory and the capacity of the hard disks become the major limitations.

10.7.1.2 Waveform Acquisition

After a peak is detected, the system samples its waveform and stores the digital values in a file. The sampling rate is adjustable. We use a sampling rate of 10 to 20 kHz, which produces 25 to 50 points per single waveform.

10.7.1.3 Recording of Stimulus Parameters and Time Marks

Our stimulation system produces a time mark at the beginning of stimulation and a stimulus code. Both signals are combined as a binary code on a separate channel on the tape recorder. The acquisition system records this binary code, together with the separated waveforms in the same output file.

10.7.2 Step 2: Sorting of Waveforms and Saving the Event Time (Timestamp) for Each Waveform as a Spike Train

Spike sorting is the most difficult part of the analysis, close to an art, and is still a developing area of engineering science. Among the most common methods are feature analysis, principal component analysis, and template matching.

10.7.2.1 Feature Analysis

In general, more features give better distinction between different spikes. One of the earliest approaches to spike sorting was to use spike height and width, or peak-to-peak amplitude. Modern methods utilize several features (Wheeler and Heetderks, 1982; Schmidt, 1984). Our sorting software, Autocut (DataWave Technologies, Inc.), is based on the feature analysis method. The software automatically scores up to 32 parameters. However, for the following cluster analysis, we can use only eight of these features. The combination of eight parameters produces 28 two-dimensional arrays of data points. The program separates, in total, up to 20 clusters of waveforms.

As an alternative to Autocut, we use Off-line sorter (Plexon, Inc.). Off-line sorter separates waveforms recorded by a single electrode into several clusters using template matching, principal component, and cluster analyses.

10.7.2.2 Principal Component Analysis

The idea behind principal component analysis is to find an ordered set of orthogonal basis vectors that capture the directions in the data of largest variation (Glaser and Marks, 1968; Glaser, 1971; Gerstein et al., 1983). The data are the recorded waveforms. In high-level mathematical languages such as Matlab, Octave (a public domain program with similar functionality), or Mathematica, the principal components can be computed with just one line of code.

10.7.2.3 Template Matching

Template matching uses model spike shapes to classify action potentials (Capowski, 1976; Millecchia and McIntyre, 1978). Earlier methods of template matching relied on the user to choose a small set of spikes that would serve as the templates. Modern methods choose the spike templates automatically (Yang and Shamma, 1988).

10.7.2.4 Cluster Analysis

The last step in sorting is cluster analysis. It is one of the most common methods for finding groups in multidimensional data set (Duda and Hart, 1973; Hartigan, 1975; Everitt, 1993). A basic assumption underlying cluster methods is that the data result from several independent classes, each of which can be described by a relatively simple model. This assumption fits the case of spike sorting rather well, as the action potentials arise from different neurons.

10.7.2.5 Controls

The sorted and selected spike trains are transferred into Neuroexplorer for further analysis. This software package reads output files from the Autocut, Off-line sorter, and several other commercial multiunit acquisition systems (CED Spike-2, Cortex, Bionic Technologies, Instrutech, Multichannel Systems, Neuralynx, Plexon, RC Electronics).

After sorting, we check the quality of the spike separation. If the spike waveform in question is indeed a single unit, then there should be no interspike interval less than the refractory period, which for most cells is no less than 1 msec (1 msec rule). For this, we use interspike interval time histogram (ISTH) or autocorrelogram. Fibers with apparent peaks on ISTH or autocorrelogram in the intervals ±1 msec or less should be excluded from further analysis.

Breathing, heart beats and AC 60 Hz (50 Hz) may create artifacts. The common feature among these artifacts is their regularity. We use spectral analysis of the spike trains to identify affected signals. The spike trains with obvious peaks of frequency at 60 Hz or its harmonics, heart rate (usually 3 to 6 Hz), or respiration (0.5 to 2 Hz) are excluded from the analysis.

10.7.3 Step 3: Analysis of the Spike Trains

The final analysis includes the building of peristimulus time histograms (PSTH), construction of raster plots, and numerical analysis of the spontaneous activity and responses to different compounds. As an extra control, we build histograms of the neuronal activity throughout the recording and trial-by-trial Raster plots to estimate consistency of the spike trains and fiber responsiveness. This helps us to exclude deteriorating recordings. At the end of the analysis, numerical values of spontaneous activity and responses are exported from Neuroexplorer to MS Excel or Matlab software packages for further statistical analysis.

10.8 DATA ANALYSIS

10.8.1 Cluster Analysis

The classification of fibers into distinct types is facilitated by hierarchical cluster analysis. It has been used in chemosensory research for more than 20 years and is today included in practically all statistical packages. We use SYSTAT for Macintosh.

Hierarchical cluster analysis is a multivariative procedure for detecting natural grouping in n fibers with m stimuli. The responses to all stimuli are taken into consideration and the analysis considers each stimulus as an independent variable and calculates Pearson correlation coefficients between the n response profiles. We use correlation measures because they are not influenced by differences in the absolute values of the responses. The whole matrix of the correlation coefficients is subjected to the analysis. Then we look for similarities between the whole pattern of response profiles. We usually use an average linkage method, although other procedures for linkage can be used (Bieber and Smith, 1986). The result is usually presented as a dendrogram. One possible extreme outcome of the analysis is that separate clusters cannot not be distinguished, although analysis of single CT or NG taste fibers usually results in a few groups or clusters of fibers.

Since hierarchical cluster analysis is a mathematical procedure, it can be considered more objective than non-mathematical methods. Furthermore, the analysis can utilize the responses to more stimuli than a classification based on the best response to four to five basic stimuli. In fact, as many stimuli as possible should be included because the analysis will then describe the response profiles more thoroughly. The stimuli can all be one concentration of many different compounds or several concentrations of a small number of compounds. In each case, the analysis considers each stimulus as an independent variable and calculates correlation coefficients between all of them. Non-stimulating compounds represent the problem, because they will increase the similarities between the *n* response profiles. This will increase the correlation between the response profiles. Thus, the analysis will distinguish fewer clusters than with compounds that give a response in at least some fibers.

Although analysis of single CT or NG taste fibers usually results in groups or clusters of fibers, one possible extreme outcome of the analysis is that separate clusters cannot be distinguished, as, for example, in an analysis of the responses of 52 superior laryngeal nerve fibers to 20 stimuli in the hamster (Dickman and Smith 1988). The study demonstrated very small differences between the response profiles, with no suggestion of distinct clusters. Another outcome is that the analysis recognizes distinct groups of neurons with very similar profiles within each cluster. Because cluster analysis identifies the taste fibers responsive to a compound or a group of compounds with similar taste, it is particularly useful, when only limited amount of compound is available. For example, we were interested in the relationship between structure and sweetness of the sweet protein brazzein. We produced 30 brazzein derivatives, but were able to provide only a limited amount of each derivative. This put a limitation on the number of single fibers we can record from. Because native brazzein is sweet only to higher primates, we had to use the rhesus monkey as our animal model. To save compound and still optimize the number of single fibers, studied with brazzein, we used the result of an earlier cluster analysis of single taste fiber in rhesus monkey, and focused on the S cluster fibers by recording predominantly from this fiber group.

Thus, we did the separation of fibers in real time based on our knowledge of response profiles of the different clusters obtained in past time.

10.8.2 Multidimensional Scaling Analysis

Multidimensional scaling has been used to describe both the human and the nonhuman taste world (Schiffman, 1971; Schiffman 1976; Takane et al. 1977; Schiffman et al., 1978; Schiffman et al., 1979; Mullen, 1991; Erickson et al., 1993) in behavioral (Ninomiya et al. 1984) and electrophysiological studies (Travers and Smith, 1984; Di Lorenzo, 1989; Girardot and Derby, 1990). The presence of clusters does not necessarily mean that there are taste fibers dedicated to particular taste qualities or taste categories. With the aid of MDS analysis we can address the question of whether the input of a group or cluster of fibers provides sufficient information for separation of its stimuli from all others, or the input from all clusters is necessary to distinguish these stimuli from the other qualities.

MDS computes coordinates of points in a multidimensional space where each point represents the response to a particular stimulus. The result is usually plotted as two- or three-dimensional diagrams. The closeness between the points reflects similarities between the stimuli, and a grouping of stimuli can be considered as an information channel for taste quality. If the separation between a particular group and all other stimuli remains, when only the responses of the presumed cluster are used in the MDS analysis, then this suggests that the activity in that fiber cluster is sufficient to convey the taste of these particular stimuli, without the need for any input from other taste fibers.

We, for example, used MDS to ask whether the S cluster could be described as a channel or line for the sweet taste quality in chimpanzee. Thus, we asked whether S fibers alone provide sufficient information for separation of sweet stimuli, or if input from all clusters was necessary to distinguish sweet stimuli from the other qualities (Hellekant et al., 1997). We computed two maps: one that was based on the correlation matrix of responses of all 48 fibers to 22 stimuli, and one based on the matrix of responses of the 21 S fibers to the 22 stimuli. In both cases, all sweeteners were positioned close to each other and, what is more important, all sweeteners were well-separated from the other stimuli. This suggests that the activity in the S-cluster fibers alone might be sufficient to separate stimuli of the sweet taste quality from other stimuli (Hellekant et al., 1997; Hellekant et al., 1998).

10.9 CONCLUSION

This chapter has dealt with methods used in electrophysiological recordings from different taste nerves, a technique which over many years has contributed important knowledge on how the sense of taste functions. However, we would like to emphasize that this technique reaches a fuller potential when it is combined with behavioral methods in animals and psychophysical observations in humans. Furthermore, the technique has and will greatly benefit from contributions of the anatomical, morphological, biochemical, molecular, and more recently genetic areas.

REFERENCES

Adrian, E.D. and Y. Zotterman (1926). The impulses produced by sensory nerve-endings, *J. Physiol.,* LXI(2), 151-171.

Bieber, S.L. and D.V. Smith (1986). Multivariative analysis of sensory data: a comparison of methods, *Chem. Senses,* 11(1), 19-47.

Capowski, J.J. (1976). *The Spike Program: A Computer System for Analysis of Neurophysiologicsl Action Potentials*, Washington, D.C., Hemisphere.

Danilova, V., G. Hellekant, et al. (1998). Gustatory responses of the hamster *Mesocricetus auratus* to various compounds considered sweet by humans, *J. Neurophysiol.,* 80, 2102-2112.

Danilova, V., G. Hellekant, et al. (1998). Behavioral and single chorda tympani taste fiber responses in the common marmoset, *Callithrix jacchus jacchus, Ann. N.Y. Acad. Sci. Olfaction and Taste XII,* 855, 160-164.

Di Lorenzo, P.M. (1989). Across unit patterns in the neural response to taste: vector space analysis, *J. Neurophysiol.,* 62(4), 823-833.

Dickman, J.D. and D.V. Smith (1988). Response properties of fibers in the hamster superior laryngeal nerve, *Brain Res.,* 450(1-2), 25-38.

Duda, R.O. and P.E. Hart (1973). *Pattern Classification and Scene Analysis*, New York, Wiley.

Erickson, R.P., J.L. Rodgers, et al. (1993). Statistical analysis of neural organization, *J. Neurophysiol.,* 70(6), 2289-300.

Everitt, B.S. (1993). *Claster Analysis*, New York, Wiley.

Frank, M.E. (1991). Taste-responsive neurons of the glossopharyngeal nerve of the rat, *J. Neurophysiol.,* 65(6), 1452-63.

Gerstein, G.L., M.J. Bloom, et al. (1983). Design of a laboratory for multineuron studies, *IEEE Trans. Systems, Man. Cybern.,* 13, 668-676.

Girardot, M.N. and C.D. Derby (1990). Independent components of the neural population response for discrimination of quality and intensity of chemical stimuli, *Brain, Behavior and Evolution,* 35(3), 129-145.

Glaser, E.M. (1971). Separation of neuronal activity by waveform analysis, *Adv. Biomed. Eng.,* 1, 77-136.

Glaser, E.M. and W.B. Marks (1968). On-line separation of interleaved neuronal pulse sequences, *Data Acquisition Process. Biol. Med.,* 5, 137-156.

Grant, R., S. Miller, et al. (1989). The effect of chorda tympani section on ipsilateral and contralateral salivary secretion and taste in man, *J. Neurol. Neurosurg. Psychiatry,* 52(9), 1058-1062.

Greene, E.C. (1968). *Anatomy of the Rat*, New York and London, Hafner Publishing.

Hartigan, J.A. (1975). *Clusterin Algorithms*, New York, Wiley.

Hellekant, G. and Y. Kasahara (1973). Secretory fibres in the trigeminal part of the lingual nerve to the mandibular salivary gland of the rat, *Acta Physiolog. Scand.,* 89(2), 198-207.

Hellekant, G., Y. Ninomiya, et al. (1997). Taste in chimpanzees. II. single chorda tympani fibers, *Physiol. Behav.,* 61(6), 829-41.

Hellekant, G., Y. Ninomiya, et al. (1998). Taste in chimpanzees. III. Labeled line coding in sweet taste, *Physiol. Behav.,* 65(2), 191-200.

Hellekant, G. and T.W. Roberts (1995). Whole nerve and single fiber recordings in non-human primates, in *Experimental Cell Biology of Taste and Olfaction,* A.I. Spielman and J.G. Brand, Eds., Boca Raton, FL, CRC Press, 277-290.

Hellekant, G., D.E. Walters, et al. (1991). Electrophysiological evaluation of sweeteners, in *Sweeteners. Discovery, Molecular Design, and Chemoreception*, D.E. Walters, F.T. Orthoefer, and G.E. DuBois, Eds., Washington, D.C., Am. Chem. Soc., 290-300.

Izumi, H. and K. Karita (1995). Salivary secretion in cat submandibular gland mediated by chorda tympani afferents, *Am. J. Physiol.,* 268(2 Pt 2), R438-44.

Langley, J.N. (1898). The salivary glands, in *Textbook of Physiology*, E.A. Schaefer, Ed., Edingburgh, Young J. Pentland, 1, 475-530.

Lewicki, M.S. (1998). A review of methods for spike sorting: the detection and classification of neural action potentials, *Network: Comput. Neural Syst.,* 9, R53-R78.

Millecchia, R. and R. McIntyre (1978). Automatic nerve impulse identification and separation, *Comput. Biomed. Res.,* 11, 459-468.

Mullen, K. (1991). A simple multivariate probabilistic model for preferential and triadic choices, *Psychometrika,* 56, 69-75.

Ninomiya, Y., T. Higashi, et al. (1984). Qualitative discrimination of gustatory stimuli in three different strains of mice, *Brain Res.,* 322(1), 83-92.

Rehnberg, B.G., T.P. Hettinger, et al. (1992). Salivary ions and neural taste responses in the hamster, *Chem. Senses,* 17(2)), 179-190.

Schiffman, S.S. (1976). Taste of Dipeptides, *Physiol. Behav.,* 17, 523-535.

Schiffman, S.S., G. Musante, et al. (1978). Application of multidimensional scaling to ratings of foods for obese and normal weight individuals, *Physiol. Behav.,* 21(3), 417-22.

Schiffman, S.S., D.A. Reilly, et al. (1979). Qualitative differences among sweeteners, *Physiol. Behav.,* 23(1), 1-9.

Schmidt, E.M. (1984). Computer separation of multi-unit neuroelectric data: a review, *J. Neurosci. Methods,* 12(2), 95-111.

Takane, Y., F.W. Young, et al. (1977). Nonmetric individual differences multidimensional scaling: an alternating least squares method with optimal scaling features, *Psychometrika,* 42(1)), 7-63.

Travers, S.P. and D.V. Smith (1984). Responsiveness of neurons in the hamster parabrachial nuclei to taste mixtures, *J. Gen. Physiol.,* 84(2), 221-250.

Wheeler, B.C. and W.J. Heetderks (1982). A comparison of techniques for classification of multiple neural signals, *IEEE Trans. Biomed. Eng.,* 29(12), 752-759.

Yang, X. and S.A. Shamma (1988). A totally automated system for detection and classification of neural spikes, *IEEE Trans. Biomed. Eng.,* 35, 806-816.

Section 3

Higher-Order Studies in Taste and Olfaction

11 Activation of Neurons in Trigeminal Subnucleus Caudalis (Vc) by Irritant Chemical Stimulation: Extracellular Single-Unit Recording and c-*fos* Immunohistochemical Methods

Earl E. Carstens, Christopher T. Simons, Jean-Marc Dressier, Makoto Sudo, Satoko Sudo, Mirela Iodi Carstens, and Steven L. Jinks

CONTENTS

0-8493-2329-0/02/$0.00+$1.50

11.1 INTRODUCTION

This chapter describes *in vivo* methods to investigate the activation of central trigeminal neurons by irritant chemical stimuli delivered to the oral mucosa. It primarily focuses on two distinct methodologies: electrophysiological single-unit recordings and c-*fos* immunohistochemistry. These two methods provide very different information but complement each other nicely. Extracellular single-unit recording is a standard method that is familiar to most chemosensory scientists. It allows one to investigate the responsiveness of individual neurons to applied stimuli on a moment-to-moment basis, and thus provides information about neural information processing that may be directly linked to sensory perception. The use of c-*fos* immunohistochemistry has increased markedly since its introduction over 10 years ago (Hunt et al., 1987) and is becoming a standard method, although caution must still be exercised in interpreting data (Harris, 1998). This approach allows one to identify an entire population of neurons that has been activated by a sensory stimulus. However, this anatomical method only gives a static picture of the locations of neurons that were activated at one given time point following the stimulus. As discussed, by combining these two methodological approaches, one can identify the locations and size of neuronal populations, as well as the response characteristics of individual neurons in the population, that are excited by a specific sensory stimulus.

A major goal in our laboratory is to understand the capacity for individual trigeminal neurons to signal chemically evoked orofacial irritant sensations, and how activity in trigeminal neurons relates to human perception of irritation. One question is: where are chemoresponsive trigeminal neurons located? Here, the c-*fos* method is ideal for demonstrating the brain stem locations of neurons that are activated by chemical stimuli delivered to the oral cavity. Another question is whether one and the same neuron can signal irritation elicited by different chemicals. This question cannot be answered by the c-*fos* method. Suppose that each trigeminal neuron is specifically activated by only one class of irritant chemical (Figure 11.1A). Use of

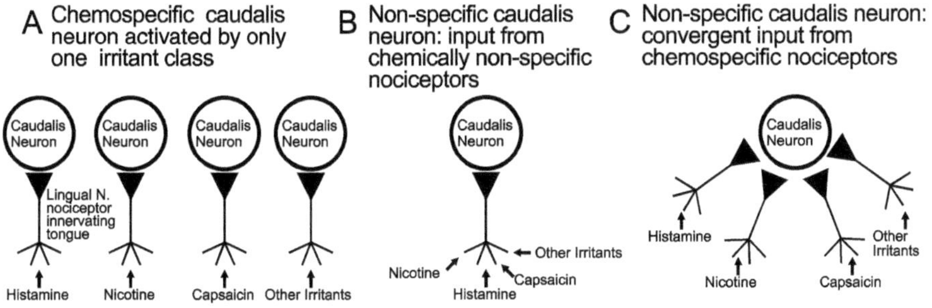

FIGURE 11.1 Schematic diagram showing chemospecific (A) and nonspecific (B, C) Vc neurons and the possible types of chemonociceptive inputs (see text for further explanation).

the c-*fos* method allows one to identify the locations of neurons activated by application of only one irritant chemical. In a separate set of animals, application of a different irritant chemical might elicit a similar pattern of activation, but one cannot be certain if the same population of neurons was activated by both chemicals (Figure 11.1B), or if two separate "chemospecific" populations of neurons having a generally similar regional distribution were activated by the two different irritants (Figure 11.1A). This question can be resolved with single-unit recording methods, which allow one to demonstrate that an individual neuron is or is not excited by the two or more different irritant chemicals. However, this method is limited in that it is unable to distinguish whether a central neuron receives input from receptors that themselves respond to multiple irritant chemicals (Figure 11.1B), or if it receives convergent, chemospecific afferent inputs (Figure 11.1C). To distinguish between these possibilities, it will be necessary to determine if individual peripheral receptors are chemospecific or if they respond to multiple classes of irritant chemicals. This chapter deals only with central trigeminal neurons, leaving open the question of chemospecificity of primary afferent fibers.

11.2 EXTRACELLULAR SINGLE-UNIT RECORDING METHODS

11.2.1 Advantages and Disadvantages

As stated in the introduction to this chapter, the major advantage of this method is that it allows a moment-to-moment analysis of the firing of a single neuron, and allows one to determine whether that neuron is activated by a variety of different stimuli. While one can learn much about the properties of individual neurons, a major disadvantage of traditional single-unit recording is that analysis is usually restricted to one, or at most a few, neurons at a time. This limits our understanding of sensory coding by populations of neurons. However, this limitation is starting to be overcome with techniques that allow multiple, simultaneous single-unit recordings; a full discussion of this topic is beyond the scope of this chapter.

A related disadvantage of single-unit recording is that one is studying an individual neuronal element that normally functions in a circuit. One cannot be certain of that neuron's interconnections within the circuit. This problem can be overcome to some extent *in vivo* using electrophysiological stimulation methods to specify the axonal projections of, and sources of input to, the neuron under study. Another, more reductionistic approach to this problem is to record *in vitro* from a slice preparation (e.g., isolated spinal cord or trigeminal slice) that contains important circuit elements. However, in this case, the reduced preparation is isolated from its wider connections.

An additional limitation of extracellular recording is that one cannot specify exactly which neuron was recorded from. Often, this is of minor concern when recording from a region where neuronal morphologies and interconnections are already known. This limitation, too, can be overcome by the more technically demanding approach of intracellular recording and subsequent staining of the neuron to reveal its morphology.

Despite all these limitations, the extracellular single-unit recording method has contributed tremendously to our understanding of neural processing and continues to be a viable approach in neuroscience.

11.2.2 Surgical and Recording Methods

The authors have conducted studies using adult male Sprague-Dawley rats anesthetized with thiopental (80 mg/kg), supplemented in recent studies with halothane (0.5% in O_2). Surgical preparation and recording methods have been previously described in detail (Carstens et al., 1998; Simons et al., 1999; Dressier et al., 2000d) and are summarized here. The caudal medulla is exposed surgically by midline incision, followed by laminectomy at C1-C2 and removal of the caudal portion of the occipital bone. The head is placed in a stereotaxic frame and ventroflexed, and the spine is fixed with a vertebral clamp at C3. To allow chemical stimulation of the oral cavity, the mouth is held in a partially opened position using a retractor, and is frequently rinsed with physiological saline to prevent dessication.

Based on c-*fos* labeling (see Section 11.3), chemonociceptive neurons receiving afferent input from the tongue are located in superficial laminae of dorsomedial trigeminal subnucleus caudalis (Vc). We routinely record from these units by advancing a tungsten microelectrode (F. Haer, 10–12 MΩ) into the superficial Vc to depths of 50 to 300 μm using a hydraulic microdrive. The microelectrode is connected to commercially available amplification equipment to filter, amplify, and monitor isolated extracellular action potentials. The conditioned signal can be viewed and heard on an oscilloscope and audio monitor, respectively, and digitized and fed to a computer for analysis and storage. A variety of software is available to store and process the recorded signals; we currently use a system developed at the University of Erlangen (Forster and Handwerker, 1990). Action potentials generated by a single unit are discriminated based on their amplitude and waveform. Some software programs allow discrimination of more than one simultaneously recorded action potential. The spike data are stored on disk for later data analysis and display, commonly in the form of a peristimulus time histogram (PSTH).

The location of the Vc neuron is marked at the conclusion of the recording session by passing current through the recording electrode to make an electrolytic lesion. The brain stem is fixed by immersion for several days in 10% buffered formalin and cut into 40 to 50-μm frozen sections on a microtome. Sections are mounted on microscope slides, counterstained as desired (e.g., Neutral Red; cresyl violet), cover-slipped, and inspected under the light microscope to identify the general recording site but not the exact cell recorded.

11.2.3 Search Strategies to Identify Chemonociceptive Vc Units

A variety of search strategies can be used to identify nociceptive neurons, including the following.

- Electrical stimulation of the lingual nerve, or of the dorsal anterior surface of the tongue using an Ag-AgCl ball electrode. The advantage of this search strategy is that it is free of any sampling bias toward units with specific types of natural inputs (low- vs. high-threshold mechanical). To identify units with nociceptive input requires that the stimulus intensity be sufficient to activate Aδ and C fibers. Once a unit is isolated in this manner, its receptive field must be determined using mechanical, thermal, and chemical stimuli. Another major advantage of this method is that it affords the possibility to identify Vc units that respond to chemical stimulation even if they do not respond to mechanical or thermal stimuli. Application of electrical stimuli to the surface of the tongue is relatively simple. However, a disadvantage of lingual nerve stimulation is the additional surgical preparation required to isolate the nerve and place it on a pair of silver hook stimulating electrodes.
- Use of mechanical (touch, pressure and pinch) stimulation of the tongue to identify responsive units. The advantage of this strategy is speed. However, a disadvantage is that an excessive number of noxious pinch stimuli may be needed to identify units, leading to swelling of the tongue and possibly the development of hyperalgesia. Also, many Vc units that respond to noxious pinch do not necessarily respond to noxious heat or chemicals.
- Use of noxious heat, such as topical application of hot (50 to 54°C) water to the tongue. The advantage of this strategy is again speed, because a large fraction of heat-responsive units additionally respond to noxious chemicals. The disadvantage is that noxious heat stimuli cannot be applied frequently because they may damage the tongue, leading to swelling and possibly hyperalgesia.
- Use of a nondesensitizing noxious chemical, such as pentanoic acid. We have recently found that pentanoic acid (300 m*M*) excites a large percentage of noxious heat-responsive Vc units that also subsequently responded to capsaicin and/or nicotine (Dressier et al., 2000d). Other nondesensitizing chemicals that might be used to search for chemonociceptive Vc units include carbonated water (Simons et al., 1999), histamine, HCl at pH < 3, and 5 *M* NaCl (Carstens et al., 1998). Chemicals that induce desensitization, such as capsaicin, are not helpful to identify chemonociceptive units because their responses to subsequent chemical stimuli may be depressed (cross-desensitized) after the initial application (see Section 11.2.6.2).

- Antidromic stimulation. This allows one to identify Vc units based on their axonal projection to a more rostral target such as ventral posterior medial (VPM) thalamus, and is unbiased in terms of the type of afferent fiber inputs the Vc units receive. This method, like the peripheral electrical search strategy (see first item on list), affords the possibility to identify units that respond only to chemical stimuli. Successful implementation of this method requires that the following criteria be met: the unit responds at a constant latency to antidromic stimulation, it follows a high (>250 Hz) stimulus frequency, and collision of antidromic and orthodromically elicited action potentials can be shown within a critical time period (Lipski, 1981). Disadvantages of this method are the additional surgery and the time required to optimally place the stimulating electrode(s) at the Vc target projection site.

11.2.4 Characterization of Receptive Field and Unit Classification

Once a single Vc unit is isolated by one of the search strategies noted above, its receptive field area is mapped with graded mechanical stimuli. Graded pressure stimuli are applied with forceps or calibrated von Frey filaments, and the threshold is taken as the force eliciting an action potential on approximately 50% of trials (or eliciting a detectable increase in firing if the unit exhibits spontaneous activity). Because of the compliance of the tongue and other oral soft tissue, the exact mechanical threshold may be difficult to establish. At the expense of greater accessibility to intra-oral tissue, we avoid methods to fix or to extend the tongue (e.g., with a suture through the tongue tip) or to expose the oral cavity (e.g., removal of the lower jaw) because these manipulations might induce an inflammatory response, leading to the development of hyperalgesia that could manifest as reductions in mechanical or thermal thresholds.

Vc unit receptive fields typically respond to mild and/or intense pressure applied to the ipsilateral tongue and tongue tip bilaterally, and sometimes also to pressure-pinch of the lower lip. Units that respond to low-intensity mechanical stimuli that are not painful when self-applied, but do not respond at a higher frequency to more intense mechanical stimuli or to noxious heat, are classified as non-nociceptive or low-threshold. Units that respond to non-painful mechanical stimuli, and that respond with a higher discharge rate to noxious levels of mechanical stimuli capable of producing pain, and to noxious heat (>40°C), are classified as wide dynamic range (WDR) or multireceptive or convergent. Units that do not respond to non-noxious mechanical stimuli, but respond to noxious mechanical and thermal stimuli, are classified as nociceptive-specific (NS) or high-threshold. A high percentage of WDR and NS Vc units additionally respond to irritant chemicals applied topically to the tongue (Carstens et al., 1998), and an example is shown in Figure 11.2. Approximately one half of WDR and NS units also respond to cooling stimuli (Carstens et al., 1998); responses to noxious cold have not been investigated.

It is not known if some Vc units are chemospecific, i.e., they respond to irritant chemicals but not to mechanical or thermal stimuli. Such units, if they exist, could

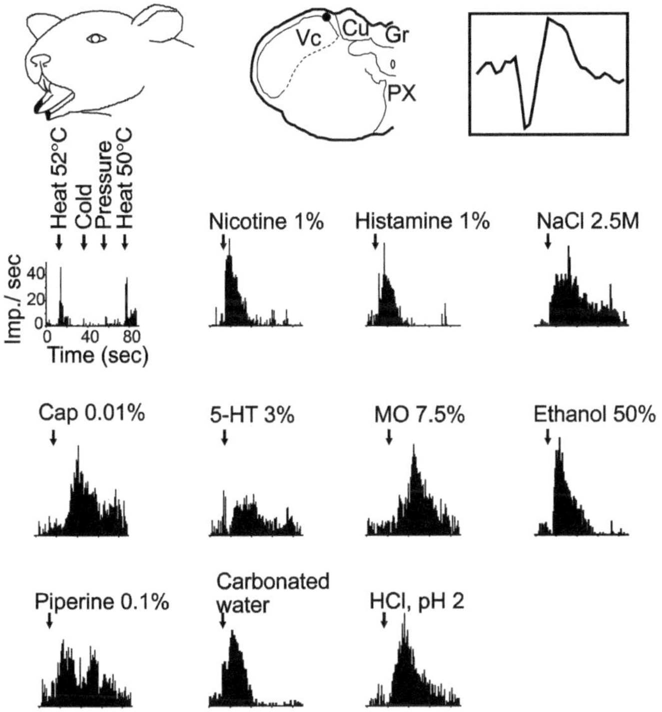

FIGURE 11.2 Example of responses of individual Vc unit to application of different stimuli to the tongue. Upper row shows, from left to right, a figurine of the rat's face with the lingual receptive field (black), the recording site (dot) in Vc, and an example of the action potential waveform. The second row shows peristimulus-time histograms (PSTHs; bin width: 1 s) of the unit's responses to various stimuli. The left-hand PSTH shows its responses to pressure and noxious heat stimuli, but not to cooling. All subsequent PSTHs have the identical time-base and *y*-axis (impulse frequency), and show the unit's response to each indicated chemical applied (at arrows) in a 0.1-ml bolus to the dorsal surface of the tongue. The chemical stimulus was left on for 60 s, at which time the tongue was rinsed with saline. Abbreviations: cap, capsaicin; 5-HT, serotonin; MO, mustard oil; CU, cuneate n.; Gr, gracile n.; Px, pyramidal decussation. Note that the unit responded vigorously to each chemical. (Adapted from Carstens, E.E. et al. *J. Neurophysiol.* 80, 465-492, 1998.)

only be isolated by one of the electrical or chemical search strategies noted above and would constitute a novel class.

11.2.5 Neuronal Outputs and Inputs

The axonal projection(s) of a single Vc neuron can be determined to some extent using antidromic stimulation of a potential target site, such as VPM thalamus or parabrachial nucleus. Criteria for the demonstration of antidromic activation were described above. A disadvantage of this method is that the inability to antidromically activate a neuron does not prove that it lacks an ascending projection because the ascending axon may be at a distance from, or was damaged by, the stimulating electrode.

Inputs to a neuron can also be assessed by electrical stimulation at a possible source. An excitatory input would be manifested by orthodromic activation of the neuron, while an inhibitory input would be manifested by a reduction in ongoing or evoked activity. It is also possible, but technically demanding, to demonstrate monosynaptic inputs to a neuron using spike-triggered averaging.

11.2.6 Responses to Irritant Chemicals

Irritant chemicals can be manually delivered to the anterior dorsal surface of the tongue by syringe, or by intermittent or continuous flow through narrow tubing placed toward the back of the tongue with flow directed toward the tip (Dressier et al., 2000d). It is advisable to avoid spread of chemical stimuli away from the tongue, even if the Vc unit's receptive field is restricted to this area, because activation of extralingual nociceptors could conceivably induce central sensitization. A strip of Parafilm can be placed beneath the tongue to prevent gingiva and other oral tissues from being stimulated. With the head in a ventroflexed position, chemicals delivered to the lingual surface will either remain there due to surface tension, or will drip off the tongue without contacting other oral tissue.

There are a number of issues to consider in determining if a Vc unit is chemonociceptive, as discussed below.

11.2.6.1 Chemospecificity

It is conceivable that some Vc units only respond to a particular class of irritant chemicals (see Figure 11.1A). Obviously, the unit's chemical responsiveness would be missed unless that particular irritant chemical was tested. Thus, several chemicals might need to be tested sequentially until the effective excitatory one is found. In this regard, the order of chemical presentation becomes important, as discussed further below.

11.2.6.2 Order of Chemical Stimulus Presentation: Cross-Desensitization

In a prior study, we found that a majority of WDR and NS unit responded to several irritant chemicals (Carstens et al., 1998). However, some irritants, including capsaicin, piperine, and mustard oil, appeared to cross-desensitize Vc unit responses to additional, subsequently tested irritant chemicals. Here the terms cross-desensitization and cross-tachyphylaxis are used as synonyms, meaning that presentation of one irritant, such as capsaicin, reduces or eliminates the ability of the unit to respond to application of a different irritant. Thus, a unit might normally respond to a variety of different irritants; but if capsaicin were tested first and prevented the unit from responding to further chemicals due to cross-desensitization, one would reach the false conclusion that the unit responded selectively to capsaicin.

Therefore, when presenting a sequence of irritant chemicals, those that do not induce cross-desensitization should be tested earlier, while cross-desensitizing chemicals such as capsaicin, piperine, and mustard oil should be tested last. A problem here is that not all irritant chemicals have been tested for their ability to induce

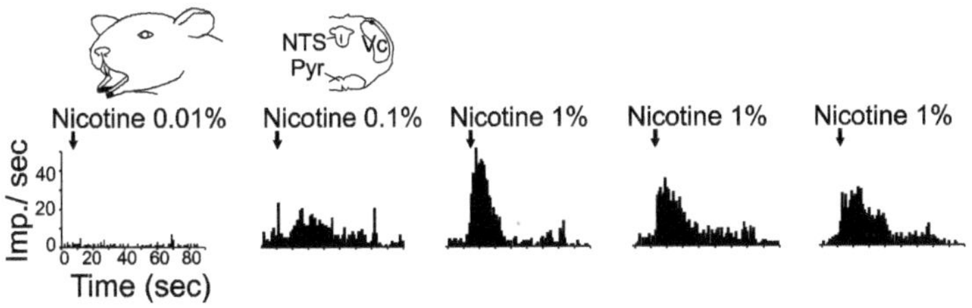

FIGURE 11.3 Tachyphylaxis in responses of Vc unit to repeated lingual application of nicotine. Shown are PSTHs (bin width: 1 s) of responses of Vc unit to successive applications of nicotine to the tongue at the indicated concentrations. Note the increase in response magnitude with increasing concentration (first three PSTHs from left), as well as the reduction in successive response magnitude to repeated application of 1% nicotine (last three PSTHs from left). Upper-left inset shows receptive field (black) and upper right inset shows recording site (dot) in Vc. Additional abbreviation: NTS, n. of solitary tract.

cross-desensitization. In human studies, capsaicin has been shown to cross-desensitize irritation elicited by capsaicin (Green, 1989; 1991b; 1993), piperine (Green, 1996), zingerone (Green, 1993); nicotine (Dressier et al., 1997), citric acid (Gilmore and Green, 1993; Dressier et al., 2000c), sodium chloride (Green, 1991a; Gilmore and Green, 1993), cinnamic aldehyde (Green, 1991a), and menthol (Cliff and Green, 1996). Piperine also cross-desensitized irritation elicited by capsaicin (Green, 1996) and nicotine (Dressier et al., 1999). Citric acid (Dressier et al., 2000b) and menthol cross-desensitized capsaicin-evoked irritation (Green and McAuliffe, 2000), as did a high dose of nicotine (Dressier et al., 2000a). Menthol also cross-desensitizes nicotine (Dressier et al., 2001b).

There are few electrophysiological studies of cross-desensitization. In previous studies of spinal dorsal horn WDR neurons, intracutaneous capsaicin (Carstens, 1997) and nicotine (Jinks and Carstens, 1999) cross-desensitized responses to histamine. Our prior study of Vc units (Carstens et al., 1998) did not provide evidence for any cross-desensitizing effect for most irritant chemicals tested, with the exceptions of capsaicin, piperine, and mustard oil. In that study, a given irritant chemical was applied several times at 5-min intervals to test for tachyphylaxis, i.e., a reduction in successive responses to repeated stimuli. The following chemicals exhibited no significant tachyphylaxis: histamine, 5 *M* NaCl, ethanol, HCl (Carstens et al., 1998), carbonated water (Simons et al., 1999) and pentanoic acid (Dressier et al., 2000d). The following chemicals exhibited significant tachyphylaxis: nicotine, serotonin, capsaicin, piperine, and mustard oil in some but not all cases (Carstens et al., 1998). An example of tachyphylaxis in a Vc unit's response to repeated application of 1% nicotine is shown in Figure 11.3 (three right-hand PSTHs). However, despite the tachyphylaxis exhibited by nicotine and serotonin, there was no evidence that they caused cross-tachyphylaxis to other irritant chemicals tested later. The evidence against cross-tachyphylaxis is indirect and is based on order of presentation. The responses of a group of Vc units to chemicals presented in one specific order (e.g., histamine followed by nicotine followed by NaCl) were not significantly different from the responses of a separate group of Vc units to the same chemicals presented in a different order (e.g., NaCl, nicotine, histamine). Had nicotine produced cross-tachyphylaxis, the responses to NaCl and histamine should have been smaller when preceded by nicotine, but they were not. This implies that nicotine and serotonin either do not induce cross-tachyphylaxis (although they exhibit tachyphylaxis), or that the cross-tachyphylaxis had worn off by the time the next irritant chemical was presented.

11.2.6.3 Cross-Sensitization

The converse problem arises if an irritant chemical causes the Vc unit to give a larger response to a different, subsequently tested chemical: cross-sensitization. There are very few examples of cross-sensitization in the psychophysical literature. It was reported that zingerone (Green, 1993) and 5 *M* NaCl (Dressier et al., 2001a) cross-sensitized irritation elicited by capsaicin. NaCl was also shown to cross-sensitize irritation elicited by ethanol (Green, 1990). At the extreme, cross-sensitization might enhance nociceptive transmission to such an extent that a normally ineffective chemical stimulus might become capable of exciting a Vc unit, producing an apparent false-positive result. An example might be the activation of silent or sleeping nociceptors that are not initially excited by the irritant chemical, but may become responsive to that chemical over time and thus excite the Vc unit. However, using the same argument presented above with different orders of chemical presentation, there is no evidence that any irritant chemical induces a consistent increase in a Vc unit's response to a subsequently tested chemical (Carstens et al., 1998). Moreover, we have not observed any Vc units that exhibited a progressive increase in their spontaneous or evoked firing over time, as might be expected if cross-sensitization were developing.

11.2.6.4 Interstimulus Interval

A critical factor in testing a Vc unit's responses to sequentially presented chemical stimuli is the interstimulus interval (ISI). The ISI has been used as the interval between the onset of one stimulus and the onset of the next (e.g., Cliff and Green, 1994); but in recent psychophysical studies, it is more consistently taken as the time from the offset of one stimulus to the onset of the next (e.g., Prescott and Stevenson, 1996b; Prescott and Stevenson, 1996a; Green and Rentmeister-Bryant, 1998). The interval between the onset of successive stimuli and the duration of each stimulus are both important factors that contribute to sensation and to the pattern of Vc unit responses to sequentially delivered stimuli.

The ISI is important when testing whether or not a Vc unit responds to different irritant chemicals. For example, a Vc unit might respond to several irritant chemicals. However, if the initially tested chemical induced cross-tachyphylaxis, the response to the next-tested chemical might be smaller or absent if it was applied at a short ISI that fell within the time period of cross-tachyphylaxis. This would result in a false-negative.

The ISI is also important in testing for tachyphylaxis to repeated presentation of one chemical. We found that responses of spinal dorsal horn WDR units to repeated intracutaneous microinjections of nicotine (10% in 1 μl) exhibited more rapid and complete tachyphylaxis at shorter (5-min) compared to longer (10-, 15-min) intervals (Jinks and Carstens, 1999).

11.2.6.5 Terminology

In addition to the term "tachyphylaxis," other terms are frequently used in the literature when referring to changes in the size of successive responses to repetitive stimuli, including habituation, self-desensitization, adaptation, and fatigue.

Habituation also refers to a progressive reduction in the magnitude of successive responses, often of a behavioral nature, to repeated application of an identical stimulus. There are several specific features associated with habituation.

- The degree of habituation is greater with weaker stimuli and weaker with strong stimuli (intensity dependence).
- The degree of habituation is greater when stimuli are presented at short ISI, and is weaker at long ISI (frequency dependence).
- After habituation is established, presentation of a strong stimulus elicits a response equal to the pre-habituated response (dishabituation).
- After habituation has been established, reapplication of the same stimulus after a sufficiently long waiting period elicits a response at the pre-habituated level (spontaneous recovery).
- When habituation has been established by repeated stimulation at one specific location, presentation of the same stimulus at a separate location (within certain spatial limits) elicits a response that is at the habituated level (transfer of habituation).
- After habituation has been established with an initial stimulus series, repetition of the stimulus series results in a more rapid re-establishment of habituation (potentiation of habituation).

The tachyphylaxis of spinal WDR neuronal responses observed with repeated intracutaneous application of nicotine differs from the expected pattern of habituation because the tachyphylaxis was much stronger at higher (10%) than lower (1%) nicotine concentrations (Jinks and Carstens, 1999). Moreover, when capsaicin is applied to the tongue, followed by a prolonged waiting period (minutes, hours, or even days) in humans, reapplication of capsaicin elicits significantly less irritation (self-desensitization) (Green, 1989; Green, 1991b; Karrer and Bartoshuk, 1991). In contrast, repetitive application of capsaicin at a short (e.g., 30 s) ISI induces sensitization. These electrophysiological and psychophysical observations indicate that the tachyphylaxis or self-desensitization induced by chemical stimulation should not be called habituation because they do not exhibit intensity dependence, spontaneous recovery, or dishabituation, and the frequency dependence observed for capsaicin has a pattern opposite to that of habituation.

Adaptation is commonly defined as a reduction in the magnitude of a response during a maintained stimulus. Adaptation may refer to a reduction in depolarization of a receptor, or in the perceived intensity of a sensation. By this definition, adaptation should not be used when referring to responses to repetitive stimuli. However, when an irritant chemical is delivered repeatedly to the tongue, it is not certain how long an effective concentration of that chemical is maintained at the level of nociceptors in the lingual epithelium. This will depend on the rate of chemical diffusion through the lingual epithelium and the rate of clearance. If the clearance rate is slower than the rate at which the concentration is replenished by repetitive application, this is akin to continual application, in which case adaptation may rightfully apply to a reduction in neural response or perceived intensity.

In the psychophysical literature, the term desensitization refers to a progressive reduction in the magnitude of sensation elicited by repeated stimuli. Desensitization also refers to a persistent reduction in irritant sensation when the stimulus is reapplied some time later. Physiologically, desensitization has been recently defined as reduction in currents evoked in individual neurons to repetitive agonist application (Liu et al., 2000). These authors further defined acute desensitization as a decrease in currents in the presence of an agonist, although it might be argued that this more strictly reflects adaptation.

11.2.6.6 Sensitization and Stimulus-Induced Recovery (SIR)

Lingual application of capsaicin at a short (30 s) ISI elicits irritation in humans that often increases across trials; this has been called sensitization (Green, 1989; 1991b). This psychophysical term should not be confused with the same word that is used to describe an increase in the sensitivity of neurons, although physiological sensitization of nociceptors or central pain-transmitting neurons may represent a mechanism underlying the psychophysical effect.

We have recently tested whether repeated application of capsaicin at the same 1-min interval elicits a progressive increase in responses of Vc neurons that might parallel psychophysical sensitization. Figure 11.4A shows a typical example in which repeated application of capsaicin at 1-min intervals caused a progressive increase in firing over approximately the first 8 min. This pattern was highly consistent in all Vc units tested (Dressier et al., 2000d).

It was recently discovered that, following application of a series of capsaicin stimuli and a rest period, reapplication of capsaicin at the same (1-min) interval (30-s ISI) eventually started to elicit irritation that increased across trials to reattain the maximal intensity achieved in the first stimulus series (Green, 1996; Green and Rentmeister-Bryant, 1998). This phenomenon has been called stimulus-induced recovery (SIR), with the implication that desensitization can be overcome by recurrent stimulus application. We also tested if Vc units exhibit a firing pattern consistent with SIR. Figure 11.4B shows the response of a Vc unit to a second series of capsaicin stimuli, delivered 1 h following the initial capsaicin stimulus series. Initially, reapplication of capsaicin did not elicit any increase in firing, consistent with desensitization. However, the unit eventually exhibited an increase in firing, albeit not to the level observed in the first series. This is consistent with partial SIR.

In contrast to capsaicin, repeated application of nicotine (at the same 1-min interval) to the human tongue elicits irritation, the intensity of which decreases across trials (Dressier et al., 1997; 1999). We were interested to determine if a similar self-desensitization could be seen in the responses of Vc units to nicotine. Figure 11.5A shows a typical example in which repeated application of nicotine initially excited the Vc unit over the first three trials, followed by a decline in responses. This pattern was observed consistently in all Vc units tested. These data indicate that repeated

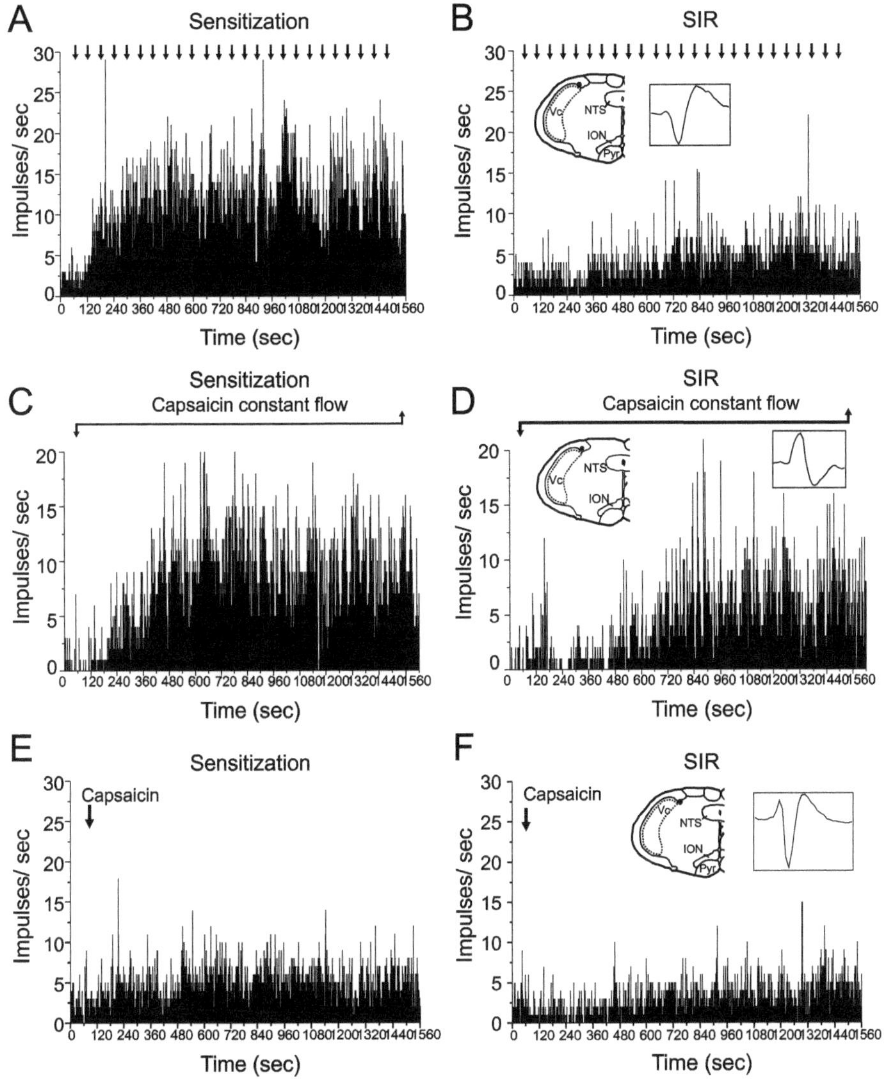

FIGURE 11.4 Sensitization and SIR in the responses of Vc units to repeated or continuous application of capsaicin. A: PSTH (bin width: 1 s) of Vc unit's response to repeated application (at arrows) of capsaicin (100 ppm; 0.1 ml) to tongue at 1-min intervals. Note buildup in response (sensitization) during initial eight trials. B: Response of same unit in A to repeated application of capsaicin after a 1-h rest period. Note partial SIR. Insets show recording site in Vc (left) and action potential waveform (right). C: Response of another Vc unit to constant-flow (0.32 ml/min) application of capsaicin to tongue. Upper bar with arrows shows on- and offset of flow stimulus. Note sensitization. D: Response of same unit in C to reapplication of capsaicin by flow after a 1-h rest period. Note SIR. Insets: recording site and action potential waveform. E: Response of another Vc unit to single application of capsaicin (100 ppm, 0.1 ml) to tongue. Note absence of sensitization. F: Response of same unit in E to reapplication of capsaicin after a 1-h rest period. Note absence of SIR. Insets: recording site and action potential waveform. (From Dressier, J.M. et al., *J. Neurophysiol,* 84, 1851-1862, 2000d. With permission.)

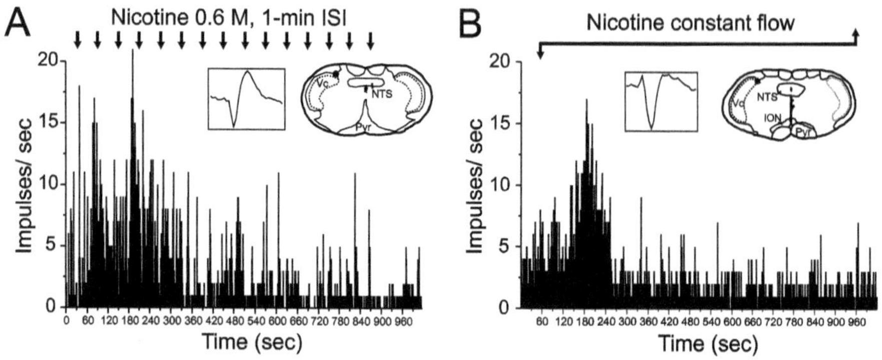

FIGURE 11.5 Adaptation or desensitization in responses of a Vc unit to repeated or continuous application of nicotine. A: PSTH of a Vc unit's response to repeated application of 0.6 *M* nicotine (arrows) at 1-min intervals. Note initial excitation followed by decline in firing. Insets: action potential waveform and recording site. B: Response of another Vc unit to constant-flow application (upper bar with arrows) of nicotine. Note initial excitation followed by decline. Insets: action potential waveform and recording site. (From Dressier, J.M. et al., *J. Neurophysiol,* 84, 1851-1862, 2000d. With permission.)

applications of capsaicin and nicotine have divergent effects on irritant sensation in humans that are reflected in the different patterns of responses of Vc units.

11.2.6.7 Continual Chemical Stimulation

The use of continuous chemical application by constant flow allows one to investigate Vc unit responses to a maintained stimulus. Application of capsaicin by constant flow (0.32 ml/min via syringe pump) to the tongue excites Vc units in a manner identical to repeated application at 1-min intervals. An example is shown in Figure 11.4C. Thus, the firing increased during the initial 8 min, consistent with sensitization, and appeared to reach a plateau. After a 1-h rest period, the capsaicin was reapplied at the same flow rate. After an initial delay consistent with desensitization, the firing rate increased again consistent with SIR (Figure 11.4D).

As a control, a single bolus (0.1 ml) of capsaicin was applied once. This also significantly elevated Vc unit firing, but at a much slower rate and to a lesser degree (Figure 11.4E) compared to continual application. Furthermore, when a single bolus of capsaicin was reapplied after a 1-h waiting period, there was no response consistent with desensitization and an absence of SIR (Figure 11.4F).

We propose that repetitive (1-min intervals) or continuous application of capsaicin increases its intra-epithelial concentration to a maximal level that is maintained throughout the stimulus period, resulting in the Vc firing pattern characterized by sensitization to a plateau. The single application presumably caused a slower rise in the intra-epithelial concentration which then declined, as mirrored by the Vc unit response. When a single bolus was reapplied after a rest period, there was no increase in Vc firing, indicating that this application procedure was incapable of inducing SIR (Figure 11.4F).

For comparison, we also tested the effect of continual application of nicotine. Similar to repetitive application, constant-flow application of nicotine initially excited Vc units, followed by a rapid adaptation or desensitization to the pre-stimulus level (Figure 11.5B). When nicotine was reapplied following a rest period, many Vc units exhibited no further increase in firing indicative of desensitization.

We are currently examining responses of Vc units to additional irritant chemicals. Like capsaicin, lingual application of 5 *M* NaCl and 250 m*M* citric acid both elicit irritation in humans that increases in intensity across trials of repeated application at 1-min intervals (Dressier et al., 2000b; 2001a). Ongoing studies indicate that similar patterns of sensitization can be elicited by constant-flow application of 5 *M* NaCl (Figure 11.6A) and 300 m*M* pentanoic acid (Figure 11.6B) in Vc units. The latter example shows an initial brief excitatory response with subsequent sensitization.

11.2.6.8 Concentration Dependence

In addition to the temporal characteristics of Vc neuronal firing, another important factor is the concentration dependence of the neural response. This is determined by recording responses to successive application of one chemical at different concentrations (i.e., doses; stimulus volume is held constant). One limitation of this approach is that responses to later stimuli in the series might be artifactually low if preceding stimuli induce tachyphylaxis (see above). We have attempted to minimize this potential confound using an ascending series of concentrations. The rationale is that lower concentrations are presumed to induce less, if any, tachyphylaxis and that this only becomes a problem at the highest concentrations tested. However, this assumption has not yet been tested and requires a systematic study of the relationship between the irritant concentration and the degree of tachyphylaxis.

Vc units exhibited concentration-dependent increases in firing to all chemicals tested (Carstens et al., 1998). In most cases, the highest firing rate was elicited by the highest concentration. An example of a concentration-related excitation of a Vc unit by nicotine is shown in Figure 11.3 (first 3 PSTHs from left). Interestingly, however, capsaicin evoked a larger response at 0.01% than 0.1%, suggesting that the latter response may have been attenuated due to desensitization induced by the prior, albeit weaker, stimulus (see Figure 6M in Carstens et al., 1998). Therefore, for those chemicals inducing tachyphylaxis (see above), one should assume that the neuronal response to the last, highest-concentration stimulus in the ascending series is smaller than it would be in the absence of any preceding stimuli.

Some of the irritant chemicals tested in Vc units have also been used in human studies of oral irritation, including capsaicin, nicotine, NaCl, ethanol, piperine, and acid. In general, those concentrations that elicited moderate levels of oral irritation were at the low end of the range of concentrations that excited Vc units (Table 11.1). For example, the human capsaicin detection threshold was reported to be 0.2 ppm (0.66 μM) (Szolcsànyi, 1990) and is usually used at concentrations of 3 to 100 ppm (33 to 330 μM) (Green, 1989, 1996; Dressier et al., 1997) or even higher (1000 ppm) (Karrer and Bartoshuk, 1991) to elicit moderate to high levels of irritation. However, Vc units are generally unresponsive to 10 ppm capsaicin but respond at 100 ppm

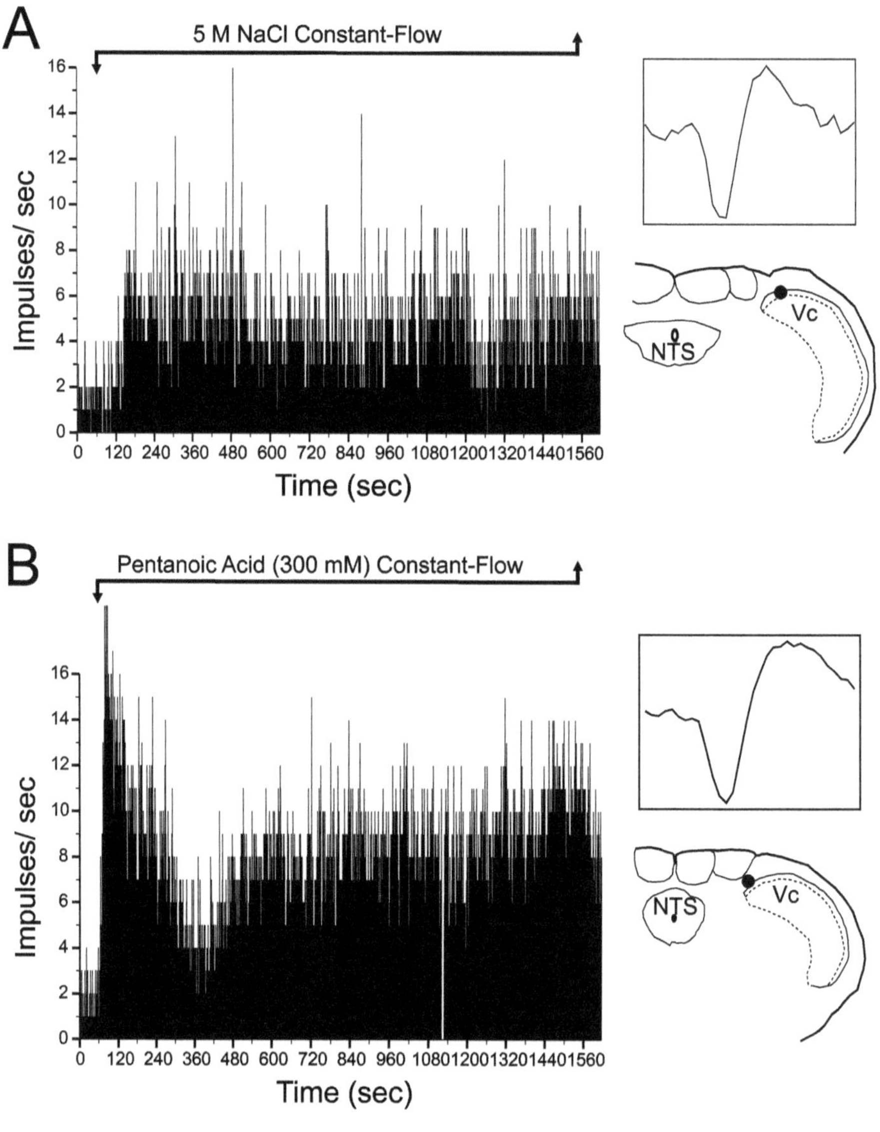

FIGURE 11.6 Sensitization in responses of Vc units to application of hypertonic NaCl or pentanoic acid by constant flow. A: PSTH of a Vc unit's response to lingual application of 5 *M* NaCl by constant flow (upper bar with arrows). Note sensitization. Insets to right: action potential waveform and recording site in Vc. B: PSTH of another unit's response to constant-flow application of pentanoic acid. Note initial excitation followed by sensitization. Insets to right: action potential waveform and recording site in Vc.

(Carstens et al., 1998; Dressier et al., 2000d). Interestingly, using a paired preference paradigm, we have observed that rodents reject capsaicin at concentrations as low as 0.5 to 1 ppm (1.65 to 3.3 μ*M*) (authors' unpublished observations). Thus, if Vc units are involved in detecting and signaling oral irritation, their response threshold

TABLE 11.1
Concentrations of Common Irritant Chemicals above Threshold for Exciting Vc Neurons and for Inducing Irritation

Irritant Chemical	Concentration: Vc (%)	Concentration: Vc (m*M*)	Humans (range) (m*M*)
Capsaicin	0.01 (100 ppm)	0.33	10×10^{-3}–0.33
Piperine	0.10	3.5	0.06–0.26
Histamine	1.0	90	
Nicotine	1.0	62	0.74–300
NaCl	14.6	2.5×10^3	400–5×10^3
Ethanol	25.0	5.4×10^3	5.4×10^3–10.8×10^3
Serotonin (5-HT)	1.0	57	
Pentanoic acid	3.1	300	200
Citric acid	4.8	250	250
HCl	0.36	100	
Mustard oil	4.0	403	
Menthol	1.0	64	3.3–19.2

is approximately two orders of magnitude higher than the behavioral detection threshold. One possible reason for this difference is that the anesthetic (thiopental) used in the electrophysiological experiments increases the thresholds of Vc units to irritant chemicals. Another possibility is a species difference in the diffusion of chemicals through the lingual epithelium.

11.2.6.9 Chemonociceptive Responses of Vc Neurons and Sensory Discrimination

That many Vc neurons show broad tuning in their responses to a variety of irritant chemicals (Figure 11.2) raises the issue of sensory discrimination. It remains an open question as to whether or not humans can accurately discriminate among different irritant sensations. Obviously, discrimination of highly acidic or saline solutions would be aided by concomitant sour and salty tastes. However, some irritants, such as capsaicin and piperine, are considered tasteless. Discrimination between their irritant sensations might be accomplished based on the time-course (rate of onset; overall duration), relative intensity, or qualitative properties (burning vs. pricking). It is difficult to understand how broadly tuned Vc neurons would contribute to such discrimination. However, this might be accomplished if different Vc neurons exhibit selectivity in their responses to different irritant chemicals, or if sub-populations of Vc neurons exist that respond only to one class of irritant.

11.2.7 Summary

Our studies to date indicate that substantial numbers of both WDR and NS classes of Vc units are chemonociceptive, generally responding to lingual application of a variety of different irritant chemicals in a concentration-dependent manner (Carstens et al., 1998). It is not currently known if some Vc units respond only to irritant chemicals and not to mechanical or thermal stimuli; unbiased electrical search strategies are needed to test for such units. Repeated application of some chemicals induces tachyphylaxis, as well as cross-tachyphylaxis (or cross-desensitization). Cross-tachyphylaxis is a potential confound when investigating the responses of a Vc unit to sequential application of different irritant chemicals and requires that non-desensitizing chemicals be tested first and desensitizing chemicals last. The interval at which a chemical is repeatedly delivered is also an important factor. Capsaicin induces sensitization when delivered continually or once per minute, but not when presented a single time. Moreover, following a rest period, reapplication of capsaicin only elicits a delayed response, consistent with desensitization, which eventually rises, consistent with SIR. In contrast, continual or repetitive application of nicotine initially excites Vc units followed by rapid adaptation to the prestimulus level. These firing patterns are consistent with human psychophysical studies showing sensitization with capsaicin and desensitization with nicotine. However, the threshold for Vc unit response to capsaicin is much higher than that required to elicit irritation, suggesting possible depressant effects of anesthetics in the electrophysiological studies.

11.3 c-*fos* IMMUNOHISTOCHEMISTRY

11.3.1 Advantages and Disadvantages

This method has seen increasing use in studies of trigeminal nociception from oral, ocular, and intranasal mucosae (e.g., Anton et al., 1992; Strassman and Vos, 1993; Bereiter et al., 1994; Coimbra and Coimbra, 1994; Carstens et al., 1995; Mineta et al., 1995; Martinez and Belmonte, 1996; Meng and Bereiter, 1996; Bereiter, 1997; Li et al., 1997; 1999; Yamashiro et al., 1997; Bereiter et al., 1998; Sugimoto et al., 1998; Oakden and Boissonade, 1998; Takeda et al., 1998; Aihara et al., 1999; Chattipakorn et al., 1999; Zhou et al., 1999; Bereiter and Bereiter, 2000; Byers et al., 2000; He et al., 2000; Leong et al., 2000; Takemura et al., 2000). As noted in the introduction, a major advantage of this method is that it provides a more global view of populations of neurons that are activated by a sensory stimulus. The immunohistochemical method is fairly straightforward, and the data can be readily quantified by counting cells with *fos*-like immunoreactivity (FLI). A major disadvantage is that the method provides only a static view of the distribution of FLI. In addition, there are still uncertainties regarding the link between sensory stimulation and neuronal c-*fos* expression (see Harris, 1998, for recent review). Expression of FLI is best related to noxious stimulation (Hunt et al., 1987; Zimmermann and Herdegen, 1994; Harris, 1998). Even here, however, there are discrepancies between results based on FLI vs. single-unit recording. For example, it has been shown that neurons in both Vc (Strassman and Vos, 1993; Carstens et al., 1998) and subnucleus oralis (Dallel et al., 1990; 1996; Raboisson et al., 1991; 1995) respond to noxious oro-facial stimuli including capsaicin. However, intra-oral application of irritant chemicals elicits FLI in Vc but has generally failed to result in FLI in oralis (Strassman and Vos, 1993;

Carstens et al., 1995; 2000), with the exception of a recent report that lingual capasicin elicited FLI in both Vc and oralis (He et al., 2000). Therefore, results obtained with the c-*fos* method should be interpreted with caution and preferably corroborated by functional data obtained using other approaches.

Our laboratory has used the c-*fos* method to target brain stem locations of neurons that are activated by oral irritant chemicals, as well as to investigate transduction pharmacology. This chapter briefly presents the methodology and gives examples showing the utility of the c-*fos* immunohistochemical approach in relation to single-unit recordings.

11.3.2 Immunohistochemical Methods

Adult male Sprague-Dawley rats are anesthetized with sodium pentobarbital (65 mg/kg, i.p.). The mouth is held in a partially opened position with a small retractor and a strip of Parafilm is placed underneath the tongue to prevent stimulation of other oral tissue. The chemical stimulus is delivered to the anterior dorsal surface of the tongue, either by topical application or by flow (up to 10-min duration), as in the electrophysiological experiments described above. We usually close the mouth 10 to 20 min after application of the aqueous stimulus to prevent desiccation.

Appropriate control groups are essential in interpreting c-*fos* data. It is important to include a vehicle control to test for nonspecific stimuli such as mechanical stimulation of lingual tissue during chemical application. We also recommend an unstimulated group to control for anesthetic, passage of time, and other factors such as dessication (Strassman and Vos, 1993; Carstens et al., 1995; 2000; Simons et al., 1999).

Two hours following the onset of chemical stimulation, the animals are perfused through the aorta with phosphate-buffered saline followed by 4% paraformaldehyde. Brain stems are removed and post-fixed for 8 h, and then transferred to a 30% sucrose solution. Two to three days later, the brain stems are cut into 50-μm frozen sections and collected in serial order in three separate 24-well containers (Costar, Corning, NY) in phosphate-buffered saline. Each container thus has serial sections at a 150-μm sampling interval. Tissue in one of these containers is then processed immunohistochemically. This involves washing the sections, blocking in 3% normal goat serum, followed by incubation in primary c-*fos* antibody (Arnel, New York, NY; 1:50,000) for 24 to 36 h. The sections are then washed and exposed to secondary biotinylated (goat-anti-rabbit) antibody, followed by an avidin-biotin-peroxidase complex (ABC; Vector, Burlingame, CA) reaction enhanced with biotinyl tyramide/H_2O_2. To visualize the reaction product, sections are finally subjected to a nickel-enhanced diaminobenzidine reaction, mounted on microscope slides, and coverslipped. They are examined under a light microscope and cell nuclei displaying black *fos*-like immunoreactivity (FLI) are counted. Our microscope is equipped with a color video camera allowing us to image individual sections, which can then be imported into graphic software (we use CorelDraw) to plot FLI directly on a tracing of the section or onto templates taken from the atlas of Paxinos and Watson (1998). We routinely count FLI bilaterally in five regions of interest: dorsomedial Vc, ventrolateral Vc, the ventrolateral medullary area dorsal to the lateral reticular nucleus, the nucleus of the solitary tract (NTS), and the area postrema. Counts of FLI are made by an investigator blind as to treatment, in all sections from the level of the pyramidal decussation caudally to the rostral extent of the area postrema. Between-treatment group comparisons of mean bilateral counts of FLI for each region of interest can be statistically analyzed by analysis of variance (ANOVA).

11.3.3 Brainstem Locations of FLI Following Irritant Chemical Stimulation of the Tongue

We initially employed this method to establish the brain stem locations of FLI elicited by irritant chemical stimulation of the tongue (Carstens et al., 1995). Each irritant tested (capsaicin, nicotine, histamine, piperine, and acetylcholine) elicited very similar distributions of FLI in the dorsomedial region of Vc, as well as in several other distinct locations, including the ventrolateral aspect of Vc, the ventrolateral medulla dorsal to the lateral reticular nucleus, nucleus of the solitary tract (NTS), and area postrema. An example of the distribution of FLI elicited by nicotine in each of these brain stem regions is shown in Figure 11.7 (from Carstens et al., 2000).

The pattern of FLI in dorsomedial Vc was similar, regardless of the irritant chemical applied. As noted in the introduction to this chapter, this method cannot determine whether one and the same Vc neuron was activated by different chemicals, or if each chemical activates a select sub-population of neurons within the same general region of Vc (Figure 11.1). However, the advantage of this method is that it provides an overall picture of the numbers and distributions of activated neurons at a specific time point.

11.3.4 Pharmacological Studies

Another arguable advantage of the c-*fos* method is that the chemical is only applied once, providing a view of the distribution of all activated neurons that is not compromised by factors, such as tachyphylaxis, associated with repeated stimulus application (see Section 11.2.6.4).

As an example, repeated application of nicotine to the tongue induces tachyphylaxis in successive Vc unit responses (Carstens et al., 1998). This makes it difficult to investigate the pharmacology of nicotinic receptors involved because a nicotinic antagonist might reduce nicotine-evoked responses, but it is uncertain how much of the reduction is due to tachyphylaxis and how much is due to true antagonism. The c-*fos* method obviates this problem because the chemicals are only applied once. Figure 11.8A shows an example of the brain stem distribution of FLI from one rat following application of nicotine to the tongue. In a separate animal, application of nicotine after pretreatment of the tongue with a nicotinic antagonist (mecamylamine) resulted in considerably less FLI (Figure 11.8B). Overall, there was significantly less FLI in Vc of the group receiving mecamylamine followed by nicotine compared to the group receiving nicotine alone (Carstens et al., 2000). Unstimulated controls exhibited no FLI in dorsomedial Vc, while vehicle- (saline) stimulated controls exhibited a small degree of FLI in this region (Figure 11.8C) which, however, was significantly less compared to the nicotine-only group.

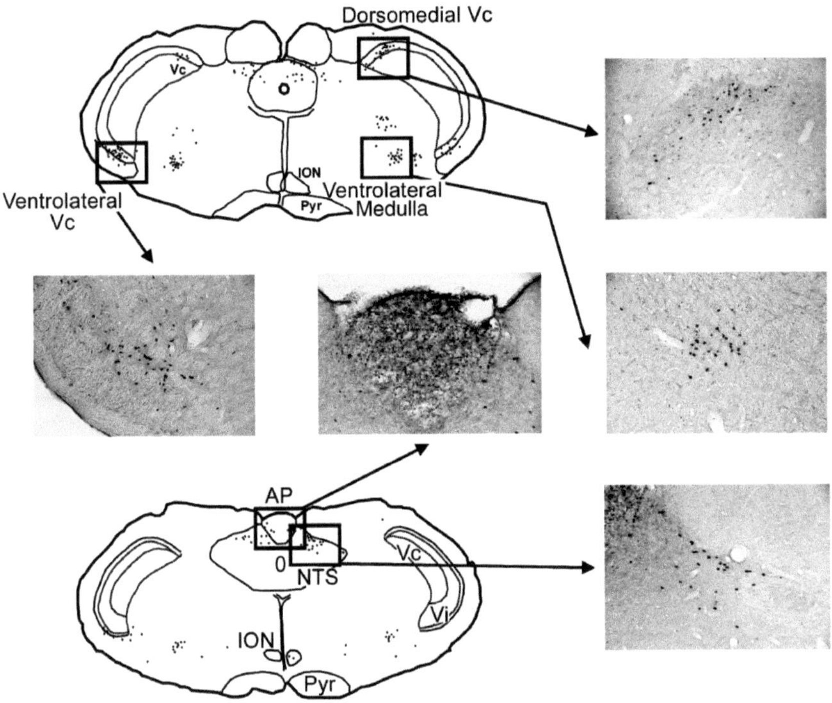

FIGURE 11.7 Medullary regions exhibiting FLI. The upper and lower line drawings are cross-sections through the medulla at the level of caudal medulla (upper section) and area postrema (lower section). Dots indicate locations of cell nuclei expressing FLI. Boxes surround areas that correspond to the photomicrographs (at 20×). Abbreviations as in Figure 11.8. (From Carstens et al., 2000, with permission of Springer Verlag.)

We have used the same approach to show that FLI elicited in Vc by application of carbonated water to the tongue was significantly attenuated by pretreatment with different blockers of carbonic anhydrase (Simons et al., 1999; Dressier et al., 2000c). In a correlative electrophysiological analysis, one of the carbonic anhydrase blockers (dorzolamide) also significantly reduced responses of Vc units to carbonated water flowed over the tongue (Simons et al., 1999). These results therefore illustrate the usefulness of the c-*fos* method in studying the pharmacology of transduction mechanisms for irritant chemical stimuli, and in this case reveal a close correlation between results obtained with the two different methods.

11.3.5 Summary

The c-*fos* method can be used to great advantage in targeting the locations of neurons that are activated by irritant chemical stimuli, and in investigating the transduction mechanisms. However, to understand the extent to which individual neurons can respond to different irritants, correlative single-unit recordings are necessary. These

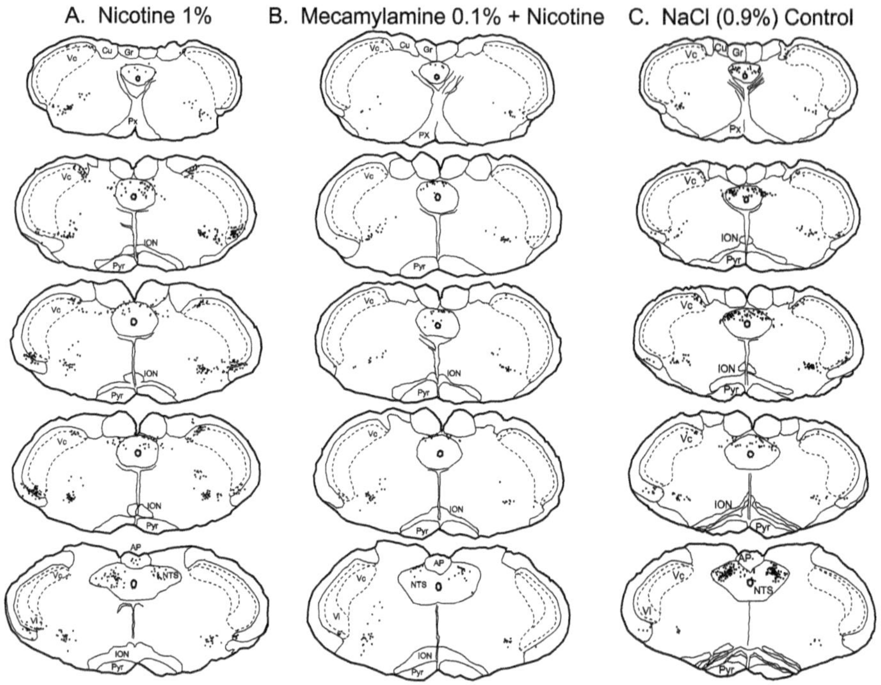

FIGURE 11.8 Brain stem distribution of FLI in rats receiving nicotine on the tongue. In each panel, dots indicate locations of cell nuclei expressing FLI, plotted on drawings of medullary sections arranged from caudal (top) to rostral (bottom). Dashed lines indicate approximate border of laminae I-II in Vc, and Vc/Vi transition in bottom-most section. A: Nicotine (1%) alone. B: Mecamylamine (0.1%) followed by 1% nicotine. C: Saline (vehicle) control. Abbreviations: AP, area postrema; Cu, n. cuneatus; ION, inferior olivary n.; Gr, n. gracilis; NTS, n. of the solitary tract; Px, pyramidal decussation; Pyr, medullary pyramid; Vc, trigeminal subnucleus caudalis; Vi, trigeminal subnucleus interopolaris.

two methods complement each other well in research directed toward identifying and characterizing neuronal circuits involved in signaling oral chemical irritation and pain.

ACKNOWLEDGMENTS

Work reported in this chapter was supported by grants from the NIH (NS 35788) and the California Tobacco-Related Disease Research Program (6RT-0231).

REFERENCES

Aihara, Y., Maeda, T., Hanada, K., and Wakisaka, S. Effects of morphine on the distribution of Fos protein in the trigeminal subnucleus caudalis neurons during experimental tooth movement of the rat molar, *Brain Res.*, 819, 48-57, 1999.

Anton, F., Herdegen, T., Peppel, P., and Leah, J.D. C-*fos*-like immunoreactivity in rat brainstem neurons following noxious chemical stimulation of the nasal mucosa, *Neuroscience,* 41, 629-641, 1992.

Bereiter, D.A. Morphine and somatostatin analogue reduce c-*fos* expression in trigeminal subnucleus caudalis produced by corneal stimulation in the rat, *Neuroscience,* 77, 863-874, 1997.

Bereiter, D.A. and Bereiter, D.F. Morphine and NMDA receptor antagonism reduce c-*fos* expression in spinal trigeminal nucleus produced by acute injury to the TMJ region, *Pain,* 85, 65-77, 2000.

Bereiter, D.A., Bereiter, D.F., Tonnessen, B.H., and Maclean, D.B. Selective blockade of substance P or neurokinin A receptors reduces the expression of c-*fos* in trigeminal subnucleus caudalis after corneal stimulation in the rat, *Neuroscience,* 83, 525-553, 1998.

Bereiter, D.A., Hathaway, C.B., and Benetti, A.P. Caudal portions of the spinal trigeminal complex are necessary for autonomic responses and display Fos-like immunoreactivity after corneal stimulation in the cat, *Brain Res.,* 657, 73-82, 1994.

Byers, M.R., Chudler, E.H., and Iadarola, M.J. Chronic tooth pulp inflammation causes transient and persistent expression of *fos* in dynorphin-rich regions of rat brainstem, *Brain Res.*, 861, 191-207, 2000.

Carstens, E. Responses of rat spinal dorsal horn neurons to intracutaneous microinjection of histamine, capsaicin, and other irritants, *J. Neurophysiol.,* 77, 2499-2514, 1997.

Carstens, E.E., Kuenzler, N., and Handwerker, H.O. Activation of neurons in rat trigeminal subnucleus caudalis by application of different classes of irritant chemicals to the oral and occular mucosa, *J. Neurophysiol.,* 80, 465-492, 1998.

Carstens, E., Saxe, I., and Ralph, R., Brain stem neurons signaling expressing c-*fos* immunoreactivity following chemical stimulation of the rat's tongue, *Neuroscience,* 69, 939-953, 1995.

Carstens, E., Simons, C.T., Dressier, J.-M., Iodi Carstens, M., and Jinks, S.L. Role of neuronal nicotinic acetylcholine receptors in the activation of neurons in trigeminal subnucleus caudalis by nicotine delivered to the oral mucosa, *Exp. Brain Res.,* 132, 375-383, 2000.

Chattipakorn, S.C., Light, A.R., Willcockson, H.H., Närhi, M., and Maixner, W. The effect of fentanyl on c-*fos* expression in the trigeminal brainstem complex produced by pulpal heat stimulation in the ferret, *Pain,* 82, 207-215, 1999.

Cliff, M.A. and Green, B.G. Sensory irritation and coolness produced by menthol: evidence for selective desensitization of irritation, *Physiol. Behav.,* 56, 1021-1029, 1994.

Coimbra, F. and Coimbra, A. Dental noxious input reaches the subnucleus caudalis of the trigeminal complex in the rat, as shown by c-*fos* expression upon thermal or mechanical stimulation, *Neurosci. Lett.*, 173, 201-204, 1994.

Cliff, M.A. and Green, B.G. Sensitization and desensitization to capsaicin and menthol in the oral cavity: interactions and individual differences, *Physiol. Behav.,* 59, 487-494, 1996.

Dallel, R., Luccarini, P., Molat, J.-L., and Woda, A. Effects of systemic morphine on the activity of convergent neurons of spinal trigeminal nucleus oralis in the rat, *Eur. J. Pharmacol.,* 314, 19-25, 1996.

Dallel, R., Raboisson, P., Woda, A., and Sessle, B.J. Properties of nociceptive and non-nociceptive neurons in trigeminal subnucleus oralis of the rat, *Brain Res.*, 521, 95-106, 1990.

Dressier, J.-M., Chang, H.-K., O'Mahony, M., and Carstens, E. Cross-desensitization of capsaicin-evoked oral irritation by high but not low concentrations of nicotine, *Neurosci. Lett.,* 290, 133-136, 2000a.

Dressier, J.-M., Nguyen, N., Sieffermann, J.-M., Carstens, E., and O'Mahony, M. Oral irritant properties of piperine and nicotine: psychophysical evidence for asymmetrical desensitization effects, *Chem. Senses,* 24, 405-413, 1999.

Dressier, J.-M., O'Mahony, M., and Carstens, E. Oral irritant effects of nicotine: psychophysical evidence for decreased sensation following repeated application and lack of cross-desensitization to capsaicin, *Chem. Senses,* 22, 483-492, 1997.

Dressier, J.-M., O'Mahony, M., Iodi-Carstens, M., Yao, E., and Carstens, E. Oral irritation by sodium chloride: sensitization, self-desensitization, and cross sensitication to capsaicin, *Physiol. Behav.,* 72, 317-324, 2001a.

Dressier, J.-M., O'Mahony, M., and Carstens, E. Oral irritant properties of menthol: sensitizing and desensitizing effects of repeated application and cross-desensitization to nicotine, *Physiol. Behav.,* 73, 25-36, 2001b.

Dressier, J.-M., O'Mahony, M., Iodi-Carstens, M., and Carstens, E. Sensory properties of citric acid: psychophysical evidence for sensitization, self-desensitization, cross-desensitization and cross-stimulus-induced recovery following capsaicin, *Chem. Senses,* 25, 769-780, 2000b.

Dressier, J.-M., Simons, C.T., Iodi Carstens, M., O'Mahony, M., and Carstens, E. Psychophysical and neurobiological evidence that the oral sensation elicited by carbonated water is of chemogenic origin, *Chem. Senses,* 25, 277-284, 2000c.

Dressier, J.-M., Simons, C.T., Sudo, M., Sudo, S., and Carstens, E. Sensitization, desensitization and stimulus-induced recovery (SIR) of responses of rat trigeminal caudalis neurons to repeated oral application of capsaicin and nicotine, *J. Neurophysiol.,* 84, 1851-1862, 2000d.

Forster, C. and Handwerker, H.O. Automatic classification and analysis of microneurographic spike data using a PC/AT, *J. Neurosci. Meth.,* 31, 109-118, 1990.

Gilmore, M.M. and Green, B.G. Sensory irritation and taste produced by NaCl and citric acid: effects of capsaicin desensitization, *Chem. Senses,* 18, 257-272, 1993.

Green, B.G. Capsaicin sensitization and desensitization on the tongue produced by brief exposures to a low concentration, *Neurosci. Lett.,* 107, 173-178, 1989.

Green, B.G. Effects of thermal, mechanical, and chemical stimulation on the perception of oral irritation, in *Chemical Senses, Vol. 2: Irritation,* B.G. Green, J.R. Mason and M.R. Kare, Eds., New York, Dekker, 171-192, 1990.

Green, B.G. Capsaicin cross-desensitization on the tongue: psychophysical evidence that oral chemical irritation is mediated by more than one sensory pathway, *Chem. Senses,* 16, 675-689, 1991a.

Green, B.G. Temporal characteristics of capsaicin sensitization and desensitization on the tongue, *Physiol. Behav.,* 49, 501-505, 1991b.

Green, B.G. Evidence that removal of capsaicin accelerates desensitization on the tongue, *Neurosci. Lett.,* 150, 44-48, 1993.

Green, B.G. Rapid recovery from capsaicin desensitization during recurrent stimulation, *Pain,* 68, 245-253, 1996.

Green, B.G. and McAuliffe, B.L. Menthol desensitization of capsaicin irritation: evidence of a short-term anti-nociceptive effect, *Physiol. Behav.,* 68, 631-639, 2000.

Green, B.G. and Rentmeister-Bryant, H. Temporal characteristics of capsaicin desensitization and stimulus-induced recovery in the oral cavity, *Physiol. Behav.,* 65, 141-149, 1998.

Harris, J.A. Using c-*fos* as a neural marker of pain, *Brain Res. Bull.,* 45, 1-8, 1998.

He, Y.F., Ichikawa, H., and Sugimoto, T. The effect of neonatal capsaicin on the c-*fos*-like immunoreactivity induced in subnucleus oralis neurons by noxious intraoral stimulation, *Brain Res.,* 860, 203-207, 2000.

Hunt, S.P., Pini, A., and Evan, G. Induction of c-*fos*-like protein in spinal cord neurons following sensory stimulation, *Nature,* 328, 632-634, 1987.

Jinks, S.L. and Carstens, E. Activation of spinal wide dynamic range neurons by intracutaneous microinjections of nicotine, *J. Neurophysiol.,* 82, 3046-3055, 1999.

Karrer, T. and Bartoshuk, L. Capsaicin desensitization and recovery on the human tongue, *Physiol. Behav.,* 49, 757-764, 1991.

Leong, S., Liu, H., and Yeo, J. Nitric oxide synthase and glutamate receptor immunoreactivity in the rat spinal trigeminal neurons expressing Fos protein after formalin injection, *Brain Res.,* 855, 107-115, 2000.

Li, J.L., Kaneko, T., Nomura, S., Li, Y.Q., and Mizuno, N. Association of serotonin-like immunoreactive axons with nociceptive projection neurons in the caudal spinal trigeminal nucleus of the rat, *J. Comp. Neurol.*, 384, 127-141, 1997.

Li, J.L., Li, Y.Q., Kaneko, T., and Mizuno, N. Preprodynorphin-like immunoreactivity in medullary dorsal horn neurons projecting to the thalamic regions in the rat, *Neurosci. Lett.,* 264, 13-16, 1999.

Lipski, J. Antidromic activation of neurones as an analytic tool in the study of the central nervous system, *J. Neurosci. Meth.*, 4, 1-32, 1981.

Liu, L, Welch, J.M., Erickson, R.P., Reinhart, P.H., and Simon, S.A. Different responses to repeated applications of zingerone in behavioral studies, recordings from intact and cultured TG neurons, and from VR1 receptors, *Physiol. Beh.*, in press, 2000.

Martinez, S. and Belmonte, C. c-*Fos* expression in trigeminal nucleus neurons after chemical irritation of the cornea: reduction by selective blockade of nociceptor chemosensitivity, *Exp. Brain Res.*, 109, 56-62, 1996.

Meng, I.D. and Bereiter, D.A. Differential distribution of *Fos*-like immunoreactivity in the spinal trigeminal nucleus after noxious and innocuous thermal and chemical stimulation of rat cornea, *Neuroscience,* 72, 243-254, 1996.

Mineta, Y., Eisenberg, E., and Strassman, A.M. Distribution of *fos*-like immunoreactivity in the caudal medullary reticular formation following noxious facial stimulation in the rat, *Exp. Brain Res.*, 107, 34-38, 1995.

Oakden, E.L. and Boissonade, F.M. *Fos* expression in the ferret trigeminal nuclear complex following tooth pulp stimulation, *Neuroscience,* 84, 1197-1208, 1998.

Paxinos, G. and Watson, C. *The Rat Brain in Stereotaxic Coordinates*, 4th ed., Academic Press, New York, 1998.

Prescott, J. and Stevenson, R.J. Desensitization to oral zingerone irritation: effects of stimulus parameters, *Physiol. Behav.,* 60, 1473-1480, 1996a.

Prescott, J. and Stevenson, R.J. Psychophysical responses to single and multiple presentations of the oral irritant zingerone: relationship to frequency of chili consumption, *Physiol. Behav.,* 60, 617-624, 1996b.

Raboisson, P., Bourdiol, P., Dallell, R., Clavelou, P., and Woda, A. Responses of trigeminal subnucleus oralis nociceptive neurones to subcutaneous formalin in the rat, *Neurosci. Lett.,* 125, 179-182, 1991.

Raboisson, P., Dallell, R., Bourdiol, P., Clavelou, P., Sessle, B.J. and Woda, A. Effects of subcutaneous formalin on the activity of trigeminal brain stem nociceptive neurones in the rat, *J. Neurophysiol.,* 73, 496-505, 1995.

Simons, C.T., Dressier, J.M., Carstens, M.I., O'Mahony, M., and Carstens, E. Neurobiological and psychophysical mechanisms underlying the oral sensation produced by carbonated water, *J. Neurosci.,* 19, 8134-8144, 1999.

Strassman, A.M. and Vos, B.P. Somatotopic and laminar organization of Fos-like immunoreactivity in the medullary and upper cervical dorsal horn induced by noxious facial stimulation in the rat, *J. Comp. Neurol.*, 331, 495-516, 1993.

Sugimoto, T., He, Y.F., Funahashi, M., and Ichikawa, H. Induction of immediate-early genes c-*fos* and *zif*268 in the subnucleus oralis by noxious tooth pulp stimulation, *Brain Res.*, 794, 353-358, 1998.

Szolcsànyi, J. Capsaicin, irritation, and desensitization: neurophysiological basis and future perspectives, in *Chemical Senses, Vol. 2: Irritation*, B.G. Green, J.R. Mason, and M.R. Kare, Eds., New York, Dekker, 141-168, 1990.

Takeda, M., Matsumoto, S., and Tanimoto, T. c-*Fos*-like immunoreactivity in the upper cervical spinal dorsal horn neurons following noxious chemical stimulation of the nasal mucosa in pentobarbital-anesthetized rats, *Arch. Histol. Cytol.*, 61, 83-87, 1998.

Takemura, M., Shimada, T., Sugiyo, S., Nokubi, T., and Shigenaga, Y. Mapping of c-*Fos* in the trigeminal sensory nucleus following high- and low-intensity afferent stimulation in the rat, *Exp. Brain Res.*,130, 113-123, 2000.

Yamashiro, T., Nakagawa, K., Satoh, K., Moriyama, H., and Takada, K. c-*Fos* expression in the trigeminal sensory complex and pontine parabrachial areas following experimental tooth movement, *Neuroreport,* 8, 2351-2353, 1997.

Zhou, Q., Imbe, H., Dubner, R., and Ren, K. Persistent Fos protein expression after orofacial deep or cutaneous tissue inflammation in rats: implications for persistent orofacial pain, *J. Comp. Neurol.*, 412, 276-291, 1999.

Zimmermann, M. and Herdegen, T. Control of gene transcription by Jun and Fos proteins in the nervous system: beneficial or harmful molecular mechanisms of neuronal response to noxious stimulation?, *Am. Pain Soc. J.*, 3, 33-48, 1994.

12 Methodological Considerations for Electrophysiological Recording and Analysis of Taste-Responsive Neurons in the Brain Stem of the Rat

Patricia M. Di Lorenzo and Christian H. Lemon

CONTENTS

0-8493-2329-0/02/$0.00+$1.50

12.1 INTRODUCTION

This chapter provides a review of some of the most relevant points and pointers related to extracellular recording of taste-responsive neurons in the brain stem of the rat. Although the focus is on the rat, most of what is discussed is relevant to other rodent species commonly used in taste research, such as the hamster, gerbil, or, most recently, mouse. In the interest of brevity, this chapter discusses only those points of the methodology for which the authors believe there is something to say apart from the standard recording procedures that might be used in extracellular single-cell recording. Thus, this chapter is far from encyclopedic in scope, and the authors apologize if any desirable details or references have been omitted. All in all, the purpose of this chapter is to discuss some issues about the relationship of methodology to the investigation of taste coding.

12.2 BACKGROUND

12.2.1 Peripheral Anatomy of the Gustatory System

The cascade of nervous system activity that results in the perception of taste begins with processes occurring at the level of taste receptor cells (TRCs) in the oral cavity. These receptor cells are differentiated, polarized epithelial cells whose basic function is to transduce chemical information into an electrochemical neural signal. Each TRC lives a very short life of about 10 to 12 days in length[2,3]; the TRC population is in a constant state of turnover within the system. Two specialized regions are found on TRCs: the apical and basolateral membranes. The apical membrane contains chemoreceptor proteins exposed to the mucosal space, while the basolateral membrane synapses with gustatory afferent fibers; TRCs do not possess their own axons.[1,55] Action potentials have been observed in TRCs in response to stimulation with some chemical stimuli.[1,9,41,59] Although one TRC may sometimes synapse with several afferent fibers, it is not known if these connections are determined by the particular sensitivity profiles of the TRCs.[59] To complicate matters, it is generally believed that TRCs are broadly tuned[77,85,87]; i.e., they respond to more than one of the four prototypical, or basic taste qualities: sweet, sour, salty, and bitter.

In mammals, TRCs reside in taste buds, which are embedded in the tongue and oral epithelium.[41] Taste buds are typically comprised of 40 to 120 cells of various types; some of these cells are TRCs, some are precursors to TRCs, and some may play a solely supportive role within the bud.[59] Taste buds are located in papillae, of which there are four types: (1) filliform papillae are distributed across the tongue and do not contain any taste buds; (2) fungiform papillae are located on the anterior two-thirds of the tongue and are innervated by the chorda tympani (CT) nerve, a branch of the facial or VIIth cranial nerve; (3) foliate papillae lie in two areas on opposite sides of the tongue, the anterior foliate being innervated by the CT while the posterior foliate is innervated by the lingual-tonsillar branch of the glossopharyngeal or IXth cranial nerve; and (4) circumvallate papillae are found on the posterior tongue and are innervated exclusively by the lingual-tonsillar nerve. The soft palate, located on the roof of the oral cavity, also contains taste buds. These are innervated by the greater superficial petrosal branch of the facial, or VIIth cranial nerve. Refer to complete reviews of papilla anatomy and innervation.[41,59]

12.2.2 Central Pathways in the Gustatory System

The first-order gustatory afferents synapse ipsilaterally onto glomeruli located within the rostral central and rostral lateral subdivisions of the nucleus of the solitary tract (NTS) in the brain stem.[121] Neurons within the gustatory portion of the NTS are quite small and very densely packed. Evidence suggests that glutamate mediates excitatory synaptic transmission between afferent gustatory input and rostral NTS neurons.[57,118] Additionally, anatomical[13,55] and physiological[47,101] studies suggest that gamma-aminobutyric acid (GABA) and substance P[15] also play a significant role in synaptic processing in the gustatory NTS. In this context, there is evidence that NTS synaptic potentials are complex mixtures of excitation and inhibition.[48]

In the rodent, second-order gustatory afferents, originating from the NTS, project ipsilaterally to the caudal parabrachial nucleus of the pons (PbN).[72,73] Originally termed the pontine taste area,[72,73] the gustatory region of the PbN is comprised of three main subdivisions: medial, which is ventral to the brachium; the "waist" area, which is located within the brachium; and lateral, which lies just dorsal to the brachium.[70] Cells located in these areas have more complex dendritic arborizations than cells in the gustatory NTS.[12] PbN neurons use primarily excitatory amino acids as neurotransmitters, such as glutamate and aspartate, although there is evidence for the presence of noradrenergic and GABAergic influences.[84]

It should be noted that the ascending gustatory projection from the NTS to the PbN is apparent only in rodents. In non-human primates, neuroanatomical data have revealed that only the caudal NTS (non-gustatory) projects to the PbN; ascending gustatory fibers extend to higher centers without traversing the PbN.[67]

Gustatory afferents originating from the caudal PbN project to a variety of forebrain areas forming two systems: the thalamocortical and ventral forebrain systems.[71] The thalamocortical route, as the name suggests, illustrates the typical ascending pathway for sensory information to the cortex. The caudal PbN projects to the parvicellular portion of the ventroposteromedial thalamus, which then projects to the gustatory neocortex.[72] The ventral forebrain system is comprised of projections from the PbN to the lateral hypothalamus, central nucleus of the amygdala, bed nucleus of the stria terminalis, and the substantia innominata.[69,72] Interestingly, the corticofugal system can influence the sensibilities of taste-driven neurons in the NTS[24] and PbN[16,18,23] by way of descending projections to the brain stem.

12.3 CODING

Any discussion of neural coding should begin with a statement of definitions. In that way, even if there is disagreement about the terms or their meanings, the starting point is clear. In fact, some would argue that the idea of coding is itself useless, or at best misleading.[49] That idea notwithstanding, for our purposes, neural code is defined as the set of rules for transformations of neural signals that permits the communication of specific information. In most studies of coding, the mere demonstration that a particular type of information can be gleaned by applying a specific transformation, i.e., the coding mechanism, to the neural signals (usually the spike train) is enough to suggest that the nervous system encodes that information in just that way. For example, by demonstrating that more intense stimuli produce higher rates of firing in a given neuron, many would conclude that stimulus intensity is encoded by the firing rate. Although this may on its surface seem reasonable, there are cases for which this conclusion may be only partially true. Consider the circumstance in which changing the intensity of a stimulus may alter its quality or add other qualities, as when very intense stimuli produce pain, for example. In that case, and very likely in many cases, different types of information can be conveyed simultaneously by the same neural elements. Moreover, there may be instances in which *what* is being encoded must be determined before *how* it is being encoded in order to understand the workings of the system. The gustatory system may represent one of those instances.

Extrapolating from Mueller's law of specific nerve energies, a taste stimulus or tastant is any stimulus that evokes activity in the taste nerves or neural pathways associated with taste. In everyday life, taste stimuli are most commonly chemicals that are dissolved or suspended in fluid (often saliva) that come in contact with the tongue. The issue of what is being encoded about a tastant can, on one level, be determined by psychophysics. The organization of the taste world into groups of stimuli that taste alike, defined as taste qualities, represents a rough description of what must be conveyed by the nervous system. It should be noted that there is disagreement about the number of taste qualities[53] and even about whether they exist at all.[30,31,33,34,88] Even so, the decision of whether stimuli taste alike might seem to be a straightforward problem for the nervous system to solve. However, in addition to its quality, there are several other aspects of taste stimuli that are encoded, or deduced, by the neural representation in the gustatory pathway. These include intensity (concentration),[26,40,94] flow rate,[99] and hedonic quality.[89,93] Furthermore, there are several variables, such as hormonal state,[21,22,44,45] sodium deprivation,[10,11,54,62,97] or prior learning experiences,[8,43,63] that can affect the response evoked by a taste stimulus.

12.3.1 Issues to Consider

In investigations of the neural code for taste stimuli, there are several assumptions that are implicit, but that may affect the conceptualization of how the brain encodes taste. First, although it is well-known that the taste receptor cells are replaced on average approximately every 10 to 12 days,[2,3] it is assumed that the system encodes a given stimulus in the same way over time. That is, the assumption is made that there is stability in the neural code in the face of ever-changing inputs. Second, it is usually assumed that the same coding mechanism is used to encode every stimulus. For example, if one proposes that salty stimuli are encoded by the activity in those cells that respond best to salty tastes, called a labeled line, then it is assumed that sweet, sour, and bitter stimuli are also encoded by activity in cells that respond best to exemplars of those stimuli. This assumption prevails, even in the face of evidence suggesting that different nerves which innervate different areas of the oropharyngeal area also serve different behavioral functions.[106-109] Third, most theories of taste coding, with the possible exception of temporal coding theories, assume that the spike activity produced by a taste-responsive cell is integrated over time. More discussion of this point comes later. Finally, it is often assumed that the same coding mechanisms are used at every level of the nervous system. This implies that if labeled lines are identified in the NTS, then in all probability labeled lines are also used in the PbN, thalamus, and neocortex. This assumption puts the investigator at risk of ignoring the potentially unique role that each neural structure might have in the coding process.

The most common strategy for studying taste coding, and one that is not unique to the study of gustation, is to record from a single cell in a single neural structure during the presentation of a variety of effective stimuli. Through the analysis of the evoked spike trains across cells and subjects, the neural code is deduced. Unfortunately, this strategy does not allow the study of the interactions among cells within or across neural structures. In effect, it is akin to trying to deduce the rules of football by examining the activities of each player one at a time. Even so, despite its limitations, the cell-by-cell strategy has borne much fruit in the form of two influential theories of taste coding, both of which have their share of predictive power. These are the labeled line[36-39,46,52,78,79,92] and the across unit pattern[27,28,40,103,122] theories.

An alternative strategy for investigating the workings of the taste system in the brain stem has been to record from more than one cell at a time. This has both advantages and disadvantages. The advantages are that the interactions among cells can be analyzed, as well as their individual response properties. That, of course, assumes that the cells that are recorded are indeed interacting and, further, that these interactions can be detected and analyzed. The disadvantages of this strategy are that it is technically very challenging to obtain such multi-site records and that not every pair of simultaneously recorded units show evidence of functional interactions. (These interactions are detected by analyses of the relationship between the timing of spikes in the two records.) This renders this type of data rare, although potentially quite valuable.[25]

12.4 ELECTROPHYSIOLOGICAL RECORDING

12.4.1 CHOICE OF SUBJECTS AND ANESTHESIA

Single-unit electrophysiological studies of the central structures of the mammalian gustatory system have used rats, hamsters, mice, gerbils, cats, rabbits, dogs, and non-human primates as subjects. The choice of subject depends on several factors, including cost and the particular question being asked. In this context, species differences can play a significant role. For example, it is well-known that cats do not have any sweet receptors on their tongues and that the CT of hamsters is far more sensitive to sweet tastes than that of the rat. Although the taste sensitivities of non-human primates most closely resemble those of humans, it is not always practical to conduct taste-related research on monkeys. It is generally assumed that the mechanisms used for coding of taste stimuli are sufficiently similar in rodents and non-human primates so as to make research on the former a reasonable pursuit.

Most of the recordings from rodents have been obtained from acute, anesthetized preparations although some chronic recording experiments have been conducted on awake, unanesthetized animals.[50,65-68]

To anesthetize the rat, we typically use urethane (ethyl carbamate, 1.5 gm/kg, i.p.); however, many investigators have used sodium pentobarbital (50 to 60 mg/kg) or, on occasion, chloral hydrate (60 mg/kg). To prevent respiratory congestion, atropine (0.25 mg/kg, i.p.) can be administered prior to anesthesia. Generally, a single dose of urethane will maintain an animal at a surgical level of anesthesia for several hours without the necessity of supplemental doses. However, the level of anesthesia should be checked periodically through heart rate monitoring or reaction to tail pinch so that supplemental doses (of usually either urethane or sodium pentobarbital) can be administered as needed. Body temperature should also be monitored throughout the experiment and maintained at approximately 37°C (range, 36 to 38°C) with a heating pad.

12.4.2 Surgery

Surgical preparation for recording from the brain stem begins with a tracheotomy and insertion of an endotracheal tube, typically of polyethylene. The endotracheal tube should be secured with suture and the wound closed with wound clips or suture. The head is then mounted in either a non-traumatic head holder[29] or a stereotaxic instrument. When using a stereotaxic instrument, only blunt ear bars should be used to prevent damage to the CT nerve that courses through the middle ear.[98] The surgical approach to the NTS involves positioning the head so that the brain stem is horizontal and exposing the Obex (the most caudal point of the fourth ventricle) and the portion of the brain stem containing the rostral NTS. To do this, some of the occipital bone must be removed along with the most caudal aspect of the overlying cerebellum. The bone can be removed with rongeurs or by drilling a deep trough around the relevant area and then gently lifting away the bone flap. The cerebellum can then be gently aspirated with suction to expose the surface of the brain stem. The surgical approach to the PbN is similar to that of the NTS, except that more of the occipital bone needs to be removed to expose the cerebellar surface that directly overlies the PbN. The amount of exposure of the brain stem is similar but the amount of cerebellar surface that needs to be exposed is slightly greater. To avoid perturbing the transverse sinus, the head should be positioned at a downward angle. Alternatively, the path of the recording electrode should be angled slightly off the vertical. Once exposed, the brain should be covered with wet cotton soaked with sterile isotonic saline or bathed in warm mineral oil.

In our laboratory, the head is oriented by positioning the tooth bar at 5.5 mm below the interaural line. This angle allows vertical access to both the NTS and PbN. Prior to exposing the brain stem surface, we place two or three skull screws in the rostral skull. We then position a large inverted bolt close to the skulls screws and embed the bolt and screws in dental acrylic cement. The bolt is held by an armature attached to the stereotaxic instrument. This technique allows the removal of the ear and tooth bars but preserves the correct orientation of the skull for recording. With these methods, the NTS is located at 2.7 mm rostral, 1.8 mm lateral to the Obex, and 0.8 to 1.0 mm below the surface of the brain stem. The PbN is located 5.7 mm rostral, 1.8 mm lateral to the Obex, and 5.5 to 6.0 mm below the surface of the cerebellum.

12.4.3 Electrodes and Amplification

Typically, *in vivo* single-unit electrophysiological studies in gustation are conducted by recording action potentials extracellularly from taste-driven neurons. The small size of taste neurons in the brain stem has made intracellular recording in intact animals a daunting challenge, although some have been successful.[81-83] In this context, Renehan et al.[81-83] have adopted the strategy of first recording extracellular taste responses from an NTS neuron and then impaling the cell with the recording electrode for the injection of labeling dyes.

Different types of electrodes have been used for single-unit extracellular recordings from the gustatory system. Pipettes pulled from glass capillary tubing would be the obvious choice if the experimenter wished to eject dyes or drugs in the vicinity of a neuron. The fine, sharp tip that can be obtained with glass microelectrodes minimizes tissue damage during probe advancement; however, it may also be easily clogged as it moves through tissue. The high impedance, which can be greater than 50 MΩ, can provide well-isolated, low-noise, single-unit recordings, but the electrode must be very close to a cell before a signal can be recorded. Because the taste-responsive areas of the brain stem are generally small (i.e., about 300 μm in any direction) the use of glass pipettes as electrodes generally precludes the detection of a background response, presumably originating from cells in the general vicinity of the electrode tip, as a harbinger of the closeness of the taste area. Perhaps for this reason, some investigators use electrical stimulation of the CT[81-83] or anodal stimulation of the tongue[101] as part of their search strategy.

Another popular choice of electrode is etched tungsten, insulated except for the tip. Some insulate with glass[67,68,113,114] and others with baked-on epoxy. Although most investigators use electrodes with an impedance of less than 5 MΩ (at 1 kHz), the epoxy-insulated tungsten electrodes that we use begin with an impedance of 18 to 20 MΩ (at 1 kHz, FHC, Inc., Bowdoin, ME). The advantage of a tungsten over a glass pipette is in its strength and ability to maintain a sharp tip geometry. Over the years we have experimented with different tip geometries; the common wisdom is that a blunt tip will enable cells to be isolated for longer periods of time while a sharper tip will promote better signal-to-noise ratios. We have settled on a generally sharp and narrow tip geometry.

Because extracellular single-unit action potentials are high-frequency events, electrode signals are typically AC-coupled during the amplification process. AC-coupling allows the experimenter to essentially high-pass filter the data prior to hard storage or online analysis. This rids the record of any unwanted low-frequency DC potentials. Additionally, low-noise, single-unit recordings are best obtained when using a differential amplification circuit, which will eliminate simultaneous, extraneous potentials (e.g., stimulus artifact) in signals arising from both the source and indifferent electrodes through common mode rejection. Additional shielding (e.g., use of a copper-shielded room or Faraday cage) may be necessary to protect the recording from radio-frequency and 60-Hz noise. There are a number of companies that sell biological amplifiers that are suitable for extracellular single-unit recording [e.g., Astro-Med, Inc. (formerly Grass; www.grassinstruments.com); WPI, Inc. (www.wpiinc.com); A-M Systems, Inc. (www.a-msystems.com)].

12.4.4 Location and Testing of Taste-Responsive Cells

The first task for an investigator of taste in the brain stem is to find the taste-responsive cells. The small dimensions of the taste-related portions of both the NTS and PbN pose a special challenge to the investigator. For that reason, a "search stimulus" is often used to forecast the nearness of taste-responsive cells. As noted, some experimenters electrically stimulate the CT[81-83] or use anodal stimulation of the tongue[101] to identify taste-responsive cells. Some bathe the tongue with a cocktail stimulus.[113-114] This is a combination of taste stimuli, usually containing representatives of the four basic taste qualities. In most of our past experiments, we have used 10 m*M* NaCl as a search stimulus because it is known that greater than 90% of the taste-responsive cells in the brain stem of anesthetized rats respond to salt. Recent studies in unanesthetized rats, however, have shown a much more restricted sensitivity to salt and a much larger sensitivity to sucrose in these animals.[65-68] In our most recent studies, we have tested each isolated unit in the NTS with all four of the basic taste stimuli and, as a result, our latest data set has included a relatively large sample of sucrose-best cells, many of which did not respond to any other taste stimuli. These results imply that the preponderance of sensitivity to salt in anesthetized rats may reflect, in part, a sampling error.

During a recording session, taste-responsive neurons are identified as units that alter (i.e., either increase or decrease) their spike discharge rate in response to bathing the tongue with a taste stimulus. There are occasions when this determination must be done by analyses conducted off-line; however, most of the time, the presence of a taste response is obvious from either the oscilloscope trace or the sound of the spikes on the audio monitor. Waveforms arising from a single cell can be identified by uniform amplitude and waveform.[4] In our experience, NTS units appear to be more densely packed and therefore more difficult to isolate and hold than units in the PbN.

Once a single unit has been isolated, the experimental paradigm is begun. The nature of this paradigm will obviously vary according to the experimental question that is being studied, but will obviously include the presentation of taste stimuli to some portion of the oropharyngeal cavity. Taste stimuli are usually presented for 10 to 20 s, followed by a distilled water rinse whose length is sufficient to return the spike discharge rate to its pre-stimulus, spontaneous level. Some experimenters opt to control for intraoral tactile stimulation induced by the taste stimulus presentation by flowing distilled water across the tongue immediately prior to tastant delivery.[5] Presentations of taste stimuli are usually spaced at least 2 min apart to minimize the effects of adaptation.[100,102]

TABLE 12.1
Taste Stimuli and Typical Concentrations Used in Experiments on Rats

Taste Stimulus	Conc. (m*M*)
Salty salts	
NaCl	100
LiCl	100
Na_2SO_4	100
$NaNO_3$	100
Bitter salts	
KCl	100
$CaCl_2$	100
NH_4Cl	100
$MgCl_2$	100
Sour	
HCl	10
Citric acid	10
Malic acid	10
HNO_3	30
NaOH	10
Sweet	
Sucrose	500 or 100
Fructose	300
Glucose	100
DL-Alanine	100
Glycine	300
Bitter	
Quinine·HCl	10
Quinine·SO_4	10
Urea	100
Sucrose octaacetate	
Other	
Monosodium glutamate	100
Maltose	300
Polycose	100–200

12.5 TASTE STIMULI

Taste stimuli used for experimentation are often selected as representative prototypes of the proposed four basic taste qualities. Solutions of sucrose, NaCl, HCl, and quinine, respectively, are the most frequently used to represent these taste qualities. A fifth taste quality, umami, has also been suggested.[123,124] Depending on the experimental question that is asked, some investigators have used an extended selection of tastants. A sample of these is presented in Table 12.1. Although the tastants are grouped according to the taste qualities that they generally evoke in humans, it can be seen that there are some tastants that are difficult to classify. Because there are known species differences in the perception of some tastants,[90] it is risky at best to specify what each of these chemicals tastes like to a rat. By choosing to restrict the stimulus array to four or five stimuli, the investigator is already making the assumption that these stimuli can account for the majority of taste sensibilities. That is, stimuli are chosen so as to be so different from each other that they can be considered in some sense independent and, further, so that there are no other stimuli that represent additional taste qualities. Independence here has sometimes been defined by cross-generalization studies using conditioned taste aversions.[74]

Tastants are usually prepared by dissolving reagent-grade stock in distilled water. Unless used as an experimental factor, it is important that all taste stimuli are presented at the same (usually room) temperature because taste responses recorded from the CT[60,75] and NTS[75] are known to be modulated by thermal cues. Stimulus concentration also influences the magnitude of a taste response: the higher the concentration, the larger the response. In an effort to standardize stimulus strength, many experimenters have chosen stimulus concentrations known to evoke half-maximal responding from the rodent CT nerve[36]: sucrose, 500 m*M*; NaCl, 100 m*M*; HCl, 10 m*M*; quinine, 10 m*M*. However, a more complete description of neural sensitivity to an array of taste stimuli can be obtained through the use of within-stimulus concentration series. Single brain-stem neurons, which appear to be narrowly tuned to one stimulus, for example, usually exhibit greater breadth of sensitivity at high stimulus concentrations.[39,61,112] The drawback of recording responses to an extended concentration series is that units must be held for extended periods of time, sometimes on the order of hours, to obtain a complete data set.

12.5.1 Taste Stimulus Delivery

Methods used to deliver taste stimuli to the rodent oral cavity vary. A gravity flow system is probably the most basic and widely used. As the name implies, gravity is used to draw solutions into the mouth: the stimulus reservoirs are simply elevated relative to the experimental preparation. A slightly more complicated approach involves the use of an air-pressurized stimulus delivery device. Pressurization allows the experimenter to ensure constancy in the stimulus flow rate and, if the device is computer actuated, precise control of stimulus onset/offset timings. See Box 12.1 for a description of three systems that have been used in our lab.

12.6 DATA ANALYSIS

12.6.1 Data Reduction — Determination of the Response Measure

The process of data analysis always begins with data reduction. The raw data consist of a compilation of waveforms associated with action potentials that in most cases are almost immediately transformed into a stream of time-stamped events. This first step can be accomplished in a variety of ways, from very basic to highly sophisticated. In the most basic system, the analog signal would be fed to a spike-amplitude

BOX 12.1
Stimulus Delivery Techniques

It seems that every laboratory that studies gustation has its own idiosyncratic way of presenting taste stimuli. Each of these methods can vary in terms of the area of the oropharyngeal cavity that is stimulated, in the flow rate of the stimulus, and in the precision of the timing of the stimulus presentation. However, because all studies of taste coding rely on the correspondence of the neural response with the physical presentation of the stimulus, it is important that any stimulus delivery system is at least consistent across presentations.

The first consideration that must be made when choosing a method of stimulus delivery is that of the area of the oropharyngeal cavity to be stimulated. Because taste buds on various portions of the tongue and palate are innervated by different nerves,[71] it is desirable to report the extent of the stimulated area. This is especially important in light of the fact that there is an orotopic map in the NTS.[113-115] Thus, the response properties of cells in that structure will reflect the particular inputs that are stimulated by any taste stimulus.

Here we describe three solutions to the problem of stimulus delivery that we have used in our lab over the years. Each has been designed with particular applications in mind.

The first of these is a simple flow chamber that fits over the rostral two-thirds of the tongue. (See Figure 12.1 top) This is designed to stimulate only the fungiform taste buds innervated by the chorda tympani nerve.[71] It consists of a 2- to 3-cm portion of the barrel of a 3-cc polyethylene syringe fitted with a latex barrier drawn across one end, either glued or tied in place. A slit is made in the middle of this barrier so that the tongue of the animal can be drawn through it. When in use, the rostral portion of the tongue is within the flow chamber and the latex barrier prevents backwash toward the caudal tongue and pharynx. At about the midway point of the flow chamber barrel, a hole is drilled and one end of a segment (about 1 cm long) of a 1-cc polyethylene syringe is fitted into the hole and glued in place with epoxy. This latter segment is placed at an angle so that it can clear the front teeth. The other end is attached to Teflon tubing for delivery of taste stimuli. This flow chamber is easily adapted to a gravity flow system of stimulus delivery. For example, the free end of the Teflon tube can be attached to a stopcock and the funnel mounted on a ring stand. Premeasured stimuli can then be delivered to the flow chamber by simply releasing the stopcock. Alternatively, stimulus reservoirs, physically positioned above the animal, can be attached to solenoids that can be controlled by a computer. Opening of the solenoids then initiates the stimulus flow. In general, flow rates in gravity flow systems are about 3 ml/sec.

The second method of stimulus delivery that we have used consists of a battery of stimulus reservoirs fitted with a polyethylene tube (PE200) leading to the tongue (Figure 12.1, middle). For stimulus presentation, compressed air is forced into an individual reservoir and the stimulus is pumped out. Computer-controlled solenoids regulate the air flow. A one-way check valve attached to the reservoir end of the polyethylene tube ensures that fluid flows only toward the tongue and that the polyethylene tubes are nearly filled with fluid at all times.

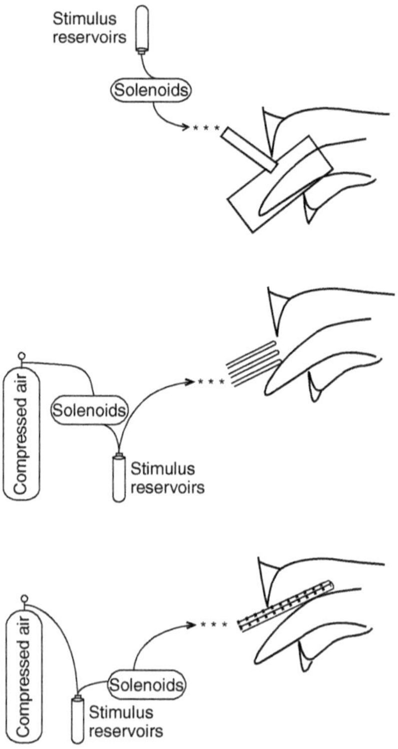

FIGURE 12.1 Stimulus delivery techniques: a stylized drawing of three different techniques for taste stimulus presentation. Top: gravity flow system controlled by a computer-actuated solenoid. Flow chamber allows stimulation of the rostral portion of the tongue. Middle: pressurized system that presents taste stimuli to the whole mouth. Each stimulus is delivered through its own stainless steel tube to prevent cross-contamination across tastants. Manifold is sometimes used with pressurized system instead of bundle of tubes. In that case, one tube is used for suction to clear out manifold between stimulus presentations. Bottom: stimulus reservoirs are kept pressurized at all times. Opening of solenoids produces spray of tastant to the whole mouth through perforated stainless steel tubes.

At the tongue, the bundle of tubes can be positioned in the mouth. Using this method, the number of stimuli that can be delivered is limited by the size of the tubing and the physical dimensions of the tongue and mouth. However, the advantage of such a system is that there is no contamination of one stimulus by another because each is delivered through a separate tube. The flow rate for this system is about 1.5 ml/sec.

The number of possible stimuli can be expanded by delivering the stimuli through a manifold. We have designed a rectangular manifold from stainless steel (dimensions 2.5 cm L × 0.5 cm W × 0.3 cm D) with seven holes, three positioned toward the tongue, two toward the soft palate, and one toward the back of the mouth. As many as 14 stainless steel tubes (14 ga. TW) can be soldered to the manifold. With this method, however, some means of clearing the manifold between stimulus presentations must be incorporated into the design. We have solved this problem by introducing 1 s of suction at the end of each stimulus (or rinse) presentation through one of the tubes. Similar to the stimulus presentation, the suction is controlled by a solenoid actuated by a computer.

The third means of stimulus delivery returns to the one-tube/one-stimulus type that we have used in the past. However, with this technique, the stimulus reservoirs are maintained under continuous pressure until a solenoid is opened (see Figure 12.1, bottom). At that point, the stimulus is sprayed throughout the mouth via a series of pinholes in one of several stainless steel tubes (16 ga. TW) mounted in the mouth. Because each stimulus is carried by its own tube, there is a limit to the number of stimuli that can be delivered. However, because the stimuli are under pressure, the delivery of the stimulus is almost immediate and can be regulated with precision down to 50-msec duration. Pressurization also means that the flow rate can be very high (e.g., 5 ml/sec); this allows the dispersion of the stimulus throughout the mouth even when very short (e.g., 100 msec or less) stimulus presentations are used.

discriminator that allows the experimenter to determine the level of signal above which any waveform will indicate the occurrence of an action potential. The device then outputs a pulse, often a TTL pulse, that can be counted and analyzed by a computer. This method of data reduction requires that the signal is well-isolated so that there can be no doubt about the presence of action potentials originating from other cells. A variation of the spike-amplitude discriminator that improves the spike-amplitude discrimination is the window discriminator, where both lower and upper limits of an acceptable signal amplitude can be set. In that case, actions potentials from other cells that are either of lesser or greater magnitude than the cell in question can be excluded.

In recent years, more sophisticated methods for spike detection have been developed. To some extent, each of these involves the measurement of various aspects of the action potential waveform, in addition to its peak amplitude. In one application (DataWave, Longmont, CO), these include, but are not limited to, peak-to-peak amplitude, width of the waveform, peak negative value, etc. The experimenter can then determine the limits of each parameter that define a given waveform as having arisen from the same cell. In other software applications (e.g., Cambridge Electronic Design, Cambridge, U.K.; Spectrum Scientific, Dallas, TX), incoming signals are compared to a template of the waveform of interest. The software then makes a decision, using a principal components analysis, as to whether the waveform is similar enough to the template to be counted as originating from the cell of interest. Most of these applications also have the capacity to acquire data from multiple channels simultaneously. This is essential when recording from more than one electrode simultaneously. It is important to note that, despite the added capacity for detecting spikes, these newer software applications are no substitute for recordings where action potentials are well-isolated from the background noise level.

Once the spike train has been digitized and time-stamped, perhaps the most common form of further data reduction is the construction of a peristimulus-time histogram from the spike train. This procedure begins with the assumption that the system integrates incoming spikes. That is, the assumption is made that the specific timing of the spikes is not relevant, at least on some, usually arbitrary, time scale. The first step in this procedure is to choose a time bin over which to sum the occurrence of spikes. The number of spikes that occur in successive time bins beginning sometime before and continuing during and after the response are counted. These numbers are then plotted as a histogram with time on the abscissa and number of spikes on the ordinate. Obviously, the choice of the magnitude of the time bin implies an integration constant for the response. Various experimenters have chosen time bins ranging from 50 to 500 msec.

Another decision that has to be made is how to measure the spontaneous or baseline rate of activity. Some experimenters use the firing rate that occurs in response to water as a baseline firing rate. Others use an average spontaneous firing rate that is calculated across some interval when there is no stimulus on the tongue. Because there is some variability in the spontaneous firing rate from trial to trial, we have chosen to use the mean firing rate in the 10 s just prior to each stimulus presentation as the baseline firing rate against which we measure a taste response. In any case, the detection of a significant response is almost always based on some statistical comparison of the firing rate when the stimulus is present on the tongue to some measure of baseline firing rate.

Methods for detecting and measuring the beginning of a significant taste response is the next decision for the investigator. This can be problematical because each type of stimulus delivery system will introduce some variability in the actual time of contact of the stimulus with the tongue and because the critical event (i.e., stimulus-receptor contact) is difficult, if not impossible to measure. As an approximation of the time of stimulus-receptor contact, some researchers utilize devices that are triggered by the flow of the stimulus just prior to its contact with the tongue, for example, a photocell[94] or thermistor[80] device. The neural response to a taste stimulus is then measured as the firing rate that occurs during some time interval following the marker of stimulus arrival at the tongue. The disadvantage of this method, which some would argue is trivial, is that this response interval includes the variability of the time for the stimulus to be distributed across the tongue and the latency of the neural response once the stimulus contacts the receptors. Other laboratories, including our own, rely on changes in the firing rate to signal the beginning of a response. Obviously, this ignores the latency of the neural response with respect to the time of stimulus-receptor contact. The problem then becomes how to determine when a change in firing rate is significantly different from the spontaneous rate. We deal with this problem by dividing the activity during a trial into 50-msec bins. We then use a sliding window of 1 s, where the number of spikes that occur within this window is compared to the spontaneous firing rate, defined as the average firing rate (spikes per second) that occurs in the 10 s just prior to stimulus delivery. If the activity in the window is statistically significantly greater than the spontaneous firing rate, then the trailing 50-msec bin of that 1 s window is considered to be the beginning of the response.

The interval of time over which a response is measured also varies across laboratories. In general, most people use between 3 and 5 s of activity as an indication of a response. For the most part, this decision is arbitrary because the stimulus presentation is almost always longer than the response interval. Furthermore, the stimulus-evoked activity often outlasts the presentation of the stimulus. This may be due to the lingering presence of the stimulus on the tongue, although not every response persists for the entire stimulus presentation. In some early work, the initial portion of the response was excluded from the response measure[40] because it was found to be variable. In fact, there is evidence that the magnitude of this initial phasic portion of the response is proportional to the flow rate of the stimulus on the tongue[99] and is generally unstable.[27]

Although the decision of the response interval may be somewhat arbitrary, it is not trivial. Because all other analyses are based on this measure, one would hope that the firing rate of each response would be stable enough that measuring say 1, 3, or 5 s of activity would make no difference in the conclusions produced by further analyses. This may not always be the case. For example, we have shown that when taste response in the PbN was divided into 0.5-s epochs, the correlations of across-unit patterns varied over the time course of the response interval.[26] Whether this is a difference that is significant is not known; however, this result points to the fact that taste responses may evolve over time and that the temporal pattern of a response may indeed convey some information.

12.6.2 Classification of Response Profiles

The determination of a response measure, however accomplished, will ultimately result in the assignment of a single number to the response of a given unit to a given stimulus. This number will then be used in all subsequent analyses of coding mechanisms. The responses of a unit to each of the taste stimuli that were tested is defined as the response profile of that unit. These profiles can then be used to determine the degree of "tuning" of the unit, i.e., the extent to which the unit responds to one or more stimuli in favor of others. The most widely used measure of the degree of tuning of a taste-responsive cell is that of the Uncertainty measure developed by Smith and Travers in 1979.[104] This metric is based on information theory[96] and provides a number between zero and one for each unit. Low numbers indicate that the unit responds to a single taste stimulus from an array, and numbers close to one indicate that the cell responds equally well to more than one stimulus. See Box 12.2 for further discussion of the Uncertainty measure.

Response profiles can also be used to classify units in terms of their best stimulus, defined as the stimulus that evokes the most robust response from a given set of test stimuli. It has been found that the units with the same best stimulus show similar relative response magnitudes to other taste stimuli.[36-39,52,78,79] This observation has fueled the assertion that groups of units with the same best stimulus constitute a neuron type, and that these groups share a common function in the taste coding process. That is, it has been argued that the collection of units with the same best stimulus are exclusively responsible for encoding that stimulus as well as others with the same quality (sweet, sour, salty, or bitter). This idea is called the labeled line theory.

BOX 12.2
The Uncertainty Measure

In 1979, Smith and Travers[104] published a paper that described a method for assessing the breadth of tuning of a taste-responsive cell. In effect, these investigators borrowed the measure of entropy from information theory[96] and applied it to the response magnitudes evoked by representatives of the four basic taste qualities: sweet, sour, salty, and bitter. This metric, called the Uncertainty measure (denoted by H), is an index of the amount of information conveyed by a message. When H = 1.0, the implication is that there is total uncertainty, i.e., no information is contained in the message. Conversely, when H = 0, the message is unambiguous. As applied to taste cells, H is close to 1.0 when the cell responds equally well to all four taste stimuli. In that case, it cannot, therefore, by itself convey the identity of a stimulus that is on the tongue. Such a cell is said to be broadly tuned. H is close to zero when a cell's response signals only one tastant from among the four possible alternatives. This type of cell is said to be narrowly tuned. The formula for the Uncertainty measure is:

$$H = -1.66\left(\sum_{i=1}^{4} P_i \log P_i\right)$$

where P represents the proportion of the total number of response spikes for a given stimulus i across four taste stimuli: NaCl, HCl, quinine, and sucrose.

At least three points about the Uncertainty metric are noteworthy. First of all, because the formula includes a log transformation, negative response values — indicative of inhibitory responses — cannot be analyzed. In their original paper on the Uncertainty measure, Smith and Travers acknowledge this point. They also note that this problem arising from negative response values cannot be solved by the inclusion of spontaneous rates of activity because it would inflate the breadth of responsiveness. The authors leave the discussion by citing the low incidence of inhibitory responses in the brain stem, thus implying that there is seldom a need to assess the breadth of tuning of cells whose response repertoire includes inhibitory responses. The second point to note is that the initial constant coefficient is derived from the fact that there are exactly four stimuli. Choosing this particular constant scales the Uncertainty measure so that when $P_i = .25$, H = 1.0. If more stimuli are used, a different coefficient must be used. However, if the additional stimuli were similar to any of the others, the Uncertainty measure would be predictably and artificially inflated. Smith and Travers[104] chose to include the responses from only four tastants based on the idea that they span the taste experience. That is, by definition, any other taste stimulus could be expressed as some combination of these four. This is an assumption that remains controversial.[30,31,33,34,88]

The third point concerning the Uncertainty measure that is worth consideration is that of its distribution when data from real cells are analyzed. Figure 12.2 plots the distribution of Uncertainty measure as the relative value of the maximum response (P_{max}) varies (after Figure 1 in Reference 104). The top curve shows H when the remaining three responses are equal. The lower curve shows

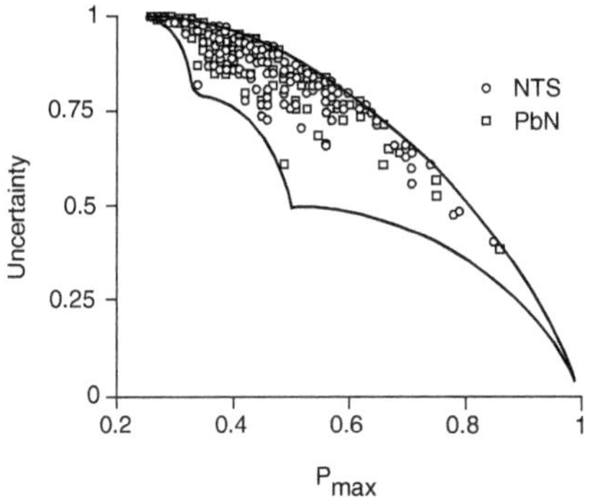

FIGURE 12.2 Uncertainty measure of cells in the NTS and PbN (see text). Solid curves represent the upper (top function) and lower (bottom function) limits of the possible values of H when there are four possible responses plotted against the value of the maximum proportion of the total response-related spikes summed across all four tastants. (From D.V. Smith and J.B. Travers, *Chem. Sens. Flav.*, 4, 215, 1979.)

the distribution that results when the responses are as narrowly distributed as possible. For example, when P_{max} is .9, the remaining *P*s are 0.1, 0, and 0. When $P_{max} = 0.3$, the remaining $P_i = 0.3$, 0.3, and 0.1. Also shown are the Uncertainty measures for 301 cells recorded from the NTS (n = 183)[20,24,25,64] and the PbN (n = 118)[18,21,22,25] of urethane-anesthetized rats. It can be seen that the Uncertainty values for most of the cells are quite high, indicating that most cells in the brain stem are generally broadly tuned. The average H = 0.85 ± 0.01 s.e.m. in the NTS and H = 0.89 = ± 0.01 in the PbN; the median H values are 0.87 and 0.93, respectively.

The best stimulus classification scheme has been criticized on several levels. First, the rule for classification can seem very arbitrary considering the fact that a unit may be placed in a given best stimulus group even if its response to the best stimulus is only one spike more than its response to the next-best stimulus. Simply replicating the presentation of the stimuli might conceivably alter the classification of such a unit. The result is that units with apparently very different response profiles can be considered to be part of the same functional group. Second, the best stimulus of a unit can vary depending on the concentration of the stimuli. That is, by manipulating the concentration of the test stimuli, it is hypothetically possible to determine that a given unit can encode any stimulus. Opponents of this view argue that if concentrations are chosen that are at the same relative point in the intensity-response curve, usually mid-range, associated with a tastant, results will be internally consistent. Finally, there are several manipulations that have been shown to alter response profiles. As noted above, these include hormonal state,[21,22,44,45] sodium deprivation,[10,11,54,62,76] or prior learning experiences.[8,43,63] The potential for modification of a unit's best stimulus calls into question the idea that this stimulus identifies the role it has in the coding process.

12.6.3 Organization of Response Profiles

Although a unit's best stimulus may not be the most stable tag for classifying units into types, many investigators have argued that there is nevertheless a non-random organization to the response profiles in the brain stem.[20] There are many other examples. To detect such an organization, two quantitative techniques have been used extensively in taste research. These are hierarchical cluster analysis (HCA) and multidimensional scaling (MDS). In both of these analyses, the input is a matrix of similarities or dissimilarities among response profiles. Across-unit patterns of responses generated by different stimuli can also be used in these analyses to study how the nervous system organizes responses to a variety of taste stimuli. The output of both is a configuration of units such that units with similar response profiles are placed close together in a hypothetical taste space in MDS, or in the same cluster in an HCA, and units with dissimilar response profiles are placed far apart in an MDS or in different clusters in an HCA. Results of these analyses have been used to argue that there are distinct neuron types in the brain stem that share similar response profiles.[105] However, there are some cautionary notes worth considering when using these analyses. Some of these have been described by Erickson, Rodgers, and Sarle.[35] In general, these criticisms derive from the fact that both of these analyses are designed to impose an organization onto the data, even if there is inherently none (i.e., if there were a continuum in the variation of response profiles).

As mentioned, the first step in both HCA and MDS is the construction of a similarity matrix among response profiles. There are several options for measuring similarity, each of which has its own implications for biasing the results. The most commonly used metric for similarity is the Pearson product-moment correlation. The formula for calculating this measure is as follows:

$$r = \frac{\sum_{i=1}^{n}(x_i - \bar{x})(y_i - \bar{y})}{\left(\sum_{i=1}^{n}(x_i - \bar{x})^2\right)^{1/2}\left(\sum_{i=1}^{n}(x_i - \bar{y})^2\right)^{1/2}} \tag{12.1}$$

where x_i and y_i are the responses to taste stimuli X and Y, and $\bar{x}$ and $\bar{y}$ are the mean response rates to X and Y, respectively. It can be seen that the Pearson correlation is scaled according to the mean response. That is, each response is expressed as a deviation from the overall mean. The effect of this normalization is that the Pearson correlation measures the similarity in the overall envelope of the response profile, apart from its absolute magnitude. The same is true when the Pearson correlation is used to index similarity of across-unit patterns of response. As a result, changes in the concentration of a tastant will produce smaller responses across all of the initially responsive cells, which will produce a high Pearson correlation value. This is desirable because one would expect, and there is evidence to support this, that stimuli that taste alike would produce across-unit response patterns that are highly positively correlated.[91] Conversely, stimuli that taste very different would show very small correlations. Erickson[32,42] has argued that this relationship only holds for stimuli that evoke relatively robust responses. When response magnitudes are small, one can also expect the correlation to be small because small variations in response magnitude are given proportionately more weight in the calculation of the correlation. This produces a potential distortion in that any two stimuli that produce weak responses can be expected to produce small across-unit correlations and would therefore be considered to be dissimilar even if they were in fact quite similar as, for example, two concentrations of the same stimulus.

The normalization of the data in the Pearson correlation carries an additional assumption. That is, it is implicit that units with small responses are equal in importance to those with large responses. This idea has been challenged in a series of papers by Erickson[32] and Gill and Erickson,[42] in which they argue that the nervous system most probably gives more weight to large responses compared with small ones. These authors further propose an alternative measure of similarity, called the Neural Mass Difference (NMD),[32,42] that is the average difference in response magnitude evoked by two stimuli across cells. The formula is as follows:

$$NMD_{S_iS_j}\left(\sum_{k=1}^{n} R_{k_{S_i}} - R_{k_{S_j}}\right)\Big/ n \qquad (12.2)$$

where S_i is the standard stimulus, S_j is the comparison stimulus, and n is the number of units tested with S_i or S_j; R_k is the response of unit k to S_i or S_j. The results of the analyses of across-unit patterns are very similar to those using the Pearson correlation, except when the stimuli evoke very weak responses. In that case, the small correlations that result imply that the two stimuli are very distinct from one another and the small NMD suggests just the opposite.

A third type of analysis of similarity/dissimilarity that has been used is called vector space analysis.[17,22] In this approach, the response of each unit to a given stimulus is considered to be one coordinate of a vector in n-dimensional space, where n is the number of units in the sample. The length of the vector is then proportional to the response magnitude evoked by the stimulus. The angle between two such vectors representing the across-unit responses evoked by two different stimuli can be used as an index of the similarity of these two patterns of response. The formula for the angle (θ) between two vectors v and w associated with the across-unit patterns evoked by two taste stimuli (V and W) is:

$$\cos\theta = \frac{\sum_{i=1}^{n} v_i w_i}{\left(\sum_{i=1}^{n} v_i^2\right)^{1/2}\left(\sum_{i=1}^{n} w_i^2\right)^{1/2}} \qquad (12.3)$$

It can be seen that this formula is essentially identical to that of the Pearson product-moment correlation, except that it is not normalized according to the mean response. This fact imbues this measure with sensitivity to both the relative and absolute values of the responses. For a complete discussion of this type of analysis, see Di Lorenzo.[17] This type of analysis has been used to evaluate taste responses in the brain stem and across-unit responses in other sensory systems as well.[110,116,119]

Once a matrix of similarity/dissimilarity has been formed, using whatever metric that the experimenter deems appropriate, HCA and/or MDS analyses can then be applied. Although the input to each of these analyses is the same, the outputs are different.

For example, the output of the HCA will, by definition, be a grouping of units (or stimuli, depending on the input) that is imposed on the data. This will most times be expressed as a dendrogram (see Box 12.3 for more about HCA). This is literally a chart of the distances, proportional to similarity, between units (stimuli) or between clusters of units (stimuli). The usual goal of this analysis is to identify functionally distinct groups or types of units. The determination of the correct number of clusters is sometimes made by comparison of the average intracluster distances to the average intercluster distances. When intracluster distances are much smaller than intercluster distances, the implication is that the members of the clusters are more similar to each other than they are to members of the other clusters. This comparison is important because the cluster analysis will force the data into groups, even if the data are not inherently divisible into groups.

The MDS analysis, in contrast to HCA, does not decide on groupings among units. Instead, the output of this type of analysis is the construction of a taste space, so called because taste-responsive units (or taste stimuli) are positioned in this mathematical space in such a way as to reflect the similarities of their response profiles (or across-unit patterns). Taste units that have similar response profiles are placed close together in the taste space and those that are very different are placed far apart. The MDS analysis utilizes the similarity matrix provided by the experimenter as input, but also requires the experimenter to decide on the number of dimensions of the resulting taste space. The analysis will then generate a configuration of the data in the taste space and calculate a goodness-of-fit measure for that configuration. It then reiterates that process. In various programs, the measure of goodness of fit can vary, as well as the number of iterations of the process of configuring the data. For each solution in a given dimensionality, a final goodness-of-fit measure is provided. The decision of the number of dimensions that are needed to adequately characterize the data is often determined by the scree test. In this test, the number of dimensions is plotted against the goodness-of-fit measure. The best number of dimensions for the data is determined where there is an elbow in that curve, that is, where the addition of another dimension does not significantly improve the goodness of fit. In general, most applications of MDS to taste responses result in taste spaces of two or three dimensions. Figure 12.4 shows an example of the results of an MDS applied to data from the PbN of a urethane-anesthetized rat.[26]

BOX 12.3
Hierarchical Cluster Analysis

One of the most persistent issues in the literature on the neural code for taste has been the question of whether there exist "types" of cells that are functionally distinct. One approach to answering this question has been to investigate the organization of neuronal response profiles, i.e., the relative response rates evoked by a variety of taste stimuli. The rationale for this strategy is that if there were separate groups of neurons that each had a distinct role in the coding process, the members of each group would respond similarly, and uniquely compared with other groups, to taste stimuli. For example, all cells that respond to NaCl > HCl > quinine > sucrose might form one group, and cells that respond to NaCl > sucrose > HCl > quinine might form another group. To identify such groups, several investigators have used hierarchical cluster analysis (HCA).[20,21,51,65-68,105,122] This is a quantitative technique that is designed to classify entities (in this case, neurons) based on the similarity of their values along one or more variables (in this case, responses to taste stimuli). (It should be noted that the initial choice of variables to be included in the analysis represents an initial classification of the data. For example, for data from taste-responsive cells, the choice of the array of taste stimuli is a direct reflection of the judgment of relevance on the part of the investigator.) This technique has also been used to classify morphological cell types in the gustatory system.[95,120]

In HCA, a matrix of similarities or distances among neurons is constructed and, depending on the particular type of HCA chosen, a specific algorithm is used to join neurons into groups beginning with the two most similar (or closest) two neurons. The algorithm is a rule that determines how the distance between cells or groups of cells will be calculated. This process is repeated by joining the next two most similar neurons or groups of neurons until the entire data set is included in a single group. A graphical representation, called a dendrogram, is often constructed to illustrate this process. In a dendrogram, the sequence of clustering (i.e., which cells or groups of cells are joined first, second, etc.) is shown, as well as the distances between clusters as they are formed. Figure 12.3 shows an example of a dendrogram resulting from an HCA of a set of data sampled from the NTS.[20]

Beyond the decision of what variable to include in the analysis, two additional decisions are made by the investigator, both of which can have a major impact on the results of the HCA. The first of these is the choice of the measure of similarity across cells. For example, one option is to simply calculate the Euclidean distance between the cells using the response rates as coordinates in a vector space. This measure is very sensitive to changes in scale; thus, if one is only interested in the relative response rates, and not the absolute values of those response rates, some sort of standardization would be necessary. Alternatively, the angle between these vectors might also be used (see Reference 17 for a detailed discussion of this approach). The most commonly used measure of similarity in gustatory research is the Pearson product-moment correlation. The formula for the correlation is essentially the same as that for the angular separation of the associated vectors, except that it is scaled according to

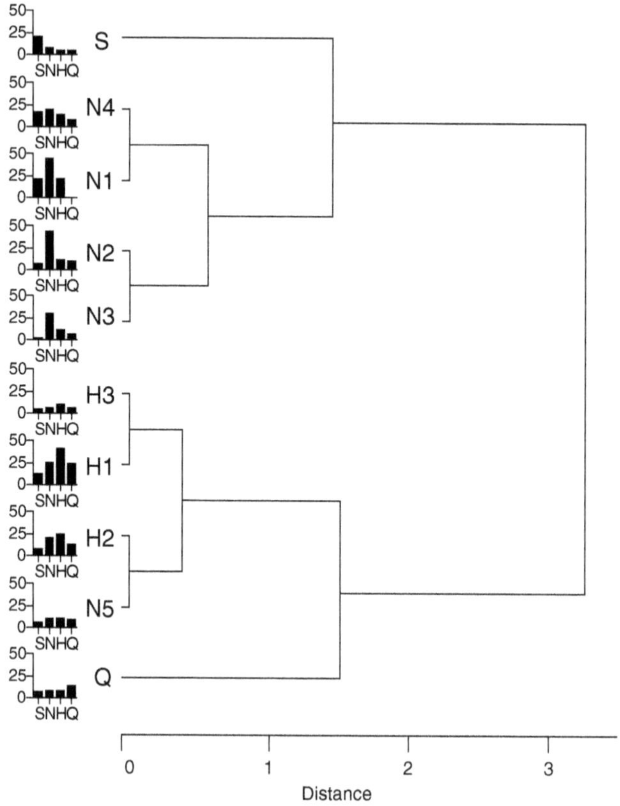

FIGURE 12.3 Dendrogram showing the results of a cluster analysis of the response profile of 10 units recorded from the NTS.[20] Response profiles of each unit are shown to the left. Ordinate shows average spikes per second above spontaneous rate. Stimuli are shown along the abscissa. Abbreviations are as follows: S, 500 m*M* sucrose; N, 100 m*M* NaCl; H, 10 m*M* HCl; Q, 10 m*M* quinine·HCl. Units are labeled according to their best stimulus (letter) and by the rank of the response magnitude to their best stimulus (number). For example, H2 is an HCl-best unit with the second largest response to HCl from among the group of HCl-best units. Dendrogram shows the organization of response profiles. Distance (plotted along the abscissa) is an arbitrary measure corresponding to the Pearson product-moment correlation (used as a measure of similarity of response profiles). Results show a division of the NaCl best units from the HCl-best units. The sole sucrose-best unit was determined to be most similar to the NaCl-best units and the one quinine-best unit was determined to be most similar to the HCl-best units.

the mean response. It is therefore relatively insensitive to absolute values, but is instead a good detector of linear dependence. There are, of course, many other possible measures of similarity that can be used.

The second decision that must be made is the method of cluster formation, i.e., the clustering algorithm. There are several choices available and each will affect the results in a different way. They differ in the way in which the closeness of two cells or groups of cells is determined. For example, one can use the minimum of the distances between the two closest members, one in each cluster. This is called the nearest neighbor method and forms the basis for single linkage HCA. Conversely, the distance between the two furthest members, one in each cluster, can be used; this is called complete linkage HCA. The centroid method involves the use of the minimum distance between the centroids of each cluster. In "Ward's method," it is assumed that at any stage in the analysis there is a loss of information that occurs as each pair of clusters is joined. This loss can be calculated as the total of the squared deviations of each member of a cluster from its mean. At each stage in the analysis, the two clusters whose fusion results in the minimum loss of information are joined. In addition to the clustering algorithms mentioned here, there are many others that can be used.

There are two problems that are inherent to the HCA technique that may affect the outcome. The first is referred to as chaining. This is the tendency of an HCA to join objects at a very low level by linking chains of intermediates. The result is a dendrogram that suggests a continuum of objects rather than a set of groups. The second problem is that, although an object may be poorly classified at an early stage of the HCA, there is no provision for its reclassification later in the analysis.

Once the HCA is complete, for many investigators determining the number of clusters present in the data is the next step. That is, one may ask whether the organization of associations revealed by the HCA suggests some natural grouping of the data into different types. Generally speaking, however, there is no standard method for making this determination. Some have used comparisons of intra vs. intercluster distances, while others have examined the dendrograms for large changes in intercluster distances. Each of these methods have their shortcomings. Because there remains the problem of defining what a "cluster" actually is, a more worthwhile enterprise might be to view the entire dendrogram as a diagram of relationships and structure, rather than a division into different neuron types.

The conclusions concerning the grouping of either units or stimuli in the results of MDS analyses are typically derived from eye-balling. For example, units can be labeled according to their best stimulus and the observation may be made that they are placed close together in the taste space. When across-unit patterns associated with various taste stimuli are used as input, the observation can be made that similar tasting stimuli are located close to one another in the taste space. The utility of the MDS analysis then is to graphically display the complex relationships among response profiles associated with units or among across-unit response patterns associated with taste stimuli. The interpretation of these relationships is at the discretion of the investigator.

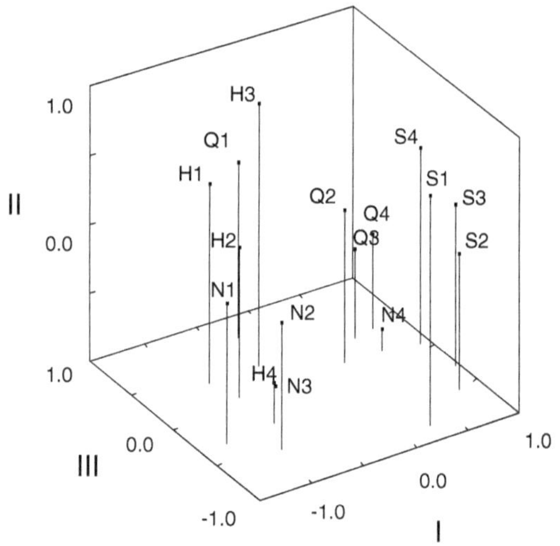

FIGURE 12.4 MDS solution in three dimensions of the analysis of across-unit response patterns from the PbN.[26] Taste stimuli are four concentrations of each of four tastants: N1–4, 100 m*M*, 10 m*M*, 1 m*M*, 0.1 m*M* NaCl; H1–4, 10 m*M*, 5 m*M*, 1 m*M*, 0.1 m*M* HCl; Q1–4, 10 m*M*, 5 m*M*, 1 m*M*, 0.1 m*M* quinine·HCl; S1–4, 500 m*M*, 100 m*M*, 10 m*M*, 1 m*M* sucrose. Across-unit response patterns are grouped roughly according to taste quality. Notable exceptions are the response patterns associated with the highest concentrations of quinine and HCl, which are placed close together, implying that they evoke similar across-unit response patterns. Stress (goodness of fit) values were: 0.342, 1 dimension; 0.162, 2 dimensions; 0.104, 3 dimensions; 0.071, 4 dimensions; 0.056, 5 dimensions.

Both HCA and MDS are aimed at the study of the organization of the responses in a particular brain area. Certainly, the knowledge of such organization can reveal or suggest the coding mechanisms that are utilized by the nervous system. However, there are other aspects of the taste responses not included in these analyses that also hold potential for conveying information and that point to other types of coding mechanisms. One example is the temporal pattern of response. This aspect is completely ignored by both HCA and MDS because, for both analyses, the number of spikes is summed over time, thus discarding information about their temporal arrangement. There is some evidence,[19] however, that the temporal pattern of a taste response can be used to produce behavioral reactions that are appropriate to the tastant that generated them.

12.7 CONCLUSION

There are some perhaps obvious, but nevertheless profoundly important points to be made about the investigation of coding in the taste system. First, as in most of scientific inquiry, the results one obtains are directly dependent on the questions asked. This can affect the types of analyses chosen and the interpretation of the results of those analyses in nontrivial ways. For example, if one believes that the taste system consists of groups of units defined by their best stimulus, then there will be little doubt that some evidence for that grouping will be apparent in the data. Conversely, if one believes that response profiles are not inherently grouped by type, then there will be evidence for a continuum. Often, there can be evidence for both conceptualizations within the same data. In the end, this fact has produced a somewhat artificial dichotomy whereby the two major theories of taste coding (i.e., labeled line and across-unit pattern theory) have vigorously vied for dominance for over 3 decades. The end result is that there is still controversy over which theory is closer to the truth. The fact that both theories have considerable predictive value only complicates the issue.

In recent times, the study of neural coding in the taste system has broadened to include questions other than those aimed at differentiating between the labeled line and across-unit pattern theories. For example, more complete characterization of the response properties of brain stem taste cells and their relationship to anatomical features has been one of the avenues explored by some investigators.[56,81,82] The study of the biophysical[6,7,111] and pharmacological[13-15,117] properties of taste cells has also added important information to our understanding of how the brain might encode taste. Exploration of how these properties might impact the response potential of taste cells has also borne fruit.[58,101] Another strategy employed in our lab has been the use of multisite recording, i.e., recording from more than one taste-responsive cell simultaneously. This allows the study of the functional relationships among cells in the brain stem.[25] As more information is gathered about the circuitry that connects taste-responsive cells in the brain stem, the use of artificial neural networks may also prove useful as a method for understanding the ways in which the population of cells in the brain stem encode taste stimuli. In all, the expansion of the field of gustatory neural coding to include a variety of experimental strategies and paradigms can be expected to yield conceptualizations of coding mechanisms that do not as yet exist. However, these new ideas may be informed by, and perhaps even be based on, results of experiments that have and will utilize the methods discussed here.

ACKNOWLEDGMENTS

The authors wish to acknowledge Ms. Swarooparani Vadlamudi for help in the preparation of this manuscript. Supported by NSF grant IBN9630326 to PMD.

REFERENCES

1. Avenet, P. and Lindemann, B., Perspectives of taste reception, *J. Membr Biol.*, 112(1), 1, 1989.
2. Beidler, L.M., Physiological properties of mammalian taste receptors, in *Taste and Smell in Vertebrates,* Wolstenholme, G.E.W., Ed., London: J. & A. Churchill, 1970.
3. Beidler, L.M. and Smallman, R.L., Renewal of cells within taste buds, *J. Cell Biol.*, 27(2), 263-272, 1965.
4. Bishop, P.O., Burke, W., and Davis, R., The identification of single units in central visual pathways, *J. Physiol.,* 162, 409, 1962.

5. Boughter, J.D., Jr., St. John, S.J., and Smith, D.V., Neural representation of the taste of NaCl and KCl in gustatory neurons of the hamster solitary nucleus, *J. Neurophysiol.,* 81(6), 2636, 1999.
6. Bradley, R.M. and Sweazey, R.D., *In vitro* intracellular recordings from gustatory neurons in the rat solitary nucleus, *Brain Res.,* 508, 168, 1990.
7. Bradley, R.M. and Sweazey, R.D., Separation of neuron types in the gustatory zone of the nucleus tractus solitarii on the basis of intrinsic firing properties, *J. Neurophysiol.,* 67, 1659, 1992.
8. Chang, F.-C. and Scott, T.R., Conditioned taste aversions modify neural responses in the rat nucleus tractus solitarius, *J. Neurosci.,* 4(7), 1850, 1984.
9. Chen, Y., Sun, X.D., and Herness, S., Characteristics of action potentials and their underlying outward currents in rat taste receptor cells, *J. Neurophysiol.,* 75(2), 820, 1996.
10. Contreras, R.J., Changes in gustatory nerve discharges with sodium deficiency: a single unit analysis, *Brain. Res.,* 121, 373, 1977.
11. Contreras, R.J. and Frank, M.E., Sodium deprivation alters neural responses to gustatory stimuli, *J. Gen. Physiol.,* 73, 569, 1979.
12. Davis, B.J., The ascending gustatory pathway: a Golgi analysis of the medial and lateral parabrachial complex in the adult hamster, *Brain Res. Bull.,* 27, 63, 1991.
13. Davis, B.J., GABA-like immunoreactivity in the gustatory zone of the nucleus of the solitary tract in the hamster: light and electron microscopic studies, *Brain Res. Bull.,* 30, 69, 1993.
14. Davis, B.J. and Jang, T., Tyrosine hydroxylase-like and dopamine b-hydroxylase-like immunoreactivity in the gustatory zone of the nucleus of the solitary tract in the hamster: light- and electron-microscope studies, *Neuroscience,* 27, 949, 1988.
15. Davis, B.J. and Smith, D.V., Substance P modulates taste responses in the nucleus of the solitary tract of the hamster, *Neuroreport,* 8(7), 1723, 1997.
16. Di Lorenzo, P.M., Taste responses in the parabrachial pons of decerebrate rats, *J. Neurophysiol.,* 59(6), 1871, 1988.
17. Di Lorenzo, P.M., Across-unit patterns in the neural response to taste: Vector space analysis, *J. Neurophysiol.,* 62, 823, 1989.
18. Di Lorenzo, P.M., Corticofugal influence on taste responses in the parabrachial pons of the rat, *Brain Res.,* 530, 73, 1990.
19. Di Lorenzo, P.M. and Hecht, G.S., Perceptual consequences of electrical stimulation in the gustatory system, *Behav. Neurosci.,* 107(1), 130, 1993.
20. Di Lorenzo, P.M. and Lemon, C.H., The neural code for taste in the nucleus of the solitary tract of the rat: effects of adaptation, *Brain Res.,* 852, 383, 2000.
21. Di Lorenzo, P.M. and Monroe, S., Taste responses in the parabrachial pons of male, female and pregnant rats, *Brain Res. Bull.,* 23, 219, 1989.
22. Di Lorenzo, P.M. and Monroe, S., Taste responses in the parabrachial pons of ovariectomized rats, *Brain Res. Bull.,* 25, 741, 1990.
23. Di Lorenzo, P.M. and Monroe, S., Corticofugal input to taste-responsive units in the parabrachial pons, *Brain Res. Bull.,* 29, 925, 1992.
24. Di Lorenzo, P.M. and Monroe, S., Corticofugal influence on taste responses in the nucleus of the solitary tract in the rat, *J. Neurophysiol.,* 74(1), 258, 1995.
25. Di Lorenzo, P.M. and Monroe, S., Transfer of information about taste from the nucleus of the solitary tract to the parabrachial nucleus of the pons, *Brain Res.,* 763, 167, 1997.
26. Di Lorenzo, P.M., Monroe, S., and Hecht, G.S., Information processing in the parabrachial nucleus of the pons, in Kurihara, K., Suzuki, N., Ogawa, H., Eds. *Olfaction and Taste XI*; New York: Springer Verlag, 1994, 402.

27. Doetsch, G.S. and Erickson, R.P., Synaptic processing of taste-quality information in the nucleus tractus solitarius of the rat, *J. Neurophysiol.*, 33, 490, 1970.
28. Erickson, R.P., Sensory neural patterns and gustation, in Zotterman, Y., Ed., *Olfaction and Taste*, Oxford, U.K., Pergamon, 1963, 205.
29. Erickson, R.P., Nontraumatic headholders for mammals, *Physiol. Behav.*, 1, 97, 1966.
30. Erickson, R.P., Studies on the perception of taste: do primaries exist?, *Physiol. Behav.*, 28(1), 57, 1982.
31. Erickson, R.P., Ohrwall, Henning, and von Skramlik: the foundations of the four primary positions in taste, *Neurosci Biobehav. Rev.*, 8(1), 105, 1984.
32. Erickson, R.P., A neural metric, *Neurosci Biobehav. Rev.*, 10, 377, 1986.
33. Erickson, R.P. and Covey, E., On the singularity of taste sensations: what is a taste primary?, *Physiol. Behav.*, 25(4), 527, 1980.
34. Erickson, R.P., Covey, E., and Doetsch, G.S., Neuron and stimulus typologies in the rat gustatory system, *Brain Res.*, 196(2), 513, 1980.
35. Erickson, R.P., Rodgers, J.L., and Sarle, W.S., Statistical analysis of neural organization, *J. Neurophysiol.*, 70(6), 2289, 1993.
36. Frank, M., An analysis of hamster afferent taste nerve response functions, *J. Gen. Physiol.*, 61, 588, 1973.
37. Frank, M., The classification of mammalian afferent taste fibers, *Chem. Senses*, 1, 53, 1974.
38. Frank, M.E., Bieber, S.L., and Smith, D.V., The organization of taste sensibilities in hamster chorda tympani nerve fibers, *J. Gen. Physiol.*, 91, 861, 1988.
39. Frank, M.E., Contreras, R., and Hettinger, T., Nerve fibers sensitive to ionic taste stimuli in chorda tympani of the rat, *J. Neurophysiol.*, 50(4), 941, 1983.
40. Ganchrow, J. and Erickson, R.P., Neural correlates of gustatory intensity and quality, *J. Neurophysiol.*, 33, 768, 1970.
41. Gilbertson, T.A., Peripheral mechanisms of taste, in Linden, R.W.A. Ed., *The Scientific Basis of Eating*, Basel: Karger, Vol. 9, 1998, 1-28.
42. Gill, J.M., II and Erickson, R.P., Neural mass differences in gustation, *Chem. Senses*, 10, 531, 1985.
43. Giza, B.K., Ackroff, K., McCaughey, S.A., Sclafani, A., and Scott, T.R., Preference conditioning alters taste responses in the nucleus of the solitary tract of the rat, *Am. J. Physiol.*, 273(4 Pt 2), R1230, 1997.
44. Giza, B.K., Deems, R.O., Vanderweele, D.A., and Scott, T. R., Pancreatic glucagon suppresses gustatory responsiveness to glucose, *Am. J. Physiol.*, 265(6 Pt 2), R1231, 1993.
45. Giza, B.K. and Scott, T.R., Intravenous insulin infusions in rats decrease gustatory-evoked responses to sugars, *Am. J. Physiol.*, 252(5 Pt 2), R994, 1987.
46. Giza, B.K. and Scott, T.R., The effect of amiloride on taste-evoked activity in the nucleus tractus solitarius of the rat, *Brain Res.*, 550, 247, 1991.
47. Grabauskas, G. and Bradley, R.M., Ionic mechanism of GABAA biphasic synaptic potentials in gustatory nucleus of the solitary tract, *Ann. N.Y. Acad. Sci.*, 30, 855, 486, 1998.
48. Grabauskas, G. and Bradley, R.M., Tetanic stimulation induces short-term potentiation of inhibitory synaptic activity in the rostral nucleus of the solitary tract, *J. Neurophysiol.*, 79(2), 595, 1998.
49. Halpern, B.P., Sensory coding, decoding, and representations. Unnecessary and troublesome constructs?, *Physiol Behav.*, 69(1), 115, 2000.
50. Hajnal, A., Takenouchi, K., and Norgren, R., Effect of intraduodenal lipid on parabrachial gustatory coding in awake rats, *J. Neurosci.*, 19(16), 7182, 1999.

51. Halsell, C.B. and Travers, S.P., Anterior and posterior oral cavity responsive neurons are differentially distributed among parabrachial subnuclei in rat, *J. Neurophysiol.*, 78(2), 920, 1997.
52. Hettinger, T.P. and Frank, M.E., Specificity of amiloride inhibition of hamster taste responses, *Brain Res.,* 513, 24, 1990.
53. Hettinger, T.P., Frank, M.E., and Myers, W.E., Are the tastes of polycose and monosodium glutamate unique?, *Chem. Senses,* 21(3), 341, 1996.
54. Jacobs, K.M., Mark, G.P., and Scott, T.R., Taste responses in the nucleus tractus solitarius of sodium-deprived rats, *J. Physiol (London),* 406, 393, 1988.
55. Lasiter, P.S. and Kachele, D.L., Organization of GABA and GABA-transaminase containing neurons in the gustatory zone of the nucleus of the solitary tract, *Brain Res. Bull.,* 21, 623, 1988.
56. Leonard, N.L., Renehan, W.E., and Schweitzer, L., Structure and function of gustatory neurons in the nucleus of the solitary tract. IV. The morphology and synaptology of GABA-immunoreactive terminals, *Neuroscience,* 92(1), 151, 1999.
57. Li, C.S. and Smith, D.V., Glutamate receptor antagonists block gustatory afferent input to the nucleus of the solitary tract, *J. Neurophysiol.,* 77(3), 1514, 1997.
58. Liu, H., Behbehani, M.M., and Smith, D.V., The influence of GABA on cells in the gustatory region of hamster solitary nucleus, *Chem. Senses,* 18(3), 285, 1993.
59. Lindemann, B., Chemo reception: tasting the sweet and the bitter, *Curr. Biol.,* 6, (10) 1234, 1996.
60. Lundy, R.F., Jr. and Contreras, R.J., Temperature and amiloride alter taste nerve response to Na^+, K^+, and NH_4 salts in rats, *Brain Res.*, 744(2), 309, 1997.
61. Maes, F.W., Improved best-stimulus classification of taste neurons, *Chem. Senses,* 10, 35, 1985.
62. McCaughey, S.A., Giza, B.K., and Scott, T.R., Activity in rat nucleus tractus solitarius after recovery from sodium deprivation, *Physiol. Behav.,* 60(2), 501, 1996.
63. McCaughey, S.A., Giza, B.K., Nolan, L.J., and Scott, T.R., Extinction of a conditioned taste aversion in rats: II. Neural effects in the nucleus of the solitary tract, *Physiol. Behav.*, 61(3), 373, 1997.
64. Monroe, S. and Di Lorenzo, P.M., Taste responses in neurons in the nucleus of the solitary tract that do and do not project to the parabrachial pons, *J. Neurophysiol.,* 74(1), 249, 1995.
65. Nakamura, K. and Norgren, R., Gustatory responses of neurons in the nucleus of the solitary tract of behaving rats, *J. Neurophysiol.*, 66(4), 1232, 1991.
66. Nakamura, K. and Norgren, R., Taste responses of neurons in the nucleus of the solitary tract of awake rats: an extended stimulus array, *J. Neurophysiol.,* 70(3), 879, 1993.
67. Nishijo, H. and Norgren, R., Responses from parabrachial gustatory neurons in behaving rats, *J. Neurophysiol.*, 63(4), 707, 1990.
68. Nishijo, H. and Norgren, R., Parabrachial neural coding of taste stimuli in awake rats, *J. Neurophysiol.*, 78(5), 2254, 1997.
69. Norgren, R., Taste pathways to hypothalamus and amygdala, *J. Comp. Neurol.,* 166(1), 17, 1976.
70. Norgren, R., Projections from the nucleus of the solitary tract in the rat, *Neuroscience,* 3, 207, 1978.
71. Norgren, R., Taste and the autonomic nervous system, *Chem. Senses,* 10, 143, 1985.
72. Norgren, R. and Leonard, C.M., Ascending central gustatory pathways, *J. Comp. Neurol.*, 150(2), 217, 1973.

73. Norgren, R. and Pfaffmann, C., The pontine taste area in the rat, *Brain Res.,* 91(1), 99, 1975.
74. Nowlis, G.H., Frank, M.E., and Pfaffmann, C., Specificity of acquired aversions to taste qualities in hamsters and rats, *J. Comp. Physiol. Psychol.*, 94(5), 932, 1980.
75. Ogawa, H., Hayama, T., and Yamashita, Y., Thermal sensitivity of neurons in a rostral part of the rat solitary tract nucleus, *Brain Res.,* 454(1-2), 321, 1988.
76. Ogawa, H., Sato, M., and Yamashita, S., Multiple sensitivity of chorda tympani fibres of the rat and hamster to gustatory and thermal stimuli, *J. Physiol. (London),* 199(1), 223, 1968.
77. Ozeki, M. and Sato, M., Responses of gustatory cells in the tongue of the rat to stimuli representing the four taste qualities, *Comp. Biochem. Physiol. A.,* 41, 391, 1972.
78. Pfaffmann, C., Gustatory nerve impulses in the rat, cat and rabbit, *J. Neurophysiol.,* 18, 429, 1955.
79. Pfaffmann, C., Frank, M., Bartoshuk, L.M., and Snell, T.C., Coding gustatory information in the squirrel monkey chorda tympani, *Prog. Psychobiol. Physiol. Psychol.,* 6, 1, 1976.
80. Pritchard, T.C., Lipton, A., and Pfaffmann, C., A noninvasive stimulus onset device for use in gustatory research, *Physiol. Behav.*, 40(2), 255, 1987.
81. Renehan, W.E., Jin, Z., Zhang, X., and Schweitzer, L., The structure and function of gustatory neurons in the nucleus of the solitary tract. I. A classification of neurons based on morphological features, *J. Comp. Neurol.,* 347, 531, 1994.
82. Renehan, W.E., Jin, Z., Zhang, X., and Schweitzer, L., Structure and function of gustatory neurons in the nucleus of the solitary tract. II. Relationships between neuronal morphology and physiology, *J. Comp. Neurol.,* 367(2), 205, 1996.
83. Renehan, W.E., Massey, J., Jin, Z., Zhang, X., Liu, Y.Z., and Schweitzer, L., Developmental changes in the dendritic architecture of salt-sensitive neurons in the nucleus of the solitary tract, *Brain Res. Dev. Brain Res.,* 102(2), 231, 1997.
84. Saleh, T.M. and Cechetto, D.F., Neurotransmitters in the parabrachial nucleus mediating visceral input to the thalamus in rats, *Am. J. Physiol.,* 266(4 Pt 2), R1287, 1994.
85. Sato, T. and Beidler, L.M., The response characteristics of rat taste cells to four basic taste stimuli, *Comp. Biochem. Physiol. A.,* 73(1), 1, 1982.
86. Sato, T. and Beidler, L.M., Dependence of gustatory neural response on depolarizing and hyperpolarizing receptor potentials of taste cells in the rat, *Comp. Biochem. Physiol. A.*, 75(2), 131, 1983.
87. Sato, T. and Beidler, L.M., Broad tuning of rat taste cells for four basic taste stimuli, *Chem. Senses,* 22, 287, 1997.
88. Schiffman, S.S. and Erickson, R.P., The issue of primary tastes vs. a taste continuum, *Neurosci. Biobehav Rev. Summer,* 4(2), 109, 1980.
89. Schwartzbaum, J.S., Electrophysiology of taste-mediated functions in parabrachial nuclei of behaving rabbit, *Brain Res. Bull.,* 11(1), 61, 1983.
90. Sclafani, A. and Abrams, M., Rats show only a weak preference for the artificial sweetener aspartame, *Physiol. Behav.,* 37(2), 253, 1986.
91. Scott, T.R., Behavioral support for a neural taste theory, *Physiol. Behav.*, 12(3), 413, 1974.
92. Scott, T.R. and Giza, B., Coding channels in the rat taste system, *Science,* 249, 1585, 1990.
93. Scott, T.R. and Mark, G.P., The taste system encodes stimulus toxicity, *Brain Res.,* 414(1), 197, 1987.

94. Scott, T.R. and Perrotto, R.S., Intensity coding in pontine taste area: gustatory information is processed similarly throughout rat's brain stem, *J. Neurophysiol.,* 44, 739, 1980.
95. Schweitzer, L. and Renehan, W.E., The use of cluster analysis for cell typing, *Brain Res. Protoc.,* 1(1), 100, 1997.
96. Shannon, C.E. and Weaver, W., *The Mathematical Theory of Communication*, Urbana, IL: Univ. Illinois Press, 1949.
97. Shimura, T., Komori, M., and Yamamoto, T., Acute sodium deficiency reduces gustatory responsiveness to NaCl in the parabrachial nucleus of rats, *Neurosci. Lett.,* 236(1), 33, 1997.
98. Slotnick, B.M., Sheelar, S., and Rentmeister-Bryant, H., Transection of the chorda tympani and insertion of ear pins for stereotaxic surgery: equivalent effects on taste sensitivity, *Physiol. Behav.,* 50(6), 1123, 1991.
99. Smith, D.V. and Bealer, S.L., Sensitivity of the rat gustatory system to the rate of stimulus onset, *Physiol. Behav.,* 15(3), 303, 1975.
100. Smith, D.V. and Bealer, S.L., Recovery of excitability after gustatory adaptation: effects of stimulus intensity, *Sens. Processes,* 1(2), 99, 1976.
101. Smith, D.V. and Li, C.S., Tonic GABAergic inhibition of taste-responsive neurons in the nucleus of the solitary tract, *Chem. Senses,* 23, 159, 1998.
102. Smith, D.V., Steadman, J.W., and Rhodine, C.N., An analysis of the time course of gustatory neural adaptation in the rat, *Am. J. Physiol.,* 229(4), 1134, 1975.
103. Smith, D.V. and St. John, S.J., Neural coding of gustatory information, *Curr. Opin. Neurobiol.,* 9(4), 427, 1999.
104. Smith, D.V. and Travers, J.B., A metric for the breadth of tuning of gustatory neurons, *Chem. Sens. Flav.,* 4, 215, 1979.
105. Smith, D.V., Van Buskirk, R.L., Travers, J.B., and Bieber, S.L., Gustatory neuron types in the hamster brain stem, *J. Neurophysiol.,* 50, 522, 1983.
106. Spector, A.C., Redman, R., and Garcea, M., The consequences of gustatory nerve transection on taste-guided licking of sucrose and maltose in the rat, *Behav. Neurosci.,* 110(5), 1096, 1996.
107. Spector, A.C., Travers, S.P., and Norgren, R., Taste receptors on the anterior tongue and nasoincisor ducts of rats contribute synergistically to behavioral responses to sucrose, *Behav. Neurosci.,* 107(4), 694, 1993.
108. St. John, S.J. and Spector, A.C., Combined glossopharyngeal and chorda tympani nerve transection elevates quinine detection thresholds in rats (*Rattus norvegicus*), *Behav. Neurosci.,* 110(6), 1456, 1996.
109. St. John, S.J. and Spector, A.C., Behavioral discrimination between quinine and KCl is dependent on input from the seventh cranial nerve: implications for the functional roles of the gustatory nerves in rats, *J. Neurosci.,* 18(11), 4353, 1998.
110. Steullet, P. and Derby, C.D., Coding of blend ratios of binary mixtures by olfactory neurons in the Florida spiny lobster, *Panulirus argus, J. Comp. Physiol. A,* 180 (2), 123, 1997.
111. Tell, F. and Bradley, R.M., Whole-cell analysis of ionic currents underlying the firing pattern of neurons in the gustatory zone of the nucleus tractus solitarii, *J. Neurophysiol.,* 71(2), 479, 1994.
112. Travers, J.B. and Smith, D.V., Gustatory sensitivities in neurons of the hamster nucleus tractus solitarius, *Sens. Processes,* 3(1), 1, 1979.
113. Travers, S.P. and Norgren, R., Coding the sweet taste in the nucleus of the solitary tract: differential roles for anterior tongue and nasoincisor duct gustatory receptors in the rat, *J. Neurophysiol.,* 65(6), 1372, 1991.

114. Travers, S.P. and Norgren, R., Organization of orosensory responses in the nucleus of the solitary tract of rat, *J. Neurophysiol.,* 73(6), 2144, 1995.
115. Travers, S.P., Pfaffmann, C., and Norgren, R., Convergence of lingual and palatal gustatory neural activity in the nucleus of the solitary tract, *Brain Res.,* 365(2), 305, 1986.
116. Vinje, W.E. and Gallant, J.L., Sparse coding and decorrelation in primary visual cortex during natural vision, *Science,* 287, 1273, 2000.
117. Wang, L. and Bradley, R.M., Influence of GABA on neurons of the gustatory zone of the rat nucleus of the solitary tract, *Brain Res.,* 616, 144, 1993.
118. Wang, L. and Bradley, R.M., *In vitro* study of afferent synaptic transmission in the rostral gustatory zone of the rat nucleus of the solitary tract, *Brain Res.,* 702(1-2), 188, 1995.
119. Wang, X., Merzenich, M.M., Beitel, R., and Schreiner, C.E., Representation of a species–specific vocalization in the primary auditory cortex of the common marmoset: temporal and spectral characteristics, *J. Neurophysiol.,* 74(6), 2685, 1995.
120. Wetherton, B.M., Leonard, N.L., Renehan, W.E., and Schweitzer, L., Structure and function of gustatory neurons in the nucleus of the solitary tract. III. Classification of terminals using cluster analysis, *Biotech. Histochem.*, 73(3), 164, 1998.
121. Whitehead, M.C. and Frank, M.E., Anatomy of the gustatory system in the hamster: central projections of the chorda tympani and the lingual nerve, *J. Comp. Neurol.,* 220, 378, 1983.
122. Woolston, D.C. and Erickson, R.P., Concept of neuron types in gustation in the rat, *J. Neurophysiol.,* 42(5), 1390, 1979.
123. Yamaguchi, S., Basic properties of umami and effects on humans, *Physiol. Behav.,* 49(5), 833, 1991.
124. Yoshida, M. and Saito, S., Multidimensional scaling of the taste of amino acids, *Jpn. Psychol. Res.,* 11, 149, 1969.

13 Electrophysiological Analysis of Olfactory Coding in the CNS

Thomas A. Christensen and John G. Hildebrand

CONTENTS

13.1 INTRODUCTION

For more than five decades, olfactory neuroscientists have relied on a range of different electrophysiological methods to help them decipher complex patterns of odor-evoked activity in the brain. As far back as the early 1940s, Lord Adrian used simple extracellular wire electrodes to begin to study the mechanisms underlying odor discrimination in the mammalian olfactory bulb and piriform cortex.[1] These remarkable records, crude as they were by today's standards, provided a first glimpse of the stunning complexity of odor-driven population responses in the brain. In comparing olfaction and vision, Adrian proposed that the process of distinguishing odors is comparable not to the discrimination of simple colors, but to the differentiation of complex visual patterns.[1] Moreover, in describing these odor-evoked activity patterns, Adrian was among the first to speculate that "...the pattern of excitation will be different for different smells." As we shall see, this basic premise has stood the test of time, and we now know that it applies not only to vertebrate animals, but also to other diverse animal taxa, including the insects and other invertebrates that we and many others study as experimental models.

0-8493-2329-0/02/$0.00+$1.50

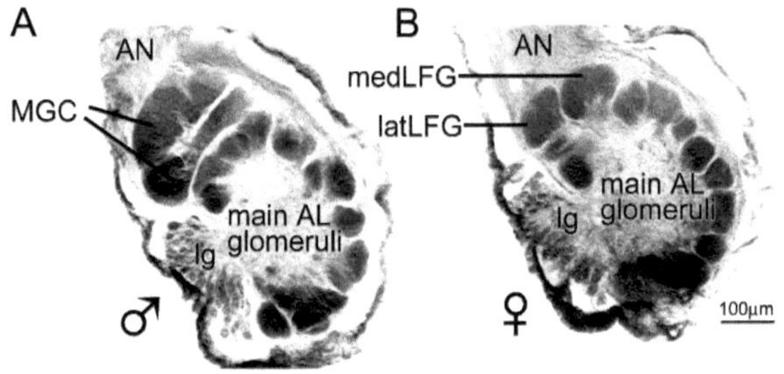

FIGURE 13.1 The first-order olfactory centers in the insect brain: the antennal lobes. Like its vertebrate counterpart, the olfactory bulb, the antennal lobe (AL) exhibits a characteristic, orderly array of synaptic glomeruli. In males (A), the macroglomerular complex (MGC) resides at the entrance of the antennal nerve (AN), outside the shell of main AL glomeruli. Occupying a similar position in females (B) are the medial and lateral large female glomeruli (medLFG and latLFG, respectively.[15] The large lateral group of AL cell bodies (lg) is visible near the lower left in each section.

Through the decades following Adrian's seminal work, as more was learned about the organization of the olfactory bulb, attention shifted to the fundamental role of olfactory glomeruli (Figure 13.1) in processing odor signals from the sensory receptor cells in the periphery.[2] By the end of the 1970s and through the 1980s, reports detailing the many similarities between the antennal lobe of insects and the olfactory bulb of vertebrates began to emerge.[3-6] The most obvious feature shared by both systems is the presence of multiple, discrete modules of synaptic neuropil — the glomeruli — but the similarities certainly do not end there.[7,8] Both sensory networks are characterized by a population of neurons that serve either excitatory or inhibitory functions, and the precise synaptic arrangements among certain types of neurons are crucial in the initial stages of odor-information processing.[9] For example, glomerular output neurons (mitral/tufted cells in vertebrates and projection neurons in insects) are involved in complex synaptic interactions with GABAergic inhibitory interneurons that participate in shaping the patterned output of a glomerulus and its neighbors. It is the precise interplay between these different cell types that underlies the initial neural representations of odors at the early stages of processing.[2,8,9]

A long-standing hypothesis asserts that each glomerulus is functionally unique and dedicated to processing specific information about a particular odor, or even a chemical determinant of the odor molecule.[2,10-12] According to this premise, different odor stimuli each evoke a unique pattern of activity that is first distributed across the glomerular array and is subsequently read by higher-order processing centers in the olfactory pathway.[9,12-15] Over the decades, olfactory scientists have developed an impressive arsenal of techniques to examine (either directly or indirectly) these activity patterns in the brain. Owing to space constraints, we will first briefly summarize some of the most powerful methods for functional imaging of odor-evoked neural activity (more detailed accounts of these methods are found in Chapters 5, 18, and 19). The remainder of the review focuses on the still-growing list of electrophysiological tools currently used to explore the neural networks that underlie the brain's dynamic olfactory maps.

13.2 METHODS FOR INDIRECT RECORDING OF ODOR-EVOKED NEURAL ACTIVITY

In the mid-1970s, several studies showed that glucose uptake in the glomerular layer of the mammalian olfactory bulb could be used as a marker to reveal distinguishable patterns of activity evoked by different odors.[3,16,17] In those studies, after receiving an intracardial injection of [^{14}C]2-deoxy-D-glucose, rats were exposed to different odors and then sacrificed in preparation for autoradiographic analysis of sectioned brain tissue. These experiments were among the first to demonstrate that specific glomerular loci (e.g., the modified glomerular complex in the dorso-medial region of the olfactory bulb of laboratory rats) are activated selectively by certain odors (e.g., rat suckling pheromone).[3] Subsequent analyses using the same method in both vertebrates and invertebrates have continued to provide support for the critical concept that glomeruli are structurally and functionally distinct modules in the brain, and that a given odor is represented by multiple sites of focal activity distributed across a subset of olfactory glomeruli.[18,19]

With the development of fluorescent activity markers such as voltage- and ion-sensitive dyes,[12,20-24] as well as other functional imaging techniques,[25-27] olfactory scientists gained powerful new tools for labeling the odor-evoked activity of groups of glomeruli that can be identified across preparations (other chapters 5, 18, and 19). In addition to offering enhanced spatial resolution, these methods can also provide information about the time-course of odor-evoked activity patterns in the brain. The role of temporal patterning in olfactory coding is currently a topic of much interest and is discussed in more detail later in this chapter.

Recent progress in molecular studies of olfaction has also led to the development of methods that can track odor-evoked expression of immediate early genes, such as c-*fos*, in activated glomeruli. The induction of c-*fos* expression can be visualized by *in situ* hybridization with c-*fos* cRNA,[28] or by examining the pattern of stimulus-evoked Fos-like immunoreactivity.[29]

All of these methods for indirect monitoring of neural activity continue to provide crucial evidence to support a role for specific subsets of glomeruli in the neural representation of different odors. The remainder of this chapter will focus on the methods used to identify the specific neural interactions that underlie these complex activity patterns.

13.3 METHODS FOR DIRECT RECORDING OF ODOR-EVOKED NEURAL ACTIVITY

13.3.1 Extracellular Microelectrodes

One of the oldest and simplest methods for recording patterns of central neural activity is to amplify extracellular signals recorded through metal wires inserted into the tissue.[30] The typical electrode of this kind, fabricated from tungsten or steel wire, is insulated to the tip, leaving only the cross-sectional surface of the wire in contact with the tissue. The wire may also be etched chemically, so that the resulting sharpened tip is ideally suited for penetrating soft brain tissue. Wire electrodes are also commonly used to help penetrate harder structures, such as insect cuticle, and this method is still one of the most effective for recording extracellular activity from sensory receptor cells housed in cuticle-covered olfactory sensilla on the surface of the antenna.[31]

Extracellular signals may also be recorded with glass microelectrodes.[32] These electrodes are fashioned from a thin glass-capillary tube (typically 1 to 1.5 mm OD) that is heated to soften the glass in the center, and then drawn apart rapidly to yield two micropipettes with gradually tapering tips. Tip diameters of such micropipettes are usually in the range of 1 to 10 μm, and the resulting microelectrodes have low resistances (typically less than 10 MΩ).

Examination of trains of action potentials (spikes) recorded extracellularly from the olfactory bulb or the antennal lobe shows that odor-evoked responses are temporally complex, often involving inhibitory as well as excitatory phases.[33-35] Olfactory physiologists employed these extracellular recording methods almost exclusively until the mid-1970s, when the need to examine the cellular interactions underlying these complex responses led to an increase in the use of intracellular recording. Nevertheless, a half-century after Adrian's experiments, the relative ease and long time-course of extracellular recordings continue to make this method valuable to olfactory neuroscientists.[36]

13.3.2 Intracellular Microelectrodes

Intracellular recording from single neurons has contributed significantly to our understanding of the functional organization of the olfactory system, and has facilitated the analysis of olfactory coding in many species. This method offers three important advantages over extracellular recording: (1) it enables one to record activity unambiguously from an individual neuron; (2) one can observe postsynaptic potentials in addition to spikes; and (3) by injection of an intracellular stain, one can correlate the anatomical and physiological properties of the same neuron (Figure 13.2).[32] The method relies on an ultra-fine (0.1-μm tip diameter) glass electrode that penetrates into the neuron so that its transmembrane potentials can be recorded. When the electrolyte solution in the recording electrode contains a suitable intracellular stain, that marker can be injected by iontophoresis or pressure and then allowed to diffuse throughout the neuron, often revealing its complete morphology (Figure 13.2B). Beginning in the mid-1970s, the application of these methods to studies of the olfactory system provided a means to test hypotheses about the synaptic organization and processing capabilities of glomerular circuits in both vertebrates and invertebrates.[4,37-42]

In numerous studies, we have taken advantage of intracellular methods to examine the functional and anatomical properties of neural circuits that process odor information in the antennal lobe of insects. Through combined intracellular recording and staining of single neurons or pairs of neurons (Figure 13.3), it was discovered that the projection (output) neurons innervating a particular glomerulus are a diverse group, exhibiting marked heterogeneity in both form and function.[14,15,42-44] This complexity reflects the fact that an olfactory glomerulus is not simply a site of convergence of olfactory receptor cells responding to the same stimuli; importantly,

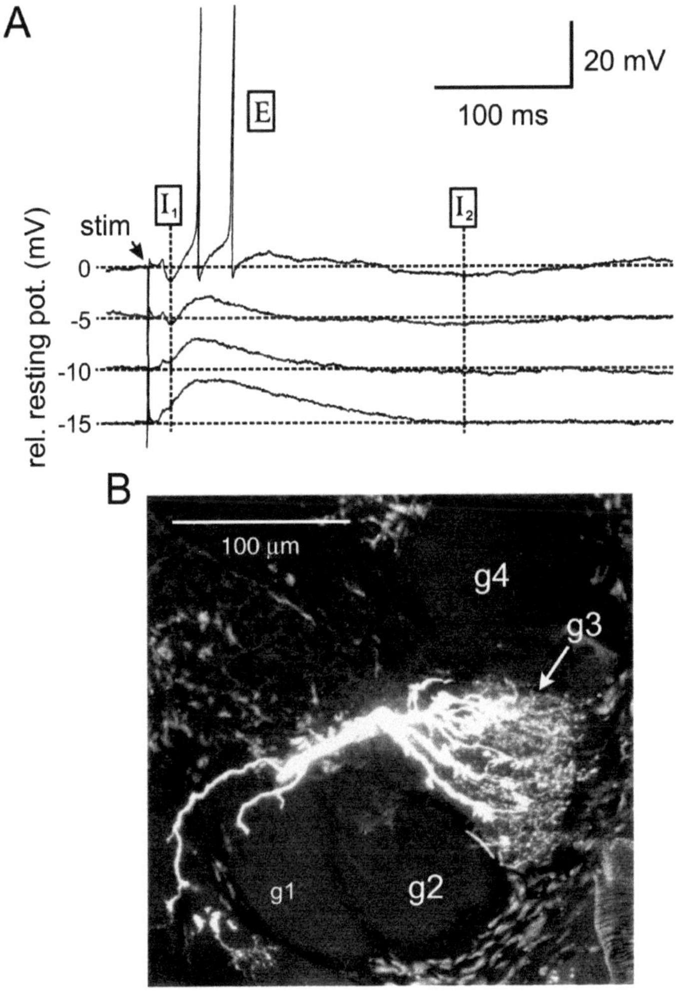

FIGURE 13.2 Intracellular recording is used extensively in olfactory research to help elucidate the complex synaptic circuits underlying odor-information processing in the brain. A: Unlike extracellular (AC-coupled) records, intracellular traces allow the investigator to monitor and analyze postsynaptic potentials (PSPs) in addition to spikes. The recording electrode can also be used to pass current into the neuron to permit further examination of its synaptic input. This series of traces shows the effect of injecting negative current steps on the PSPs recorded in a moth antennal lobe neuron. Electrical stimulation of the antennal nerve (stim) evoked a characteristic triphasic response in the neuron, and each phase responded differently to the negative shift in membrane potential. Note the different reversal potentials for the two inhibitory phases, I_1 and I_2, reflecting the different ionic mechanisms driving these two potentials. B: Another advantage of intracellular methods is that a tracer dye can be used to characterize the morphology of the neuron being analyzed. After staining with Lucifer Yellow, a laser-scanning fluorescence confocal microscope was used to visualize the fine anatomical details of the neuron, thereby also revealing the type of neuron and the identity of its target glomerulus. This method showed that the extensive branches of this output neuron were all confined to only one of the four glomeruli (g1–g4) situated in this small region of the antennal lobe.

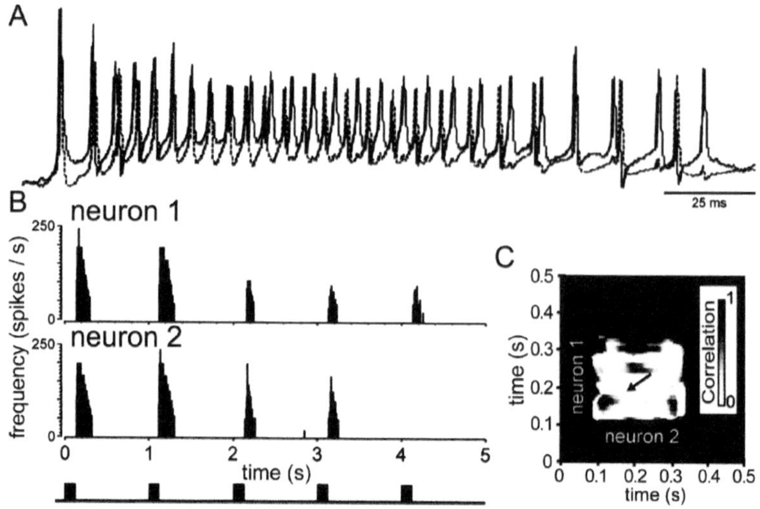

FIGURE 13.3 Patterns in time: temporal relationships between two olfactory output neurons that innervate the same glomerulus and respond to the same odor. A: Superimposed traces showing the spiking patterns of the two neurons in response to the key odor stimulus (a component of the sex pheromone; data provided by Dr. H. Lei). B: Instantaneous frequency plots illustrating the responses to five 100-ms odor pulses delivered at 1-s intervals. Notice how neuron 2 responded to only four of the five pulses. C: Dynamic cross-correlation analysis reveals the time course of synchrony between the two neurons. An incidence of high correlation is shown as a black spot in this three-dimensional surface plot. The highest correlation occurs at an expected latency just after stimulus onset (arrow). In sum, this method not only measures co-activity between neurons, but also provides information about the time course of co-activity within a coding ensemble.

it also encapsulates the synaptic circuitry that is responsible for significant amplification, integration, and coding of incoming odor signals at the first central relay in the olfactory pathway.[9]

13.3.3 Patch Microelectrodes

A more recent development in micropipet recording is the patch-clamp technique,[45] and its application in studies of the olfactory system is increasingly common. This method has the advantage that only a small area of the cell membrane (or patch) is in contact with the recording pipet, and thus the functional properties of ion channels in the patch can be examined by clamping the membrane potential at fixed voltages and measuring the induced currents. Alternatively, the membrane patch can be ruptured, resulting in direct contact between the cell cytoplasm and the solution in the electrode. This quasi-intracellular (or whole-cell) configuration can be used to measure the overall responses of the neuron to injected current or voltage steps, and it furthermore allows the experimenter to study the neuron's geometry if a tracer dye is included in the recording electrode.[46,47] This method has been used to examine olfactory-bulb neurons in brain-slice preparations[46-48] and in cell culture,[49] as well as to study insect antennal-lobe neurons *in vitro*.[50]

13.3.4 Multichannel Microelectrode Arrays

Building on and retaining the power of single-unit recording techniques, a method has been developed to probe neural function and information processing at the level of multineuronal circuits. This relatively recent development in neurophysiological activity monitoring employs multichannel extracellular microelectrode arrays that enable unambiguous recordings from many neurons simultaneously (Figure 13.5).[51] Originally developed for use in the mammalian brain, we now use this method extensively in our laboratory to study odor-evoked activity patterns in the insect olfactory system.[52]

The first multichannel recording arrays were simply multiple wire electrodes, inserted separately or as groups into the cortex.[30] Since their introduction in the 1950s, these electrode arrays are regularly being redesigned and improved. A variety of different electrode and insulation materials, as well as different wire configurations, have been tested to help improve the separation of signals from different neurons.[51] Enhanced methods for stabilizing the arrays once they have been implanted in the nervous system have also greatly increased data yields. In addition to these changes to the electrodes themselves, and because of the widespread use of computers in neurophysiological research, there are also steady improvements in the hardware and software used for multichannel data acquisition and analysis.[30,51]

In our research on the olfactory system of the moth *Manduca sexta*, we use multichannel silicon microprobes developed using thin-film microfabrication methods at the Center for Neural Communication Technology at the University of Michigan.[53] In most experiments, we use a 12-channel, 3-pronged array of iridium electrodes with dimensions that are matched to those of the moth's antennal lobe (Figure 13.5).[52] Although we have been using this technology in the moth brain for only a relatively short time, it is clear that ensemble recording methods provide a number of important advantages over methods used in the past. First, one now has the ability to monitor the responses of *groups* of olfactory information-processing neurons (as opposed to one or two), and this approach is now revealing information about cellular interactions in these circuits that could not be obtained with other recording methods (Figure 13.4).[52] Second, as long as the insect's head is fixed firmly in place, multichannel silicon-iridium electrode arrays are extremely stable, and these probes have provided long-term population recordings continuously for nearly 24 h. In contrast, our previous studies using single-unit intracellular electrodes (often from 5 to 10 μm neurites) showed that intracellular impalements were seldom stable for 30 min. Third, the spatial arrangement of the recording array (see Figure 13.5) allows one to record odor-driven responses simultaneously from multiple sites in the antennal lobe, thus providing a means to monitor activity patterns both within and across the olfactory glomeruli.[52] An alternative approach is to sample different sites sequentially using a single microelectrode, but this method does not permit the investigator to examine the interactions between the functionally diverse types of neurons that participate in these odor-encoding ensembles. Imaging studies can suffer from the same limitation because they often reveal only spatial patterns of activity without revealing which neurons are firing in response to a given stimulus. Moreover, the temporal resolution of an optical recording is usually slower than that

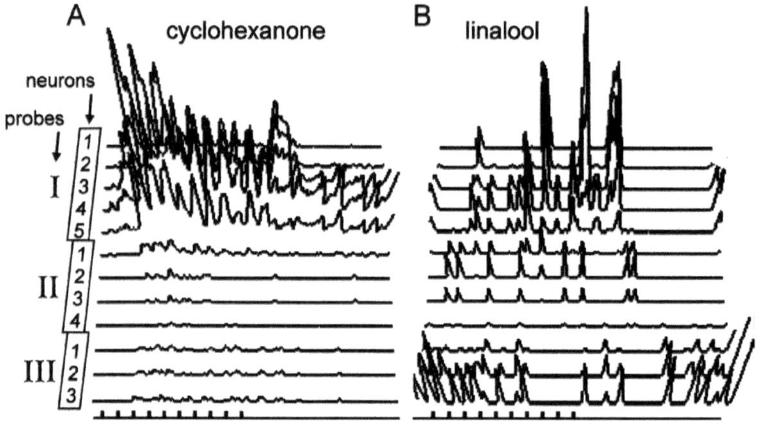

FIGURE 13.4 Neural ensemble recordings in the moth antennal lobe. We use a fork-like silicon multielectrode array (Figure 13.5) to analyze dynamic network interactions within and between olfactory glomeruli. These records are the post-stimulus-time histograms representing the firing patterns of 12 neurons recorded simultaneously across the three probes of the array, in response to an organic solvent (A) and a common plant odor (B). Stimulus markers beneath records illustrate the stimulus time-course; ten 100-msec pulses delivered at 2 Hz. Notice that while some neurons are activated by both odors, the two odors nevertheless evoke distinctly different spatiotemporal patterns of activity across the ensemble.

of a microelectrode; thus, most conventional imaging methods cannot resolve the rapid changes in neuronal firing that underlie odor recognition. Fourth, multichannel recording methods are increasingly being adapted for use in awake animals, which allows the experimenter to look for neural correlates of specific behaviors.[51] In sum, the technical advances that led to the current resurgence in the use of multichannel recording methods in behaving mammals will certainly continue, and it is expected that this approach will be increasingly adapted for use in smaller animal models, including insects and other invertebrates. It is also important to note, however, that the use of any one particular recording method over another is not advocated. In our view, using several different methods to examine a particular neurobiological question is a good strategy for testing the validity of one's hypothesis. Thus, we have not abandoned the use of the alternative recording methods discussed above, but we consider ensemble recording to be a powerful new addition to the list of tools used to study olfactory coding.

Results from experiments using ensemble recording methods in the moth brain have shown that the computations performed in olfactory circuits in the antennal lobe are considerably more complex than previously thought. For example, there is now evidence that ensembles of glomerular output neurons use multiple and overlapping coding strategies to process olfactory information, and that these strategies are matched to the particular circumstances surrounding odor presentation.[52] In *Manduca*, odor identity is encoded spatially, in the pattern of glomerular outputs that are activated (or inhibited), whereas the temporal pattern of firing in the ensemble encodes information about other stimulus variables, such as intensity, intermittency, and the quality of odor blends.[52] By delivering sequences of brief odor pulses

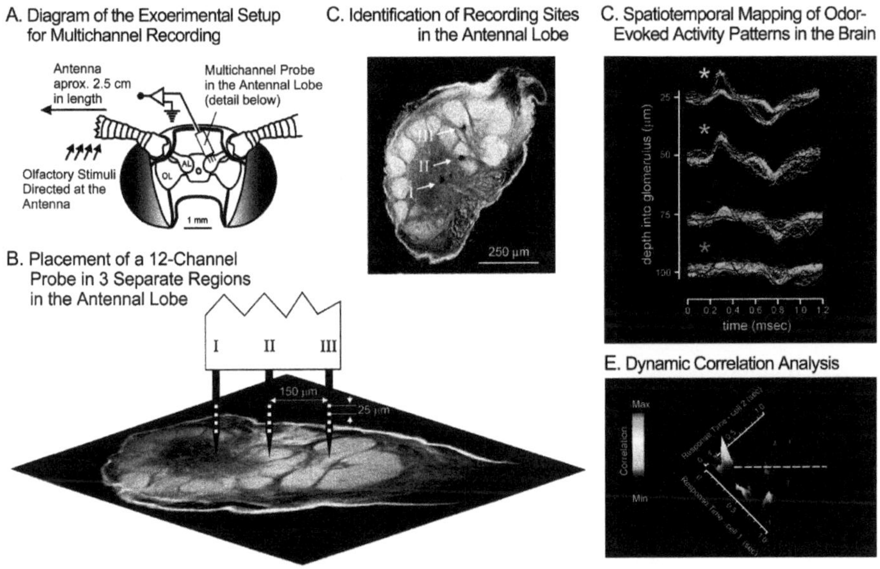

FIGURE 13.5 (see color insert following page 238) Method for obtaining simultaneous multineuronal recordings from the antennal (olfactory) lobe of the moth *Manduca sexta*. A: Diagram showing the placement of a silicon microelectrode array in the antennal lobe. B: A three-pronged probe configuration, with four recording sites per prong, is shown to scale with a plane projection of the antennal lobe (with glomeruli visible as the lighter structures). C: The exact locations of the three prongs in the glomerular neuropil can be identified because their penetration sites are visible in a confocal micrograph. D: This method is well-suited for studying the dynamic and distributed representations of odors across groups of olfactory glomeruli. For example, these records show the superimposed spike waveforms from three neurons at four depths through a glomerulus. The trace showing the maximum spike amplitude recorded for each neuron is indicated by an asterisk: the neuron represented by the green traces was closest to the top, while the blue neuron was localized to the base of the glomerulus. E: Dynamic cross-correlation analysis is then used to examine the temporal activity relationships between neurons in the ensemble. A 100-msec odor pulse was presented at time = 0. Note how these two neurons were synchronized only during a brief window of time following the stimulus presentation.

that simulate natural olfactory environments, it was discovered that these temporal patterns of firing are neither oscillatory nor odor specific, in contrast to the situation reported in other olfactory systems.[35,54] In moths, temporal patterns of spike synchrony among the members of a given neural ensemble are context dependent, illustrating that odor maps in the brain are not static representations, but dynamic patterns subject to alterations in the animal's environment.

In addition to encoding information about the odor itself, new evidence from multi-neuron recordings shows that odor-evoked activity patterns in the antennal lobe can be further modified by experience. By combining behavioral conditioning with long-term ensemble recording in moths, we are now examining the evolution of odor-evoked ensemble responses before, during, and after olfactory reinforcement training in the same animal.[55] With this powerful combined approach, we hope to better understand the specific neural interactions underlying the acquisition, retention, and extinction of these learned associations in the brain.

13.4 CONCLUDING REMARKS

This chapter provides an overview of the many different methods and experimental strategies that have contributed to characterizing the activity of neurons and networks that process olfactory information in the brains of vertebrates and invertebrates. These methods, adopted from fields as distant as electrophysiology and optical sciences, have enabled brisk progress toward understanding the specific neural mechanisms that encode the different attributes of an olfactory stimulus. Importantly, these methods are not only beginning to tell us how neural networks discriminate one odor from another, but they also indicate that odor-evoked activity patterns (even at the very earliest stages of processing in the brain!) are not static but context dependent, and they can be modulated by numerous factors unrelated to the molecular identity of the odor.[56] Moreover, parallel studies in vertebrates and invertebrates have revealed numerous similarities as well as interesting differences across taxa, and this diverse approach will surely continue to improve our understanding of how olfactory information is mapped in the brain.

ACKNOWLEDGMENTS

We wish to thank the many members of our lab who over many years have contributed immeasurably to the work and ideas reported here. Research from our lab reported in this chapter has been supported by grants from NIH (NIDCD) and a contract from DARPA (CBS).

REFERENCES

1. Adrian, E.D., Olfactory reactions in the brain of the hedgehog, *J. Physiol. (London)*, 100, 459, 1942.
2. Shepherd, G.M., Principles of specificity and redundancy underlying the organization of the olfactory system, *Microsc. Res. Tech.*, 24, 106, 1993.
3. Teicher, M.H., Stewart, W.B., Kauer, J.S., and Shepherd, G.M., Suckling pheromone stimulation of a modified glomerular region in the developing rat olfactory bulb revealed by the 2-deoxyglucose method, *Brain Res.*, 194, 530, 1980.
4. Matsumoto, S.G. and Hildebrand, J.G., Olfactory mechanisms in the moth *Manduca sexta*: response characteristics and morphology of central neurons in the antennal lobes, *Proc. R. Soc. Lond. B*, 213, 249, 1981.
5. Christensen, T.A. and Hildebrand, J.G., Functions, organization and physiology of the olfactory pathways in the lepidopteran brain, in *Arthropod Brain: Its Evolution, Development, Structure and Functions*, Gupta, A.P. (Ed.), Wiley, New York, 1987, 457-484.
6. Boeckh, J., Distler, P., Ernst, K.D., Hösl, M., and Malun, D., Olfactory bulb and antennal lobe, in *Chemosensory Information Processing*, Schild, D. (Ed.), Springer-Verlag, Berlin, 1990, 201-227.

7. Oland, L.A. and Tolbert, L.P., Multiple factors shape development of olfactory glomeruli: insights from an insect model system, *J. Neurobiol.,* 30, 92, 1996.
8. Christensen, T.A., Heinbockel, T., and Hildebrand, J.G., Olfactory information processing in the brain: encoding chemical and temporal features of odors, *J. Neurobiol.,* 30, 82, 1996.
9. Christensen, T.A. and White, J., Representation of olfactory information in the brain, in *The Neurobiology of Taste and Smell,* Vol. 2, Finger, T.E., Silver, W.L., and Restrepo, D. (Eds.), Wiley, New York, 2000, 201-232.
10. Allison, A.C. and Warwick, R.T., Quantitative observations on the olfactory system of the rabbit, *Brain*, 72, 186, 1949.
11. Mori, K., Fujita, S.C., Imamura, K., and Obata, K., Immunohistochemical study of subclasses of olfactory nerve fibers and their projections to the olfactory bulb in the rabbit, *J. Comp. Neurol.*, 242, 214, 1985.
12. Kauer, J.S., Contributions of topography and parallel processing to odor coding in the vertebrate olfactory pathway, *Trends Neurosci.,* 14, 79, 1991.
13. Sullivan, S.L., Ressler, K.J., and Buck, L.B., Spatial patterning and information coding in the olfactory system, *Curr. Opin. Genet. Devel.*, 5, 516, 1995.
14. Vickers, N.J., Christensen, T.A., and Hildebrand, J.G., Combinatorial odor discrimination in the brain: attractive and antagonist odor blends are represented in distinct combinations of uniquely identifiable glomeruli, *J. Comp. Neurol.*, 400, 35, 1998.
15. King, J.R., Christensen, T.A., and Hildebrand, J.G., Response characteristics of an identified, sexually dimorphic olfactory glomerulus, *J. Neurosci.,* 15, 2391, 2000.
16. Sharp, F.R., Kauer, J.S., and Shepherd, G.M., Local sites of activity-related glucose metabolism in rat olfactory bulb during olfactory stimulation, *Brain Res.*, 98, 596, 1975.
17. Lancet, D., Kauer, J.S., Greer, C.A., and Shepherd, G.M., High resolution 2-deoxyglucose localization in olfactory epithelium, *Chem. Senses*, 6, 343, 1981.
18. Rodrigues, V., Spatial coding of olfactory information in the antennal lobe of *Drosophila melanogaster*, *Brain Res.*, 453, 299, 1988.
19. Johnson, B.A., Woo, C.C., Hingco, E.E., Pham, K.L., and Leon, M., Multidimensional chemotopic responses to n-aliphatic acid odorants in the rat olfactory bulb, *J. Comp. Neurol.*, 409, 529, 1999.
20. Kleinfeld, D., Delaney, K.R., Fee, M.S., Flores, J.A., Tank, D.W., and Gelperin, A., Dynamics of propagating waves in the olfactory network of a terrestrial mollusc: an electrical and optical study, *J. Neurophysiol.*, 72, 1402, 1994.
21. Friedrich, R.W. and Korsching, S.I., Combinatorial and chemotopic odorant coding in the zebrafish olfactory bulb visualized by optical imaging, *Neuron*, 18, 737, 1997.
22. Joerges, J., Küttner, A., Galizia, C.G., and Menzel, R., Representations of odours and odour mixtures visualized in the honeybee brain, *Nature*, 387, 285, 1997.
23. Kimura, T., Toda, S., Sekiguchi, T., Kawahara, S., and Kirino, Y., Optical recording analysis of olfactory response of the procerebral lobe in the slug brain, *Learning & Memory*, 4, 389, 1998.
24. Lam, Y-W., Cohen, L., Wachowiak, M., and Zochowski, M.R., Odors elicit three different oscillations in the turtle olfactory bulb, *J. Neurosci.*, 20, 749, 2000.
25. Lieke, E.E., Optical recording of neuronal activity in the insect central nervous system: odorant coding by the antennal lobes of honeybees, *Eur. J. Neurosci.*, 5, 49, 1993.
26. Rubin, B.D. and Katz, L.C., Optical imaging of odorant representations in the mammalian olfactory bulb, *Neuron*, 23, 499, 1999.

27. Sobel, N., Prabhakaran, V., Hartley, C.A., Desmond, J.E., Glover, G.H., Sullivan, E.V., and Gabrieli, J.D.E., Blind smell: brain activation induced by an undetected airborne chemical, *Brain*, 122, 209, 1999.
28. Guthrie, K.M., Anderson, A.J., Leon, M., and Gall, C., Odor-induced increases in c-*fos* mRNA expression reveal an anatomical "unit" for odor processing in olfactory bulb, *Proc. Natl. Acad. Sci. U.S.A.*, 90, 3329, 1993.
29. Guthrie, K.M. and Gall, C.M., Functional mapping of odor-activated neurons in the olfactory bulb, *Chem. Senses*, 20, 271, 1995.
30. Schmidt, E.M., Electrodes for many single neuron recordings, in *Methods for Neural Ensemble Recordings*, Nicolelis, M.A.L., (Ed.), CRC Press, Boca Raton, FL, chap. 1, 1999.
31. Hansson, B.S. and Christensen, T.A., Functional characteristics of the antennal lobe, in *Insect Olfaction*, Hansson, B.S., (Ed.), Springer, Berlin, ch. 5, 1999.
32. Purves, R.D., *Microelectrode Methods for Intracellular Recording and Ionophoresis*, Academic Press, London, 1981.
33. Boeckh, J. and Boeckh, V., Threshold and odor specificity of pheromone-sensitive neurons in the deutocerebrum of *Antheraea pernyi* and *A. polyphemus* (Saturnidae), *J. Comp. Physiol. A*, 132, 235, 1979.
34. Meredith, M., Patterned response to odor in mammalian olfactory bulb: the influence of intensity, *J. Neurophysiol.*, 56, 572, 1986.
35. Stopfer, M., Wehr, M., MacLeod, K., and Laurent, G., Neural dynamics, oscillatory synchronisation, and odour codes, in *Insect Olfaction*, Hansson, B.S., (Ed.), Springer, Berlin, 1999, chap. 6.
36. Chalansonnet, M. and Chaput, M.A., Olfactory bulb output cell temporal response patterns to increasing odor concentrations in freely breathing rats, *Chem. Senses*, 23, 1, 1998.
37. Getchell, T.V. and Shepherd, G.M., Synaptic actions on mitral and tufted cells elicited by olfactory nerve volleys in the rabbit, *J. Physiol. (London)*, 251, 497, 1975.
38. Mori, K. and Takagi, S.F., An intracellular study of dendrodendritic inhibitory synapses on mitral cells in the rabbit olfactory bulb, *J. Physiol. (London)*, 279, 569, 1978.
39. Mori, K., Nowycky, M.C., and Shepherd, G.M., Analysis of a long duration inhibitory potential in mitral cells in the isolated turtle olfactory bulb, *J. Physiol. (London)*, 314, 311, 1981.
40. Nicoll, R.A. and Jahr, C.E., Self-excitation of olfactory bulb neurones, *Nature (London)*, 296, 441, 1982.
41. Burrows, M., Boeckh, J., and Esslen, J., Physiological and morphological properties of interneurones in the deutocerebrum of male cockroaches which respond to female pheromone, *J. Comp. Physiol. A*, 145, 447, 1982.
42. Christensen, T.A. and Hildebrand, J.G., Male-specific, sex pheromone-selective projection neurons in the antennal lobes of the moth *Manduca sexta, J. Comp. Physiol. A*, 160, 553, 1987.
43. Hansson, B.S., Christensen, T.A., and Hildebrand, J.G., Functionally distinct subdivisions of the macroglomerular complex in the antennal lobes of the sphinx moth *Manduca sexta, J. Comp. Neurol.*, 312, 264, 1991.
44. Heinbockel, T., Christensen, T.A., and Hildebrand, J.G., Temporal tuning of odor responses in pheromone-responsive projection neurons in the brain of the sphinx moth *Manduca sexta, J. Comp. Neurol.*, 409, 1, 1999.
45. Neher E. and Sakmann, B., The patch clamp technique, *Sci. Am.*, 266, 28, 1992.

46. Bufler, J., Zufall, F., Franke, C., and Hatt, H., Patch-clamp recordings of spiking and nonspiking interneurons from rabbit olfactory bulb slices: membrane properties and ionic currents, *J. Comp. Physiol. A*, 170, 145, 1992.
47. Castillo, P.E., Carleton, A., Vincent, J.-D., and Lledo, P-M., Multiple and opposing roles of cholinergic transmission in the main olfactory bulb, *J. Neurosci.*, 19, 9180, 1999.
48. Shen, G.Y., Chen, W.R., Midtgaard, J., Shepherd, G.M., and Hines, M.L., Computational analysis of action potential initiation in mitral cell soma and dendrites based on dual patch recordings, *J. Neurophysiol.*, 82, 3020, 1999.
49. Trombley, P.Q., Hill, B.J., and Horning, M.S., Interactions between GABA and glycine at inhibitory amino acid receptors on rat olfactory bulb neurons, *J. Neurophysiol.*, 82, 3417, 1999.
50. Mercer, A.R., Kloppenburg, P., and Hildebrand, J.G., Serotonin-induced changes in the excitability of cultured antennal-lobe neurons of the sphinx moth *Manduca sexta*, *J. Comp. Physiol. A*, 178, 21, 1996.
51. Nicolelis, M.A.L., *Methods for Neural Ensemble Recordings*, CRC Press, Boca Raton, FL, 1999.
52. Christensen, T.A., Pawlowski, V.M., Lei, H., and Hildebrand, J.G., Multi-unit recordings reveal context-dependent modulation of synchrony in odor-specific neural ensembles, *Nat. Neurosci.*, 3, 927, 2000.
53. Anderson, D.J., Najafi, K., Tanghe, S.J., Evans, D.A., Levy, K.L., Hetke, J.F., Xue, X., Zappia, J.J., and Wise, K.D., Batch-fabricated thin-film electrodes for stimulation of the central auditory system, *IEEE Trans. Biomed. Eng.*, 36, 693, 1989.
54. Gelperin, A., Oscillatory dynamics and information processing in olfactory systems, *J. Exp. Biol.*, 202, 1855, 1999.
55. Daly, K., Christensen, T.A., Pawlowski, V.M., Smith, B.H., and Hildebrand, J.G., Olfactory conditioning in *Manduca sexta*: evolution of neural ensemble patterns in the antennal lobe before, during, and after learned association to odors, *Soc. Neurosci. Abstr.,* 26, 1208, 2000.
56. Vickers, N.J., Christensen, T.A., Baker, T.C., and Hildebrand, J.G., Odour-plume dynamics influence the brain's olfactory code, *Nature,* 410, 466, 2001.

14 Electrophysiological Studies of Gustation in Awake Rats

Donald B. Katz, S.A. Simon, and Miguel A.L. Nicolelis

CONTENTS

14.1 INTRODUCTION

To study coding in a sensory system, an electrophysiologist typically records the activity of neurons while presenting stimuli assumed to be relevant to neural coding, ideally when the animal is awake. The neural responses, which may include tonic or phasic firing rate changes, or changes in correlations between neurons, are used to refute or support the proposed neural coding scheme.

0-8493-2329-0/02/$0.00+$1.50

Gustation is an ideal modality with which to examine some general principles of perceptual function. Gustation is linked to an animal's acquisition of appropriate sustenance in a manner that is privileged among sensory systems. Taste and smell directly inform the organism of the ingestability of incoming stimuli; and whereas objects emit odors across relatively large distances, tastes must be inside the mouth to be perceived. The gustation/ingestion coupling requires that the animal be exquisitely sensitive to the taste of a stimulus, that the gustatory system be quick to learn from its mistakes, and that the gustatory system be able to train the other sensory systems such that they can participate in the animal's recognition of appropriate foods (Garcia et al., 1985). The rapidity with which animals learn associations between taste and the consequences of ingestion serves as testament to the importance of understanding gustatory coding, as does the tight relationship between gustatory stimuli and emotional/hedonic processing.

Even under the most favorable of circumstances, however, the study of neural coding in gustation poses challenges for the behavioral electrophysiologist. It is difficult to isolate single neurons in the gustatory axis because much gustatory processing involves small clumps of small cells that are located deep beneath the surface of the brain. Moreover, even one application of an extended array of tastants requires that neurons be held for extended periods of time. When successfully isolated and held, these neurons — even those in "primary gustatory cortex" — tend to be multimodal, and thus difficult to characterize. Delivering tastant stimuli in a controlled manner is also a particularly complicated problem in gustation. The spatial distribution of gustatory receptors across the surface of the tongue and the physical properties of solutions in which tastants are dissolved make it difficult to determine the precise time-course of stimulus delivery. Controlling the stimulus is even more difficult when the animal is awake, as a rat may move a liquid tastant about the oral cavity in ways that vary from trial to trial. In addition, each tastant has an intrinsic palatability that is reflected in tastant-specific oral responses. Thus, true gustatory neural activity must be distinguished from taste-specific activity caused by differential oral responses. The lability of these responses, which change with learning, satiety, and a host of other mechanisms, further compounds the behavioral physiologist's difficulties.

Fortunately, many of the problems that face the perceptual physiologist can and have been addressed. This chapter describes the ways in which gustatory researchers have dealt with the various challenges. Following a brief review of the relevant anatomy, the chapter focuses on the different methods that researchers have used to obtain neural data from various gustatory areas using anesthetized and awake rats, and the implications of those techniques and datasets for our understanding of gustation. Also discussed are problems of stimulus control that are inherent in delivering tastants to an awake rat, how stimulus delivery issues can affect the interpretation of neural activity in gustatory structures, and techniques that are being used to maximize stimulus control. All sections focus extensively (but not exclusively) on the techniques that we use to obtain recordings of small neural ensembles from awake rats.

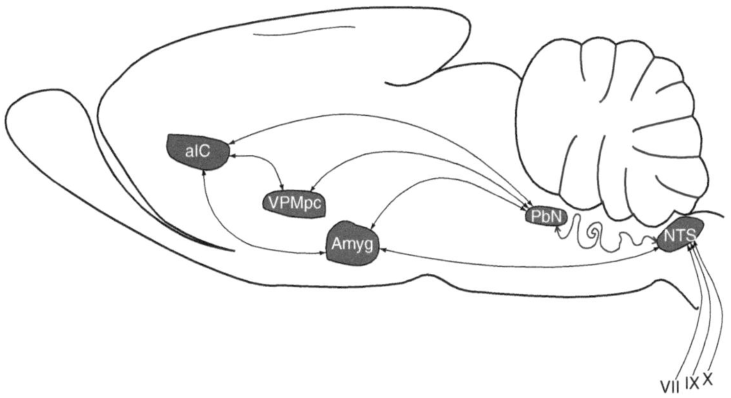

FIGURE 14.1 Schematic (not to scale) of the main gustatory axis in the central nervous system. Note both the small size of many of the gustatory areas, as well as their locations deep to the dorsal surface of the brain.

14.2 A BRIEF DESCRIPTION OF THE GUSTATORY SYSTEM

Gustatory processing begins with the activity of chemical transducers, called taste receptor cells (TRCs), that are found in onion-shaped structures called taste buds. Taste buds are distributed nonuniformly across the tongue, soft palate, and throat. Textbook characterizations of this distribution suggest that taste buds on particular parts of the tongue are specialized for detection of particular tastes. While this may be an oversimplification, the primary afferents carrying taste information to the CNS do tend to vary in their responsiveness to different tastants, depending on whether they originate in taste buds on the anterior tongue (chorda tympani), posterior tongue (hypoglossal nerve), or soft palate/throat (greater superficial petrossal nerve) (Smith and Frank, 1993).

All three gustatory nerves carry signals to the nucleus of the solitary tract (NTS) in the medulla (see Figure 14.1). In the rat, taste information travels from the NTS to the parabrachial nuclei (PbN) in the pons, and from there in parallel paths to the amygdala and the parvocellular region of the ventroposterior medial thalamus (VPMpc), which projects directly to the insular cortex (IC, primary gustatory cortex, which merges smoothly into orbital cortex, an associative area for taste, toward the frontal pole).

Figure 14.1 describes this circuitry in greater detail, approximating the layout of the rat gustatory anatomy. Most of the CNS gustatory nuclei are small and deep to the dorsal surface, and most neurons in them have small somae (see Figure 14.2). This combination of deep recording sites and small neurons increases the difficulty of recording from these areas in awake rats. Techniques involving the chronic implantation of either blunt microwires or sharp metal electrodes are unlikely to

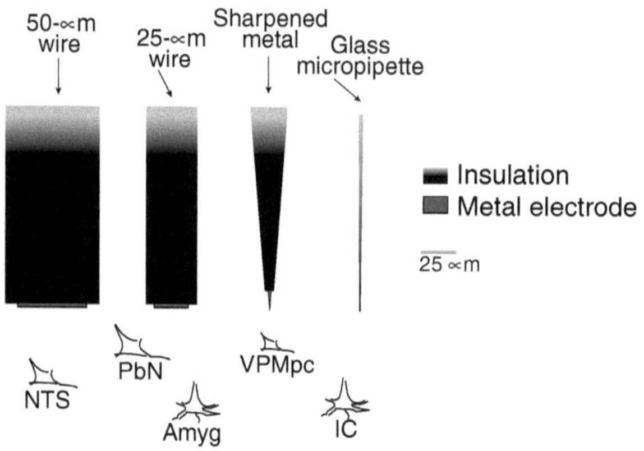

FIGURE 14.2 Electrophysiological techniques and approximate average neuron sizes. The small sizes of neurons in certain structures, most notably in VPMpc, will make isolation exceedingly difficult using certain types of electrodes.

resolve single neurons in many of the subcortical structures because the electrode tips may not remain close enough to the single neuron to isolate it from its neighbors without injuring or killing it in the days following surgery. A brief discussion of recording from each structure follows.

14.3 RECORDING FROM GUSTATORY NUCLEI

14.3.1 VPMpc

Substantial single-unit datasets have been obtained from taste thalamus in only a handful of studies, all using anesthetized animals (Nomura and Ogawa, 1985; Ogawa and Nomura, 1988; Scott and Erickson, 1971; Verhagen et al., 1999; Yamamoto et al., 1980). These studies have revealed, within a subset of VPMpc neurons, both narrowly and broadly tuned responses to gustatory stimuli. Many of these neurons also respond to somatosensory stimuli, although a neuron's somatosensory receptive field may not be the same spot on the tongue from which the largest taste responses were elicited (Nomura and Ogawa, 1985).

Because VPMpc somae probably average no larger than 13 to 15 μm in diameter (Dr. Thomas Scott, personal communication), isolation of VPMpc neurons has only been possible using glass micropipettes (see Figure 14.2) and stereotaxically mounted, intubated rats. Yamamoto et al. (1980) externalized and stimulated the chorda tympani as well to aid in localization of the VPMpc neurons. As they lowered their micropipette, the experimenters listened for responses to single-pulse stimulation of the nerve. Only isolated neurons that responded strongly to chorda tympani stimulation were considered in the dataset.

The animals used for these studies were anesthetized with either urethane or sodium amobarbital, the only anesthetics that do not completely suppress cortical and thalamic gustatory activity (it should be noted that urethane itself may cause cyclic alterations of responsiveness in cortex and thalamus; see Contreras and Steriade,

1996). In these experiments, it is common to further ensure lack of tongue movement either by cutting the hypoglossal nerve (Yamamoto et al., 1980) or by injecting the animal with a paralytic agent such as d-tubocurarin (Nomura and Ogawa, 1985; Ogawa and Nomura, 1988).

14.3.2 PbN and NTS

The terrain in the NTS and PbN is somewhat more hospitable than that in the VPMpc because the brain stem somae are slightly larger (in the hamster, these average 18 μm in the NTS and 23 μm in the PbN; see Davis, 1991). Researchers have recorded from NTS and PbN single units using pipettes, multibarrel pipettes, or sharp metal electrodes (Adachi et al., 1989; Davis and Smith, 1997; Di Lorenzo and Monroe, 1992, 1995, 1997; Di Lorenzo and Schwartzbaum, 1982a and 1982b; Hermann and Rogers, 1985; McCaughey et al., 1997; Perrotto and Scott, 1976; Shimura et al., 1997; Smith and Li, 1998). Norgren and colleagues have acutely isolated single units from both structures in awake rats (Hajnal et al., 1999; Nakamura and Norgren, 1991 and 1993; Nishijo and Norgren, 1990 and 1997). In each of these nuclei, somatosensory and gustatory regions have been found to overlap, such that isolated neurons code tastant quality, location of stimulus application, or both (Halsell et al., 1993; Hermann and Rogers, 1985). In addition, neurons in either of these regions may be sensitive to nociceptive and olfactory stimuli, or to some combination of gustatory and nongustatory stimuli (Di Lorenzo and Garcia, 1985; Van Buskirk and Erickson, 1977). The association cortex-like responsivity recorded in these structures provides rich grist for the theoretical mill, but complicates matters for the electrophysiologist.

Norgren and colleagues have noted several differences between NTS/PbN taste responses in anesthetized and awake animals. Most notable among these is a difference in average response tuning: a larger proportion of the recorded taste neurons responded significantly to only one tastant in awake rats than in anesthetized rats. Anesthesia presumably inhibits the function of some response sharpening mechanism, such that control of depth of anesthesia becomes an important variable in determining sharpness of tuning. These findings highlight the importance of recording from awake animals, particularly when studying stimuli with intrinsic or learned hedonic value.

14.3.3 IC and Amygdala

The relatively large neurons in the IC cortices and amygdalar nuclei (where the somae may be as large as 30 μm in diameter; see Washburn and Moises, 1992; Yajeya et al., 1997) are less difficult to isolate than those in the gustatory thalamus and brain stem. We discuss them together (but starting with IC) because the added ease of isolation expands the repertoire of available recording techniques. In fact, researchers interested in a variety of circuits and systems have recorded from these regions in a variety of contexts (a fact that serves as testament to their involvement in a wide range of behaviors). Remarkably few studies, however, have investigated the responses of cortical and amygdalar neurons to controlled application of tastant stimuli in awake rats.

The data that have been collected demonstrate that a small subset of neurons in IC (less than 10% by some estimates; Kosar et al., 1986; Ogawa et al., 1990) respond differentially to certain tastants, and that these same gustatory neurons may also demonstrate somatosensory, nociceptive, and olfactory responses (Hanamori et al., 1998a and 1998b). Most researchers have found taste neurons only in dysgranular and agranular IC (Kosar et al., 1986; Yamamoto et al., 1985), but Ogawa and colleagues (1992a and 1992b) found neurons in the granular portion of IC that were more specifically tuned to single tastants than those in dysgranular IC and argued that this difference relates to a granular IC specialization for tastant discrimination. The majority of these data have been collected using sharp metal electrodes acutely advanced through the brains of urethane-anesthetized animals. Some researchers approach the IC from the side of the skull, with the rats' heads tilted at 45° from the vertical axis (Ogawa et al., 1990). Access to the gustatory region is easier and faster from this angle, but the approach is difficult to use with awake animals.

Taking another approach, Yamamoto et al., (1988 and 1989) recorded units from microwires chronically implanted into IC. A similar approach is currently being used to study the gustatory and rewarding properties of alcohol (Anstrom et al., 1999). The chronic implantation of microwires offers several advantages over acute preparations. First, chronic implantation means that the rat need not be restrained during testing. Second, chronic implantation of slender microwires typically means more stable isolation of single units than with acute implantation — thus, a neuron's responsivity can be tested across a longer time course, and even throughout a one-trial learning procedure (Yasoshima et al., 1995; Yasoshima and Yamamoto, 1998). A longer isolation also allows multiple trials to be delivered per stimulus, which may allow researchers to overcome response variability. This variability, intrinsic to the responses of an awake rat, limits the power to recognize stimulus-specific activity, to the extent that some gustatory responses may be missed when only one trial can be analyzed. Larger samples of trials also greatly expands the arsenal of analytic techniques at the researcher's disposal (Chapin, 1999).

A subset of central and basolateral amygdalar neurons, recorded in anesthetized (Azuma et al., 1984) rats or with acute penetrations (Nishijo et al., 1998), or chronically implanted microwires (Yasoshima et al., 1995; Yasoshima and Yamamoto, 1998) in awake rats, also respond to some subset of stimuli in a tastant array. As with all other gustatory regions studied, multimodal responses abound here. That is, amygdalar neurons can be found that respond to any modality of stimulus delivered, including somatosensory, auditory, and olfactory. There is some indication that central nucleus neurons are likely to respond similarly to multiple tastants of similar palatability, suggesting this nucleus' involvement in evaluative processing (Nishijo et al., 1998; Azuma et al., 1984).

14.4 CHRONIC MICROWIRE IMPLANTATION INTO GUSTATORY CORTEX

In our laboratory, we have begun to use chronically implanted microwires to record IC activity. We implant bundles of 16 wires (50-µm diameter with an ~12-µm coating of teflon, cut flat; NBLabs, Denison, TX) into the brains of rats anesthetized with either a ketamine/xylazine cocktail or with pentobarbital (50 mg/kg). The bundles are divided into two parallel bundles of eight wires because experience suggests that the trauma caused by insertion of a single bundle of 16 50-μm wires reduces the quality of subsequent recordings in this region. A coating of sucrose, applied to the bundles before implantation, keeps the wires stiff during penetration; the sucrose dissolves as the bundle enters the brain.

As anesthetics suppress gustatory neural activity, the implantation is done stereotaxically (i.e., no attempt is made to elicit neural responses to tastant administration during surgery). The bundles are centered on the coordinates used for recording and intra-insular infusion (Kosar et al., 1986; Rosenblum et al., 1997; Yamamoto et al., 1989) — AP ~1.2 to 1.5, ML ~ 5.1 to 5.2, DV 4.5-5.0 from pia — and descend slowly (100 to 200 μm/min) to avoid distressing and compressing the tissue, a particular problem when implanting where the medial-lateral slope of the brain is clearly visible (and thus where the electrodes may push the tissue both ventrally and medially). Once the electrodes are properly placed, the hole is sealed with petroleum jelly or agar (4% in normal saline). Dental acrylic (Patterson Dental Supply, Raleigh, NC) is then applied around the exposed skull, covering ground and skull screws and burying the electrode assembly up to the top of the plug, and the wound is covered with antibiotic ointment. Subcutaneous injections of penicillin are also applied to reduce the risk of infection.

After approximately 1 week of recovery, single neuron activity can be discriminated from the individual wires. In our laboratory, the signals pass first through a high-impedence bank of field effect transistors (FETs) at the connector on the head, then through a programmable gain preamplifier controlled from a PC, and finally through a Multi-neuronal Acquisition Processor that digitizes as many as 96 (or as few as 16) signals simultaneously at 40 kHz (all products from Plexon Inc., Dallas, TX). Discrimination can be accomplished online, using pairs of time-voltage windows or a wave template algorithm, or offline, using templates or factors extracted from the waveforms by principle component or cluster analysis. For more details on the recording technique, the reader is referred to Nicolelis et al. (1997).

This approach, which has been used extensively to study coding in the somatosensory system (for extensive details, see Nicolelis et al., 1997), permits us to simultaneously record the activity of multiple single neurons. We typically isolate five to ten neurons (see Figure 14.3 for a screen capture of a representative session) and hold these single neurons for more than one session, as verified by a combination of consistently high S/N ratio, consistent wave shape, and consistent spiking statistics (interspike interval histograms).

14.5 CONSTRUCTION OF THE DRIVEABLE BUNDLES USED IN OUR LABORATORY

Recently, we have begun to use a driveable bundle technique for use with our recording system — a necessary advance that permits us to fine-tune our placements when the rat is awake (and thus when the taste neurons are responsive). This

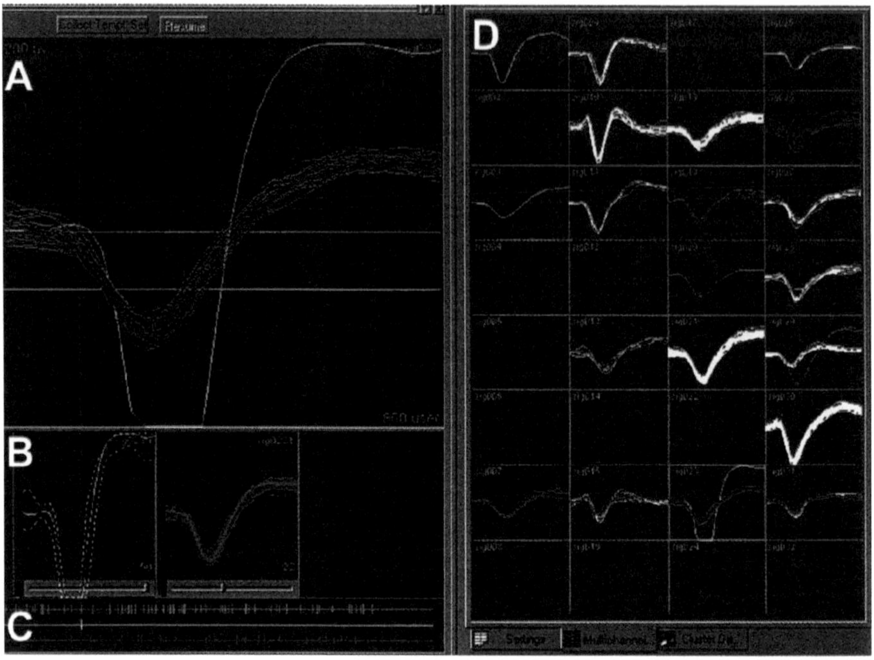

FIGURE 14.3 (see color insert following page 238) An example of the acquisition of IC units used in our laboratory (Plexon Inc., Dallas, TX). A, 800-μm segments of the channel 23 waveforms that have cleared the amplitude threshold. Note the two well-isolated single units (in green and yellow). B, below the large window, the units are presented individually. C, below these are running rasterplots of unit activity. D, the right half of the screen presents the waveforms from all 32 wires (16 in each IC). Several well isolated units can be seen, and others may be isolable using offline resorting.

technique is similar to that described by Schoenbaum for use in the amygdala and orbital regions (this volume; see also Kentros et al., 1998; Kubie, 1984), with a few adaptations that enable us to fit more than one assemby on a single skull. A bundle-carrying cannula is soldered to a 21-gauge tube. The tube is then slid into a microdrive. The microdrive, essentially a small tablet of dental acrylic with a hole running lengthwise through it, is made in a teflon mold (see Figure 14.4): a 0-80 screw projects into the mold from one end and is met halfway across by a 21-gauge stainless steel tube; after dental acrylic has been poured into the mold and allowed to harden, both screw and tube can be cracked out with gentle torque. A 0-80 set screw (all products available from Small Parts, Inc; Miami Lakes, FL) is screwed into the threaded half of the hole, and the bundle/cannula/21-gauge tube assembly is then slid in from the bottom such that it meets the set screw (Figure 14.4). Upon further turning the set screw, the tube is slowly ejected from the other end, until the set screw rests at the end of the tapped portion of the hole.

The microwires are glued both to the cannula and to the top end of the microdrive, leaving a short length of taut wire between the two glued points. The set screw is then turned back by ~2 mm, and the cannula assembly again pushed in to meet it;

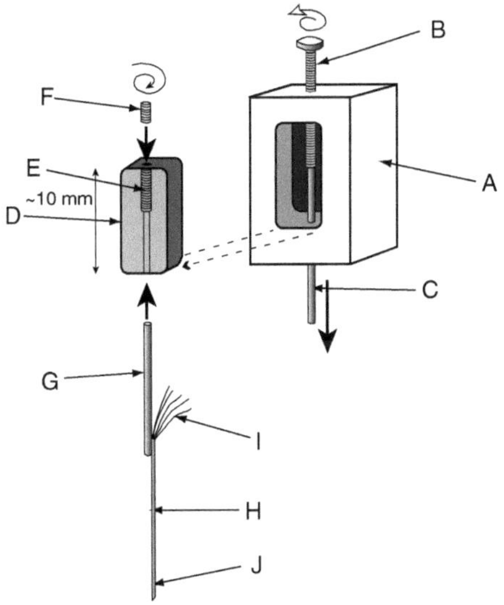

FIGURE 14.4 Construction of the dental acrylic microdrive currently being used in our laboratory. A, Teflon mold; B, 0-80 hex head screw; C, 21-g tube; D, microdrive, popped out of mold; E, hole through microdrive, with 0-80 threads molded in from the top to the midway point; F, 0-80 set screw; G, 21-g guide cannula; H, 27-g electrode cannula, soldered to the bottom of the drive cannula; I, loose top ends of microwires, to be attached to pins on assembly plug; J, 2 to 3 mm of wires emerging from bottom of electrode cannula. See text for further details.

this causes a small amount of slack to be introduced into the length of wire between the two glued points (see Figure 14.5). Finally, the microdrive is glued to the side of a connector plug (we use slim linear array connectors available from Microtech, Boothwyn, PA), and the microwires are individually glued onto the pins with a dot of silver print paint. Connection points are covered with a protective coating. For insertion, the cannula and length of slack wire are covered with petroleum jelly, such that both are free to move and penetrate further into the brain after the entire assembly has been cemented to the skull.

14.6 THE VALUE OF CHRONIC ISOLATION OF MULTIPLE SINGLE UNITS

Using driveable bundles, we gain flexibility and yield, sampling four to ten neurons at a time while advancing through IC, without giving up the advantage of chronic isolations. The use of chronic multiple-electrode physiology, particularly when room is made on the skull for multisite (i.e., more than one brain region) recording, will hopefully extend our understanding of gustation into new domains. At the most general level, this technique brings the study of gustatory dynamics within reach. Stability of recording across a single, long session enables the researcher to investigate more spatiotemporally complex single-unit responses by making reliably time-locked collection of multiple trials per tastant a fairly routine matter. Given findings

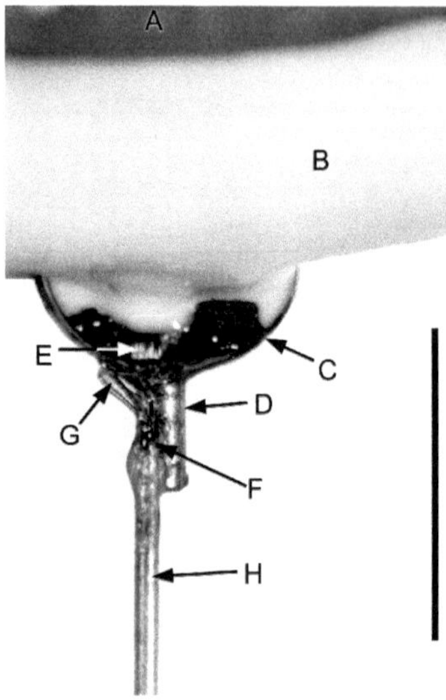

FIGURE 14.5 Detail of electrode assembly, ready for implantation. A, plug for connecting the implanted electrodes to amplifiers, etc.; B, protective covering over connections between the pins (on plug) and microwires; C, bottom of microdrive, glued to side of plug; D, 21-g drive cannula; E, point at which the bundle of microwires is cemented to the bottom of the microdrive; F, point at which the microwire bundle is cemented into the electrode cannula; G, bend in the wires, between the two cement points, that provides enough slack for 1 to 3 mm of movement following implantation; H, 27-g electrode cannula. Note that F through H are covered with a blob of petroleum jelly during implantation to ensure freedom to move. See text for further details. Scale bar = 5 mm.

from other sensory systems regarding the relative nonstationarity of the awake brain (DeAngelis et al., 1993; deCharms et al., 1998; Ghazanfar and Nicolelis, 1999; Ringach et al., 1997; Wehr and Laurent, 1996), it is reasonable to suppose that gustatory responses will prove to have a range of temporal profiles even beyond those already described (Di Lorenzo and Schwartzbaum, 1982b; Erickson et al., 1994).

The presence of multiple electrodes within a structure paves the way for expanded inquiry into how interactions between neurons may relate to gustatory coding. The few extant studies of taste neuron pairs suggest, in fact, that cross-correlations between neurons may be tastant specific (Adachi et al., 1989; Nakamura and Ogawa, 1997; Yokota et al., 1996; 1997). Examinations of neural activity in gustatory structures before and after infusions of GABA blockers furthermore suggest that intranuclear interactions play a role in the construction of tastant response profiles (Kobashi and Bradley, 1998; Ogawa et al., 1998; Smith and Li, 1998). Taken together, these findings support a view of gustatory coding that emphasizes the importance of neural populations. Ensemble recording techniques, coupled with recently developed analyses of such data, will allow researchers to investigate a variety of population coding schemes.

The collection of data from multiple neurons and structures pertains directly to the well-known labeled-line vs. across-fiber pattern debate (see Smith and St. John, 1999); first inquiries using multisite recordings suggest that the response profiles of "Na-best" neurons in the PbN may not merely be the result of feedforward transmission from NTS "Na-best" neurons (Di Lorenzo and Monroe, 1997). It is to be hoped that ensemble recordings, paired with appropriate multivariate analyses, will further support and specify our theories of functional connectivity and neural processes of perception.

Finally, the ability to isolate one ensemble of neurons for even two consecutive sessions offers even more opportunity for further theoretical insights. Currently, it can only be assumed that taste responses represent a fairly stable aspect of neural coding. With these new techniques it will be possible to test the boundaries of this stable response; that is, investigators will be able to assess what manipulations perturb single-neuron and population response profiles. Furthermore, the speed with which conditioned taste aversions form makes it possible to record a single ensemble of neurons as a rat progresses from naïveté to learnedness (Yasoshima et al., 1995; Yasoshima and Yamamoto, 1998). Such experiments may not only greatly increase our understanding of gustation, but also of the neural bases of learning.

14.7 THE BEHAVING RAT: THE CHALLENGE OF CONTROLLING THE STIMULUS

Isolation of single units is of no use to the sensory physiologist studying coding if stimuli cannot be delivered in a controlled manner. Careful control of the behavioral preparation is full partner to electrophysiological technique and therefore deserves lengthy consideration when the researcher sets out to study gustatory neural coding. The remainder of this chapter is devoted to discussing the issues inherent in dealing with the interaction between the rat and the tastant.

When squirted into the mouth or licked at by a rat, a gustatory stimulus in liquid form will strike some subset of taste receptor cells. Fluid flow and/or mouth movements will then spread the stimulus around the entire oral cavity. Stimulus delivery apparati can be designed to bathe the anterior oral cavity of an anesthetized rat in a continuous fashion. Awake rats, however, can greatly influence the time-course in which tastants hit different parts of the receptor surface by virtue of their oral behaviors.

This fact creates a variety of problems for the behavioral electrophysiologist. Primary among them is the difficulty of presenting precisely the same stimulus at the same location, trial after trial. It appears likely that a tastant that activates TRCs in one order during one trial and in another order during another trial will produce different patterns of CNS activation in the two trials. In essence, this means that the two stimuli will be measureably different in a way that may or may not have anything to do with gustatory coding. The vast majority of gustatory studies deal with this limitation by assessing only the broadest details of the neural response: taste responsivity is usually measured in terms of a single number — integrated firing across a 5-s period (but see Di Lorenzo and Schwartzbaum, 1982; Erickson et al., 1995), despite findings from a wide variety of systems suggesting that sensory stimuli induce spatiotemporally complex neural responses (DeAngelis et al., 1993; deCharms et al., 1998; Ghazanfar and Nicolelis, 1999; Ringach et al., 1997; Wehr and Laurent, 1996).

If accurate assessments of the time-course of gustatory neural responses are desired, then some effort must be made to control the initial placement of tastant on the tongue, and to limit the impact of rat behavior on tastant delivery. Researchers have sought to accomplish these tasks by monitoring a freely moving rat's contact with a lick tube, via either an infrared detector placed directly across the spout, or a low-current circuit, a capacitance circuit (Mundl and Malmo, 1979), or with a Hall-effect sensor (Kay and Laurent, 1999). Any of these will allow the researcher to know precisely when the tastant hits the tongue. Placement of that initial sample on the tongue will tend to vary between bouts of licking, however, as the rat may swivel its head around the tube or nose-poke. Also, the fact that the tastant will hit the tongue in a pulsatile fashion (rats lick according to a 6 to 10 Hz rhythm; Grill and Norgren, 1978a) may also affect neural responses. Although Nishijo and Norgren (1991) have reported that gustatory responses recorded during licking are qualitatively similar to those recorded during intra-oral infusion, this comparison was made with the rats restrained — such that the angle of access to the lick tube was controlled — and was largely (although not entirely) based on responses integrated over a 5-s period.

14.8 INTRA-ORAL CANNULATION

A more reliable stimulus delivery has been achieved using an intra-oral cannula (IOC) that deposits a sample of tastant directly onto the tongue (Grill and Norgren, 1978a; Phillips and Norgren, 1970). A PE-100 cannula is pulled up through the masseter muscle attached to a 19-gauge needle, exiting the mouth immediately anterior and lateral to the first maxillary molar, and is cemented into a dental acrylic headcap. A flared tip and washer keep the mouth end flush against the cheek, while a short length of 19-gauge stainless steel tubing attached where the PE tubing emerges from the muscle stabilizes the headcap end. Tastants can then be delivered through a PE-100 tube connected to the stainless steel top of the IOC, or through a PE-10 tube snaked down its length (Nishijo and Norgren, 1990); keeping the solutions under low or moderate pressure via a regulated nitrogen tank (a nonreactive gas that will not compromise the chemical composition of the fluids) will help to overcome resistence in the narrow tube and will limit the time delay between (solenoid-controlled) initiation of delivery and actual tongue-tastant contact to ~30 msec.

Use of the IOC ensures that a brief spurt of tastant can achieve reliable coverage of the entire tongue. While it is still theoretically possible that the rat's tongue might elude the initial coverage, the oral cavity is quickly filled by even a 30-μl dose of fluid. In our laboratory, we combine IOC implantation with a variant of the restraint technique (Nishijo and Norgren, 1990; Bermejo et al., 1998; Welsh et al., 1995) to limit behavioral variability in our rats. IOCs are implanted just prior to microwire placement. Later in the surgery, a 4-40 bolt and nut are attached to the back of the dental acrylic cap. During recovery, rats are acclimated to handling (to maintain patency, the IOCs must be reamed out with a PE-10 tube each day) and are introduced to a polyethylene restraint box. Motivated by a mild water deprivation regimen, the rats learn to self-administer small doses of water via lever press, while their heads remain bolted to the front of the restraint box.

The rats typically locate the lever during their very first session in the restraint box and quickly learn that pressing the lever results in delivery of a water reward. After only a few days of this training, they will sit patiently, awaiting rewards spaced out by seconds or even minutes. Tastant arrays can be delivered in a controlled fashion, with the rat comfortably immobilized.

14.9 THE IMPACT OF OROFACIAL BEHAVIORS: DISTINGUISHING GUSTATORY ACTIVITY FROM TASTE-RELATED ACTIVITY

The tastant is beyond the experimenter's control once delivered to the tongue, at which time the rat is free to move it around the oral cavity, and to swallow or expel it at whim. Rats exhibit a sizable repertoire of orofacial reactions to the presence of tastants in the oral cavity. How quickly the rat licks, whether the rat spits or swallows, and how the rat moves the stimulus around before doing so will depend on how palatable the rat judges the tastant to be (this phenomenon is referred to as taste reactivity; see Grill and Norgren, 1978a). Some orofacial behaviors, such as lateral tongue protrusion, are indicative of high palatability, while others, such as gapes, indicate that the rat finds the tastant distasteful (Grill and Norgren, 1978a; Berridge, 2000).

Researchers have used taste reactivity to good effect in evaluating taste quality (Breslin et al., 1992; Spector et al., 1988) and in dissociating the neural bases of liking a tastant and wanting a tastant (Berridge, 2000; Berridge and Fentress, 1985); but for the electrophysiologist such behaviors can be a source of interpretational confounds. Because orofacial behaviors cause particular patterns of somatosensory input, they will drive particular patterns of activity in somatosensory-responsive neurons. Thus, neurons with somatosensory receptive fields in the mouth region may masquerade as gustatory neurons. Although the precise nature of a tastant may not be determinable by the taste reactivity (i.e., , there is no specific "sucrose pattern"; see Berridge, 2000; Breslin et al., 1992), differences between orofacial behaviors may be sufficient to drive distinguishable neural responses.

For the most part, this seems to be a difficulty that must be lived with, a necessary price for studying awake rats. Much of taste reactivity seems, to some degree, reflexive (Grill and Norgren, 1978b), and eliminating such behaviors entirely requires that the animal be kept from behaving at all. It is possible that the best option for control of taste reactivity involves careful handling of data. Efforts can be made to test the somatosensory responsiveness of isolated neurons by infusing tasteless stimuli such as water (or better yet artificial rat saliva; DeSimone and Heck, 1993), or by attempting to stimulate the inside of the rat's mouth with a probe. Neurons that respond to somatosensory stimulation can then be considered non-gustatory and eliminated from subsequent analysis. This method will also eliminate multimodal (somatosensory-gustatory) neurons from analysis, however, and as such will likely skew conclusions reached on the basis of the dataset.

Another way that taste reactivity can be handled involves incorporating it as a factor in the analysis. The occurrence of orofacial behaviors can be videotaped and coded, and/or recorded with EMG electrodes implanted bilaterally in the temporal and masseter muscles in rats during the initial surgery. EMG recording systems (Grass Instrument Co.) are available to convert these signals into analyzable output. The onset times of these behaviors can be synchronized to the neural record, such that the neural response to gapes can be analyzed in the same way that the neural response to tastant administration itself is analyzed. The different responses can then be compared, and the activity that is tightly linked to orofacial movements can be subtracted from that evoked by the taste stimulus; any response that remains following this subtraction procedure is a true gustatory response.

Both this and the previously discussed approach are based, unfortunately, on the implicit assumption that somatosensory and gustatory coding are linearly separable (and, as such, independent). Substantial evidence is accruing, however, to suggest that somatosensation and temperature influence taste perception (see Todrank and Bartoshuk, 1991; Cruz and Green, 2000). Thus, both this and the approach above may obscure important facets of gustatory neural coding.

Taste reactivity changes as the rat learns new associations between the tastant and consequences of consumption. For example, the orofacial responses to sucrose change drastically during taste aversion conditioning, in which the taste of sucrose is associated with malaise (Breslin et al., 1992). These changes, moreover, can occur over an extremely rapid time-course — one trial or 10 minutes (Spector et al., 1988). Satiation, a process that occurs naturally within a session as the animal receives repeated trials of tastants, also reduces a tastant's palatability. It is likely that taste reactivity changes in concert with reduction of need/desire for the tastant or fluid within a single session. Such behavioral lability adds another layer to the interpretation of the above problems. Because taste reactivity can change quickly, it is as difficult to interpret neural responses in a gustatory task as it is to interpret basic responses. Care must also be taken with interpretation of data concerning the neural correlates of learning: as the animal learns to associate a tastant with outcomes, it is tempting to attribute changes in neural gustatory coding to changes in some aspect of gustatory perception. Once again, however, the phenomenon may be only indirectly related to gustation, instead being a faithful tracking of changes in somatosensory experience.

It may be possible, however, to use the lability of taste reactivity to the electrophysiologist's advantage. The potential silver lining behind the cloud is seen in the possibility of conditioning away differences in orofacial behaviors. Rats can be trained such that two dissimilar tastants lead to identical food rewards. The paradigm, sometimes referred to as the "Morrison task," involves pairing each tastant with reward only when the appropriate lever is pressed — the left lever may be mapped to the taste of sucrose, for example, and the right lever to quinine (Morrison, 1969; St. John and Spector, 1998). Through training, the rats become exquisitely sensitive to differences in taste quality, which are the keys to appropriate performance. It is possible that differences in hedonic quality, which are contradicted by the reward structure, become less salient as the rat learns.

As orofacial behaviors reflect differences in intrinsic or learned palatability, and as the conditioning leaves the rat similarly eager for either tastant, it may be that the training could reduce the differences in taste reactivity between the two stimuli. This leads to an interesting discriminative prediction: if, using chronically implanted microwires, single neuron activity is recorded as the animal learns this task, gustatory coding of the tastants used should become more distinct; as the animal recognizes the tastants' similarity with regard to ingestive consequences, however, the hedonic and somatosensory-related coding of the two tastants would converge.

14.10 SUMMARY

The challenges facing the gustatory physiologist seem daunting, but they are similar to the challenges faced by other sensory physiologists. The methods described in this and other chapters in this volume make the recording of neural responses to tastants in awake rats an achievable goal (see Figure 14.3). Neurons can be isolated and held for long enough to collect substantial numbers of trials, and the trials can be delivered in a controlled-enough fashion to enable confident interpretation. Emerging technology greatly expands the analytic and interpretive machinery that can be brought to bear on the data.

Furthermore, the very aspects of gustation that make it challenging make the rewards worth the effort. The tight coupling between taste and organismic processes such as ingestion is unique among sensory systems, and the obvious hedonic quality that is intrinsic to tastants makes gustation ideal for the study of perception and meaning. The lability of taste-related responses highlights the speed with which learning occurs in the gustatory system, and opens up unparalleled opportunities for the study of perceptual learning.

ACKNOWLEDGMENTS

The authors wish to thank Drs. Ralph Norgren and Alan Spector for their time and advice, and Drs. Asif Ghazanfar, Geoff Schoenbaum, and Marshall Shuler for critical readings. This work is supported by NIH grants DC-01065 and DC-00403, and by the Philip Morris Corporation.

REFERENCES

Adachi, M., Ohshima, T., Yamada, S., and Satoh, T. (1989). Cross-correlation analysis of taste neuron pairs in rat solitary tract nucleus, *J. Neurophysiol.,* 62, 501-509.

Anstrom, K.K., Janak, P.H., and Woodward, D.J. (1999). Changes in phasic neural activity are reinforcer-specific in an ethanol/sucrose dual-reward paradigm, *Soc. Neurosci. Abstr.,* 25, 1079.

Azuma, S., Yamamoto, T., and Kawamura, Y. (1984). Studies on gustatory responses of amygdaloid neurons in rats, *Exp. Brain Res.,* 56, 12-22.

Bermejo, R., Houben, D., and Zeigler, H.P. (1998). Optoelectronic monitoring of individual whisker movements in rats, *J. Neurosci. Methods,* 83, 89-96.

Berridge, K.C. (2000). Measuring hedonic impact in animals and infants: microstructure of affective taste reactivity patterns, *Neurosci. Biobehav. Rev.*, 24, 173-198.

Berridge, K.C. and Fentress, J.C. (1985). Trigeminal-taste interaction in palatability processing, *Science,* 228, 747-750.

Breslin, P.A.S., Spector, A.C., and Grill, H.J. (1992). A quantitative comparison of taste reactivity behaviors to sucrose before and after lithium chloride pairings: a unidimensional account of palatability, *Behav. Neurosci.,* 106, 820-836.

Bures, J., Bermúdez-Rattoni, F., and Yamamoto, T. (1998). *Conditioned Taste Aversion: Memory of a Special Kind,* Oxford: Oxford University Press.

Chapin, J.K. (1999). Population-level analysis of multi-single neuron recording data: multivariate statistical methods, in M.A.L. Nicolelis, Ed., *Methods for Neural Ensemble Recordings* Boca Raton, FL: CRC Press, 193-228.

Contreras, D. and Steriade, M. (1996). Spindle oscillation in cats: the role of corticothalamic feedback in a thalamically generated rhythm, *J. Physiol. (London),* 490, 159-179.

Cruz, A. and Green, B.G. (2000). Thermal stimulation of taste [in process citation], *Nature,* 403, 889-892.

Davis, B.J. (1991). The ascending gustatory pathway: a Golgi analysis of the medial and lateral parabrachial complex in the adult hamster, *Brain Res. Bull.,* 27, 63-73.

Davis, B.J. and Smith, D.V. (1997). Substance P modulates taste responses in the nucleus of the solitary tract of the hamster, *Neuroreport,* 8, 1723-1727.

DeAngelis, G.C., Ohzawa, I., and Freeman, R.D. (1993). Spatiotemporal organization of simple-cell receptive fields in the cat's striate cortex. I. General characteristics and postnatal development, *J. Neurophysiol.,* 69, 1091-1117.

deCharms, R.C., Blake, D.T., and Merzenich, M.M. (1998). Optimizing sound features for cortical neurons [see comments], *Science,* 280, 1439-1443.

DeSimone, J.A. and Heck, G.L. (1993). Cell biology of the lingual epithelium. In Simon, S.A. and Roper, S.D., Eds., *Mechanisms of Taste Transduction,* Boca Raton, FL: CRC Press, 201-223.

Di Lorenzo, P.M. and Garcia, J. (1985). Olfactory responses in the gustatory area of the parabrachial pons, *Brain Res. Bull.,* 15, 673-676.

Di Lorenzo, P.M. and Monroe, S. (1992). Corticofugal input to taste-responsive units in the parabrachial pons, *Brain Res. Bull.,* 29, 925-930.

Di Lorenzo, P.M. and Monroe, S. (1995). Corticofugal influence on taste responses in the nucleus of the solitary tract in the rat, *J. Neurophysiol.,* 74, 258-272.

Di Lorenzo, P.M. and Monroe, S. (1997). Transfer of information about taste from the nucleus of the solitary tract to the parabrachial nucleus of the pons, *Brain Res.,* 763, 167-181.

Di Lorenzo, P.M. and Schwartzbaum, J.S. (1982a). Coding of gustatory information in the pontine parabrachial nuclei of the rabbit: magnitude of neural response, *Brain Res.,* 251, 229-244.

Di Lorenzo, P.M. and Schwartzbaum, J.S. (1982b). Coding of gustatory information in the pontine parabrachial nuclei of the rabbit: temporal patterns of neural response, *Brain Res.,* 251, 245-257.

Erickson, R.P., Di Lorenzo, P.M., and Woodbury, M.A. (1994). Classification of taste responses in brain stem: membership in fuzzy sets, *J. Neurophysiol.,* 71, 2139-2150.

Erickson, R.P., Schiffman, S.S., Doetsch, G.S., Di Lorenzo, P.M., and Woodbury, M.A. (1995). A fuzzy set approach to the organization of the gustatory system, *Primary Sensory Neuron.,* 1, 65-80.

Gallo, M. and Bures, J. (1991). Acquisition of conditioned taste aversion in rats is mediated by ipsilateral interaction of cortical and mesencephalic mechanisms, *Neurosci. Lett.,* 133, 187-190.

Garcia, J., Lasiter, P.S., Bermudez-Rattoni, F., and Deems, D.A. (1985). A general theory of aversion learning, *Ann. N.Y. Acad. Sci.,* 443, 8-21.

Ghazanfar, A.A. and Nicolelis, M.A. (1999). Spatiotemporal properties of layer V neurons of the rat primary somatosensory cortex, *Cereb. Cortex,* 9, 348-361.

Grill, H.J. and Norgren, R. (1978a). The taste reactivity test. I. Mimetic responses to gustatory stimuli in neurologically normal rats, *Brain Res.,* 143, 263-279.

Grill, H.J. and Norgren, R. (1978b). The taste reactivity test. II. Mimetic responses to gustatory stimuli in chronic thalamic and chronic decerebrate rats, *Brain Res.,* 143, 281-297.

Hajnal, A., Takenouchi, K., and Norgren, R. (1999). Effect of intraduodenal lipid on parabrachial gustatory coding in awake rats, *J. Neurosci.,* 19, 7182-7190.

Halsell, C.B., Travers, J.B., and Travers, S.P. (1993). Gustatory and tactile stimulation of the posterior tongue activate overlapping but distinctive regions within the nucleus of the solitary tract, *Brain Res.,* 632, 161-173.

Hanamori, T., Kunitake, T., Kato, K., and Kannan, H. (1998a). Neurons in the posterior insular cortex are responsive to gustatory stimulation of the pharyngolarynx, baroreceptor and chemoreceptor stimulation, and tail pinch in rats, *Brain Res.,* 785, 97-106.

Hanamori, T., Kunitake, T., Kato, K., and Kannan, H. (1998b). Responses of neurons in the insular cortex to gustatory, visceral, and nociceptive stimuli in rats, *J. Neurophysiol.,* 79, 2535-2545.

Hermann, G.E. and Rogers, R.C. (1985). Convergence of vagal and gustatory afferent input within the parabrachial nucleus of the rat, *J. Auton. Nerv. Syst.,* 13, 1-17.

Kay, L.M. and Laurent, G. (1999). Odor- and context-dependent modulation of mitral cell activity in behaving rats, *Nat. Neurosci.,* 2, 1003-1009.

Kentros, C., Hargreaves, E., Hawkins, R.D., Kandel, E.R., Shapiro, M., and Muller, R.U. (1998). Abolition of long-term stability of new hippocampal place cell maps by NMDA receptor blockade, *Science,* 280, 2121-2126.

Kobashi, M. and Bradley, R.M. (1998). Effects of GABA on neurons of the gustatory and visceral zones of the parabrachial nucleus in rats, *Brain Res.,* 799, 323-328.

Kosar, E., Grill, H.J., and Norgren, R. (1986). Gustatory cortex in the rat. I. Physiological properties and cytoarchitecture, *Brain Res.,* 379, 329-341.

Kubie, J.L. (1984). A driveable bundle of microwires for collecting single-unit data from freely-moving rats, *Physiol. Behav.,* 32, 115-118.

McCaughey, S.A., Giza, B.K., Nolan, L.J., and Scott, T.R. (1997). Extinction of a conditioned taste aversion in rats. II. Neural effects in the nucleus of the solitary tract, *Physiol. Behav.,* 61, 373-379.

Morrison, G.R. (1969). Relative discriminability of sugars for the rat, *J. Comp. Physiol. Psychol.,* 68, 45-49.

Mundl, W.J. and Malmo, H.P. (1979). Capacitive sensor for lick-by-lick recording of drinking, *Physiol. Behav.,* 22, 781-784.

Nakamura, K. and Norgren, R. (1991). Gustatory responses of neurons in the nucleus of the solitary tract of behaving rats, *J. Neurophysiol.,* 66, 1232-1248.

Nakamura, K. and Norgren, R. (1993). Taste responses of neurons in the nucleus of the solitary tract of awake rats: an extended stimulus array, *J, Neurophysiol.,* 70, 879-891.

Nakamura, T. and Ogawa, H. (1997). Neural interaction between cortical taste neurons in rats: a cross-correlation analysis, *Chem. Senses,* 22, 517-528.

Nicolelis, M.A.L., Ghazanfar, A.A., Faggin, B.M., Votaw, S., and Oliveira, L.M.O. (1997). Reconstructing the engram: simultaneous, multisite, many single neuron recordings, *Neuron,* 18, 529-537.

Nishijo, H. and Norgren, R. (1990). Responses from parabrachial gustatory neurons in behaving rats, *J. Neurophysiol.,* 63, 707-724.

Nishijo, H. and Norgren, R. (1991). Parabrachial gustatory neural activity during licking by rats, *J. Neurophysiol.,* 66, 974-985.

Nishijo, H. and Norgren, R. (1997). Parabrachial neural coding of taste stimuli in awake rats, *J. Neurophysiol.,* 78, 2254-2268.

Nishijo, H., Uwano, T., Tamura, R., and Ono, T. (1998). Gustatory and multimodal neuronal responses in the amygdala during licking and discrimination of sensory stimuli in awake rats, *J. Neurophysiol.,* 79, 21-36.

Nomura, T. and Ogawa, H. (1985). The taste and mechanical response properties of neurons in the parvicellular part of the thalamic posteromedial ventral nucleus of the rat, *Neurosci. Res.,* 3, 91-105.

Ogawa, H., Hasegawa, K., and Murayama, N. (1992a). Difference in taste quality coding between two cortical taste areas, granular and dysgranular insular areas, in rats, *Exp. Brain Res.,* 91, 415-424.

Ogawa, H., Hasegawa, K., Otawa, S., and Ikeda, I. (1998). GABAergic inhibition and modifications of taste responses in the cortical taste area in rats, *Neurosci. Res.,* 32, 85-95.

Ogawa, H., Ito, S., Murayama, N., and Hasegawa, K. (1990). Taste area in granular and dysgranular insular cortices in the rat identified by stimulation of the entire oral cavity, *Neurosci. Res.,* 9, 196-201.

Ogawa, H., Murayama, N., and Hasegawa, K. (1992b). Difference in receptive field features of taste neurons in rat granular and dysgranular insular cortices, *Exp. Brain Res.,* 91, 408-414.

Ogawa, H. and Nomura, T. (1988). Receptive field properties of thalamo-cortical taste relay neurons in the parvicellular part of the posteromedial ventral nucleus in rats, *Exp. Brain Res.,* 73, 364-370.

Perrotto, R.S. and Scott, T.R. (1976). Gustatory neural coding in the pons, *Brain Res.,* 110, 283-300.

Phillips, M.I. and Norgren, R. (1970). A rapid method for permanent implantation of an intraoral fistula in rats, *Behav. Res. Methods Instrument,* 2, 124.

Ringach, D.L., Hawken, M.J., and Shapley, R. (1997). Dynamics of orientation tuning in macaque primary visual cortex, *Nature,* 387, 281-284.

Rosenblum, K., Berman, D.E., Hazvi, S., Lamprecht, R., and Dudai, Y. (1997). NMDA receptor and the tyrosine phosphorylation of its 2B subunit in taste learning in the rat insular cortex, *J. Neurosci.,* 17, 5129-5135.

Scott, T.R., Jr. and Erickson, R.P. (1971). Synaptic processing of taste-quality information in thalamus of the rat, *J. Neurophysiol.,* 34, 868-883.

Shimura, T., Tanaka, H., and Yamamoto, T. (1997). Salient responsiveness of parabrachial neurons to the conditioned stimulus after the acquisition of taste aversion learning in rats, *Neuroscience,* 81, 239-247.

Smith, D.V. and Frank, M.E. (1993). Sensory coding by peripheral taste fibers, in Simon, S.A. and Roper, S.D., Eds., *Mechanisms of Taste Transduction* Boca Raton, FL: CRC Press, p. 295-338.

Smith, D.V. and Li, C.S. (1998). Tonic GABAergic inhibition of taste-responsive neurons in the nucleus of the solitary tract, *Chem. Senses,* 23, 159-169.

Smith, D.V. and St John, S.J. (1999). Neural coding of gustatory information, *Curr. Opin. Neurobiol.,* 9, 427-435.

Spector, A.C. (1995). Gustatory function in the parabrachial nuclei: implications from lesion studies in rats, *Rev. Neurosci.,* 6, 143-175.

Spector, A.C., Breslin, P., and Grill, H.J. (1988). Taste reactivity as a dependent measure of the rapid formation of conditioned taste aversion: a tool for the neural analysis of taste-visceral associations, *Behav. Neurosci.,* 102, 942-952.

St. John, S.J. and Spector, A.C. (1998). Behavioral discrimination between quinine and KCl is dependent on input from the seventh cranial nerve: implications for the functional roles of the gustatory nerves in rats, *J. Neurosci.,* 18, 4353-4362.

Todrank, J. and Bartoshuk, L.M. (1991). A taste illusion: taste sensation localized by touch, *Physiol. Behav.,* 50, 1027-1031.

Van Buskirk, R.L. and Erickson, R.P. (1977). Odorant responses in taste neurons of the rat NTS, *Brain Res.,* 135, 287-303.

Verhagen, J.V., Giza, B.K., and Scott, T.R. (1999). Taste in the rat thalamus, *Soc. Neurosci. Abstacts,* 25, 2185.

Washburn, M.S. and Moises, H.C. (1992). Electrophysiological and morphological properties of rat basolateral amygdaloid neurons *in vitro, J. Neurosci.,* 12, 4066-4079.

Wehr, M. and Laurent, G. (1996). Odour encoding by temporal sequences of firing in oscillating neural assemblies, *Nature,* 384, 162-166.

Welsh, J.P., Lang, E.J., Sugihara, I., and Llinás, R. (1995). Dynamic organization of motor control within the olivocerebellar system, *Nature,* 374, 453-457.

Yajeya, J., de la Fuente Juan, A., Merchan, M.A., Riolobos, A.S., Heredia, M., and Criado, J.M. (1997). Cholinergic responses of morphologically and electrophysiologically characterized neurons of the basolateral complex in rat amygdala slices, *Neuroscience,* 78, 731-743.

Yamamoto, T., Matsuo, R., and Kawamura, Y. (1980). Corticofugal effects on the activity of thalamic taste cells, *Brain Res.,* 193, 258-262.

Yamamoto, T., Matsuo, R., Kiyomitsu, Y., and Kitamura, R. (1988). Sensory inputs from the oral region to the cerebral cortex in behaving rats: an analysis of unit responses in cortical somatosensory and taste areas during ingestive behavior, *J. Neurophysiol.,* 60, 1303-1321.

Yamamoto, T., Matsuo, R., Kiyomitsu, Y., and Kitamura, R. (1989). Taste responses of cortical neurons in freely ingesting rats, *J. Neurophysiol.,* 61, 1244-1258.

Yamamoto, T., Shimura, T., Sako, N., Yasoshima, Y., and Sakai, N. (1994). Neural substrates for conditioned taste aversion in the rat, *Behav. Brain Res.,* 65, 123-137.

Yamamoto, T., Yuyama, N., Kato, T., and Kawamura, Y. (1985). Gustatory responses of cortical neurons in rats. III. Neural and behavioral measures compared, *J. Neurophysiol.,* 53, 1370-1386.

Yasoshima, Y., Shimura, T., and Yamamoto, T. (1995). Single unit responses of the amygdala after conditioned taste aversion in conscious rats, *Neuroreport,* 6, 2424-2428.

Yasoshima, Y. and Yamamoto, T. (1998). Short-term and long-term excitability changes of the insular cortical neurons after the acquisition of taste aversion learning in behaving rats, *Neuroscience,* 84, 1-5.

Yokota, T., Eguchi, K., and Satoh, T. (1996). Correlated discharges of two neurons in rat gustatory cortex during gustatory stimulation, *Neurosci. Lett.,* 209, 204-206.

Yokota, T., Eguchi, K., and Satoh, T. (1997). Sensitivity of rat cortical neurons in distinguishing taste qualities by individual and correlative activities, *Chem. Senses,* 22, 363-373.

15 Recording from Single Neurons in the Primary Taste Cortex of the Alert Macaque

Thomas R. Scott and Carlos R. Plata-Salamán

CONTENTS

15.1 INTRODUCTION

Receptor processes in gustation can be investigated effectively in non-mammalian and rodent models. However, as interest shifts to higher-order relays of the central nervous system, it is increasingly important to the relationship among neurophysiology, psychophysics, and clinical practice that neural activity be assessed in animals that share sensory values closely with humans. This was the primary impetus for

0-8493-2329-0/02/$0.00+$1.50

the development of a primate model of gustation. The value of this model is further enhanced by the animal's being alert. This offers several advantages, among which are the following.

- The depressing effects of anesthetic on cortical neurons is spared, along with the hyperglycemia that accompanies anesthetic administration,[33] both of which have been shown to affect taste responses in the cortex and ventral forebrain.[1,19,20,34]
- The subject participates actively in tasting stimuli. Therefore, flow rates, stimulus volume, and area of stimulation are maintained in the physiological range. Moreover, the normal interaction between stimulus molecules and receptor sites is permitted by the active manipulation of solution in the mouth.
- Neural and behavioral experiments can be conducted simultaneously. The monkey's behavioral responses may offer a direct measure of the hedonic value of a tastant or of his level of satiety while the activity of the neurons that may mediate these processes is monitored.
- The number of subjects is minimized. Chronic recording from alert animals permits the use of one monkey in daily sessions for perhaps a year, offering an enormous database. In a typical preparation we record from 2000 to 3000 cells, of which 150 to 200 are gustatory.
- The rate of data collection is high. Two major impediments to rapid data collection — performance of surgical procedures and familiarization with the idiosyncracies of each animal's anatomy—are obviated in the chronic, alert preparation. Data collection begins minutes after the monkey enters the recording chair, and the familiarity between experimenter and subject tends to sustain the momentum of the previous session.

The approach described below was adapted from those developed in the laboratories of Edmund Rolls and of Yutaka Oomura for investigating the neural control of feeding. It has been applied to gustation at the level of the nucleus of the solitary tract,[28,35] the primary taste cortex in the frontal operculum[29] and anterior insula,[14-18,22,25-27,32,36] the secondary taste cortex in caudolateral orbitofrontal cortex,[21] the amygdala,[11,23,34] the hypothalamus[1,9,10,12] and the basal ganglia.[7]

15.2 METHODS

15.2.1 Subjects

Our subjects are male cynomolgus or rhesus macaques typically weighing 4 to 7 kg. Their cranial implants require that they be housed individually, but we offer them diversions in the form of toys and videotapes. They have pellet food and tap water available *ad lib.* Several weeks before surgery, we begin adapting the macaque to his primate chair in which we offer him fruits and fruit juice. When he enters the chair voluntarily and appears comfortable in it, we deem him prepared for surgery.

15.2.2 Surgery

15.2.2.1 Implant Construction

We implant a fixed pair of sterile insect pins, insulated except at their tips, into the brain at a site remote from the gustatory regions of interest. These serve as an electrical reference for differential recording and as a constant feature against which to compare the location of each recording track through X-ray photography. The reference guides are constructed of 00 gauge, 7.5-cm stainless steel insect pins coated with Insulex. We fix them with cyanoacrylate and dental acrylic to the stainless steel recording chamber described below.

15.2.2.2 Surgical Procedure

We observe full sterile precautions throughout surgery. We sedate each macaque with ketamine HCl (10 mg/kg, intramuscularly [i.m.]) and anesthetize him to a surgical level with sodium pentobarbital (30 to 40 mg/kg, i.m.). We monitor depth of anesthesia by testing frequently for the presence of flexion reflexes and active muscle tonus and administer supplemental doses of pentobarbital as indicated. We also administer atropine (0.1 mg/kg, i.m.) to reduce mucus secretion and salivation, and apply glycerine ophthalmically to maintain the moistness of the eyes. We monitor heart rate and respiration throughout surgery, and maintain core temperature between 35 and 37°C.

We secure the macaque in a Kopf stereotaxic instrument. We lower a 30-mm (outside diameter) stainless steel ring with its reference pins to the skull surface, drilling two 2-mm diameter holes for admission of the insect pins. We center the ring 5 mm lateral to the midline and at the probable anteroposterior level of the insular-opercular cortex as predicted from preliminary X-rays and from magnetic resonance imaging performed during the acclimation period. We originally centered the ring at the midline to afford access to both hemispheres. However, this created the requirement that the electrode be angled quite sharply to reach the insular-opercular cortex, which is 16 to 19 mm lateral to the midline in a 5-kg macaque.[25] By centering the ring 5 mm lateral to the midline and committing our search to one hemisphere, we can comfortably reach the most lateral extent of the operculum using an angle of 12°. This approach brings the electrode down approximately parallel to the long axis of the taste cortex, maximizing the number of taste cells encountered. Neuroimaging data from human subjects initially suggested that, at the level of primary taste cortex, gustation is more heavily represented in the non-dominant hemisphere.[30,31] Thus, to determine which hemisphere will become our focus in each monkey, we offer him 100 raisins from a variety of positions, and note which hand he prefers to use in reaching for them. We then implant the ring on that side of the skull. It should be noted that the growing literature on human neuroimaging of taste-induced activity[2,5] as well as a clinical report[6] have introduced contradictory findings on laterality.

We typically drill ten holes in the exposed skull, five for skull screws and five for I-shaped pieces that are inserted through slots and rotated 90° to provide a secure grip beneath the skull. We attach the stainless steel ring to the skull through a matrix of dental acrylic that flows under and around these ten anchors. To ensure full support and stability for the macaque's head during recording sessions, we fix two stainless steel tubes to the acrylic skull cap, one in front of and one behind the ring. These are 12-mm o.d., 7-mm i.d., and 67-mm long. They must be small enough to be incorporated into the skull cap, yet with a large enough inner diameter that the horizontal support bars inserted through them during recording remain rigid as the macaque shifts his weight. The remainder of the skull is left intact except for a 2-mm diameter hole drilled at a promising position over the primary taste cortex. It is through this hole that the search for gustatory neurons will begin, and it will be expanded cautiously as the exploration widens. We apply 0.2% nitrofurazone topically and 100,000 U/kg penicillin G benzathine and G procaine i.m. for several days as a prophylaxis against post-operative infection.

15.2.2.3 Management of the Dura

The connective tissue above the dura begins to thicken immediately after surgery. The tissue is loosely organized and imbued with cerebrospinal fluid at first; but after repeated penetrations by the guide tube and electrode, it may become dense and impenetrable. To avoid this, we treat the recording chamber daily with sterile saline and nitrofurazone, and each 3 to 4 weeks with 3% hydrogen peroxide solution. By keeping the chamber clean and preventing infection at the wound margins, the dura remains soft and penetrable for months. If penetration does become an obstacle, we surgically reduce the dura to its original layer under local anesthesia.

15.2.2.4 Maintenance of Bone Integrity

The durability of the implanted ring is a function of the integrity of the skull. It is worthwhile to conduct a presurgical X-ray survey of the skull to plan the location of the ten anchors so as to take advantage of the thickest bone and to distribute the physical stress as evenly as possible. During surgery, we disturb the periosteum as little as possible and cool the drill burr with a constant stream of sterile saline. Following surgery, we maintain the implants and wound margins carefully to resist the infection that leads to bone necrosis.

15.2.3 Electrophysiology

15.2.3.1 Recording Sessions

Following at least 10 days of recovery from surgery, we transfer the macaque to the primate chair with which he has been familiarized and insert the horizontal support bars that hold his head in place. He is otherwise free to move, and typically sits in a relaxed manner. When he has settled into a comfortable position, we make suitable adjustments to the chair to fit it optimally to his size and shape, then dedicate the tailored chair to that monkey for the duration of his tenure. The successful conduct of these studies depends on a cooperative relationship between experimenter and subject. Therefore, we attend to the macaque's comfort and desires on a regular basis, including offering him fruits and nuts on occasion. The duration of a recording session is normally 6 hr, although we shorten it if the monkey becomes restless, and may lengthen it by up to 50% if the recording track is productive and the macaque shows no signs of distress.

15.2.3.2 Electrodes

We have experimented with a variety of electrodes and settled on glass-coated tungsten as offering the best combination of sturdiness necessary to penetrate 25 mm of tissue and delicacy needed to isolate the activity of 15-μm somas in gustatory cortex. Another virtue of these electrodes is that the damage they cause with each track is minimal, which permits chronic recording from a circumscribed area over a period of months. The electron-dense tungsten also appears prominently in the X-ray photographs taken at the end of each track. We construct the electrodes in three stages: coarse etching of the tungsten to create a pencil-shaped tip, insulation by melting and drawing glass over the tip of the tungsten, and fine etching under microscopic control to expose a recording tip of about 2 × 2 μm.

15.2.3.3 Positioning and Penetration

We attach a Kopf x-y positioner to the implanted stainless steel ring and set it to the desired coordinates for each track. We anesthetize the dura with 0.1 ml xylocaine and then use a sterile 21-gauge stainless steel guide tube to penetrate it while monitoring the monkey's reaction to ensure he is not experiencing discomfort. The sterile microelectrode is protected within the guide tube. Once within the parenchyma, the electrode is lowered manually to within 5 mm of the insular-opercular cortex and advanced using a Trent-Wells hydraulic microdrive and chronic adapter system.

15.2.3.4 Recording

We use standard electrophysiological recording techniques for differential amplification, display, and recording of the neural signal. We attach the indifferent lead either to ground or to the deep, fixed insect pins, whichever offers the better signal-to-noise ratio for that track.

15.2.4 Stimulus Delivery

One compromise that must be accepted in working with an alert animal is that he has some control over stimulus flow within the mouth. This is true whether the stimulus is delivered directly to the tongue through a syringe or by way of an intraoral cannula. The macaque's reaction to quinine will, in either case, prevent that solution from contacting the same receptor surface that his response to sucrose will permit. Investing some degree of control in the subject has the advantage that the dynamic interaction between stimulus molecules and receptor sites approximates the normal condition. It has the disadvantage that the independent variable—the administration of a tastant to the gustatory receptor surface—may not be rigidly replicated from one application to the next.

Two methods of stimulus delivery are used, depending on the actions of the monkey and the requirements of the experiment. Normally, with a strong positive relationship between experimenter and subject, the two enter into a cooperative cadence. With the approach of the syringe and the word ready, the monkey opens his mouth and may extend his tongue. We place 1 ml of solution on the center of the tongue. We are careful to maintain a high ratio of appetitive to aversive tastes and are vigilant to indications of reluctance by the macaque. Under these circumstances, we are forgiven the occasional application of $MgCl_2$, tannic acid, or quinine. We assess the reliability of our stimulation and recording procedures by reapplying the stimulus series and comparing the response of each neuron elicited by the first and the second application of every stimulus. For various datasets, the correlation between the two columns of discharge rates thus generated ranges from +0.82[22] to +0.96.[16] This high level of agreement offers assurance that, in addition to being simple, humane and interactive with the monkey, manual delivery is also reliable.

When we encounter the occasional reluctant animal, we implant an intraoral cannula for stimulus delivery.[8] We draw a length of PE60 tubing from the margin of the skull cap subcutaneously to exit through the gum just behind the first molar. We place a Teflon washer over the distal end of the tube, which we then heat-flare to secure the washer's position. We retract the tube so that the washer rests snugly against the upper gum. We fit the central end of the cannula with a syringe needle and incorporate it into the headpiece with dental acrylic leaving only a short length of the needle exposed. It is by this access that we inject solutions into the mouth. Finally, we impregnate surgical swabs with antibiotic and place them on the skin along the course of the cannula, from headpiece to cheek. This serves as a matrix along which we build an extension of the headpiece in dental acrylic to protect the cannula. This system allows us to deliver tastants to the same location regardless of the monkey's willingness to accept them, although it still leaves to him the distribution of stimuli over the receptor surface. Disadvantages of the intraoral cannula are the requirement of additional surgery, the danger of infection, and the possible introduction of spurious tastes by bacterial growth in the cannula or by reflux of previously administered solutions back into the cannula.

Following each stimulus application, we wash 5 ml distilled water into the mouth and allow 30 to 120 sec of rest, depending on the intensity of the preceding tastant. If there are either electrophysiological (altered discharge levels) or behavioral (licking, chewing, facial expressions) indications that baseline has not been reestablished, we extend the rinse and rest periods.

With either method of stimulus delivery, we mark the moment of stimulus application. We attach a wire, shielded except at its tip, to the syringe needle through which stimuli are delivered. We can mark stimulus onset with precision through a circuit that is completed when the solution contacts the mouth.[3] Non-ionic stimuli (sugars) should be mixed in 5% tap water to provide the conductance necessary to activate the system. The circuit contains sufficient impedance to restrict current levels to two orders of magnitude below the threshold for electric taste.[13] The output of the high-impedance integrated circuit is a square wave that is processed by additional circuitry to offer continuous audiovisual event indicators during stimulus delivery.

Because some degree of multimodality characterizes cortical taste neurons, we assess the responses of each cell to olfactory, visual, and tactile stimulation and to motor activity. For olfaction, we saturate a cotton-tipped applicator with each stimulus and present it under the monkey's nose, out of his view. For vision, we expose the stimulus syringe to the monkey at a distance of 60 cm for 3 sec, and then bring the syringe toward his mouth over another 3 sec. These events will often elicit activity from cortical taste cells. To test for tactile sensitivity, we stroke the teeth, gums, gingivae, palate, and lips with a cotton-tipped applicator. Spontaneous actions, such as chewing, gaping, or extending the tongue, are noted and correlated with neural activity. However, we cannot determine with certainty whether the responses that accompany such actions are related to their tactile, proprioceptive, or motor components.

15.2.5 Data Analysis

We begin counting action potentials 3 sec before stimulus application to gain a measure of spontaneous rate, and continue for 5 sec following stimulus onset. We subtract the mean spontaneous rate from the gross evoked rate to determine the net evoked response to each stimulus. This is the basic unit of information derived from each stimulus application. From the net rate, we calculate the profile of activity generated across neurons by each stimulus —originally termed the across-fiber pattern because it was derived from the responses of chorda tympani axons —as well as the profile given by each neuron across the stimulus array as a prelude to classifying cells. Both pairs of profiles serve as the bases for calculating matrices of Pearson product-moment correlation coefficients, the former among stimuli, the latter among neurons. Each matrix of correlations then serves as the basis for generating a cluster analysis and a multidimensional space through which the degree of similarity among stimulus qualities and discrete categories of neurons, respectively, can be revealed. The responses of a single cortical taste cell to a range of stimulus qualities and concentrations are shown in Figure 15.1.

It is evident that the time-course of neural activity also carries information relevant to the quality code.[4,24] Therefore, we divide the 5-sec response period into fifty 100-msec bins and generate poststimulus time histograms that characterize the time course of activity evoked by each stimulus. We treat these temporal profiles in the same manner as the across-fiber patterns by computing a correlation matrix of similarity and using that to generate a cluster analysis and multidimensional space of temporal similarity.

15.2.6 Localization of Recording Sites

In addition to ceding some level of control over stimulus flow, there is one other compromise that must be accepted in working with a chronic, alert animal: recording locations cannot be identified as precisely as in the acute preparation. Daily assessments are approximations, and by the time histology is performed, the tissue may have undergone changes and will have been subjected to the physical effects of multiple electrode tracks.

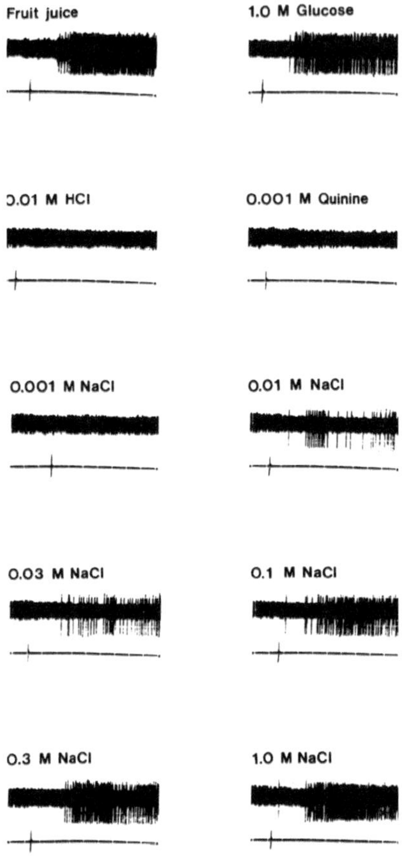

FIGURE 15.1 Extracellular action potentials evoked from a single cortical taste neuron. Stimuli include fruit juice, representatives of the basic taste qualities sweet (glucose), sour (HCl) and bitter (quinine), and a concentration series of NaCl extending from 0.001 *M* to 1.0 *M*. Stimulus onset is marked by the pulse in the lower trace. The entire sweep represents 2 sec of activity. The cell discharged with a latency of about 420 msec.

We determine the position of the recording sites in two ways. First, we take X-ray photographs from frontal and lateral perspectives following each track. We can then infer recording locations to within about 250 μm by reference to the fixed insect pins and to skull landmarks. Second, during the final recording session, we make marking lesions (100 μA for 20 sec, electrode negative). We administer a lethal dose of sodium pentobarbital and perfuse the macaque transcardially with physiological saline followed by a 10% formalin-saline solution. We remove the brain and place it in 10% formalin-30% sucrose solution until it sinks, after which we cut 50-μm transverse serial frozen sections and stain them with Neutral Red or Cresyl Violet. A transverse section taken at the anteroposterior level of sphenoid is shown in Figure 15.2. Also shown are the presumed locations of 17 taste cells identified according to the basic stimulus that evoked that cell's maximum response.

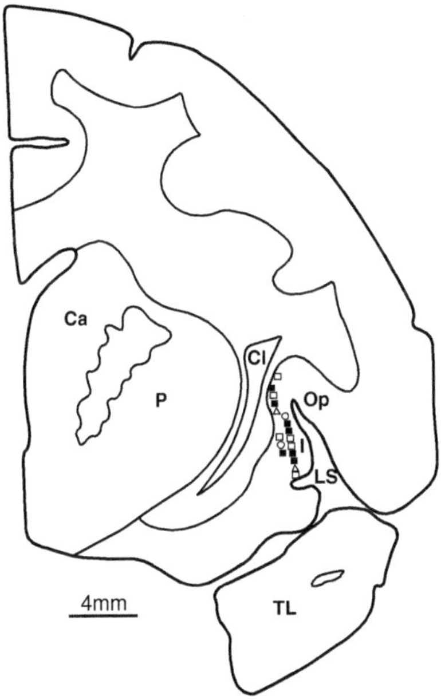

FIGURE 15.2 Tracing of a coronal section through the left hemisphere of a macaque at the anteroposterior level of the anterior clinoid process of the sphenoid bone (0.0 AP). Taste-responsive neurons encountered during three successful tracks through this section are labeled according to the basic stimulus that elicited the maximum response. Glucose-best neurons (■); NaCl-best neurons (□); HCl-best neurons (Δ); quinine-best neurons (○). Abbreviations: Ca, caudate nucleus; Cl, claustrum; I, anterior insula; LS, lateral sulcus; Op, frontal operculum; P, putamen; TL, temporal lobe.

15.2.7 Time Considerations

Our typical involvement with a macaque extends through some 18 months, 12 of which are spent gathering electrophysiological data. The anticipated time line is:

- Purchase, performance of blood tests, quarantine . . . . 2–3 months
- Adaptation to monkey chair, familiarization of monkey with experimenter, MRI, initial X-rays, preparation for surgery . . . . 2 months
- Surgery and recovery. . . . . 2 weeks
- Recording period (extended if implant is stable and monkey healthy). . . . . 12 months
- Perfusion and histology . . . . 2 weeks

REFERENCES

1. Burton, M., Rolls, E., and Mora, F., Effects of hunger on the response of neurons in the lateral hypothalamus to the sight and taste of food, *Exp. Neurol.*, 51, 668, 1976.
2. Cerf, B., Lebihan, D., Van de Mortele, P.F., and Faurion, A., Functional lateralization of human gustatory cortex related to handedness disclosed by fMRI study, *Ann. N.Y. Acad. Sci.*, 855, 575, 1998.
3. Chang, F.-C.T. and Scott, T.R., A technique for gustatory stimulus delivery in the rodent, *Chem. Senses*, 9, 91, 1984.
4. DiLorenzo, P.M. and Schwartzbaum, J.S., Coding of gustatory information in the pontine parabrachial nuclei of the rabbit: temporal patterns of neural response, *Brain Res.*, 251, 245, 1982.
5. Faurion, A., Cerf, B., Le Bihan, D., and Pillias, A.M., fMRI study of taste cortical areas in humans, *Ann. N.Y. Acad. Sci.*, 855, 535, 1998.
6. Fujikane, M., Itoh, M., Nakazawa, M., Yamaguchi, Y., Hirata, K., and Tsudo, N., Cerebral infarction accompanied by dysgeusia — A clinical study on the gustatory pathway in the CNS, *Rinsho Shinkeigaku*, 39, 771, 1999.
7. Karádi, Z., Faludi, B., Lénárd, L., Czurko, A., Niedetsky, C., Vida, I., and Nishino, H., Glucose-sensitive neurons in the globus-pallidus. II. Complex functional attributes, *Brain Res. Bull.*, 37, 157, 1995.
8. Karádi, Z., Nishino, H., Oomura, Y., Scott, T.R., and Aou, S., A method for gustatory stimulus delivery in awake rhesus monkeys, *Brain Res. Bull.*, 25, 665, 1990.
9. Karádi, Z., Oomura, Y., Nishino, H., and Scott, T.R., Characteristics of lateral hypothalamic neurons in the awake monkey, *Ann. Rev. NIPS,* 9, 256, 1988.
10. Karádi, Z., Oomura, Y., Nishino, H., Scott, T.R., Lénárd, L., and Aou, S., Responses of lateral hypothalamic glucose-sensitive and glucose-insensitive neurons to chemical stimuli in behaving rhesus monkeys, *J. Neurophysiol.*, 67, 389, 1992.
11. Karádi, Z., Scott, T.R., Oomura, Y., Nishino, H., Aou, S., and Lénárd, L., Complex functional attributes of amygdaloid gustatory neurons in the rhesus monkey, in *Olfaction and Taste, Vol. XII,* Murphy, C., Ed., N.Y. Acad. Sci., New York, 855, 488, 1988.
12. Oomura, Y., Nishino, H., Karádi, Z., Aou, S., and Scott, T.R., Taste and olfactory modification of feeding related neurons in behaving monkey, *Physiol. Behav.*, 49, 943, 1991.
13. Pfaffmann, C. and Pritchard, T.C., Ion specificity of electric taste, *Olfaction and Taste, Vol. VII,* Van der Starre, H., Ed., IRL Press, London, 1980, 175.
14. Plata-Salamán, C.R. and Scott, T.R., Taste neurons in the cortex of the alert cynomolgus monkey, *Brain Res. Bull.*, 28, 333, 1992.
15. Plata-Salamán, C.R., Scott, T.R., and Smith-Swintosky, V.L., Gustatory neural coding in the monkey cortex: L-amino acids, *J. Neurophysiol.*, 67, 1552, 1992.
16. Plata-Salamán, C.R., Scott, T.R., and Smith-Swintosky, V.L. Gustatory neural coding in the monkey cortex: the quality of sweetness, *J. Neurophysiol.*, 60, 482, 1993.
17. Plata-Salamán, C.R., Scott, T.R., and Smith-Swintosky, V.L., Gustatory neural coding in the monkey cortex: acid stimuli, *J. Neurophysiol.*, 74, 556, 1995.
18. Plata-Salamán, C.R., Smith-Swintosky, V.L., and Scott, T.R., Gustatory neural coding in the monkey cortex: mixtures, *J. Neurophysiol.*, 75, 2369, 1996.
19. Rolls, E.T., The orbitofrontal cortex, *Phil. Trans. R. Soc. Lond.,* 351, 1433, 1996.
20. Rolls, E.T., Sienkiewicz, Z.J., and Yaxley, S., Hunger modulates the responses to gustatory stimuli of single neurons in the caudolateral orbitofrontal cortex of the macaque monkey, *Eur. J. Neurosci.*, 1, 53, 1989.

21. Rolls, E.T., Yaxley, S., and Sienkiewicz, Z.J., Gustatory responses of single neurons in the orbitofrontal cortex of the macaque monkey, *J. Neurophysiol.,* 64, 1055, 1990.
22. Scott, T.R., Giza, B.K., and Yan, J., Gustatory neural coding in the cortex of the alert cynomolgus macaque: the quality of bitterness, *J. Neurophysiol.*, 81, 60, 1999.
23. Scott, T.R., Karádi, Z., Oomura, Y., Nishino, H., Plata-Salamán, C.R., Lénárd, L., Giza, B.J., and Aou, S., Gustatory neural coding in the amygdala of the alert macaque monkey, *J. Neurophysiol.*, 69, 1810, 1993.
24. Scott, T.R. and Mark, G.P., The taste system encodes stimulus toxicity, *Brain Res.*, 414, 197, 1987.
25. Scott, T.R. and Plata-Salamán, C.R., Taste in the monkey cortex, *Physiol. Behav.*, 67, 489, 1999.
26. Scott, T.R., Plata-Salamán, C.R., Smith-Swintosky, V.L., and Giza, B.K., Gustatory neural coding in the monkey cortex: stimulus intensity, *J. Neurophysiol.*, 65, 76, 1991.
27. Scott, T.R., Plata-Salamán, C.R., and Smith-Swintosky, V.L. Gustatory neural coding in the monkey cortex: the quality of saltiness, *J. Neurophysiol.*, 71, 1692, 1994.
28. Scott, T.R., Yaxley, S., Sienkiewicz, Z.J., and Rolls, E.T., Gustatory responses in the nucleus tractus solitarius of the alert cynomolgus monkey, *J. Neurophysiol.*, 55, 182, 1986.
29. Scott, T.R., Yaxley, S., Sienkiewicz, Z.J., and Rolls, E.T., Gustatory responses in the frontal opercular cortex of the alert cynomolgus monkey, *J. Neurophysiol.*, 56, 876, 1986.
30. Small, D.M., Jones-Gotman, M., Zatorre, R.J., Petrides, M., and Evans, A.C., A role for the right anterior temporal lobe in taste quality recognition, *J. Neurosci.*, 17, 5136, 1997.
31. Small, D.M., Zald, D.H., Jones-Gotman, M., Zatorre, R.J., Pardo, J.V., Frey, S., and Petrides, M., Human cortical gustatory areas: a review of functional neuroimaging data, *Neuroreport*, 10, 7, 1999.
32. Smith-Swintosky, V.L., Plata-Salamán, C.R., and Scott, T.R., Gustatory neural coding in the monkey cortex: stimulus quality, *J. Neurophysiol.*, 66, 1156, 1991.
33. Taylor, M.H. and Howard, E., Impaired glucose homeostasis in adult rats after corticosterone treatment in infancy, *Endocrinology*, 88, 1190, 1971.
34. Yan, J. and Scott, T.R., The effect of satiety on taste cells in the amygdala of the alert macaque, *Brain Res.*, 740, 193, 1996.
35. Yaxley, S., Rolls, E.T., Sienkiewicz, Z.J., and Scott, T.R., Satiety does not affect gustatory activity in the nucleus of the solitary tract of the alert monkey, *Brain Res.*, 347, 85, 1985.
36. Yaxley, S., Rolls, E.T., and Sienkiewicz, Z.J., Gustatory responses of single neurons in the insula of the macaque monkey, *J. Neurophysiol.*, 63, 689, 1990.

16 Olfactory Learning and the Neurophysiological Study of Rat Prefrontal Function

Geoffrey Schoenbaum

CONTENTS

16.1 INTRODUCTION

The rat olfactory system comprises a set of structures well-suited to the study of learning and memory. This chapter discusses the study of olfactory processing in the rat and the role it plays in learning, memory, and behavior. In particular, this chapter describes the privileged role olfaction appears to hold for rats, briefly reviews the system of structures involved in processing olfactory information, and then focuses on the use of olfaction within a subset of these structures as a tool for the study of learning and behavior. This discussion emphasizes the use of olfaction to

0-8493-2329-0/02/$0.00+$1.50

study prefrontal function in rats, stressing parallels with prefrontal function in other species, and in the process describing in detail the methods used in several experiments.

16.2 RAT OLFACTORY PROCESSING: IS IT SPECIAL?

It has been argued that odor-based information plays a special role in rat cognition.[1,2] Unlike primates, which rely heavily on their highly developed visual processing system, rats utilize olfactory cues to a much greater extent. The importance of olfactory processing is evident in the very low thresholds at which rats can detect odors[3] and their preferential attention to odors in learning tasks.[4] When odor cues are used in discrimination training, rats develop a learning set that is evident in facilitated learning of new discrimination problems[9] and reversals.[4] This facilitation is less readily apparent when other modalities are used.[4,10] Rats are also able to perform a variety of odor-based tasks that are analogs of those used in primate research.[4-9] For example, rats are proficient at delayed non-match to sample,[6] paired-associate testing,[7] and even tests of transitive inference[8] when odor cues are substituted for visual cues.

Kolb has suggested that common functions can be examined in different species provided the appropriate species-specific behaviors are assessed.[11] In rats, species-specific behaviors may utilize olfactory information preferentially. Indeed, olfactory-based learning has been used as a tool to assess hippocampal function in rats. This strategy has been productive, confirming and extending many observations regarding hippocampal function made in primate species.[12] This chapter describes attempts to employ a similar strategy to investigate prefrontal function in rats.

16.3 THE RAT OLFACTORY SYSTEM: ANATOMY AND STAGES OF PROCESSING

Olfactory input progresses through several stages of processing. Unlike most other sensory input to the brain, olfactory information does not reach cortical regions through a thalamic relay. Instead, input from chemoreceptors in the nasal mucosa is processed in the olfactory bulb and from there is sent primarily to the piriform cortex.[13,14] Although the primary role of these early structures is likely to be odor identification, it is clear from both recording and behavioral studies that these regions play a larger role. Immediately within the piriform cortex and indeed even within the olfactory bulb, descending influences begin to integrate the olfactory input with information from other modalities.[15,16]

After these early stages of processing, olfactory information is relayed to a number of structures in which processing linked to several different brain functions may occur. For example, the piriform cortex sends projections to the medial temporal regions[17,18] and amygdala[20] as well as the mediodorsal thalamus.[19] In these structures, olfactory information is integrated with input from other modalities. Many of these structures, including the piriform cortex, send afferents to the orbitofrontal cortex (OFC), a prefrontal area in the rat and primate. Through these connections, networks in the OFC gain access to many different types of information, but chief among these appears to be information regarding the olfactory cues in the environment.

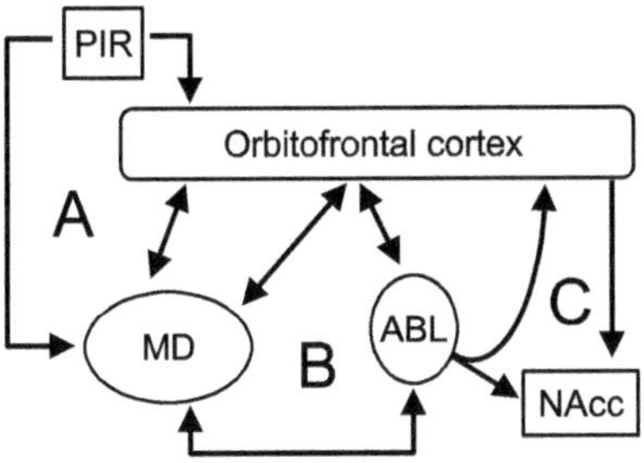

FIGURE 16.1 Schematic representation of three triangular circuits involving rat OFC. OFC receives input from piriform cortex (A), amygdala (B), and sends output to nucleus accumbens (C). Input from the piriform cortex and amygdala comes directly and also after processing in the mediodorsal thalamus (MD). These connections are reciprocated, allowing descending input from the OFC to modify processing in these areas. In addition, the OFC can influence the MD via output to the nucleus accumbens (NAcc). Note that NAcc also receives input from the same cells in ABL that project to the OFC. Many of these connections are similar in rat and primate, although olfactory connections appear to receive greater proportional representation in rat OFC than in primate. These connections suggest that the OFC may be preferentially involved in the processing of olfactory information in the rat and information regarding the motivational value of cues across species.

16.4 MODELING PREFRONTAL FUNCTION IN RAT ORBITOFRONTAL CORTEX

As in primate,[21] the rat prefrontal cortex is composed of subdivisions that appear to be differentiated, in part, by their pattern of connectivity.[22] The orbitofrontal subdivision of the rat prefrontal cortex is situated above the rhinal sulcus and anterior to the genu of the corpus callosum. This region includes the medial, ventrolateral, and lateral orbital regions and, depending on the connectivity one wishes to include, also the ventral and dorsal divisions of the agranular insular cortex. The OFC may also be extended into the posterior agranular region; however, this includes regions devoted to visceral and taste input (see Chapter 14 of this book) that may differ functionally from the more anterior areas. For that reason, this discussion is restricted to the olfactory and associated areas. These prefrontal areas are notable for their common involvement in olfactory processing and for their reciprocal connections with other areas similarly involved.

Several anatomical relationships likely to have functional significance are apparent (see Figure 16.1). These three circuits, as well as many others involving the OFC, are evident both in the rat and in primate species (see References 23, 31, and 32 for a discussion of primates). For this reason, the suggestion has been made that the OFC may represent the most likely rat homolog of primate prefrontal cortex.[23,24]

These three circuits implicate the OFC in the processing of olfactory and motivational information. As outlined above, the OFC receives olfactory input both directly from the piriform cortex (PIR) and also indirectly through projections from the mediodorsal thalamus (MD). The direct and indirect pathways converge on the same regions within the OFC; however, the neurons from which the two pathways originate in the PIR differ in their morphology.[25] Cells that send output to the MD are large multipolar neurons, whereas those that project directly to the OFC are much smaller and more superficially located. The significance of this dichotomy is unclear, but it suggests that the two pathways carry different information regarding olfactory processing. In addition, projections between the OFC and these structures are reciprocal,[11,19] allowing top-down modulation of processing in these two structures. A similar relationship is present between the OFC, MD, and amygdala. Neurons in the basolateral amygdala (ABL) project heavily to the dorsal agranular insular cortex,[24,26] and the ABL also projects to the MD,[19] which sends afferents back to the dorsal agranular insular region.[19] Again, neurons in the ABL that project indirectly to the OFC through the MD appear to be larger and more integrative than those that project directly to the OFC,[25] and the direct projections to the OFC also are reciprocated.[11,19] Finally, there exists a triangular relationship between the OFC, ABL, and nucleus accumbens. Neurons in the ABL project both to the accumbens and the OFC, and the OFC sends output to the same neurons in the accumbens that receive the ABL collaterals.[20,27,28] The accumbens is proposed to integrate input regarding cognitive and emotion-related information via these projections[29] in the selection of action repertoires.[30] This is one of the few projections from OFC that is not reciprocated, suggesting it may play a special role in the output of this region.

Subdivisions of the prefrontal cortex are proposed to perform somewhat similar functions within specific informational domains defined by their connections.[33] The connectivity of the rat orbitofrontal cortex described above suggests that motivational and olfactory information are crucial to the role of this region, and behavioral findings support this notion. OFC lesions in primates and rats result in an inability to correctly use the motivational or incentive value of cues to guide behavior.[34-37] Similar impairments are also seen following lesions of the amygdala,[35,38-41] suggesting that the OFC and ABL may form a system for the manipulation of incentive-based information similar to other prefrontal circuits. Parallel studies in rats[42,43] and more recently in primates[44] show that lesions of the OFC/ABL system prevent rats from decreasingly responding to a conditioned stimulus when the associated reinforcer is "devalued" by pairing with illness or over-feeding (see Box 16.1 — *Representation of Incentive Value in OFC and ABL* — at end of the chapter).

Olfaction also plays a significant role in the orbital region in both rats and primates. In rats, in particular, olfactory information is prominent in the pattern of connectivity, as outlined above. The significance of olfactory processing in the OFC is apparent in behavioral studies, which report that lesions of the OFC impair performance in olfactory-based tasks for rats[6,45] in much the same manner that orbital lesions in other species impair performance in tasks based on other modalities.[35,46] Notably, these impairments do not extend universally to all olfactory behavior. For example, rats with OFC lesions are able to detect odors[6] and, in fact, respond correctly to them under some circumstances.[6,45] The impairment caused by OFC lesions appears to result from an inability to appropriately utilize the significance of the cues. Although olfaction may be particularly well-represented in the rat OFC,[47] a more general function related to the appropriate use of motivational information may be common to other species. Investigation of this common function in the rat might best be conducted using a task that stresses the processing of olfactory and motivational information.

16.5 LEARNING AND BEHAVIOR IN ORBITOFRONTAL CORTEX

This author's research into the function of OFC has utilized both recording and behavioral methods. The following chapter sections describe the findings in two of these experiments, with particular emphasis on the methods used in both the olfactory behavior and in electrophysiology. The first experiment[15,48] describes recordings of neural activity in the OFC (and piriform cortex) in rats that were well-trained on an odor discrimination task. The second experiment[49-51] describes firing activity in the OFC (and ABL) in rats learning new odor discrimination problems in each session. Contrasting firing activity in different regions of this prefrontal system within a single task — a task that requires processing of olfactory and motivational information — provides a picture of how different structures within the circuit might contribute to performance in this task. Together, the two studies demonstrate the importance of considering the contributions of learning and behavior when studying neural encoding.

16.5.1 Odor Discrimination Training

Rats will readily learn to perform behaviors in response to different odors. This author uses synthetic odor compounds obtained from International Flavors and Fragrances (IFF). Alternatively, spices and flavorings from McCormick, which are available at most grocery stores, can be used. The odors from IFF are grouped into categories according to a system of subjective classifications[5] and discrimination problems are composed of odors from different categories (fruity, woody, herbal, etc.). Categories used in successive discrimination problems are counter-balanced by association with reward and arranged so that they do not repeat from one day to the next.

Rats are trained within a shielded aluminum chamber 45 cm on each side with sloped walls (Figure 16.2A). Inside the chamber, two panel lights are mounted on the right wall 30 cm above the floor. Below the panel lights is a Plexiglas insert on which are mounted an odor sampling port and a fluid delivery well (Figure 16.2B). Odors are contained in isolated odor-dilution chambers on a removable cartridge (see Box 16.3 — *Olfactory Stimulus Control* — at end of the chapter) and can be individually delivered to the odor port through a system of solenoids and flowmeters that terminates at the back of the port (Figure 16.2C). In the discrimination task, the rats sample an odor delivered at the odor sampling port and then must decide, based on the identity of that odor, whether to make a go response at the fluid well for reward or to withhold that response (no-go) to avoid a penalty (see Box 16.2 — *Odor Discrimination Training* — at end of the chapter).

Behaviors are monitored and events are controlled within the task by computer using a behavioral program written in C++. This computer program interacts with the equipment through input and output registers on a card in the computer. These

FIGURE 16.2A

programmable registers are used to send out and receive electrical signals in the form of TTL (transistor-transistor logic) pulses. In the recording setup, input and output registers are available on the A/D cards that are part of most commercially available data acquisition systems (Datawave Technologies, Cambridge Electronic Design, Plexon Inc.). These systems have the advantage of integrating timestamps for events into the neural data record. The behavioral control can also be implemented separately using inexpensive I/O cards (Data Translation, DT2817) to run training boxes that are not used for recording. In this case, a programmable clock timer board (Computer Boards, CIO-DIO24/CTR3) is also necessary to allow events to be timed

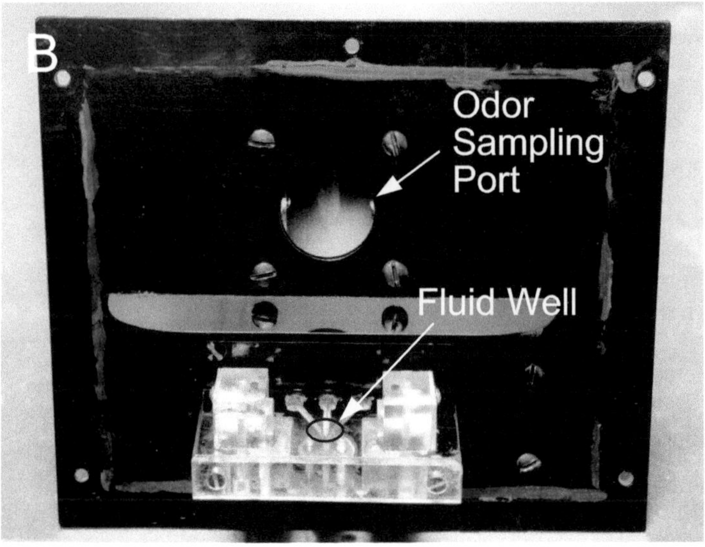

FIGURE 16.2B

FIGURE 16.2 Photographs of the odor discrimination training chamber. A: Outside of behavioral box, showing aluminum chamber with the power supply below, relay switches above, and the flowmeters used to control the air flow, the vacuum lines, and the delivery rate for each of the odors that can be delivered. The reinforcers and removable odor cartridge are located behind the flowmeters on a shelf above the behavioral chamber. B: Removable Plexiglas insert that contains the odor port, fluid well, and the electrical circuitry to detect responses. This insert is bolted into an opening in the right wall of the aluminum training chamber shown in A. The 2.5-cm diameter opening is the odor sampling port. An infrared emitter is located in the left rim, and a detector is located in the right rim directly opposite the emitter. The port will easily fit an adult rat's snout. Behind the opening is a half cylinder with a 2.5-cm internal diameter into which the odor is delivered after a response. Below the odor sampling port is a Plexiglas ledge that contains the fluid well; the well is visible in the center. Three lines enter the well from the back wall of the Plexiglas insert, and a drain line exits from the bottom of the well block and through the Plexiglas wall. These lines allow delivery of any of three different fluids to the well. One fluid is typically water and can be used to flush the well between trials. A vacuum solenoid attached to the drain line closes to occlude the drain, thereby preventing leakage out of the well, and opens to facilitate flushing of the well. An infrared emitter is located at the base of the square Plexiglas block to the left of the well, and a detector is located opposite this emitter in the block located to the right of the well. C: Back of Plexiglas insert described in B (now attached to the training chamber), showing the system of air lines that allows the precise control of odor delivery. An air/odor line delivers odorants at 1.5 l/min to the odor delivery chamber behind the odor port (see B). Air is drawn out of the odor delivery chamber by a vacuum line at the top at 2 l/min. A vacuum dump allows the odor to be diverted away from the delivery chamber when not required. This dump draws air away at 2 l/min. Three configurations are described in the text for flow rates in each line corresponding to the setup during the inter-trial intervals, during preparation for odor delivery, and during odor sampling. These configurations allow for the rapid delivery and clearance of the odor cues and prevent odorized air from entering the behavioral chamber.

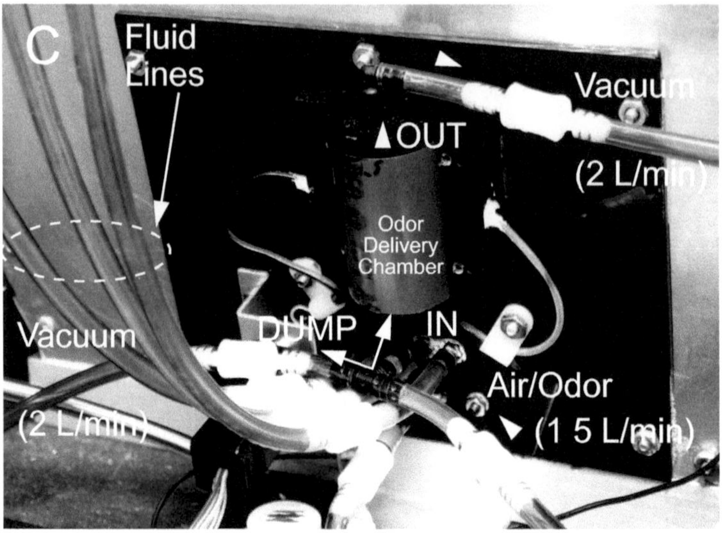

FIGURE 16.2C

with a greater precision than is possible with the timer functions available in most programming languages.

One input register on the I/O cards is used to monitor responses at the odor sampling port and the fluid well in the training chamber. Infrared detectors (Newark Electronics, 84F233) and emitters (Newark Electronics, 84F229) are positioned across the openings of the port and the well (Figure 16.2B). Output from the detector is connected to an optically isolated input relay from Crydom (Newark Electronics, 90F6055) mounted on a Potter and Brumfield board (Newark Electronics, 75F004), which is visible in Figure 16.2A. When appropriately used, these switches serve to isolate the computer boards from the detection circuitry, thereby reducing noise artifacts. When the emitter beam contacts the detector, the output of the detector is approximately equal to the input voltage of 5 V. When the rat interrupts the infrared beam from the emitter, the resistance in the detector increases and the output voltage falls to near 0 V. This transition switches the input relay, which is detected by the computer.

The output registers on the I/O cards control odor and fluid delivery. Again, control is implemented through a set of optically isolated relays to minimize noise artifacts (Newark Electronics, 75F004, 81F9285). The same Potter and Brumfield relay board can be used with both Potter and Brumfield output relays and with the Crydom input relays discussed above. The output relays operate Clippard solenoid valves that regulate the odor and fluid lines and the panel lights that start a trial. Odor delivery is accomplished through these solenoid valves (McMaster-Carr, 4916K13) and a system of flowmeters (Minuteman Controls, Dwyer MMA-21). Before entering this system, clean air is filtered to remove any contaminants (Fischer Scientific, 09-746A) at 5 psi. This air stream is then regulated by a main flowmeter to 1.5 l/min before being split into a clean and an odorized pathway. The odorized pathway is diverted to a manifold attached to a removable polycarbonate odor cartridge containing 16 isolated odor dilution chambers. The odor cartridge is designed to allow independent activation of any one of the 16 different odors without mixing or cross-contamination of the delivery lines with different odors (see Box 16.2 — *Odor Discrimination Training* — at end of the chapter). The "clean" pathway is diverted around this cartridge and through a second regulating flowmeter designed to approximate the resistance of one of the odor dilution chambers. The two pathways converge at the back of the odor delivery chamber in a second manifold that leads to a single line (Figure 16.2C). The odor delivery chamber is a hemicylinder located behind the odor port. The air line ends just outside this hemicylinder where it is again divided. One branch, the odor delivery line, goes to the bottom of the odor delivery chamber, opening less than 1 cm from the odor port. The other branch is connected to a solenoid-controlled vacuum dump. The vacuum dump has a flow rate of 2.0 l/min away from the odor delivery line. In addition, a second vacuum line provides a 2.0 l/min negative air flow at the top of the odor delivery chamber at all times. All connections are made using ⅛-in. i.d. polyurethane tubing and barbed connectors (Minuteman Controls, 3814-6-color-RL, F-3120-85).

During the inter-trial intervals, the 16 solenoids controlling the odors on the removable cartridge remain closed, and the clean air stream is diverted around the odor cartridge and past the odor delivery chamber to the vacuum dump. This serves to remove any odor remaining from the prior trial from the air lines. The clean air flows at 1.5 l/min, and the vacuum dump draws at 2.0 l/min; the difference in flow rates results in a negative flow out of the odor delivery chamber through the odor delivery line of 0.5 l/min. In addition, there is negative flow from the top of the odor delivery chamber of 2.0 l/min, resulting in a net flow out of the behavioral chamber through the odor sampling port of 2.5 l/min.

Before a trial can be initiated, an odor is selected from the cartridge by activation of the appropriate solenoid. This allows clean air to flow through one of the odor dilution chambers containing air saturated with the odor. Because the clean air line bypassing the cartridge is regulated to exactly match the resistance of each line in the odor cartridge, the total air flow is split evenly between the odor and clean air paths, resulting in a 50% saturated air stream at the convergence point of the second manifold. Check valves prevent backflow before and after each odor dilution chamber on the cartridge (see Box 16.3 — *Olfactory Stimulus Control* — at end of the chapter). Sufficient time is allowed for the odorized air stream to reach the odor delivery line before the rat is allowed to initiate the trial. The odor does not enter the port because the vacuum dump remains activated at 2.0 l/min, drawing the odorized air stream plus 0.5 l/min of air from the box into the vacuum dump.

When the rat initiates a trial by poking its nose into the odor sampling port from inside the box, the odorized air stream is delivered by deactivation of the solenoid controlling the vacuum dump. This results in the immediate diversion of the entire 1.5 l/min stream of odorized air into the odor delivery chamber behind the odor port. The length of tubing is minimal, resulting in a delivery time of 25 to 40 msec. As long as the rat remains in the port, the odor continues to be delivered. Odor never enters the box, however, because of the second vacuum line at the top of the odor sampling port. This vacuum line remains activated at 2.0 l/min, resulting in a negative draw of 0.5 l/min out of the behavioral chamber throughout odor sampling.

When the rat leaves the odor port, the vacuum dump solenoid is reactivated to turn off the odor. Odor remaining in the chamber is rapidly cleared by the vacuum exhaust at the top of the delivery chamber and the vacuum dump operating at the bottom of the delivery chamber. The process repeats on the next trial. This system allows for the rapid delivery of multiple odor cues with minimal contamination between cues. Odor mixing can only occur in the common line between the second manifold and the odor delivery chamber, and this line can be changed regularly to prevent any contamination between odor problems.

Fluid delivery is also accomplished via a system of solenoid valves. The fluid well consists of a small depression in a Plexiglas ledge (Figure 16.2B). The ledge extends 2.5 cm from the chamber wall. The well is cut into the middle of the ledge and is flanked on either side by 1-cm-high Plexiglas blocks that contain the circuitry for the infrared emitter and detector. The infrared beam passes immediately above the lip of the well. The depression is large enough to hold 0.05 ml fluid, which is the size of the bolus typically delivered. Three concealed fluid lines enter a common area at the bottom of the well. Fluid in each line is delivered by gravity-flow control via a stainless-steel, large-bore Clippard solenoid valve (Minuteman Controls, Special Order CS-2325). This model of the Clippard valve has the advantage of being rapidly activated, and the large diameter allows fluid to be delivered quickly. The stainless-steel resists corrosion and rust. Although these valves are expensive, spare internal parts can be purchased to defray some of the cost of replacement (Minuteman Controls, S6220-25-IBU, 15412-3). The well is drained by a fourth line located in an opening at the bottom of the well, below the fluid lines. The vacuum-assisted drain line is also controlled via the same type of solenoid valve. In this manner, the well can be rapidly flushed between trials with water (<50 msec) to remove any trace of the prior reinforcer.

Before the start of training, rats are placed on a moderate water-deprivation schedule for 1 to 2 days during which time they receive access to water for 30 min per day. The rats are then shaped to perform the basic task. Shaping consists of reinforcing variants of the rats' normal behavior. The behavior that is reinforced then serves as the basis upon which to build a more complicated behavior that is then subsequently reinforced. Through successive approximations, the rat is reinforced only for behavioral sequences that are progressively closer to the behavior that is required for the experiment. The discrimination task requires the rat to sample an odor at the odor port and then make a response at the fluid well to receive a fluid reward. Rats will naturally investigate the odor sampling port when placed into the behavioral chamber, so this behavior forms the basis of the initial shaping procedure. Shaping is conducted by a computer program that monitors the infrared beam across the odor port and delivers 0.05 ml water to the fluid well for any break in the beam. The shaping program gradually requires longer nosepokes into the port, coupled with a quicker response at the fluid well to trigger reward. In addition, nosepoke is restricted to periods when the panel lights are on, thereby defining the trials in the task. Over the course of 200 to 300 trials, the rat can easily be trained to hold its snout in the odor port for up to 1000 msec (the time varied for the experiments below) and make a response at the fluid well within 3000 msec of exiting the odor port (odor unpoke). Further shaping can be carried out to add variable intervals between nosepoke into the odor port and odor onset, or between a response at the well and fluid delivery. These intervals allow for analysis of neural activity after a behavior, but before the outcome of that behavior is evident to the rat.

Once the rats are trained to nosepoke for fluid reward, odor discrimination training can begin. This author generally trains rats with either two or four odors at a time, combining odor sets to create larger discrimination problems after initial training. In the first experiment described here, rats were trained to discriminate four different odor pairs, which were then combined in a single eight-odor discrimination, and neural firing was examined during accurate performance in these composite sessions. In the second experiment described here, rats were trained on two- or four-odor discrimination problems. A new discrimination problem was introduced in each session. Neural activity was examined as the rats learned each new problem, during accurate performance, and, in some sessions, during acquisition of the reversed discrimination problem.

16.5.2 The Electrode

Cells were recorded in different brain regions in each experiment using similar recording methods. In each case, recordings were made using a chronically implanted, moveable bundle of fine-wire microelectrodes, modeled after the design of Kubie et al.[52] This microelectrode array consists of ten 25-μm wires (California Fine Wire, Stablohm 675) threaded through a 27-gauge thin-wall cannula (Figure 16.3A; Small Parts, A-HTX-27TW). The cannula is mounted on a 10-pin Augat connector (Figure 16.3B; Newark Electronics, 8058-1G34), and each wire is mated to a pin using silver print paint from GC Electronics (Newark Electronics, 002869). The wires are made from NiCrFe and are insulated with Formvar except at the tips where they are cut with fine scissors. A small diameter at the exposed end is necessary for isolation of the signals from individual neurons and minimizes the sampling bias inherent in unit recording.[53] The optimal diameter varies by experimenter but reflects the requirements of a particular brain region (neuron density, soma size, and activity level). Wires with a diameter of 25 μm have worked well in recording from orbitofrontal cortex and several related areas in the rat. However, the small surface area at the tip of these wires does result in high impedance, which increases noise in the signal (see Box 16.4 — *Noise in Electrophysiology* — at end of the chapter). Typically, these wires have impedances of about 1 MΩ at 100 Hz after cutting, which yields recordings with an unacceptable amount of noise. To obtain a useful signal, the impedance of each wire is lowered by one of two methods (see Box 16.5 — *Metal Microelectrodes for Recording in Behaving Animals* — at end of the chapter). The exact value is a matter of experimenter preference, but this author has found that wires with an impedance of approximately 250 kΩ at 100 Hz work well in experiments.

The exposed pins where the wires are wrapped are covered in CutterSil Light, which is an elastic protective material (Thompson Dental Supply), and then embedded in dental cement with three 0-80 hex head screws (Figure 16.3C; Small Parts,

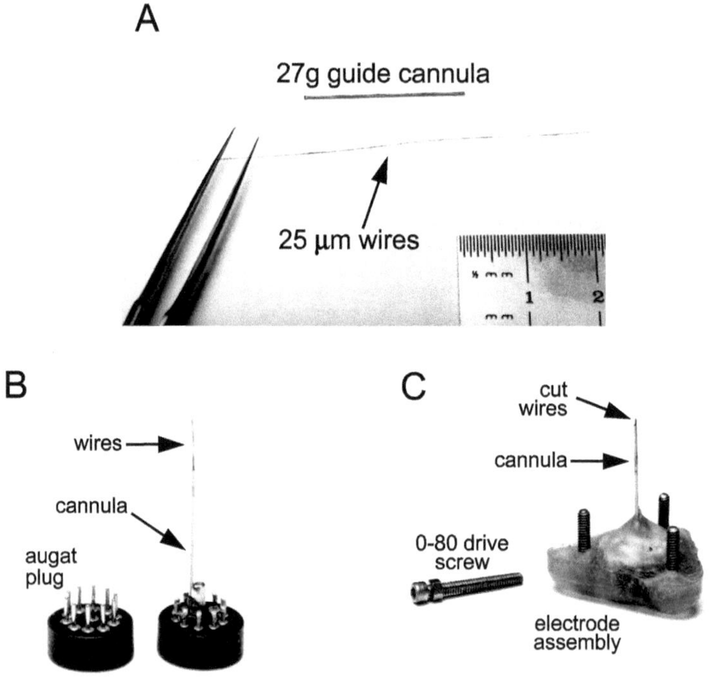

FIGURE 16.3 Close-up photo of the electrode array. A: 25-μm NiFeCr microwires beside a 27-gauge thin-walled cannula. B: 10-pin Augat plug and 10-pin Augat plug with electrode array mounted on additional center pin, and individual wires mated to individual pins by silver print. C: 0-80 hex head screw, with round place-holder nut soldered to shaft, beside molded electrode assembly consisting of the mounted array in B and three 0-80 machine screws similar to the one pictured. D: tripod mount alone (left) and with an electrode assembly attached and ready for implantation (right). This drive assembly (tripod and 0-80 drive screws) can be replaced with a smaller model that relies on a single 0-80 screw for stability and movement when a smaller implantation footprint is required.[69] E: modular JFET pre-amplifier headstage at the end of the mesh-shielded recording cable.

ASHCX-080-8). The screws are prepped to the desired length to reach a particular brain structure and have nuts soldered to the shaft just below the hex heads. These nuts have been machined round and are embedded in the dental cement next to the Augat plug base, thereby fixing the screws into the dental cement mold containing the electrode assembly. This entire assembly is then mounted by the screws on a tripod machined from a disk of aircraft alloy 7075 (McMaster-Carr), an aluminum alloy chosen for its hardness and corrosion resistance (Figure 16.3D). The tripod is cemented to the rat's skull, and the screws can then be used to advance the electrode assembly through the center hole in the tripod and into the rat's brain. The screws allow the electrode array to be advanced in 40-μm increments. This allows small populations of neurons to be sampled from multiple sites in the same animal, typically separated by 80 to 120 μm.

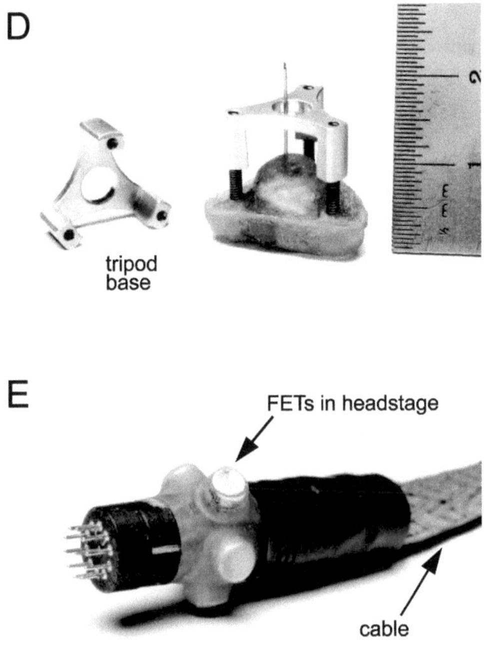

FIGURE 16.3 (continued)

16.5.3 Surgical Methods

During surgery, the electrode array described in Section 16.5.2 is chronically implanted into the brain. Rats weigh 325 to 375 g at the time of surgery. These rats are anesthetized using either an injection of Nembutal (50 mg/kg, i.p.) or more recently by isoflurane inhalation (Isovet, Mallinckrodt). Isoflurane anesthesia is more costly but has several advantages. Among these advantages are ease of induction and maintenance. The short duration of the effect allows much better control of the depth of anesthesia than is possible with Nembutal, which nearly eliminates the risk of fatality even with the rather long surgeries required for electrode implantation. After induction of anesthesia, the rat is placed into a stereotaxic apparatus (David Kopf Instruments), and standard surgical procedures are used. A midline incision is made from several millimeters anterior to bregma to a point posterior to lambda. A wide field is necessary for anchoring the tripod to the skull. The surface of the skull is then cleaned of all tissue and allowed to dry. In addition, fascia and muscle are removed from the lateral edge of the parietal ridge so that anchoring screws can be positioned there. Four 1-mm holes are drilled for anchoring screws and the ground screw, and a larger opening (~2 mm) is made over the electrode implantation site. The dura overlying the implantation site is removed using a sterile scalpel blade and small forceps so that the electrode will not be damaged during insertion. Again, the skull is cleaned and dried. Then the electrode is implanted using the following coordinates: OFC: 3.0 mm anterior to bregma, 3.0 or 3.2 mm lateral to midline, 3.0 to 4.0 mm below the surface of the brain; ABL: 3.0 mm posterior to bregma, 5.0 mm lateral, <7.5 mm below the surface of the brain. The piriform cortex can be reached by implanting in the OFC and penetrating down until piriform cells are encountered. Electrode position proximal to the piriform cortex can be ascertained, both by the lack of signal in the rhinal sulcus and also by the shape of the evoked potential in the piriform cortex elicited by olfactory bulb stimulation.[15,54-55] In other areas, electrode position can be marked by passage of a 15-μA current through each microelectrode at the conclusion of recording. The small iron deposit that results can be visualized by staining the brain sections with a 3% potassium ferrocyanide solution. The Prussian blue reaction creates a small mark that can be used to extrapolate back to the recording position in each session.

Typically, the electrode array is lowered to a position about 1 mm above the region of interest, the hole in the skull is filled with bone fragments or gel-foam, and the opening through the tripod is filled with CutterSil Light (Thompson Dental Supply). This elastic material serves to seal this route of entry into the brain once the electrode is implanted. It is important to bring the base of the tripod into contact with the skull surface; thus, the ventral position is typically somewhat variable. Anchoring screws are placed in the parietal bone ridge and the entire assembly is cemented to the skull using Grip Cement (Dentsply Caulk, 675570). Polymerization of the cement is a highly exothermic process. This heat can cause bone damage or even damage to the underlying brain. To avoid this damage, the cement should be applied in small quantities and then allowed to cure. The amounts can be determined by monitoring the temperature of the drying cement. In addition, it may be helpful to mold the edges of the anchoring cement as it dries to avoid sharp corners. These sharp edges irritate the wound edges and appear to cause the rat to scratch at the electrode assembly later. The wound is allowed to heal to the edges of the cement.

During the surgical procedure, it is crucial to maintain sterile techniques because the assembly will be chronically in-dwelling. One of the biggest impediments to successful chronic recording is the risk of infection below the implant. Infection below the implant can result in formation of a brain abscess tracking along the electrode cannula or in dissolution of the skull, causing the electrode assembly to become loose or detach from the skull. To prevent infection, the electrode assembly is sterilized with 70% EtOH prior to implantation, rats are given a pediatric suspension of the antibiotic cephalexin (30 mg/kg/d, po bid) for 2 weeks post-operatively, and they receive triple-antibiotic ointment with the anesthetic lidocaine to the wound edges for several days post-operatively.

In addition, it is important that the electrode assembly be implanted under stereoscopic guidance and at a rate slow enough to prevent bending of the wires or movement laterally. A fine wire microelectrode array is designed to "float" in the extracellular matrix. This is accomplished using very fine wires sufficiently exposed from the carrier cannula to permit some movement of the carrier independent from the wires. This author exposes 1 to 1.5 mm of the wires beyond the cannula when preparing the array for implantation. This array of floating wires must remain separated, and this requires that they enter the brain surface spread slightly and not as a single mass of wires. Stereoscopic guidance allows visualization of the electrode tips as they enter the brain to ensure this spread and also to guard against catching or bending on stray pieces of bone or dura over the implantation site. Once the tips have entered the brain, the rate of advancement is important to good outcome. A rate that is too rapid results in bending of the electrode tips back against the cannula and seems to cause the electrode track to be offset laterally, perhaps indicating that the brain has been rotated by pressure of the electrode during implantation. This problem of rotation is more acute the more laterally the electrode is implanted. (A rate of 100 microns per minute is adequate to avoid these problems.[56]

Rats are allowed to recover for 2 weeks after surgery. During this time, they receive prophylactic antibiotic treatment as noted above. This time period is also used to introduce them to the recording cable by connecting them to the cable for brief periods over several days in the training chamber. In addition, the electrode array is advanced 200 to 300 μm over this time period to remove it from the immediate vicinity of insertion where some gliosis might occur.

16.5.4 Recording Methods

During recording, activity on each microwire is amplified by passage through a preamplifier headstage located immediately above the electrode (Figure 16.3E). This preamplifier, comprised of either junctional field effect transformers (JFETs) or operational amplifiers (see Box 16.6 — *Headstage Amplifiers for Recording in Behaving Animals* — at end of the chapter), is required to convert the high-impedance neural signal into low-impedance output. A low-impedance signal is more resistant to noise artifacts caused by the cabling from the recording chamber and other sources in the local environment. For this purpose, the headstage should be located as close to the rat as possible. In our case, it is located at the end of the recording cable; an alternative method would be to include the electronics on the electrode assembly itself. This approach might result in less noise but would subject the components to the difficulties and potential damage associated with chronic implantation on the rat's head.

The resultant electrical signals from the microwire electrodes are brought by shielded cable to a patch panel outside the training chamber where the activity on each wire can be viewed and selected for recording. The signal on each microwire is referenced to a ground signal from the rat, provided by either a screw in the rat's skull or the steel guide cannula within the brain. In addition, a differential amplifier is employed for signal processing. The differential amplifier subtracts the signal of an "indifferent" microwire from the signal on the microwire of interest. By subtracting the signal on the indifferent microwire, the differential amplifer eliminates distant sources of noise impacting both wires simultaneously. Pairs of wires are selected during a short period prior to the daily session. During this period, the rat investigates the behavioral chamber or may be engaged by a practice version of the behavioral task. The neural signal is filtered at 300 to 3000 Hz and amplified at 5000×.

In the experiments described in this chapter, four amplifiers (Grass P5 Series Pre-Amplifiers, Grass Instruments) were used and the software package used was capable of discriminating units on either of the wires contributing to each channel, allowing capture of the activity on up to eight microwires. The resultant four channels of neural data were recorded in parallel with TTL channels containing marker data identifying important behavioral events in the task. Individual neurons were isolated during offline analysis with a system developed by Cambridge Electronic Design (Spike2) utilizing a template matching algorithm. For template matching, the signal from each of the four amplifiers is digitized at 25 kHz prior to template matching. Template design is accomplished by reviewing a 5- to 20-min interval of the session to be analyzed in order to establish templates that can then be used to isolate each individual neuron's activity over the course of the entire session. Templates are validated through examination of the oscilloscope trace to subjectively identify the number of individual waveforms that appear to be present on each channel, and template design is terminated when the existing templates appear to capture the waveforms on the oscilloscope trace. In this manner, it is possible to identify one to three individual neurons on each individual microwire when activity is present. Occasionally, when several microwires have activity, template matching yields a population of over 10 neurons.

Although this setup was certainly adequate, it should be noted that this author has recently migrated to the Enhanced Discovery Data Acquisition System by Datawave Technologies. This system has the advantage of integrating control of our behavioral equipment with acquisition of the neural data in real-time through the use of so-called clocked sequences. These clocked sequences are C^{++} code that is written by the experimenter and compiled into the Discovery runtime module. This code is accessed by the Discovery program to send and receive input from I/O cards and thereby control the experiment. These events and the behaviors of the rat are simultaneously labeled, timestamped, and inserted into the neural data record for later analysis. We currently combine data acquisition in this system with spike sorting and analysis using the Offline Spike Sorter and Neuroexplorer programs available from Plexon Inc. These programs isolate the single units and provide customizable graphical analysis routines that allow us to screen the data. Output can then be generated for customized statistical routines written in C^{++}. References for these companies are provided at the end of the chapter in the Appendix 16A.

16.5.5 Recording in the Context of Behavior

Behavioral neurophysiology imposes unique requirements not faced in neurophysiological nor behavioral experiments alone. These unique features and our solutions are worth noting. An incomplete list would include the attachment of the recording cable to the rat in a manner that permits relative freedom of movement, protection of the recording cable itself, shielding of the recording environment from electrical interference, and the control of behavior in such a manner as to allow for identification of neural correlates.

This author uses a homemade recording cable consisting of a modular headstage (see Box 16.6 — *Headstage Amplifiers for Recording in Behaving Animals* — at end of the chapter) composed of JFETs mounted between two Augat plugs and encased in molded dental cement (Figure 16.3E). This modular piece is connected by a second Augat plug to a cable composed of 13 to 25 wires (Cooner Wire, CZ-1105) chosen for their small gauge and low weight. These particular wires are very flexible, which allows for repetitive movement without creating breaks in the cable. The entire cable attaches to a 25-channel, low-torque commutator (Crist Instruments). The small, low-profile headstage design, coupled with the outward slant of the recording chamber (see Figure 16.2) allows the rat to make responses in the task without bringing the headstage or cable into contact with the walls of the chamber, which would introduce noise into the recording. In addition, the modular design allows easy reuse of the durable preamplifier headstage when the cable itself becomes damaged. Damage, in general, occurs when the rats become bored or frustrated and bite the recording cable where it runs between the commutator and the headstage.

Biting of the cable is minimized by several strategies. First, the cabling is tensioned by rubber bands. The tension permits the rat full range of movement inside the box but partially retracts the cable out of reach when the rat tries to bite it. Second, the cable is ensheathed in a protective spring in several places (Instech Laboratories, PS-135). This spring deflects the teeth from the wires and is very effective, although the added weight is a drawback. Finally, the cable itself is coated or sprayed with a bitter solution (denatonium benzoate, apple bitter, quinine, etc.). If this is done early in the rat's training with the cable, the rat will quickly learn not to bite the cable. This issue is of potential interest not only to minimize damage to your equipment, but also to protect the rat. Short-circuits in the recording cable can result in electrical damage to the commutator or lesions in the rat's brain.[57]

The behavioral environment and neural signal from the rat are shielded as much as possible from electrical interference (see Box 16.4 — *Noise in Electrophysiology* — at end of the chapter). The behavioral chamber is constructed of aluminum and is grounded back to the main ground point for the recording setup, thereby shielding the interior of the box from outside interference. To prevent ground loops from forming through the rat, the inside of the chamber is painted to be nonconductive and the mesh floor is coated with a rubber compound (U.S. Plastics, Plasti-Dip, 38057). The rat is always grounded by means of a wire from the main ground that enters the chamber in the recording cable and attaches to either a skull screw on the rat or to the recording cannula. In addition, the bundle of wires carrying the signal from the rat is encased in wire mesh that is grounded back to the main ground point for the rack. This design ensures that the rat remains properly grounded at all times and ground loops are eliminated.

Finally, recording places special demands on rat behavior. Analysis of neural activity requires that behavioral variability be minimized and that data be gathered from a sufficient number of trials to overcome the inherent variability in neural firing. For example, to minimize behavioral variability, rats are first trained on the recording task. Depending on the experiment, this training may involve actual experience with the odors to be used or it may involve shaping and learning with different odors. The goal of this approach is to ensure that the rats are thoroughly familiar with the training environment and with all of the behaviors required in the task, so that their actions on each trial differ only according to their current state of knowledge regarding the odor cues and not because of attentional changes or because of improvements in their understanding of the importance (or lack thereof) of other events. To accomplish this goal, most recording experiments are conducted in subjects that are well-trained on the task, including the cues used in the behavior. In these experiments, the influences of learning and of changes in behavior that result are also minimized.

In addition to pre-training the rats, this author tries to gather data from at least 20 trials for each behavioral condition that will be of later interest in the analysis of neural activity. In a straightforward experiment on odor selectivity, that might mean 20 trials involving each odor. If other events (e.g., a response by the rat) are also important, this author would try to gather data from sufficient repetitions involving a particular odor and that response. Finally, the duration of neural data gathered on each trial must be considered. A very short duration of analyzable data will, of course, increase the number of repetitions required to minimize variability. An alternative approach is to increase the duration of the data acquisition window. In a normal odor discrimination task, a rat is capable of sampling an odor in perhaps 200 msec. Our recording experiments require that the rat sample the odor for a period of 500 msec or longer. Other time periods in the task are also prolonged for similar reasons. Variable delays are inserted between responses at the odor port or fluid well and delivery of the odor or fluid reinforcer. These delays allow separation of neural activity due to the motor behavior from neural activity occurring in response to the cues and create a prolonged period — a delay — in which activity reflecting the internal expectations of the rat can be examined. These types of considerations are crucial in the success of a behavioral recording experiment, as will be discussed below.

16.6 EXPERIMENT 1: ENCODING IN OFC IN WELL-TRAINED RATS

These principles were used to investigate encoding in rat OFC (and piriform cortex) during performance of an olfactory-based task. In this experiment,[15,48] rats were shaped to nosepoke to obtain water reward, and then the animals were trained to discriminate four different odor pairs. In each pair, one odor was positive and one odor was negative. Response at the fluid well after sampling of the positive odor was reinforced by delivery of a sweetened sucrose solution to the well. The same response after sampling of the negative odor was not reinforced and was followed by a significantly prolonged inter-trial interval to further discourage these errors of commission. Once the rat had demonstrated acquisition of each two-odor discrimination problem by meeting a behavioral criterion (i.e., 90% performance in a moving block of 20 trials) the next problem was introduced. After the rats acquired all four discrimination problems, they were then trained for ten sessions in which all eight odors were presented together in a pseudo-random sequence. The rats rapidly learned to discriminate between the two types of odors during these composite sessions (Figure 16.4A). Recording began after session 10. As the performance in Figure 16.4A demonstrates, this regimen of prerecording training resulted in highly accurate (>90%) and very stable performance, both within and across sessions.

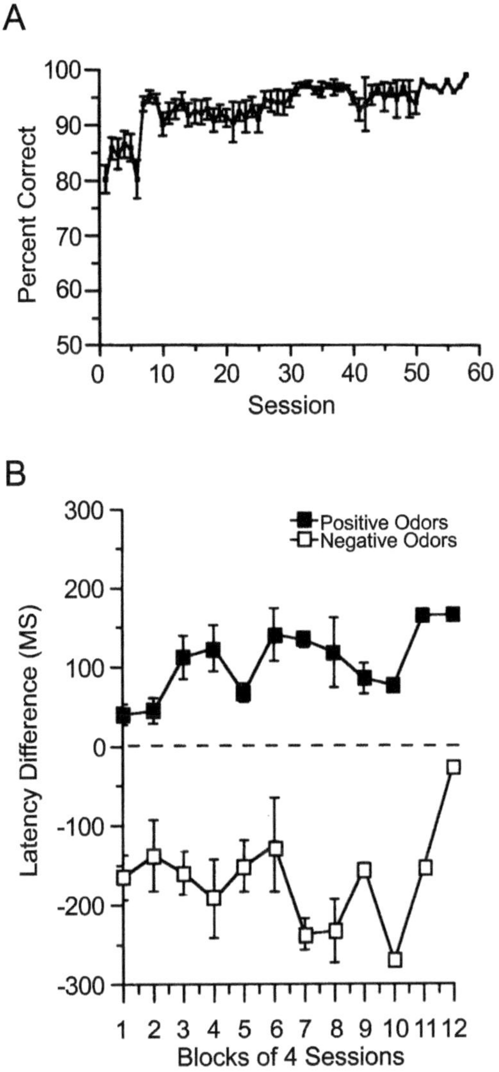

FIGURE 16.4 Measures of performance in the well-learned, eight-odor discrimination task. A: average daily performance on the eight-odor discrimination task. Each point represents the average performance across all rats. Recording began on session 10. Sessions generally consisted of 400 trials. B: difference between latency to enter the odor port on trials that were predicted vs. trials that were unpredicted (unpredicted-predicted). Latency differences were calculated from light onset to nosepoke into the odor port and controlled for current and prior trial valence. Each point represents the average difference between these measurements over a block of four recorded sessions for all of the animals, with standard error bars indicating the variability between animals. Recording began during the tenth training session; thus, for all points shown, the animals were well-trained. Significant differences were present throughout recording (see Schoenbaum and Eichenbaum, *J. Neurophysiol.,* 74, 733-750, 1995).

Stable performance within each session presented the opportunity to examine the influence of information other than the significance of the odor cues. For example, because performance was relatively consistent and the rats were content to complete a large number of trials (~400) in each session, it was possible to acquire data from trials involving many different odor cues without reducing the number of trials per odor to an unacceptable number. The sequence ensured that each of the eight odors occurred a minimum number of times within a 400-trial sequence and also required that approximately 50% of the trials be positive. Furthermore, no more than three successive trials of the same valence could occur together, and each odor was preceded with equal frequency by positive and negative trials and also by the other odors in the discrimination. Exception to these rules was made for several specific odor pairs that always occurred together. For example, odors 5+ and 6– were always followed by odor 7+, and odors 1+ and 2– always repeated. Odor 7+ also occurred unexpectedly. This arrangement provided additional information that the rat could use to guide behavior. The use of the information contained in the predictive odor pairs was evident in the rats' latencies to initiate predicted vs. unpredicted trials (Figure 16.4B). The rats initiated predicted positive trials more rapidly than nonpredicted trials of the same odor, whereas they initiated predicted negative trials more slowly. The odor sequence ensured enough trials of each odor and each type of occurrence — expected vs. unexpected, for example — to allow the analysis of neural activity in sets of trials defined by odor identity, the significance of the odor cue, and also aspects of the odor sequence such as the predictive pairings.

Peri-event time histograms synchronizing the firing activity in each cell to various behaviorally relevant events occurring during each trial were constructed to search for firing correlates. These histograms were screened for firing that differed from baseline activity during the inter-trial intervals. This initial screening process revealed that, contrary to the view that OFC serves as an odor association cortex, cells there responded to a variety of events not related to odor processing. Notably, similar patterns were found for firing selectivity in cells recorded in piriform cortex, suggesting that reciprocal connections from the OFC and other structures may act upon networks in the piriform cortex to prepare the network for more rapid or accurate identification of anticipated cues. These findings have recently been extended by recordings during odor discrimination made in olfactory bulb.[16]

Cells that differed from baseline during odor sampling were further examined for firing selectivity related to the identity and learned significance of the odor cues. The features of our odor sequence — multiple odors with differing meaning within a fixed sequence — and the large number of trials completed in each session allowed a reasonable assessment of the effect of both of these variables. The results of that analysis are evident in Figure 16.5, which shows the firing activity in 20 OFC neurons that exhibited phasic activity during odor sampling. Even a casual glance at this figure reveals the variety of patterns across odors and the distributed nature of the representation of the odors. On closer examination, several specific influences on firing become more apparent. For example, many of the neurons fired differentially based on the valence or associative significance assigned to the odor cues (Figure 16.5M, N, and O). Within these categories, there also appeared to be some

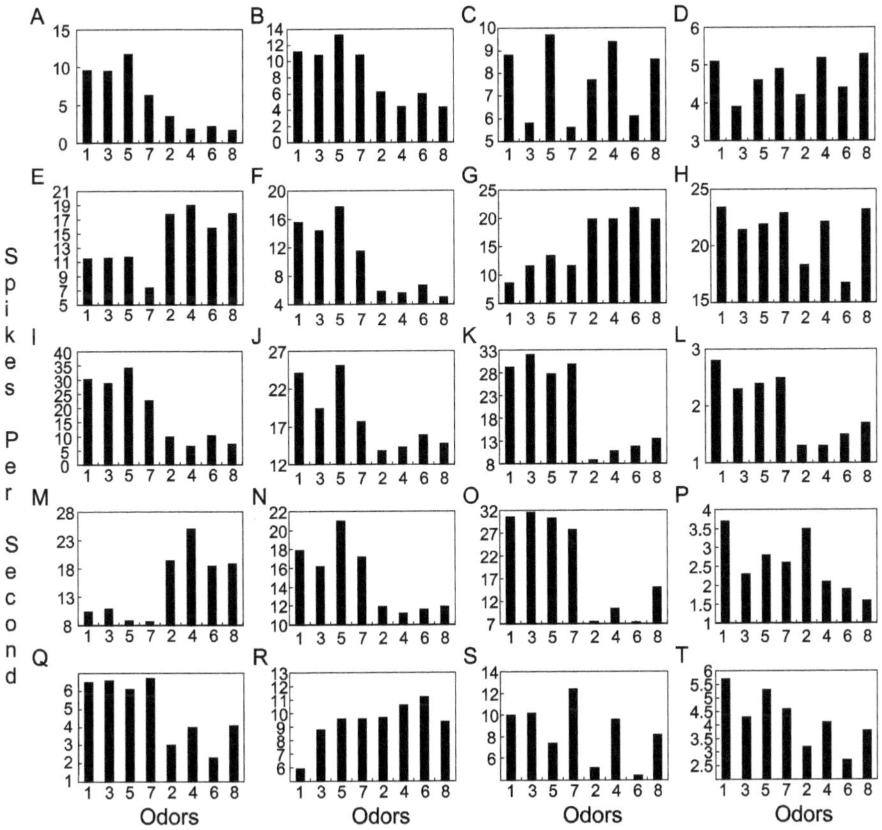

FIGURE 16.5 Activity in spikes/second (y-axis) during odor-sampling related to odor identity (x-axis) in OFC cells chosen at random from the ensemble of a particular animal. Each panel shows activity in a different neuron to each of the eight different odors presented in each session. Odor identity is noted by a number (1–8) on the x-axis of each graph. Positive odors 1^+, 3^+, 5^+, and 7^+ are on the left of each panel, and negative odors 2^-, 4^-, 6^-, and 8^- are on the right of each panel. Note the absence of true specificity in the majority of the panels, and also the variety of patterns evident consistent with distributed coding.

neurons that fired selectivity based on the identity of the odor cues (Figure 16.5J and M). ANOVA ($p < 0.05$) confirmed that substantial numbers of odor-responsive neurons in the OFC (and also the piriform cortex) exhibited these patterns.

In addition, there were neurons such as C and D (Figure 16.5) that clearly exhibited differential firing during sampling of the odor cues, but their activity was not clearly related to either odor identity or the significance of the cues. Closer examination of activity in these cells revealed the influence of information contained within the odor sequence across trials. Recall that the rats used the predictive pairs to bias their latency to initiate new trials (see Figure 16.4B). There were sufficient trials within the pseudo-random sequence to ask whether these specific odor pairings influenced firing activity. This analysis indicated that many many neurons fired differentially based on this predictive information. Finally, some cells fired differentially based on random features within the odor sequence, such as whether the current trial had been preceded by a positive or negative trial.

Many neurons participated in encoding more than one category of information, and few if any neurons were specific to a single category or cue. Rather, the encoding exhibited various degrees of selectivity and was highly distributed, particularly when activity on any given trial was considered. In other words, each cell was an unreliable predictor of the information it appeared to represent. To determine what information was represented reliably on a trial-by-trial basis by the neurons in the OFC, a modified linear discriminant analysis[58] was employed to ask how well the population response on each trial discriminated the information represented by the single neurons. Here again, the approach of recording in well-trained animals presented unique advantages. The repetitive nature of each session — where the rats were presented with the same odors in the same pseudo-random sequence each day — and the stable performance across sessions (see Figure 16.4A) allowed ensembles composed of neurons recorded from different sessions. Although this approach ignored the contribution of interactions between cells (see Reference 48 for more detailed discussion of limitations), it is analogous to the analysis of single-unit encoding discussed above. The following equations summarize the method and its application to our data.

This approach began with the definition of variables such as odor identity, reward association, and past events. The discriminant analysis creates an n-dimensional space representing the response of each neuron on a single dimension so that the population response on each trial of a session can be represented as a point in this space. We asked whether groupings of population responses in this space appeared to reflect the variables we had defined. The populations were constructed of sets of odor-sampling neurons recorded across sessions. As in the single-cell analyses, mean firing rate during the odor sampling window was used to represent information coding. Using the data from a composite recording session constructed in this manner for each animal, we measured the normalized Euclidean distance between the point within the n-dimensional activity space representing the ensemble's response to a single repetition of a given condition and the mean for the ensemble's responses to all of the repetitions of a condition. This distance was measured to the mean of each condition of the variable under consideration, and the repetition was classified as belonging to the condition with the closest mean population response. Calculation of these distances required the determination of the mean response of each neuron to each condition and the associated variance. The mean response for neuron, k, to presentation of condition, i, is:

$$\bar{y}_{ik} = \frac{\sum_{j=1}^{J} y_{ijk}}{J} \tag{16.1}$$

where J is the number of repetitions of condition i and y_{ijk} is the firing rate of neuron k to the jth repetition. The associated variance for this mean response is:

$$\sigma_{ik} = \sqrt{\frac{\sum_{j=1}^{J}\left(y_{ijk} - \bar{y}_{ik}\right)^2}{J-1}} \tag{16.2}$$

The distance between the population response to the jth presentation of condition i and the mean population response for all presentations of condition s, normalized to reflect the standard deviation of the mean, is then calculated as:

$$d_{ij}^{s} = \sqrt{\sum_{k=1}^{K}\left(\frac{y_{ijk} - \bar{y}_{sk}}{\sigma_{sk}}\right)^2} \tag{16.3}$$

where K is the number of neurons in the population. This entire procedure was repeated for each repetition of each condition; in each case, the mean for each condition and the associated variance were recalculated using a jackknife procedure and excluding the repetition of interest in order to achieve a stable estimate of the mean response to each condition unbiased by inclusion of the repetition being evaluated.[59] By comparing the classification of each repetition to its actual identity, the ensemble's ability to discriminate the conditions of a given variable was scored. This resultant performance score was evaluated by constructing a contingency table of joint probabilities to test the statistical association between condition occurrences and classifications and to measure mutual information transmission as previously described. A X^2 test of association was used to determine whether the ensemble performed significantly better than chance at classifying events at $p < 0.001$.

Our analysis of populations of orbitofrontal neurons revealed encoding of variables according to their importance to task performance (Figure 16.6). Neural ensembles performed well at discriminating odor identity, associative significance of the odor, and its predictive value for the following trial. Other variables related to context and past events, unimportant to correct performance of the task but with prominent influence on single neuron firing patterns, were poorly discriminated by the activity within the ensembles, not reaching signficance.

These findings suggest that neural ensembles in OFC encode features available from the individual neurons in a selective manner, consistent with task requirements. Ensembles in the OFC coded the most directly useful variables of odor identity and reward value under conditions that required the animal to identify odors and recall past associations with reinforcement. Our expectation would be that a task with different requirements would select different variables, a prediction that is supported by recent reports of activity in OFC in other odor-based tasks.[60,61] Thus, ensembles of neurons in the OFC appear to discriminate variables more reliably than individual neurons and also extract aspects of the representation of these variables provided by the individual neurons in a task-specific manner.

The demonstration of selective coding, driven by task requirements, becomes more significant when considered in light of the connectivity of this region in the rodent. The individual neurons in the OFC likely have access to a great variety of

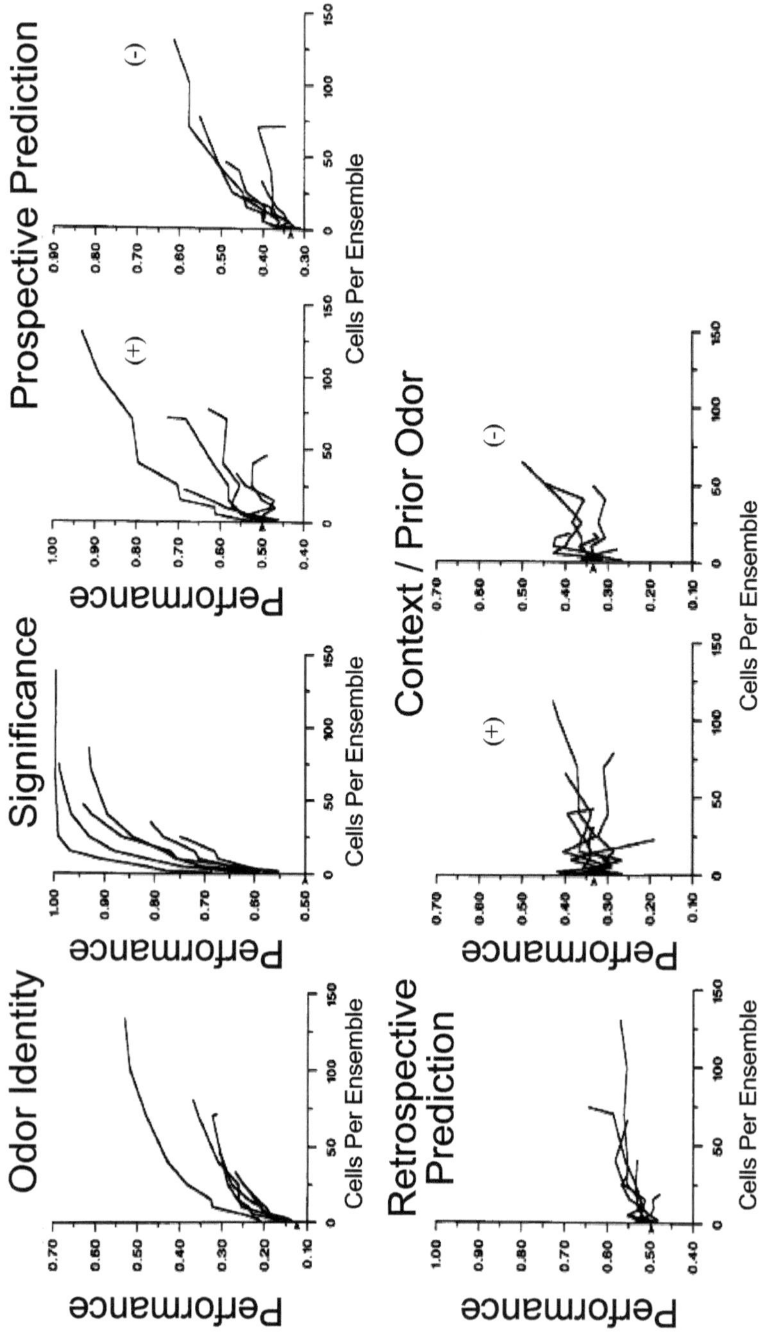
Odor Identity
Significance
Prospective Prediction
(+)
(-)
Retrospective Prediction
Context / Prior Odor
(+)
(-)
Performance
Cells Per Ensemble

FIGURE 16.6

information through reciprocal connections to a circuit of structures involved in olfactory processing, including the piriform cortex, parahippocampal areas, and amygdala. Through these connections, cells in the OFC are uniquely positioned to interact with these structures and thereby access the results of processing in them. Several of these structures are thought to be crucially involved in learning, memory, and the processing of olfactory and motivational information. A neural population located in the OFC would be able to access information regarding sensory processing and odor recognition from the piriform cortex, information regarding an odor's significance from amygdala, and information regarding context and recent events from the parahippocampal areas — and in fact, our data indicate that cellular activity in the OFC does reflect this information. Only at the level of the neural ensemble, presumably established during learning and active during performance, is the information required by the specific task represented in a useful fashion.

16.7 EXPERIMENT 2: ODOR SELECTIVITY AND THE IMPORTANCE OF LEARNING AND BEHAVIOR

The preceding experiment (Chapter 16.6) demonstrated that encoding in the OFC shared many of the features exhibited by prefrontal subdivisions in other species. For example, neural activity in the OFC reflected a diversity of events within a task designed to engage systems specific to the OFC region. In addition, correlates reflected processing within those informational domains in a task-specific manner. These features are shared by neurons in primate prefrontal cortex.[62-68] However, recording in well-trained rats confounded the influence of odor identity and learned significance, and the importance of the rats' use of the information provided by the odor cues is also inaccessible. A second experiment[49-51] was designed to address these shortcomings by recording in OFC neurons as rats learned new odor discrimination problems.

FIGURE 16.6 (opposite) Ensemble ability to discriminate information contained in the responses of the individual neurons in the OFC during odor sampling. The graphs show performance in each ensemble as a function of the number of cells included in the calculations. Each point represents the average of ten random samplings of the ensemble. Chance performance at discriminating each type of information varies according to the number of items and is marked on the y-axis of each graph by an arrow. In some cases, calculations were performed independently for data on positive and negative trials to eliminate encoding due to valence. These cases are indicated by (+) and (–) designations on the graphs. The top row of figures shows ensemble performance at identifying odor identity on the current trial, the learned significance of the odor cues, and the prediction that could be made from the current trial regarding the likelihood of reward on the next trial. This information is useful in task performance and is well encoded by the ensembles (see Schoenbaum and Eichenbaum, *J. Neurophysiol.*, 74, 751-762, 1995 for details). The bottom row of figures shows ensemble performance at identifying whether the current odor was predicted and what non-predictive odor had been presented on the prior trial. This information is not useful in task performance and is poorly encoded by the ensembles.

As in the prior study, these rats were first pre-trained to perform the behaviors involved and to discriminate odors in the task. However, they were presented with new odors in each recording session. In this manner, the firing activity during odor sampling and in other parts of the task could be examined in response to the odors when they were novel, as the rats learned about the cues, and also during a subsequent period of highly accurate performance similar to that in the earlier experiment. In addition, in many sessions, reversals were conducted in which the reinforcement contingencies of two odors were switched. Recording during learning and reversals allowed us to dissociate the influences of odor identity, significance, and behavior that were confounded in the prior study.

Pre-training consisted of shaping as in the prior experiment, followed by training on a series of three or four two-odor discrimination problems, followed by reversals. The task was similar to that used in our earlier study with two exceptions. First, rather than no reinforcement, quinine was delivered when the rat made erroneous responses after sampling of negative odors. Second, a variable delay period of 300 to 800 msec was added after a fluid well response but before delivery of the reinforcer (sucrose or quinine). During this delay, the rat was required to remain in the well, which allowed us to analyze neural activity immediately following a decision to respond, but before the rat knew the outcome of that response. The quinine and variable delay were introduced during the shaping procedure and the subsequent pre-recording discriminations.

As has been described previously,[2,4,9] rats trained on multiple odor discrimination problems develop a learning set that is reflected in facilitated acquisition of novel odor discrimination problems. Our pre-training protocol was designed to promote the formation of this learning set, for both initial learning and reversal, before recording started. A learning set develops even when the rats receive extensive pre-training on the operant behaviors involved. For example, rats that receive a large number of positive shaping trials still show a benefit of experience across multiple discrimination problems.[2] This indicates that improved performance on later problems does not result simply from improved attention to the steps in the task, but rather reflects the use of the odors to guide behavior in the task. Therefore, our pre-recording training involved both extensive shaping and experience with several discrimination problems. As expected, the rate at which our rats acquired new problems increased substantially during this pre-training phase. The rats reached criteria more rapidly on later discriminations during initial training and after reversal. This had two advantages. First, our results were not contaminated by changes in learning strategy across successive odor problems. Second, more rapid acquisition allowed us to gather sufficient data for analysis of initial learning, accurate performance, and reversal learning within a single session for recording purposes.

Although the rats did show a marked improvement in their ability to acquire discriminations and to perform reversals as more problems were presented, they did not exhibit the near-perfect learning seen in studies of learning-set formation employing discrimination paradigms in which errors of commission have no consequence.[2,4,9] Delivery of quinine on negative trials appeared to cause the rats to

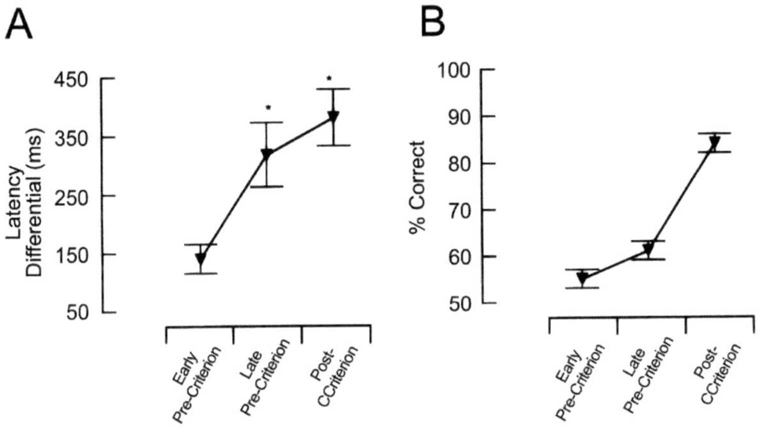

FIGURE 16.7 Changes in behavioral measures of learning during initial training. A: Changes in differential response latency in the early and late pre-criterion and the post-criterion phases. The latency difference increased significantly from the early to the late phase of pre-criterion training (see Reference 51 for details). Latency to respond on positive trials remained near 350 msec across training; thus, the increase was largely due to increases in latency to respond on negative trials. B: Go, no-go performance in the early and late pre-criterion and the post-criterion phases (chance = 50%).

continue to commit errors even after the rats had learned the reinforcement contingencies of the new odors. Evidence for this contention is available in response latencies obtained from the recording sessions, which showed a significant increase in latency to respond on negative trials during a training phase that preceded the shift to accurate performance (Figure 16.7). At the outset of each session, the rats responded on every trial, irrespective of which odor was presented, electing to make errors rather than forego rewards. Response latencies typically ranged between 250 and 500 msec and were similar on both positive and negative trials. As the rats learned the reinforcement contingencies of the odor cues, responses after positive odors became slightly faster but were limited by a floor effect due to the rapidity of the default go responses at the start of each session. Responses after negative cues became much slower, often occurring as long as 1000 msec after termination of odor sampling. The development of this response latency difference between negative and positive trials is shown in Figure 16.7 along with the go, no-go performance. Note that the difference in latency is initially very small but increases significantly in the late part of pre-criterion training. This difference emerges before the rat has become accurate at the go, no-go discrimination during the post-criterion phase. The slow acquisition allowed us to separate learning from the shift in the go, no-go decision that occurred in post-criterion performance. In many sessions, there were 40 to 60 trials separating initial evidence of learning from a shift to reliable performance.

Analysis of these trials was used to examine changes in firing rate to see whether they corresponded to learning or changes in go, no-go performance. The early pre-criterion phase was composed of those trials at the start of each session, when the new odors were introduced to the rat and before changes in either response latency or go, no-go performance were observed. The late phase of pre-criterion training consisted of the remaining trials before the rat reached the behavioral criterion. As before, the behavioral criterion was defined as 90% performance in a moving block of 20 trials. During this block of trials, the rats exhibited a significant increase in the response latency difference but continued to make errors on most negative trials (Figure 16.7). The post-criterion phase consisted of the trials after this criterion was met. Note that although no specific performance was required post-criterion, the rats typically performed well during this block of trials (Figure 16.7) and, when errors were made, continued to exhibit a latency difference between positive and negative trials. Neural activity was analyzed within each phase and then compared across phases for each neuron that showed selectivity during odor sampling to separate encoding of odor identity, associative significance, and behavioral strategy. The importance of these variables was probed further in the analysis of activity after reversal of the odors' reinforcement contingencies in many sessions.

Our analysis of these neurons focused on the cells similar to those in earlier studies that had selectivity during odor sampling in well-trained animals.[15,68] Such neurons have been proposed to encode odor identity and the learned significance of the odor cues. By examining the development of selectivity in these neurons as the rats learned about the reinforcement contingencies of novel odors and then altered their behavior to reflect that information, we discovered that encoding in these neurons is determined by the specific conjunctions between odor cues and the motivational significance of the associated reinforcers. Moreover, encoding also depends critically on the rat's use of that information to guide behavior. Figure 16.8A shows an example of a neuron in the OFC that demonstrated these characteristics. This neuron fired selectively during odor sampling in the post-criterion phase of the session, but not during either the early pre-criterion phase when the odors were novel or during late pre-criterion trials when the rat showed evidence of learning in the form of longer latency errors. Selectivity emerged in this neuron only once the rat adopted a go, no-go response strategy based on the significance of the odor cues during the post-criterion phase.

Greater than 90% of the neurons in the OFC that were selective post-criterion exhibited this pattern. Figure 16.8B shows the development of selectivity in this population of cells. This pattern suggests that few of these cells were responding based simply on the identity of the odor cues. Rather, as was suspected from earlier studies, these neurons appeared to reflect the learned significance of the cues. However, these cells only represented that information once the rat began to use it to guide go, no-go behavior. As was evident in the first experiment, where OFC neurons exhibited phasic responses as the rat performed various behaviors in the task and the population reflected the behaviorally relevant information available during odor sampling, the selectivity in the OFC neurons in this study was dependent on behavior.

The influence of the identity of the cues was evident in the effect of reversal. Firing selectivity in these OFC neurons was severely disrupted after reversal. For example, the neuron in Figure 16.8A became non-selective during reversal. Note that the selective response does not re-emerge after the rat attains criterion again; thus, the activity is not simply related to discrimination performance. Of the OFC neurons that exhibited selectivity post-criterion during initial training, 65% lost selectivity after reversal, 23% reversed odor selectivity, and only 12% maintained the same correlate. Despite the neurons that reversed, Figure 16.8B shows that, as a whole, this population of neurons became non-selective. In addition, approximately 25% of the neurons that were non-selective pre-reversal became selective when the contingencies were switched. That is, changes in the specific conjunctions between the odor cues and the associated reinforcers resulted in a turnover in the population of cells in OFC that actively represented that information. Recording during learning and reversal training in the OFC revealed the importance of identity, significance, and behavior to representations in this part of rat prefrontal cortex.

16.8 SUMMARY

The purpose of this chapter has been to provide a detailed account of how this author uses olfactory-guided behavior as a tool to study encoding in a prefrontal system. Olfaction is an excellent modality for the study of learning and behavior in the rat because rats are well-adapted to use olfactory input. Rats are skilled at the detection and identification of odors and are able to develop learning sets in tasks that utilize olfactory cues. Moreover, the OFC and the system of structures that interact with this prefrontal region in the rat are crucially involved in olfactory processing. This task and the appropriate manipulation of the cues and cognitive demands of the behavior allowed observations to be made regarding encoding in OFC that may not have been possible in another modality.

ACKNOWLEDGMENTS

The author would like to thank Dr. Kevin Austin for technical assistance on this manuscript and for preparation of some of the Web-based content (refer to www.psy.jhu.edu/~schoenbaum). A K08-AG00882-01 to GS supported this work.

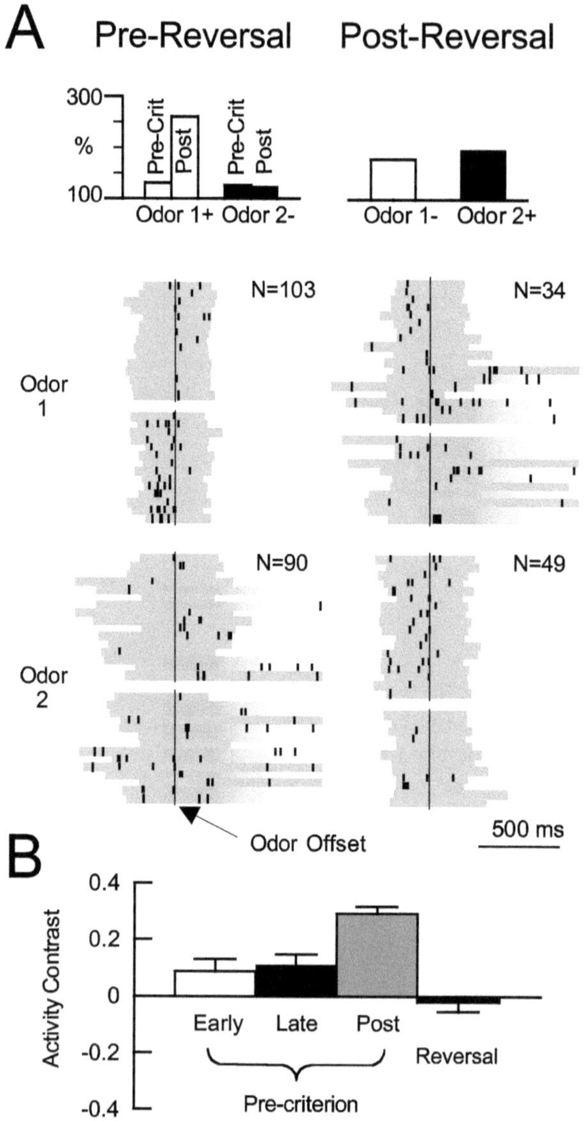
A
Pre-Reversal
Post-Reversal
300
%
100
Pre-Crit
Post
Pre-Crit
Post
Odor 1+
Odor 2-
Odor 1-
Odor 2+
N=103
N=34
Odor
1
N=90
N=49
Odor
2
Odor Offset
500 ms
B
0.4
0.2
0
-0.2
-0.4
Activity Contrast
Early
Late
Post
Reversal
Pre-criterion

FIGURE 16.8

FIGURE 16.8 (opposite) The development of selective activity to odors in the OFC during odor discrimination learning and reversal. A: Neural activity during evaluation of odor 1 (open bars) and odor 2 (closed bars) represented as a percentage of the pre-trial baseline firing rate in an OFC neuron (1.35 spikes/sec). This neuron fired more strongly during evaluation of the positive odor (odor 1) during post-criterion (post) training. That selectivity was not evident during pre-criterion (pre) training when the rat had not yet adopted a reliable response strategy to reflect the learned significance of the odors. Furthermore, the selectivity disappeared during reversal training. Raster displays below the summary bar graphs show neural activity on 30 representative trials (n = total trials) during evaluation of each odor before and after reversal, presented in the left and right panels, respectively. Neural activity, with spikes shown as black tic marks within the shaded regions, begins with odor onset and is synchronized to odor offset corresponding to withdrawal from the odor port (thin vertical line). Activity is truncated at the go response when a response was made or after 1500 msec in the event of a no-go. Trials in which a no-go occurred are evident in fading of the shaded region at the end of each Raster. Pre-criterion and post-criterion trials are separated by a small empty space in the displays of both the initial training and reversal training. The rat begins the session responding on every trial but gradually begins to withhold responses on the negative trials. During the pre-criterion phase, the rat makes several intermittent no-go responses, but the neuron fires very little during evaluation of either odor. The selective activity of this neuron develops only after a reliable change in the rat's behavioral strategy during post-criterion performance. During accurate performance, the neuron fires strongly during evaluation of the positive odor (odor 1). After reversal, the activity during odor evaluation is no longer selective either before or after the rat achieves criterion on the reversed discrimination problem. B: Contrast in activity during evaluation of positive and negative odors during the early (open bars) and late (closed bars) segments of pre-criterion training, during post-criterion performance (gray bars), and during reversal training (striped bars) for the neurons with post-criterion selectivity in the OFC. The activity contrast was calculated as the difference in firing rate during the evaluation of positive and negative odors divided by the sum of those rates, yielding values that ranged from –1 to +1. The calculation was referenced to the selectivity established during post-criterion training. Firing activity between the trials was used to calculate a baseline activity contrast of 0.04 (not shown). The degree of selectivity changed significantly in OFC during training. Post-hoc tests revealed that the degree of selectivity was not significantly different between the early and late segments of pre-criterion training, increasing significantly only during the post-criterion performance phase. During reversal training, the activity contrast decreased significantly relative to post-criterion performance, returning to a value not significantly different from baseline or either the early or the late segment of pre-criterion training. (see G. Schoenbaum et al., *J. Neurosci.,* 19, 1876–1884, 1999 for details).

REFERENCES

1. Eichenbaum, H. and Otto, T. Odor-guided learning and memory in rats: is it 'special'?, *TINS,* 16, 22-23, 1993.
2. Slotnick, B.M. The enigma of olfactory learning revisited, *Neuroscience,* 58, 1-12, 1994.
3. Slotnick, B.M. and Ptak, J.E. Olfactory intensity-difference thresholds in rats and humans, *Physiol. Behav.,* 19, 795-802, 1977.
4. Nigrosh, B.J., Slotnick, B.M., and Nevin, J.A. Olfactory discrimination, reversal learning, and stimulus control in rats, *J. Comp. Physiol. Psych.,* 89, 285-294, 1975.
5. Staubli, U., Fraser, D., Faraday, R., and Lynch, G. Olfaction and the "data" memory system in rats, *Behav. Neurosci.,* 101, 757-765, 1987.
6. Otto, T. and Eichenbaum, H. Complementary roles of the orbital prefrontal cortex and the perirhinal-entorhinal cortices in an odor-guided delayed-non-matching-to-sample task, *Behav. Neurosci.,* 106, 762-775, 1992.
7. Bunsey, M. and Eichenbaum, H. Conservation of hippocampal memory function in rats and humans, *Nature,* 379, 255-257, 1996.
8. Dusek, J.A. and Eichenbaum, H. The hippocampus and memory for orderly stimulus relations, *PNAS,* 94, 7109-7114, 1997.
9. Slotnick, B.M. and Katz, H.M. Olfactory learning-set formation in rats, *Science,* 185, 796-798, 1974.
10. Slotnick, B.M. and Brosvic, G.M. Failure of rats to acquire a taste reversal learning set, *Chem. Senses,* 12, 333-339, 1987.
11. Kolb, B. Functions of the frontal cortex of the rat: a comparative review, *Brain Res. Rev.,* 8, 65-98, 1984.
12. Eichenbaum, H. Using olfaction to study memory, *Ann. N.Y. Acad. Sci.,* 855, 657-669, 1998.
13. Scott, J.W., McBride, R.L., and Schneider, S.P. The organization of projections from the olfactory bulb to the piriform cortex and olfactory tubercle in the rat, *J. Comp. Neurol.,* 194, 519-534, 1980.
14. Slotnick, B.M. and Schoonover, F.W. Olfactory pathways and the sense of smell, *Neurosci. Biobehav. Rev.,* 16, 453-472, 1992.
15. Schoenbaum, G.M. and Eichenbaum, H.E. Information coding in the rodent prefrontal cortex. I. Single-neuron activity in orbitofrontal cortex compared with that in pyriform cortex, *J. Neurophysiol.,* 74, 733-750, 1995.
16. Kay, L.M. and Laurent, G. Odor- and context-dependent modulation of mitral cell activity in behaving rats, *Nature Neurosci.,* 2, 1003-1009, 1999.
17. Kosel, K.C., Van Hoesen, G.W., and West, J.R. Olfactory bulb projections to the parahippocampal area of the rat, *J. Comp. Neurol.,* 198, 467-482, 1981.
18. Swanson, L.W. and Kohler, C. Anatomical evidence for direct projections from the entorhinal area to the entire cortical mantle in the rat, *J. Neurosci.,* 6, 3010-3023, 1986.
19. Groenewegen, H.J. Organization of the afferent connections of the mediodorsal thalamic nucleus in the rat, related to the mediodorsal-prefrontal topography, *Neuroscience,* 24, 379-431, 1988.
20. McDonald, A.J. Cortical pathways to the mammalian amygdala, *Prog. Neurobiol.,* 55, 257-332, 1998.
21. Rosenkilde, C.E. Functional heterogeneity of the prefrontal cortex in the monkey: a review, *Behav. Neural Biol.,* 25, 301-345, 1979.
22. Kesner, R. Subregional analysis of mnemonic functions of the prefrontal cortex in the rat, *Psychobiology,* 28, 219-228, 2000.

23. Preuss, T.M. Do rats have prefrontal cortex? The Rose-Woolsey-Akert program reconsidered, *J. Cogn. Neurosci.,* 7, 1-24, 1995.
24. Krettek, J.E. and Price, J.L. The cortical projections of the mediodorsal nucleus and adjacent thalamic nuclei in the rat, *J. Comp. Neurol.,* 171, 157-192, 1977.
25. Ray, J.P. and Price, J.L. The organization of the thalamocortical connections of the mediodorsal thalamic nucleus in the rat, related to the ventral forebrain-prefrontal cortex topography, *J. Comp. Neurol.* 323, 167-197, 1992.
26. Krettek, J.E. and Price, J.L. Projections from the amygdaloid complex and adjacent olfactory structures to the entorhinal cortex and to the subiculum in the rat and cat, *J. Comp. Neurol.,* 172, 723-752, 1977.
27. McDonald, A.J. Organization of amygdaloid projections to the prefrontal cortex and associated striatum in the rat, *Neuroscience,* 44, 1-14, 1991.
28. Kita, H. and Kitai, S.T. Amygdaloid projections to the frontal cortex and the striatum in the rat, *J. Comp. Neurol.,* 298, 40-49, 1990.
29. O'Donnell, P. Ensemble coding in the nucleus accumbens, *Psychobiology,* 27, 187-197, 1999.
30. Graybiel, A.M. The basal ganglia and chunking of action repertoires, *Neurobiol. Learn. Mem.* 70, 119-136, 1998.
31. Groenewegen, H.J., Berendse, H.W., Wolters, J.G., and Lohman, A.H.M. The anatomical relationship of the prefrontal cortex with the striatopallidal system, the thalamus and the amygdala: evidence for a parallel organization, *Prog. Brain Res.,* 85, 95-118, 1990.
32. Uylings, H.B.M. and Van Eden, C.G. Qualitative and quantitative comparison of the prefrontal cortex in rat and in primates, including humans, *Prog. Brain Res.,* 85, 31-62, 1990.
33. Goldman-Rakic, P.S. The prefrontal landscape: implications of functional architecture for understanding human mentation and the central executive, *Philos. Trans. R. Soc. Lond. B. Biol. Sci.,* 351, 1445-1453, 1996.
34. Harlow, J.M. Passage of an iron bar through the head. *Publications of the Massachussets Medical Society,* 2, 329-346, 1868.
35. Jones, B. and Mishkin, M. Limbic lesions and the problem of stimulus-reinforcement associations, *Exp. Neurol.,* 36, 362-377, 1972.
36. DeCoteau, W.E., Kesner, R.P., and Williams, J.M. Short-term memory for food reward magnitude: the role of the prefrontal cortex, *Behav. Brain Res.,* 88, 239-249, 1997.
37. Bechara, A., Damasio, H., Tranel, D., and Damasio, A.R. Deciding advantageously before knowing the advantageous strategy, *Science,* 275, 1293-1294, 1997.
38. Kluver, H. and Bucy, P.C. Preliminary analysis of functions of the temporal lobes in monkeys, *Arch. Neurol. Psychiatry,* 42, 979-1000, 1939.
39. Tranel, D. and Hyman, B.T. Neuropsychological correlates of bilateral amygdala damage, *Arch. Neurol.,* 47, 349-355, 1990.
40. Killcross, S., Robbins, T.W., and Everitt, B.J. Different types of fear conditioned behavior mediated by separate nuclei within amygdala, *Nature,* 388, 377-380, 1997.
41. Kesner, R.P. and Williams, J.M. Memory for magnitude of reinforcement: dissociation between the amygdala and hippocampus, *Neurobiol. Learn. Mem.,* 64, 237-244, 1995.
42. Hatfield, T., Han, J.-S., Conley, M., Gallagher, M., and Holland, P. Neurotoxic lesions of basolateral, but not central, amygdala interfere with pavlovian second-order conditioning and reinforcer devaluation effects, *J. Neurosci.,* 16, 5256-5265, 1996.
43. Gallagher, M., McMahan, R.W., and Schoenbaum, G. Orbitofrontal cortex and representations of incentive value in associative learning, *J. Neurosci.,* 19, 6610-6614, 1999.

44. Baxter, M.G., Parker, A., Lindner, C.C.C., Izquierdo, A.D., and Murray, E.A. Control of response selection by reinforcer value requires interaction of amygdala and orbital prefrontal cortex, *J. Neurosci.,* 20, 4311-4319, 2000.
45. Eichenbaum, H., Clegg, R.A., and Feeley, A. Reexamination of functional subdivisions of the rodent prefrontal cortex, *Exp. Neurol.,* 79, 434-451, 1983.
46. Meunier, M., Bachevalier, J., and Mishkin, M. Effects of orbital frontal and anterior cingulated lesions on object and spatial memory in rhesus monkeys, *Neuropsychologia,* 35, 999-1015, 1997.
47. Price, J.L., Carmichael, S.T., Carnes, K.M., Clugnet, M.-C., Kuroda, M., and Ray, J.P. Olfactory input to the prefrontal cortex, in *Olfaction: A Model System for Computational Neuroscience,* J. Davis and H. Eichenbaum, Eds., Bradford Book, MIT Press, Cambridge MA, 1991.
48. Schoenbaum, G. and Eichenbaum, H. Information coding in the rodent prefrontal cortex. II. Ensemble activity in orbitofrontal cortex, *J. Neurophysiol.,* 74, 751-762, 1995.
49. Schoenbaum, G., Chiba, A.A., and Gallagher, M. Orbitofrontal cortex and basolateral amygdala encode expected outcomes during learning, *Nature Neurosci.,* 1, 155-159, 1998.
50. Schoenbaum, G., Chiba, A.A., and Gallagher, M. Neural encoding in orbitofrontal cortex and basolateral amygdala during olfactory discrimination learning, *J. Neurosci.,* 19, 1876-1884, 1999.
51. Schoenbaum, G., Chiba, A.A., and Gallagher, M. Changes in functional connectivity in orbitofrontal cortex and basolateral amygdala during learning and reversal training, *J. Neurosci.,* 20, 5179-5189, 2000.
52. Kubie, J.L. A driveable bundle of microwires for collecting single-unit data from freely-moving rats, *Physiol. Behav.,* 32, 115-118, 1984.
53. Lemon, R. and Prochazka, A. *Methods for Neuronal Recording in Conscious Animals,* John Wiley & Sons, New York, 1984.
54. Haberly, L.B. Summed potentials evoked in opossum prepyriform cortex, *J. Neurophysiol.,* 36, 775-788, 1973.
55. Haberly, L.B. and Shepherd, G.M. Current-density analysis of summed evoked potentials in opossum prepyriform cortex, *J. Neurophysiol.,* 36, 789-802, 1973.
56. Nicolelis, M.A.L. and Chapin, J.K. Spatiotemporal structure of somatosensory responses of many-neuron ensembles in the rat ventral posterior medial nucleus of the thalamus, *J. Neurosci.,* 14, 3511-3532, 1994.
57. Weijnen, J.A.W.M. and Chehade, N. Recording of brain potentials with FET-circuits: hazard of inadvertent lesions, *Brain Res. Bull.,* 19, 617-618, 1987.
58. Gochin, P.M., Colombo, M., Dorfman, G.A., Gerstein, G.L., and Gross, C.G. Neural ensemble coding in inferior temporal cortex, *J. Neurophysiol.,* 71, 2325-2337, 1994.
59. Mosteller, F. and Tukey, J.W. *Data Analysis and Regression,* Addison-Wesley, Reading MA, 1977.
60. Lipton, P.A., Alvarez, P., and Eichenbaum, H. Crossmodal associative memory representations in rodent orbitofrontal cortex, *Neuron,* 22, 349-359, 1999.
61. Ramus, S.J. and Eichenbaum, H.B. Neural correlates of olfactory recognition memory in the rat orbitofrontal cortex, *Society for Neuroscience Abstracts,* 25, 352.3, 1999.
62. Fuster, J.M., Bauer, R.H., and Jervey, J.P. Cellular discharge in the dorsolateral prefrontal cortex of the monkey in cognitive tasks, *Exp. Neurol.,* 77, 679-694, 1982.
63. Kubota, K. and Komatsu, H. Neuron activities of monkey prefrontal cortex during the learning of visual discrimination tasks with go/no-go performances, *Neurosci. Res.,* 3, 106-129, 1985.

64. Watanabe, M. Prefrontal unit activity during associative learning in the monkey, *Exp. Brain Res.,* 80, 296-309, 1990.
66. Wilson, F.A.W., Scalaidhe, S.P.O., and Goldman-Rakic, P.S. Dissociation of object and spatial processing domains in primate prefrontal cortex, *Science,* 260, 1955-1958, 1993.
67. Rainer, G., Asaad, W.F., and Miller, E.K. Selective representation of relevant information by neurons in the primate prefrontal cortex, *Nature,* 393, 577-579, 1998.
68. Rolls, E.T., Critchley, H.D., Mason, R., and Wakeman, E.A. Orbitofrontal cortex neurons: role in olfactory and visual association learning, *J. Neurophysiol.,* 75, 1970-1981, 1996.
69. Hetherington, P.A. and Shapiro, M.L. Hippocampal place fields are altered by the removal of single visual cues in a distance-dependent manner, *Behav. Neurosci.,* 111(1), 20-34, 1997.

Appendix — Suppliers

Aldrich Chemical
P.O. Box 355
Milwaukee, WI 53201
www.sigma-aldrich.com
414-273-3850

California Fine Wire
P.O. Box 446
Grover Beach, CA 93483
www.calfinewire.com
805-489-5144

Cambridge Electronic Design Limited
Science Park
Milton Rd.
Cambridge CB4 4FE
England
www.ced.co.uk
+44 223 420186

Cooner Wire
9265 Owensmouth Ave.
Chatsworth, CA 91311
www.coonerwire.com
818-882-8311

Computer Boards
2 Commerce Park Blvd.
Middleboro, MA 02346
www.computerboards.com
508-946-5100

Crist Instrument Co.
10200 Moxley Rd.
P.O. Box 128
Damascus, MD 20872
301-253-2184

Dentsply Caulk
38 West Clarke Ave.
P.O. Box 359
Milford, DE 19963
www.dentsply.com
800-LD-CAULK

Data Translation
100 Locke Dr.
Marlboro, MA 01752
www.datatranslation.com
508-481-3700

Datawave Technologies
380 Main St.
Suite 209
Longmont, CO 80501
www.dwavetech.com
303-776-8214

Eclectic Engineering Studio
6 Blackman Rd.
Canton, MA 02021
www.eclecticstudio.com

Fischer Scientific
585 Alpha Dr.
Pittsburgh, PA 15238
800-766-7000

Grass Instruments
Astro-Med, Grass-Telefactor Product Group
600 East Greenwich Ave.
West Warwick, RI 02893
www.grassinstruments.com
877-472-7779

Instech Laboratories
5209 Militia Hill Rd.
Plymouth Meeting, PA 19462
www.instechlabs.com
610-941-0132

Intl Flavors and Fragrances
521 West 57th St.
New York, NY 10019
212-765-5500

David Kopf Instruments
7324 Elmo St.
P.O. Box 636
Tujunga, CA 91042
www.kopfinstruments.com
818-352-3274

Mallinckrodt
675 McDonnell Blvd.
Hazelwood, MO 63042
www.mallinckrodt.com
314-654-2000

McMaster-Carr Supply Co.
6100 Fulton Industrial Blvd.
Atlanta, GA 30336
www.mcmaster-carr.com
404-346-7000

Minuteman Controls
7 Foster St.
P.O. Box 1559
Wakefield, MA 01880
www.minutemancontrols.com
800-343-8633

Newark Electronics
7272 Park Circle Dr.
Hanover, MD 21076
www.newark.com
1-800-NEWARK

Plexon Incorporated
6500 Greenville Ave.
Suite 730
Dallas, TX 75206
www.plexoninc.com
214-369-4957

Small Parts
13980 N.W. 58th Ct.
P.O. Box 4650
Miami Lakes, FL 33014
www.smallparts.com
800-220-4242

Thompson Dental Company
2720 Discovery Dr.
P.O. Box 17805
Raleigh, NC 27619
919-790-9995

US Plastics Corp
1390 Neubrecht Rd.
Lima, OH 45801
419-228-2242

BOX 16.1
Representation of Incentive Value in OFC and ABL

While olfactory information appears to be central to processing in rat OFC, processing of other types of information may also be crucial to understanding the role of this region. For example, OFC is also implicated in the processing of information regarding the motivational or incentive value of cues via connections with basolateral amygdala. Lesions of either ABL or OFC in rats result in impairments in discriminating cues associated with different magnitudes of reinforcement (Kesner and Williams, 1995; DeCoteau et al., 1997), and more recently Gallagher and colleagues have found that appropriate responding to conditioned stimuli in a devaluation paradigm depends on intact function in both regions. In this set of two studies, food-deprived rats were trained to associate a light-CS with delivery of a food-US. Prior to training, the test subjects received bilateral neurotoxic lesions of either ABL (Hatfield et al., 1996) or OFC (Gallagher et al., 1999). Rats with lesions of either structure acquired the conditioned response (food cup behaviors) normally as depicted in Figure 1.

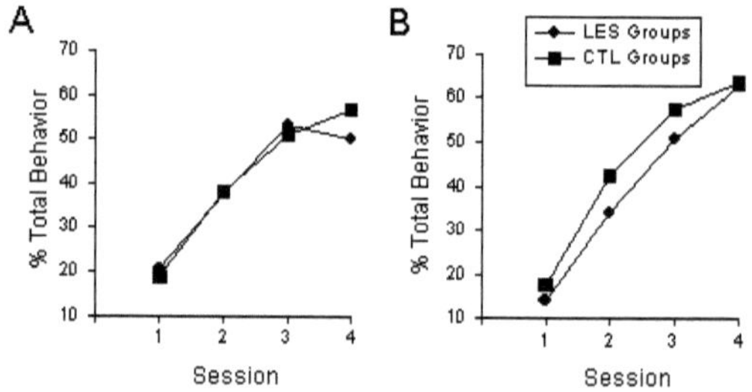

FIGURE 1 Conditioning of the food cup response in lesioned (LES) and control (CTL) subjects. Both ABL (A) and OFC (B) lesioned rats acquired the food cup response normally when compared to controls. (Figures adapted from Hatfield et al., 1996 and Gallagher et al., 1999).

After the end of training, both test and control subjects were returned to their home cages where half of the rats in each group received paired presentations of the food-US and illness induced by LiCl injections. The other half of the rats received unpaired presentations of the food and illness. Rats that received paired presentations of food and illness developed a conditioned taste aversion as a result of training. The results are shown in Figure 2 for both experiments. By the third trial (day 3), none of the rats (lesion or control) would consume the food in the paired groups; rats that received unpaired presentations continued to consume the food normally.

The rats were then returned to the training environment and exposed to the light-CS. Control subjects that received paired presentations of the food and illness exhibited fewer conditioned responses to the light-CS when compared to rats that received unpaired presentations. These data are shown in Figure 3. In contrast, rats with lesions of either OFC or ABL exhibited the same amount

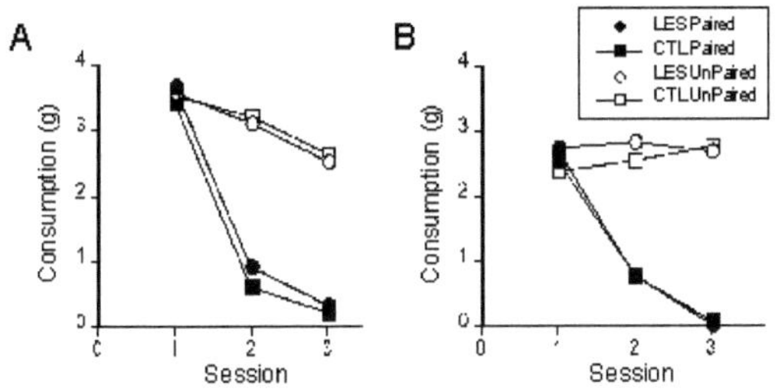

FIGURE 2 Devaluation of the food reinforcer by pairing with LiCl-induced illness in lesioned (LES) and control (CTL) subjects. Filled symbols represent rats that received paired presentations of the food and illness, and unfilled symbols represent rats that received unpaired presentations. Both ABL (A) and OFC (B) lesioned rats showed normal devaluation when presentations were paired. (Figures adapted from Hatfield et al., 1996 and Gallagher et al., 1999).

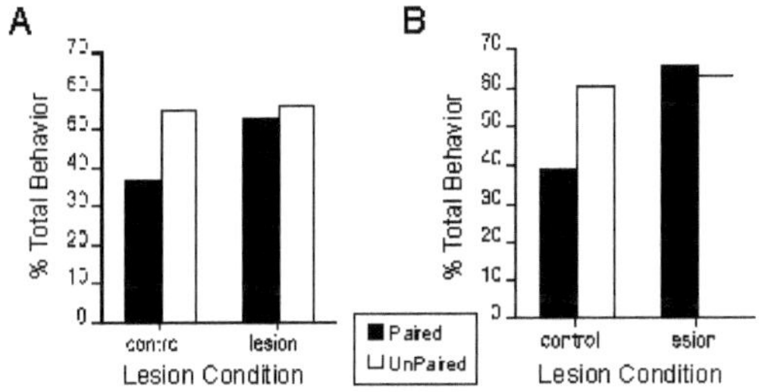

FIGURE 3 Food cup responses during devaluation testing in lesioned and control subjects. Filled bars represent rats that received paired presentations of the food and illness, and unfilled symbols represent rats that received unpaired presentations. Normal rats exhibit significantly less conditioned responding if they received paired presentations of food and illness; lesions of either ABL (A) or OFC (B) prevent the normal decrease in conditioned responding in the paired group. (Figures adapted from Hatfield et al., 1996 and Gallagher et al., 1999).

of conditioned responding to the light-CS as did rats that received unpaired presentations. These rats continued to make food cup responses despite the observation that they would not consume the food-US in either the home cage or the training environment. In other words, rats with lesions of either OFC or ABL were unable to alter their behavior to reflect the altered motivational significance of the CS. Similar results have recently been reported in primates with crossed lesions of OFC and ABL (Baxter et al., 2000). These experiments show that connections between OFC and ABL are crucial to the appropriate use of motivational information in some contexts.

Kesner RP, Willams JM. Memory of magnitude of reinforcement: Dissociation between the amygdala and hippocampus. *Neurobiol Learn Mem* 64:237-244, 1995.

DeCoteau WE, Kesner RP, Williams JM. Short-term memory for food reward magnitude: The role of the prefrontal cortex. *Behav Brain Res* 88:239-249, 1997.

Gallagher M, McMahan RW, Schoenbaum G. Orbitofrontal cortex and representations of incentive value in associative learning. *J Neurosci* 19:6610-6614, 1999.

Hatfield T, Han J-S, Conley M, Gallagher M, Holland P. Neurotoxic lesions of basolateral, but not central, amygdala interfere with pavlovian second-order conditioning and reinforcer devaluation effects. *J Neurosci* 16:5256-5265, 1996.

Baxter MG, Parker A, Lindner CCC, Izquierdo AD, Murray EA. Control of response selection by reinforcer value requires interaction of amygdala and orbital prefrontal cortex. *J Neurosci* 20:4311-4319, 2000.

Web content available at www.psy.jhu.edu/~schoenbaum

BOX 16.2
Odor Discrimination Training

A rat will readily learn to discriminate between non-biological odors to receive rewards. Figure 1 shows the behavioral apparatus used in the odor discrimination task. This apparatus consists of an odor sampling port (white circle) and a fluid delivery well (black circle). On each trial, the rats sample odors presented at the odor port (Figure 2a) and then must decide, based on the identity of the odor, whether to make a 'go' response at a nearby fluid well (Figure 2b) to receive a sucrose reward or whether to withhold that same response (Figure 2c) to avoid a penalty (quinine or a prolonged time-out period). The video clip below shows the behavior of a rat during learning regarding new odor cues.

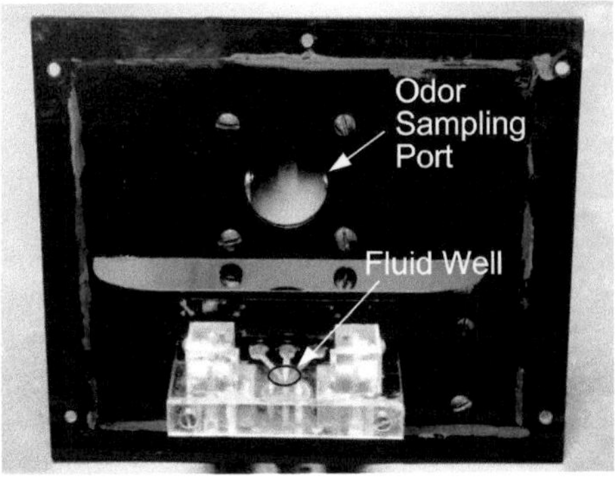

FIGURE 1 Photograph of odor port and fluid well on polycarbonate insert that the rat below is using in the odor discrimination task. This insert is bolted into the wall of a cubical aluminum training chamber pictured below.

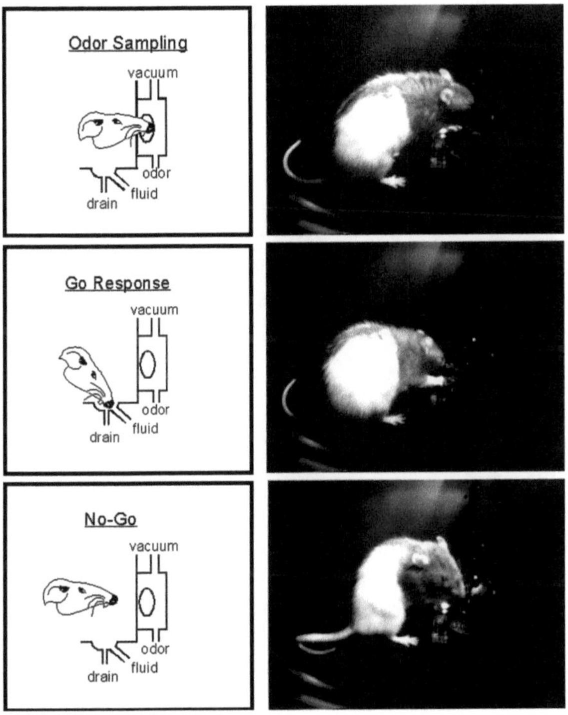

FIGURE 2 Still photographs and schematics showing the behaviors in the discrimination task. The photographs were taken by a low-light video camera mounted in the side-wall of the behavioral chamber. The camera is located on the wall to the left of the photograph in Figure 1.

Video Clip of Rat Learning New Discrimination

Media file available at www.psy.jhu.edu/~schoenbaum

BOX 16.3
Olfactory Stimulus Control

Cues must be strictly controlled in behavioral neurophysiology experiments. Crucial information is available in the onset latency and duration of neural firing patterns, so it is crucial that the experimenter be able to precisely control cue onset and offset. Moreover firing selectivity for cues would be compromised if cues were allowed to mix together. These requirements are easily met with visual, auditory or tactile cues. Onset, offset and separation of different items is simplified because the light, sound and touch are essentially instantaneous compared to the speed of neural transmission. Odors on the other hand present a challenge to the experimenter. Unlike cues in these other modalities, odors must travel via diffusion through the air and then into the nasal mucosa before being detected by the olfactory chemoreceptors. This diffusion requires time and makes control of onset and offset difficult, increasing the likelihood that one cue could mix with another.

To prevent or minimize these possibilities, it is important to exercise rigorous control of the odor cues. This control is accomplished via a system of air and vacuum lines, solenoids and flowmeters. Central to this system is a removable odor cartridge that isolates the odor cues from each other (Figure 1).

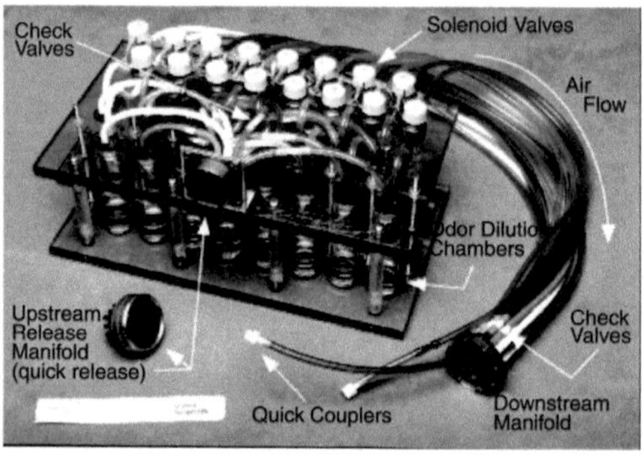

FIGURE 1 Removable 16-odor cartridge. The solenoid valves allow independent delivery of any one of the 16 different odors. Quick release connections at manifolds upstream and downstream from the odor dilutions chambers allow different cartridges to be used in the behavioral setup, so that many different odors can be used in a learning experiment. Odors are isolated into separate lines that are connected only at the manifolds. Backflow is prevented by the solenoids and also by checkvalves located at both manifolds.

This cartridge is machined from 2 pieces of 1/2" polycarbonate. The bottom piece contains depressions to hold each of 16 different odor dilution chambers made from glass sample tubes (Fischer Scientific, 0333922M). Each chamber contains a unique odor (International Flavors and Fragrances) diluted in propylene glycol. The top piece is machined with 16 circular depressions. Each circular cutout fits an o-ring, which seals the opening of the glass tube (Figure 2) when the two polycarbonate pieces are bolted together. Two holes are drilled in the polycarbonate inside each circular cutout (Figure 2). These openings are tapped with 10-32 threads to accept barbed fittings (Minuteman Controls, F312085). Air entry to each chamber is controlled by a solenoid valve (McMaster-Carr, 4916K13, 4916K81) attached to the polycarbonate top next to the openings to the odor dilution chamber (Figure 2). This valve is activated via computer-controlled relays (Newark, 81F9285, 75F004). The entire cartridge, containing 16 different odors, may be rapidly attached to the behavioral system via quick release connections/manifolds (Minuteman Controls, 2BH-19 and Small Parts, 01003312) located at either end of the lines (Figure 1). Odor mixing at these manifolds is prevented by the presence of check valves (Minuteman Controls, F2804403B85). Connections are made using 1/8" ID polyurethane tubing (Minuteman Controls, Special Order Item), which is color-coded for ease of use.

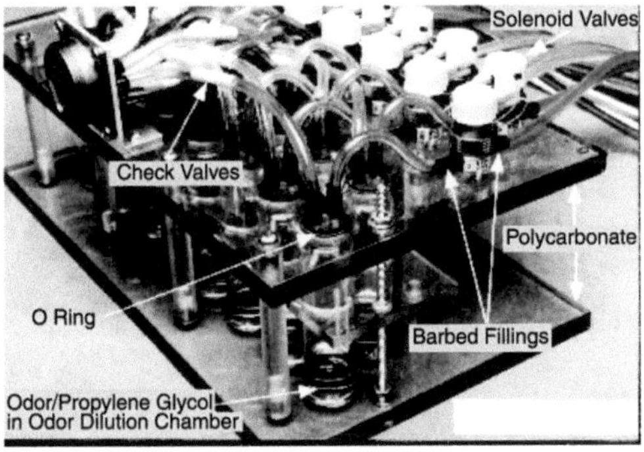

FIGURE 2 Close-up view of odor-dilution chamber. This view shows how each odor is isolated. Polyurethane tubing brings air from the manifold to the solenoid valve. Opening of the valve allows a one-way flow of air through barbed fittings in the polycarbonate top piece into the odor chamber, displacing odor-saturated air out the second barbed fitting and on to the downstream manifold. Note the circular depression fitted with an o-ring sealing the top of the glass odor chamber to the polycarbonate.

Note that all lines and equipment, such as the control solenoids, checkvalves, and the downstream manifold, are dedicated to a particular odor on the cartridge. Additional components, such as the flowmeters or upstream lines, that control the odors but do not come in contact with the odorized air flow are located off the cartridge. The only component not on the cartridge that comes into contact with multiple odors is the final delivery line into the odor delivery chamber on the behavioral box. This line is common to all odors and may be easily changed as necessary.

Suppliers

- International Flavors and Fragrances, 521 West 57th Street, New York, NY 10019, 212-765-5500.
- Fischer Scientific, 585 Alpha Drive, Pittsburgh, PA 15238, 800-766-7000 McMaster-Carr Supply Company, 6100 Fulton Industrial Boulevard, Atlanta, GA 30336, 404-346-7000.
- Minuteman Controls, 7 Foster Street, PO Box 1559, Wakefield, MA 01880, 800-343-8633.

Web content available at www.psy.jhu.edu/schoenbaum

BOX 16.4
Noise in Electrophysiology
Introduction

Electrical noise is present in all electrophysiological recordings. Unique technical challenges are presented by the freely-behaving preparation used in behavioral neurophysiology. The electrodes usually consist of fine wires which are quite different from the conventional saline-filled pipette used *in-vitro*. In addition, because the preparation is chronically implanted and behaving, the recording system must be designed to minimize artifacts associated with movement. Knowledge of basic electrical principles is essential for an understanding of electrical noise. Thus, our discussion begins with some electrical concepts.

Electrical Concepts

In an electrical circuit, current, voltage and resistance are all related as expressed by Ohm's Law,

$$V = IR$$

where V represents voltage (in Volts), I represents current (in Amperes) and R represents resistance (in Ohms) with polarity and direction as shown (Figure 1). Current flows from postive (+) to negative (–).

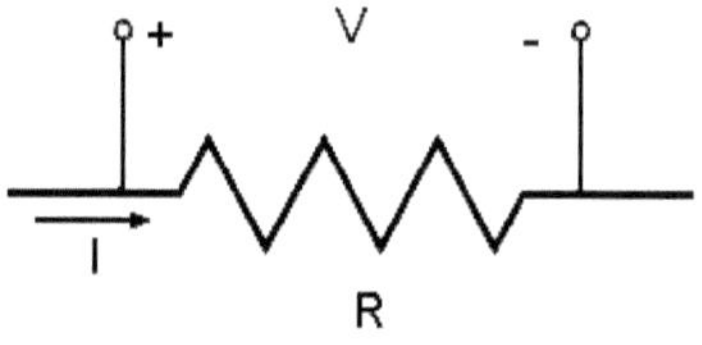

FIGURE 1

In an extracellular electrophysiological recording, changes in ionic concentration cause small currents to flow through the electrode. The electrode is connected to a preamplifier with a very high input impedance. Impedance (Z) is measured in Ohms, like resistance. However, impedance is a complex quantity that describes how a circuit element responds to both steady-state and time-varying signals. Ohm's law also applies to impedances as shown in the equation:

$$V(t) = I(t)\ Z$$

and also as depicted in Figure 2.

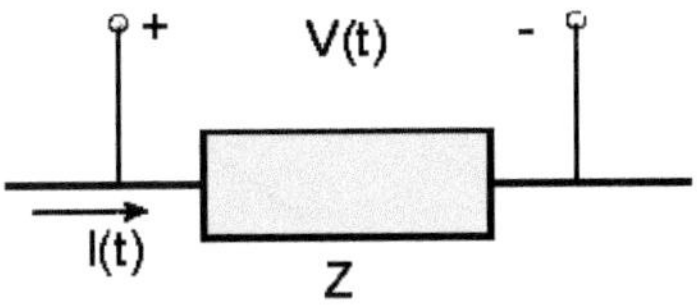

FIGURE 2

Connecting the electrode to a preamplifier with a high input impedance allows small currents induced at the electrode tip to be seen as measurable voltages (50-300 μV for spontaneous brain activity). A simplified equivalent circuit of an electrode / preamplifier recording system is shown (Figure 3). When recording from freely-behaving animals, the preamplifier is often implemented on a headstage and generally has unity gain ($A_v = 1$). For more about the headstages see movement artifact, below.

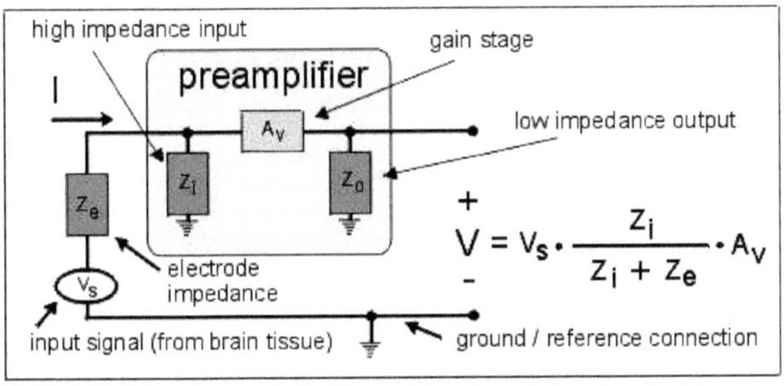

FIGURE 3

Types of Noise

High input impedance preamplifiers and high impedance electrodes are necessary for electrophysiological recording. The small tip diameter required for proper unit isolation results in electrodes with high impedances. Unfortunately, high impedances also allow small currents that are unrelated to brain function to compete with the signals of interest. Signals that interfere with measurement of neural activity are called noise. Generally, noise signals can be characterized as either intrinsic or external. Intrinsic noise is generated in the resistive components of your system by the random movement of charge carriers and is known as Johnson noise. In the figure above, the resistive portion of each impedance contributes to Johnson noise. In general, higher resistive impedances generate more noise. In the laboratory, Johnson noise is controlled by appropriate electronic (bandpass) filtering and minimization of electrode impedance (Metal Microelectrodes for Electrophysiology in Behaving Animals).

External noise is introduced through connections between elements of your recording system. The important elements and connections between them are shown (Figure 4). All of these connections are prone to noise induced by electromagnetic radiation from various sources called inductive pickup. Common sources for

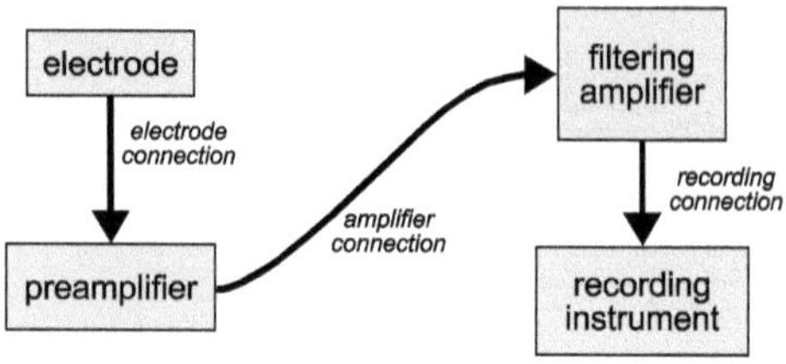

FIGURE 4

inductive pickup are computer monitors and light fixtures. In general, inductive pickup can be avoided by using shielded cables to make connection between different components. It is also helpful to keep the cable lengths as short as is practical. In particular, the length of the electrode connection should be kept to a minimum because of the high impedances involved (more about this later when we discuss movement artifact).

In addition, when electrical components make voltage measurements from different reference points a condition called a ground loop may result in a noisy recording. To avoid ground loops, all elements of the recording system should use a single point as a reference (Figure 5A) or the preparation should be electrically isolated (Figure 5B). An additional consideration for the freely-behaving preparation is that any conductive surface accessible inside the recording chamber can inadvertantly provide an additional path to ground. Thus, it is advisable to paint the interior of the recording chamber to make it non-conductive.

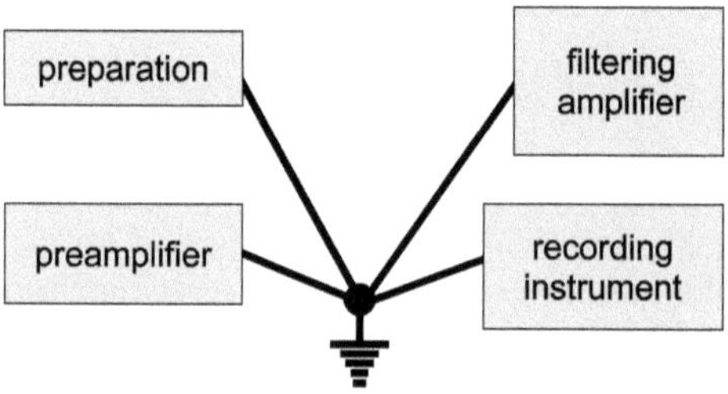

FIGURE 5A

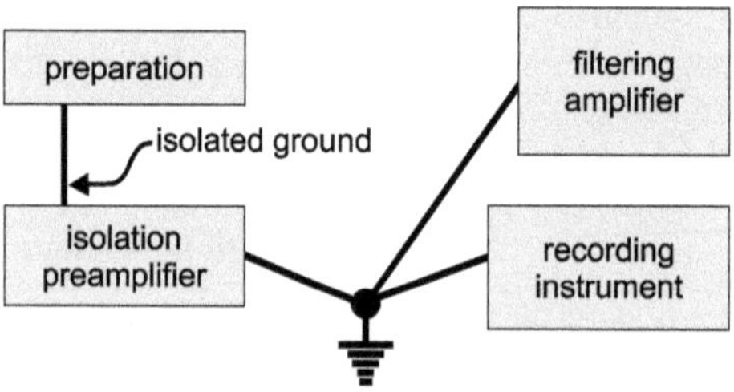

FIGURE 5B

Movement artifact is a type of noise specific to the freely-behaving preparation. Flexing of the cable connecting the animal to the recording equipment results in changes in capacitance. This change in capacitance causes the flow of charge (current) through the cable. Under high impedance recording conditions, particularly in the electrode–preamplifier connection, this movement-induced current can be many times greater than the electrophysiological signal. To minimize movement artifact, the electrode–preamplifier connection should be made as short and as rigid as possible by mounting the preamplifier directly to the skull of the preparation (called a headstage). Thus, the preamplifier presents a low impedance to the flexing cable which minimizes the movement-related artifact. The headstage effectively converts the high impedance signal from the electrode into a low impedance signal at the level of the cable (Headstage Amplifiers for Electrophysiology in Behaving Animals).

Web content available at www.psy.jhu.edu/schoenbaum

This web page coauthored by Kevin B. Austin, Ph.D., Eclectic Engineering Studio, www.EclecticStudio.com

Definitions

Current: The rate that charge passes through an electrically conductive medium. One Ampere is equal to one Coulomb per second. In a wire, current is carried by electrons. In extracellular fluid, current can be carried by any ion in solution.

Voltage: The force per unit of charge. Voltage causes charge (i.e., current) to flow in a particular direction. One Volt is equal to one Joule per Coulomb.

Resistance: The property of a material to resist current flow. One Ohm is equal to one Volt per Ampere.

Reactance: The property of an elelectical circuit element to resist changes in voltage or current. Like resistance, it is measured in Ohms. The reactance of a circuit changes with frequency. Reactance also alters the phase relationship between current and voltage. It applies to all time-variant signals. See also: impedance, resistance, capacitive reactance.

Impedance: A complex quantity representing the property of a circuit to resist both steady-state current flow (resistance) and changes in current or voltage (reactance). See also: reactance, resistance, capacitive reactance.

Complex numbers: Quantities that are expressed as a vector having both magnitude and phase information.

Johnson noise: Noise generated by the random movement of charge in a resistive medium. This noise is often described by electrophysiologists as "hiss" or "hash". It has a wide frequency spectrum, thus, it can be limited by electronic filters. Johnson noise increases with temperature and is sometimes called "thermal noise". Johnson noise also increases in proportion to resistive impedance.

Inductive pickup: Noise induced by external sources of electromagnetic radiation in unshielded connections between elements of the recording system. Electrophysiologists often describe this noise as "hum" or "buzz".

Input bias current: The current that an op-amp must draw through its inputs to operate properly.

Movement artifact: When the cable connecting the electrode to the preamplifier is bent, the change in cable capacitance causes charge to move. This charge movement generates an appreciable voltage across high impedances. This type of noise is experienced as an abrupt change in baseline that can sometimes result in amplifier saturation.

Gate leakage current: Current drawn through the input (gate) of a junctional FET circuit for proper operation.

Ground loop: A condition that occurs when the components of your recording system use different references. A ground loop is most often experienced as a prominent 60 Hz "hum" in your recording. Thus, it is indistinguishable from some types of inductive pickup. Ground loops are avoided by making sure each component of your recording system is tied to one (and only one) reference point.

Capacitive Reactance: The component of impedance that is associated with capacitance. Capacitve reactance (Xc) is inversely proportional to frequency and capacitance. When driven by a sinusoidal source, capacitive reactance is given by:

$$X_c = 1 / 2 \pi f C$$

where f is the frequency of the sinusoid in Hertz (Hz), C is the capacitance in Farads (F) and X_c is the capacitive reactance in Ohms (Ω). See also: reactance, impedance.

Operational Amplifier (op-amp): A general-purpose building-block for constructing analog circuits. Common circuits that can be built with op-amps are amplifiers, filters and comparators. Op-amps are integrated circuits and are usually grouped in packages of one, two or four. The design of op-amp circuits is beyond the scope of this document.

Field-effect transistor (FET): A discrete semiconductor component used to build integrated circuits such as operational amplifiers. A FET is a 3-terminal device with terminals labeled gate, source and drain. Generally, the input signal is connected to the gate and the output is taken from either the source or the drain. The design of FET circuits is quite involved and beyond the scope of this document.

Web content available at www.psy.jhu.edu/schoenbaum

This web page coauthored by Kevin B. Austin, Ph.D., Eclectic Engineering Studio, www.EclecticStudio.com

BOX 16.5
Metal Microelectrodes for Recording in Behaving Animals

Electrode impedance is a primary factor in the performance of any electrophysiological recording system. Electrode impedance describes the electrical characteristics of the complex interface between the metal wire microelectrode and the extracellular recording medium.

Modeling Electrode Impedance

The equivalent electrical circuit of a metal microelectrode immersed in an electrolyte solution is shown in Figure 1. It consists of both resistive and capacitive elements. The resistive portion represents the mobility of charge carriers on each side of the solution/metal interface plus the weak exchange of ions across the so-called "double-layer".

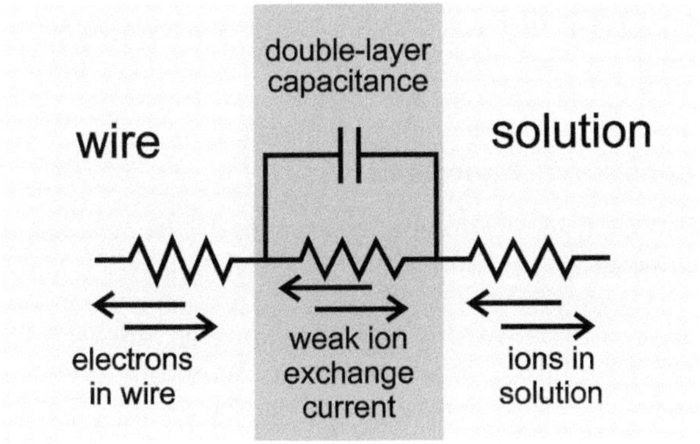

FIGURE 1

The double-layer is created by polarized water molecules at the metal/brain interface. These water molecules form a thin dielectric (Figure 2). Thus, the double-layer is modeled by a capacitor. The exchange of ions across the double layer is weak and the interface is considered polarizable. That is, ions do not physically pass into and out of the electrode. Instead, changes in ionic concentration in the extracellular space attract (or repel) electrons to (or from) the interface.

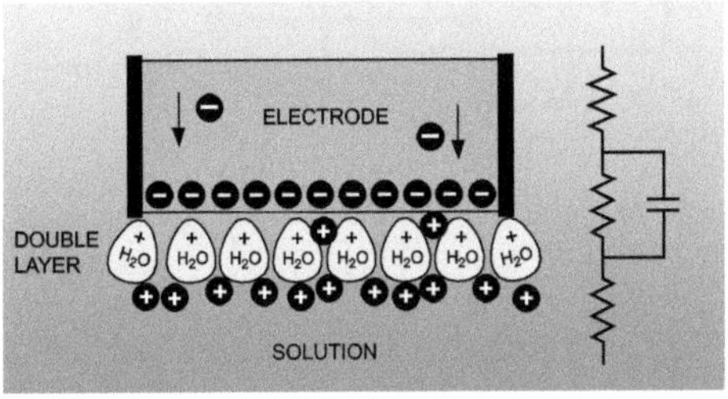

FIGURE 2

Both the capacitance and the resistance of this interface (Figure 2) are dependent on the size of the surface area of the metal in contact with the solution. In general, more surface area results in lower contact resistance and higher double-layer capacitance. Recall that higher capacitance values have smaller capacitive reactance at any given frequency. Thus, the overall effect of increasing surface area is to reduce electrode impedance.

Electrode impedance is related to Johnson (or thermal) noise. Johnson (or thermal) noise is generated by random movement of electrons in all resistive impedance elements. In general, at any given temperature, Johnson noise is proportional to resistive impedance. If the resistive impedance of the electrode is too high, these random fluctuations will interfere with the electrophysiological recording. Because electrode impedance is largely determined by surface area of the electrode tip, increasing tip diameter is one way to reduce the Johnson noise inherent to the electrode. Unfortunately, many applications require small tip diameters to obtain sufficient single-unit isolation. An alternative approach to reducing Johnson noise is to increase the surface area at the electrode tip without increasing tip diameter. This is the strategy employed by both "bubbling" and electroplating, two methods that reduce electrode impedance without sacrificing the selectivity of the microelectrode.

Measuring Electrode Impedance

The impedance of an electrode is measured by passing a small AC current through the electrode, measuring the voltage drop across a known resistance placed in series with the electrode and using this information to calculate the impedance. In our lab we use a battery-powered sine-wave generator, an oscilloscope and a beaker of saline arranged as shown in Figure 3. The sine wave generator is set to 100 mV and 100 Hz. The oscilloscope sweep rate and sensitivity is adjusted to display several cycles of the test waveform. The reference electrode should have a large surface area.

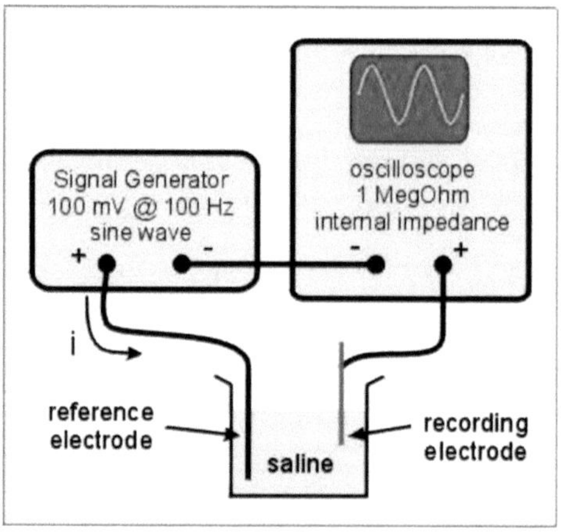

FIGURE 3

Electrode impedance may be calculated directly from the oscilloscope reading using the equation in Figure 4 below. In our lab, we use a table to make quick determinations of electrode impedance from measured voltages. In this analysis, R_m is the internal impedance of the oscilloscope (1 megohm in our example), V_0 is the output of the sine-wave generator (100 millivolts in our example) and V_m is the value displayed on the oscilloscope. The electrode impedance measured will be dependent on the frequency of the signal generator. Electrode impedance will decrease as frequency is increased consistent with the model shown in Figure 1.

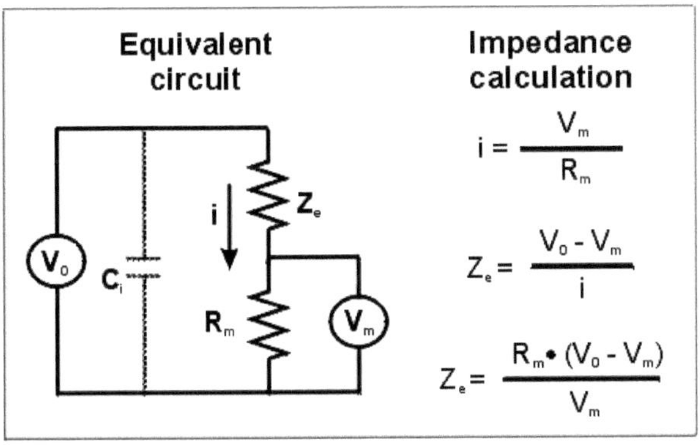

FIGURE 4

See Also: Table for Computation of Impedance

V_m (mV)	Z_e (kW)
50	1000
52	923
54	852
56	786
58	724
60	667
62	613
64	563
66	515
68	471
70	429
72	389
74	351
76	316
78	282
80	250
82	220
84	190
86	163
88	136
90	111
92	87
94	64
96	42

Calculating Electrode Impedance Using a Table

In our lab, we use this table (right) to compute electrode impedance (Z_e) from a measured voltage (V_m). The table applies to measurements made using the circuit in Metal Microelectrodes for Recording in Behaving Animals and these parameters:

Source voltage: V_0 = 100 mV pk-pk, sinusoid (100 Hz)

Oscilloscope internal impedance: R_m = 1 megohm

You may wish to adapt this technique for your particular situation. To do this, create a spreadsheet using an equation to compute impedance from your setup-specific parameters. One possible variation from our setup would use a digital voltmeter instead of an oscilloscope. This substitution would most likely require you to change R_m from 1 megohm to 10 megohms as most digital voltmeters have 10 megohms internal impedance. If you decide to use a digital voltmeter, make sure you are measuring "AC volts".

Practical Considerations

Immerse the electrode to a consistent depth. When a metal microelectrode is immersed in a conductive electrolyte, the wire and the solution form the plates of a capacitor (shown as C_i, above). The insulation coating the electrode wire is the dielectric and the capacitance depends on the depth of immersion. Thus, it is important to immerse the electrode to a consistent depth for each test so as to keep C_i invariant between tests. I typically immerse the tips to a depth of 2 mm.

Check the voltage output of the signal generator. Make sure the sine wave is devoid of DC offset (zero mean) by checking it with an oscilloscope or DC voltmeter.

Check the frequency of the sine wave. Impedance is dependent on the frequency of the signal used to measure it. For impedance measurements to be comparable and consistent, they must be made using the same frequency signal each time.

Procedures to Reduce Electrode Impedance

A small diameter tip is necessary to detect the local ionic changes that are generated extracellularly by an action potential. Wire diameter is chosen based on the cellular properties of the particular brain area being recorded. The surface area of the tip of the electrode is a strong determinant of the impedance of the electrode. Both capacitive reactance and contact resistance are reduced by increasing the surface area of the exposed electrode tip. Thus, procedures that increase surface area will result in electrodes with lower impedance. The impedance reduction procedures described below change the profile of the exposed tip without changing the diameter of the electrode tip. Thus, reducing impedance without sacrificing unit selectivity.

Method 1: The "bubbling" technique. One simple and inelegant method that has worked for us empirically is the byproduct of a procedure known as "bubbling" used to test the continuity of each wire. This procedure involves passing anodal current through each wire from a DC source, such as a 9 V or 12 V battery, and back through a saline solution. The current causes bubbles to form in the saline solution at the tip of the wire. This "bubbling" process confirms the integrity of the connections inside the electrode assembly and also results in a lowering of the impedance measured at the electrode tip. Impedance is reduced because the strong current that produces the bubbles in solution also causes etching of the electrode tip. This etching process increases the exposed surface area of the tip without changing the diameter, which determines the recording characteristics of the electrode. Damage to the insulation may also occur however, so the goal of this method would be to minimize insulation damage, which can be viewed under a stereomicroscope, while etching the tips sufficiently to achieve a usable impedance on each wire. Too much damage to the insulation will result in a non-selective electrode that is unable to isolate single cells from their neighbors.

Method 2: Plating. Another more conventional method of reducing electrode impedance is through electrolytic deposition of an inert metal onto the electrode tip (electroplating). Electroplating also reduces impedance by increasing the effective surface area without increasing the tip diameter. Platinum electroplating is accomplished by placing the electrode tips into a solution of platinum chloride and applying a small current such that the platinum in solution is reduced, causing platinum deposition at the tip of the metal electrode. I plate our electrodes using a solution of hydrogen hexachloroplatinate (8% PtC14 by weight; Sigma Chemicals) with a multi-channel, constant-current plating device available from Eclectic Engineering Studio.

A simple constant-current plating circuit is formed using a large resistance (R) and a DC voltage source (V) as in Figure 5. The resistor used in this circuit must be at least ten times larger than the largest anticipated DC resistance of the electrode. The plating current is computed using Ohm's law.

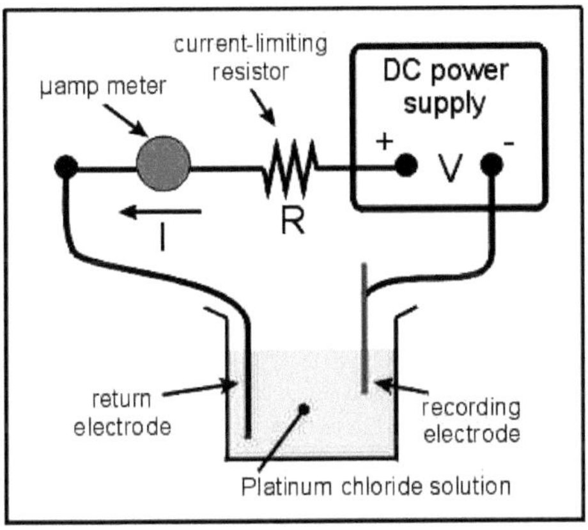

FIGURE 5

In constructing such a circuit, the current (I) should be kept relatively small (1 to 10 microamps), and the return electrode should be either graphite or platinum to prevent contamination of the platinum chloride solution.

Immersion procedure for platinum electrodeposition. Both the plating current and the concentration of the plating solution will affect the immersion time required for sufficient plating. Typically we plate our electrodes at 5 microamps for 4 to 6 seconds in two separate immersions. Prior to each immersion, the electrodes are dipped into a 90% EtOH solution to reduce small air bubbles and remove dirt particles that can have a negative impact on electrolytic deposition. Wetting the tips with EtOH and repeating the process several times serves to both reduce and distribute these factors and results in a more uniform impedance across the different wires. It is important to visually examine the tips of the wires when determining the optimal parameters for time, current, and platinum concentration. Prepare the electrodes for plating by cutting the tips using fine surgical scissors. Prior to plating, the exposed electrode tip should be "shiny" and the insulation should be intact. After plating to the desired impedance, the tips will have a rough "matte" appearance and impedance is typically reduced by 50 to 75%. If the insulation appears ragged or degraded, then the wires have been damaged, indicating that the current, duration or concentration of the plating solution should be reduced.

Variability of results. Finally it is worth noting that the exact parameters change slightly between days and on different electrodes, so we typically practice plating each electrode several times before making a final determination of the correct values. This approach is possible because extra wire is fed through the guide cannula to be cut off prior to surgery, thus several plating attempts are made before the wires are cut to their final length. Some variation in the final impedance between wires is normal and likely results from local variables in the plating environment or the mating of each wire through the electrode assembly.

Web content available at www.psy.jhu.edu/schoenbaum

This web page coauthored by Kevin B. Austin, Ph.D., Eclectic Engineering Studio, www.EclecticStudio.com

BOX 16.6
Headstage Amplifiers for Recording in Behaving Animals

Recording electrophysiological signals from a freely-behaving preparation is technically challenging. Unlike anesthetized or *in-vitro* preparations, the system used to record from a behaving animal must accommodate movement. As the preparation moves, the cable connecting the animal to the instrumentation flexes. Cable flexion results in capacitance changes which cause small currents to flow through the cable. These small currents can generate sizable voltages when they flow through high impedances. A voltage generated by cable flexion that is large enough to interfere with the electrophysiological signals you are trying to measure is called a movement artifact. Movement artifacts can be eliminated by reducing the impedance seen by the cable. This is accomplished by mounting a special preamplifier called a headstage between the electrode and the cable as shown in Figure 1. The headstage should be as close to the electrode as possible. The connection between the electrode and the headstage must be rigid.

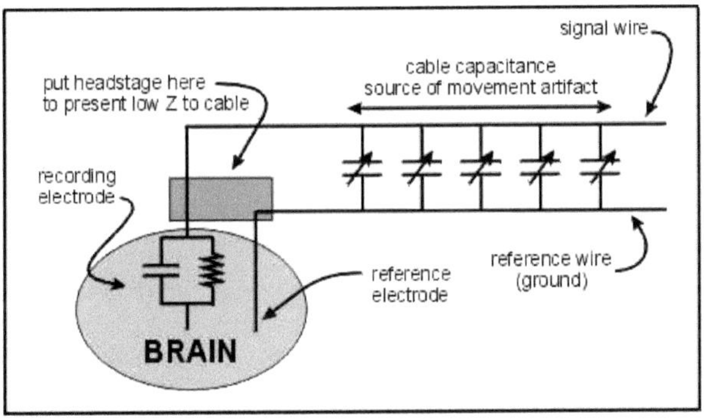

FIGURE 1

Headstage amplifiers are generally voltage-follower circuits built around an operational amplifier (op-amp) or field-effect transistor (FET). The output voltage of the headstage circuit follows the input voltage precisely. However, the output impedance of the headstage circuit is low. Thus, a headstage amplifier can be described as an impedance transformer which takes a signal from a high impedance source (the electrode) and presents it to the cable as the same signal voltage but at a lower impedance. Practical op-amp and FET realizations of voltage-follower circuits are shown below (Figure 2). Actual circuits used in particular laboratories may vary slightly from the configurations shown. However, all headstages that successfully eliminate movement artifact will act as impedance transformers and present a low impedance to the cable.

Both of these circuits are actively used in laboratories recording from freely-behaving animals. The selection of one circuit over another is largely based on legacy issues. Both circuits will successfully reduce movement artifacts. A comparison of the two types of headstage circuit is presented in the following table.

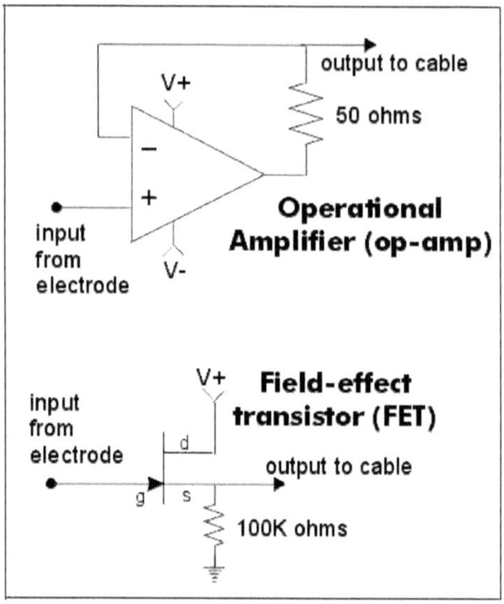

FIGURE 2

TABLE 1
FET/Op-Amp Comparison

ISSUE	FET	OP-AMP
Power	Discrete FET circuits will usually draw more power than op-amp circuits and depend on the resistance used by the circuit. Thus, battery life is reduced and more heat is generated when compared to low-power op-amps. Increased circuit temperature will result in increased wideband thermal (Johnson) noise.	Micropower op-amp versions are available (<200 microamps/channel). Power consumption is generally unaffected by the resistor value used in the headstage circuit.
Circuit	Each circuit requires at least one resistor. Power consumption and output impedance depend on the resistance value used. There is a trade-off between power consumption and output impedance. Increased power consumption will also generate more heat.	Some circumstances may require no resistors at all. If the cable has sufficiently low capacitance, the 50 ohm resistor in Figure 1 can be replaced by a wire.
Size	Increased circuit complexity results in larger sized multi-channel headstages. Individually packaged FETs can result in smaller headstages for low channel-count applications. The 2N3958 that we use contains two FET devices.	Reduced circuit complexity results in smaller headstages, especially in multi-channel situations. Op-amp ICs can have as many as 4 devices in a single package.
Technology	Discrete FET technology is relatively static.	Op-amp technology is actively developed. New and improved op-amps become available several times a year.

As the table suggests, FETs have significant drawbacks when compared to op-amps, particularly in newer preparations in which activity from a large number of wires (stereotrodes, tetrodes) will be recorded. In these cases, op-amp circuits would be far superior because of their low power consumption and reduced circuit complexity, characteristics which serve to minimize the size of the headstage mounted on the rat and also reduce the heat generated by the circuitry.

Nevertheless the lab currently uses the FET-based circuit shown in Figure 2. Each channel is built around one half of a 2N3958 dual FET device (Arrow Advantage). This is an n-channel junction FET. We use five devices to construct a 10 channel headstage. The 100 k-ohm resistor limits power consumption and heat dissipation but presents an impedance to the cable that is many times larger than that of the op-amp circuit. This trade-off appears to be reasonable for extracellular single unit recordings. However, this headstage design may not be as effective at removing movement artifact for other types of electrophysiological recording such as spontaneous EEG or evoked potentials because these signals use a different passband.

It is likely that our lab will soon adapt our headstage design to utilize an op-amp circuit. In principle, all op-amps work similarly. A key parameter to look for when seeking an op-amp to use in a headstage circuit is input bias current, which should be as low as possible (less than 50 picoamperes). Size, power consumption, operating voltage and input offset voltage are other parameters to consider. Specific examples of op-amps used by us would include the LF444, LF442 and LF441 for four-, two- and one-channel headstages, respectively. These could be combined for larger designs. Smaller packages are also available. Other IC's that have been used for unit recording include the TI2274AC, TLC2274, LT1490, and TL084. For more information and specs on different ICs, go to www.questlink.com or call Newark.

Suppliers

- Newark Electronics, 7272 Park Circle Drive, Hanover MD 21076, 1-800-NEWARK (general electronics)
- Arrow Advantage, 18640 Lake Drive East, Chanhassen MN 55317, 800-833-3557 (for 2N3958 Siliconix FET)

Web content available at www.psy.jhu.edu/schoenbaum

This web page coauthored by Kevin B. Austin, Ph.D., Eclectic Engineering Studio, www.EclecticStudio.com

17 Olfactory Event-Related Potentials

Thomas Hummel and Gerd Kobal

CONTENTS

0-8493-2329-0/02/$0.00+$1.50

17.1 A BRIEF HISTORY OF BRAIN POTENTIALS TO CHEMICAL STIMULI

In 1883, Fleischl von Marxow[1] was the first to observe that ammonia produced electrical brain potentials when presented to a rabbit's nose. Berger[2] assumed in 1929 that this phenomenon could also be found in the human electroencephalogram (EEG), but, probably due to the electrode setup, failed to see them. Then, in 1934, Adrian and Mathews first described human brain potentials to visual stimuli.[3] Finkenzeller[4] first recorded event-related potentials (ERP*) to odorous stimulation followed by Allison and Goff[6]; this work was continued by Giesen and Mrowinski,[7] Herberhold,[8] and Cianfrone and co-workers.[9] Due to technical difficulties in analyzing the large amount of data, this development was slow. In addition, data presented by Smith et al.[10] indicated that olfactory ERP cannot be found in patients who had lost their trigeminal sensitivity. As this was found to be due to the experimental conditions, this discussion was brought to an end with the development of more sophisticated stimulation techniques[11,12] (see below).

According to the sequence of their appearance after the onset of an event, ERPs are grouped into early, intermediate, late, and ultra-late potential components. An additional way of classification is to determine the distances of their generators from the recording electrode. The generators of near-field potentials are located in the cortex, relatively close to the recording electrode. In far-field recordings, generators are located subcortically at much greater distances from the recording scalp electrode.[13] With reference to olfactory ERP, to date only late near-field ERPs have been recorded.

Another way to classify ERPs is to look at the information coded in them. By definition, exogenous ERPs are determined by characteristics of stimulus. In contrast, endogenous ERPs vary with the state of the subject, or the meaning of the stimulus. ERP components are usually the less subject to endogenous processes the earlier they appear.[14,15] However, in many situations, exogenous and endogenous phenomena cannot be clearly separated. Considering the olfactory ERP, it can be assumed that all components are subject to modulation by cognitive processes[16]; that is, control of experimental conditions and contextual situations, including instructions given to the subjects, is of utmost importance to the outcome of ERP measurements.

* Event-related potentials can be elicited by both external and internal stimuli. For example, when a series of stimuli is presented with a constant interstimulus interval, the omission of one stimulus may trigger an event-related potential (although no physical external stimulus is present). In contrast, very early components of the response to auditory stimuli are termed evoked potentials (or exogenous or obligatory) as they are largely independent of the subject's state.[5]

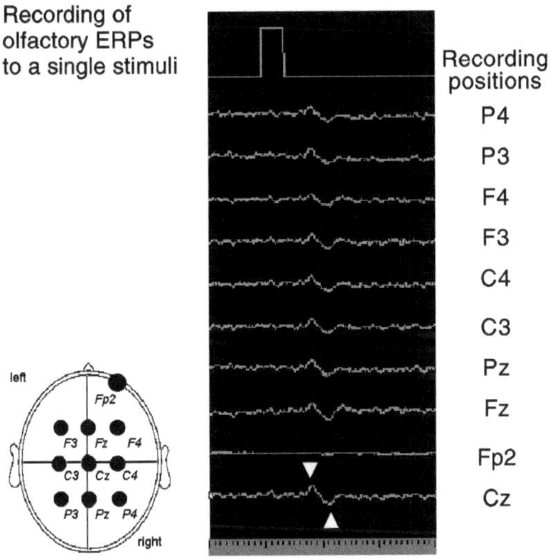

FIGURE 17.1 Examples of olfactory ERP to single phenyl ethyl alcohol stimuli recorded at different sites (stimulus duration 200 msec as indicated in top trace, stimulus onset after 500 msec; recording positions indicated on insert). In this case the response can be easily discriminated from background noise; arrows in bottom trace indicate peaks N1 and P2.

17.2 HOW TO OBTAIN OLFACTORY ERP

Event-related potentials are EEG-derived polyphasic signals. They are due to the activation of cortical neurons that generate electromagnetic fields.[15] The more neurons activated or synchronized, the larger the amplitude of the signal obtained at the surface of the scalp (Figure 17.1). Because the EEG is a noisy signal that contains activity from many cortical neurons (in addition to activity of non-neuronal origin, Figure 17.2), ERPs need to be extracted from the background activity; that is, the signal-to-noise ratio needs to be improved. The classical approach to this problem involves the averaging of individual responses to olfactory stimuli such that random activity would cancel itself out while all non-random activation would remain. In addition, stimuli are typically presented with a steep onset (<100 msec) in an extremely well-controlled, monotonous environment such that stimulus onset synchronizes the activity of as many cortical neurons as possible.

While widely used, this approach creates a number of problems, including:

- Stimuli must have a steep onset. A shallow slope of the stimulus onset may lead to a sensation; however, this sensation may not be reflected in an event-related potential because the cortical activity drowns in background noise (compare Reference 17).
- Stimuli need to be presented repetitively. This requires precise temporal control of stimulus onset as fluctuations in the timing of stimulus onset

will lead to differences in the peak latencies of individual ERPs (jitter). This jitter will lead to the modification/cancellation of peaks in the averaged response. In addition, desensitization to repeated stimuli becomes an issue.

- To properly interpret the response, it is necessary to know whether intranasal chemical stimulation activates the trigeminal or the olfactory system.*

17.3 WHY RECORD OLFACTORY ERP?

Major points why the recording of olfactory ERPs is attractive to research include that olfactory ERPs (1) are direct correlates of neuronal activation, unlike the signals that are seen, for example, in functional MR imaging; (2) have an extremely high temporal resolution in the range of microseconds; (3) allow the investigation of the sequential processing of olfactory information; and (4) can be obtained independent of the subject's response bias, i.e., they allow the investigation of subjects who have difficulties to respond properly (e.g., children, aphasic patients).

17.4 STIMULATION TECHNIQUES FOR THE RECORDING OF OLFACTORY ERP

How is it possible to produce odorous stimuli that have a rectangular shape with rapid onset; are precisely controlled in terms of timing, duration, and intensity; and the presentation of which does not simultaneously activate sensory systems other than the olfactory? Based on the principles of air-dilution olfactometry[18] such a system has been developed by Kobal.[11,12,19] Odors are applied intranasally by means of a cannula that typically has an inner diameter of 2 to 3 mm. This cannula is inserted approximately 1 cm into the nostril such that its opening lies beyond the nasal valve. Presentation of odor stimuli does not simultaneously activate mechano- or

* To illustrate this, the situation can be compared to a soccer stadium, where, similar to an EEG electrode, a microphone is positioned over the middle of the field to record all sounds. The stadium is filled with tens of thousands of people, symbolizing the millions of neurons underlying a recording electrode on the scalp. During the longest sequences of the game, only noise is recorded coming from the soccer fans, talking to each other, laughing, commenting on the quality of the game, etc. When the game becomes more exciting, there is more noise; when it is less exciting, people often, but not necessarily, become quieter. When a goal is scored, talking between soccer fans becomes synchronized with many of them shouting — which compares to the synchronization of cortical neurons by the sudden onset of a stimulus. However, similar to the electrical fields of the EEG, it is a difficult task to localize the site where the goal has been scored. If the noise comes from the left side, this does not mean that the goal has been scored on the left. It may also be due to the fact that most of the fans of the scoring team sit on the left side, but the goal was actually scored on the right side (and this situation may change during the game when teams switch sides). Finally, it may also be that the goal is not scored in our stadium, but elsewhere. Specifically, the fans of one of the two teams may have brought receivers to listen to the broadcast of a different, extremely important game; it may be that in this different stadium a goal is scored that turns the situation in the league in favor of one of the teams. This would then also create a synchronized outcry of the fans although nothing much has happened in the play we actually observe. Similarly, pinpointing the source of an event-related potential can be a difficult task.

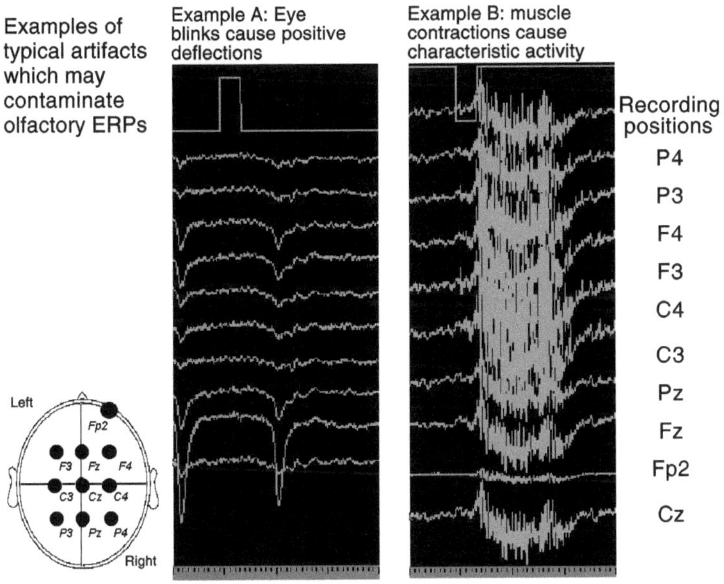

FIGURE 17.2 Examples of typical artifacts that need to be discarded before averaging olfactory ERPs. Example A: positive deflections produced by eye-blinks; amplitudes of these artifacts are typically largest at recording position Fp2; the artifact amplitudes typically decrease as a function of the distance to the frontal leads; i.e., they are largest at frontal sites, smaller at central sites, and barely detectable at parietal sites. Example B: characteristic activation produced by muscular contraction; muscle contractions cause activation with high amplitude, high frequency waves. In example B, polarity is inversed for convenience of the gaphical display.

thermoreceptors in the nasal mucosa because odor pulses are embedded in a constantly flowing air stream (typically 6 to 8 l/min).

This is achieved by directing two air streams toward the outlet of the olfactometer. Both air streams have the same flow rate, temperature, and humidity. One of the two air streams contains odor in a certain concentration (odor = O plus dilution = D); the other air stream is odorless (control = C). Immediately in front of the stimulator's outlet, a precisely tuned vacuum is attached such that during the interstimulus interval (ISI), the odor-containing air stream is sucked away and only odorless air enters the subject's nose. For stimulation, the vacuum is attached to the other side such that the odorless air is sucked away and the odor-containing air is led into the subject's nose. Thus, it is possible to switch between odorless and odor-containing air within less than 20 msec (see Figure 17.3). In this system, different odor concentrations are generated by means of air dilution, meaning that a preestablished, fully odor-saturated air-stream (odor = O) is mixed with an odorless airstream (dilution = D). While the sum of the two air streams is always constant (equal to the control air stream C), different O:D ratios produce different stimulus concentrations.

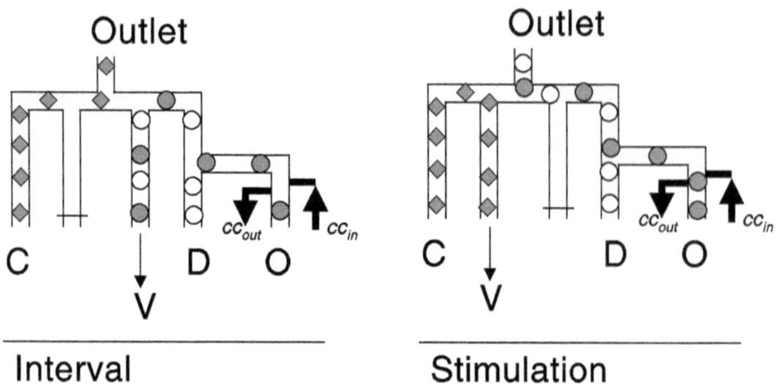

FIGURE 17.3 Schematic drawing of olfactory stimulation device. Two air streams are directed toward the outlet of the olfactometer (control [C], dilution [D] + odor [O]). Both air streams have the same flow rate, temperature, and humidity (≥80% relative humidity, 36°C). One of the two air streams contains odor in a certain concentration (D+O); the other air stream is odorless (C). A precisely tuned vacuum (V) is attached such that during the ISI, D+O is sucked away and C enters the subject's nose. For stimulation, the vacuum is switched such that C is sucked away and D+O is led into the subject's nose. Different odor concentrations are generated by mixing an odor-saturated air stream (O) with an odorless air stream (D). While D+O = C = constant, different O:D ratios produce different stimulus concentrations. Cross-contamination of odors is prevented by means of finely tuned pressure (cc_{in}) and vacuum (cc_{out}).

The constant air flow directed into the subject's nose asks for humidification (≥80% relative humidity) and thermostabilization (36°C) because dry, cool air produces nasal congestion, mucus discharge, and pain,[20-23] which interferes with olfaction in many ways.[24] When the constant air stream is warmed and humidified, subjects completely adapt to the air stream within a few seconds, meaning that the presence of an intranasal air stream is no longer felt. Furthermore, because the system is open, cross-contamination of odors needs to be prevented. A separate system of finely tuned pressure and vacuum is applied such that, similar to an air curtain, a small current of odorless air prevents molecules from O tubings to be drawn to other tubings of the open system (Figure 17.3) — which would finally lead to contamination.

In commercially available olfactometers, valves and air flows (using mass-flow controllers) are typically under computer control, and recording of stimulus-linked EEG segments is integrated in the same software that controls the olfactometer, and thus stimulus presentation. This equipment also allows the setup of sequences of stimuli with different quality, intensity, duration, or interstimulus interval. Thus, the recording of olfactory ERP becomes a routine procedure that can be carried out by any technician, a fact that is of special importance in a clinical setting.

What happens when olfactory stimuli are puffed into the nasal cavity? It is possible to obtain ERPs not only in anosmic subjects,[8,9,25-28] but also in normosmics as the result of a mixed activation of both the trigeminal and olfactory systems. Because the combined activation of the trigeminal and olfactory systems leads to numerous interactions at various levels of neuronal processing (e.g., References 29 to 33; for review, see Reference 34), this cannot be remedied by simple mathematical procedures.[26,35,36] For example, the average response to individual stimulation with carbon dioxide and hydrogen sulfide is significantly different from the response obtained after stimulation with the binary mixture of carbon dioxide and hydrogen sulfide.[31] Thus, it is difficult to interpret responses to olfactory stimuli contaminated by mechanical stimulation in patients with olfactory disorders. In addition, because the interactions between the trigeminal and olfactory systems are difficult to predict, it is misleading to interpret responses to mixed olfactory or trigeminal stimuli as being the result of predominantly (and, implicitly, more or less exclusively) olfactory or trigeminal activation.[37]

Other approaches to olfactory ERP include the electrical activation of the olfactory epithelium (OE).[38,39] This appears to be an extremely attractive technique, especially when considering that this would allow one to investigate the integrity of axonal connections from the OE to the olfactory bulb (OB). However, considering that electrical stimulation also activates trigeminal nerve endings of the OE,[40] it is uncertain whether the observed cortical responses to peripheral electrical stimulation reflect olfactory or somatosensory activation.

17.5 SHAPE AND NOMENCLATURE OF OLFACTORY ERP

The first positive peak of olfactory ERP typically occurs at latencies of >250 msec (see Table 17.2). It is then followed by at least two other peaks, a major negativity and the late positive complex. As there has been confusion in the past regarding the nomenclature of olfactory ERP, it has been suggested[41,42] to name an individual peak according to its polarity and its mean latency at position Cz. For example, a negativity at a mean latency of 340 msec would be called N_{340}. For the sake of comparison within this chapter, peaks are named P1, N1, and P2 (compare Reference 41). In terms of a nomenclature based on peak latencies at position Cz, these peaks could also be addressed as P_{383}, N_{484}, and P_{649} (using data published in Reference 19 for olfactory ERP to vanillin). Finally, the P2 peak has also been described as P1[43] or P3[44,45] (for a discussion of the P2 peak, see below).

Why is it that olfactory ERPs appear later than auditory or visual ERPs? One explanation may be that in the case of chemical stimulation, approximately 100 to 200 msec utilization time at the receptor site have to be subtracted from the latencies.[46-48] Thus, peak latencies of olfactory ERPs are comparable to N100 and P200/P300 of other sensory modalities (e.g., the auditory system). Other explanations may relate to certain ways olfactory information is processed. It can be hypothesized that, due to temporal differences in the neuronal activation produced by different components of an odor,[46] the processing of olfactory stimuli might require subcortical buffering before individual bits of olfactory information can be interpreted as an odor.[49,50]

Other than for audition or vision, to date no early ERPs have been recorded in response to olfactory stimuli (for review, see Reference 51), but only late near-field ERPs (i.e., responses from cortical neurons). Peaks of the late near-field ERPs fall into two groups. Earlier peaks such as N_{340} encode exogenous stimulus characteristics

TABLE 17.1
Typical Conditions for the Recording of Olfactory ERP in Clinical Research

Recording sites	Fp2, Fz, Cz, Pz, etc.
Reference	Pseudo-monopolar recording against linked earlobes
Filtering	AC, time constant 1 sec, low pass 15 to 30 Hz, notch filter (hum)
Sampling frequency	250 Hz
Pre-trigger period	0.5–2 sec
Post-trigger sampling	1–2 sec
Number of averages	10–20 (minimum 8)
Eye control, posture	Open eyes, sitting

to a larger extent than later, so-called endogenous components. That is, earlier components encode stimulus intensity or stimulus quality (e.g., "What is the nature of this stimulus?"), whereas later components are more related to the frequency, or the salience of the stimulus ("What is the meaning of this stimulus?")[15,52-54] (see Section 17.13).

17.6 RECORDING OF OLFACTORY ERP

Typical conditions for the recording of olfactory ERPs are presented in Table 17.1. Because the frequency spectra of late near-field ERPs range between 1 and 8 Hz, both filtering and sampling frequency must be set accordingly. Regarding the number of averages per ERP, eight records are regarded as the minimum.[12] However, others have used approximately 200 records to average a single response.[53] Given an ISI of 30 to 40 sec in single trial experiments, recording of such large numbers of responses would accumulate to at least 100 min. This appears to be impractical, especially when considering previous research indicating that improvement of the signal-to-noise ratio starts to reach a ceiling effect at 40 averages. In addition, the long times required for investigation introduces other sources of artifacts, for example, changing levels of vigilance during recording. On the other hand, while producing meaningful results,[12,55,56] a number of eight stimuli appears to invite noise (which, in turn, may only be counterbalanced by an increase in the number of investigated subjects).

Olfactory ERPs are recorded all over the scalp. In terms of the topographic distribution, olfactory ERP amplitudes exhibit characteristic patterns with a centro-parietal maximum (Figure 17.4) for both amplitudes N1 and P2[57] (compare References 58 to 60). This specific topographical distribution can be used to differentiate between trigeminal- and olfactory-induced cortical activation (compare Reference 61). The odor of nicotine may serve as an example for this change in the topographical distribution of olfactory ERP amplitudes.[56] Nicotine produces odor, burning, and stinging at increasing sensations. When nicotine is presented at a concentration where it mostly produces an odor, maximum amplitudes N1 are found at centro-parietal sites. However, when nicotine is used in concentrations that produce trigeminally mediated sensations, amplitudes are significantly larger at position Cz compared to all other

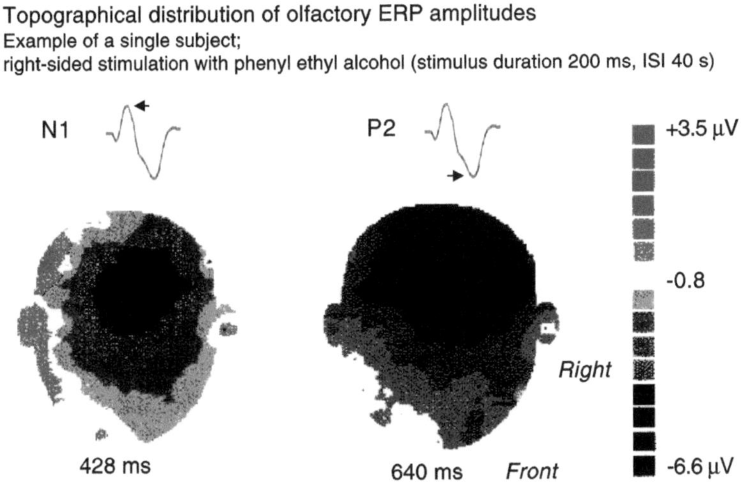

FIGURE 17.4 Topographical distribution of olfactory ERP amplitudes N1 and P2 in response to right-sided stimulation with phenyl ethyl alcohol; examples from a single subject (200-msec stimulus duration, ISI 40 sec). The darker the gray shade, the more negative (N1) or positive (P2), respectively, the amplitude (color maps have been adapted for printing purposes). For a schematic drawing of ERP components, see inserts. Maximum amplitudes N1 and P2 of olfactory ERP are typically found over the right-sided parietal cortex.

recording sites. This relates, at least partly, to different cortical areas activated by olfactory and trigeminal stimuli (see Section 17.9).

17.7 SPECIFIC CONDITIONS DURING RECORDING OF OLFACTORY ERP

During recording, stable environmental conditions are important. This includes the visual and acoustical shielding of subjects. Visual shielding can be performed using drapes; for acoustical shielding, white noise (a mixture of all frequencies at the same loudness) is typically used. Further, a defined task has proven most helpful. Specifically, many labs use a tracking task (idea/programming: Kobal/Hummel) in which subjects are requested to keep a small square controlled by a joystick inside a larger one which randomly moves on a screen at eye level, at a distance of approximately 1.5 to 2 m from the subject's eyes. This task fulfills a number of purposes:

- Eye movements are stabilized. Because eye movements are a major source of artifacts in EEG (Figure 17.2), one way to control this is to ask subjects to keep their eyes open. This task is made much easier when subjects are asked to look at a moving object, and, as the movements of the larger square on the screen require relatively small eye movements, this does not lead to significant artifacts.

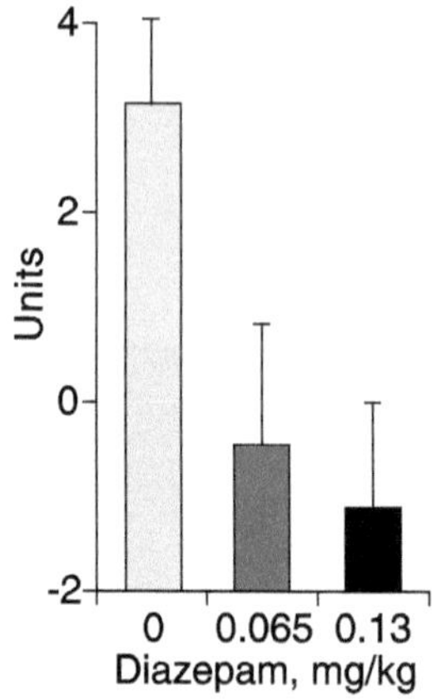

FIGURE 17.5 Results of tracking performance (in units; means ± SEM, n = 20) in relation to sedation with diazepam (intravenous administration of 0, 0.065, or 0.13 mg/kg diazepam; differences between data obtained 20 min after and before drug administration; compare Thürauf, N. et al., *Br. J. Clin. Pharmacol.,* 38, 545-555, 1994.) During administration of placebo, the subject's performance increased while it decreased in relation to the dose of the sedative diazepam.

- The task helps to stabilize the subject's vigilance/attention. Subjects are involved in a defined task that is not overly demanding to draw attention away from the olfactory stimuli.[37,53] Employing this task in combination with one or two breaks, it is possible to run single sessions over periods as long as 45 to 60 min.
- The task allows one to quantify how well subjects are able to track, which, in turn, is an indirect measure of the subject's attention. Typically, during an experiment, the subject's performance increases. The tracking performance also allows the assessment of subtle differences in vigilance, for example, following administration of the sedative diazepam, or the analgesic ibuprofen (Figure 17.5). However, data provided by Nordin et al.[61] indicate that stabilization of vigilance does not have to be of significance in ERP investigations of short duration; specifically, when investigating naïve, healthy subjects of different ages the authors did not find major differences between ERP obtained when subjects performed the tracking performance and when subjects simply had their eyes closed. However, this situation might change when the duration of sessions increases, when subjects are hyposmic, or when repetitive sessions are required.

Subjects are typically asked to relax and sit as still as possible, which is often monitored by video. In turn, this also requires that subjects be seated as comfortably as possible in a chair equipped with rests for arms, legs, and head. Finally, whenever this seems adequate, subjects are trained in a specific breathing technique, the velopharyngeal closure.[63,221] This technique helps to prevent respiratory flow through the nose by lifting the soft palate, which is under voluntary control. Simple biofeedback techniques help to instruct subjects on this technique and approximately 95% of all subjects master this technique in less than 5 min. Alternative to velopharyngeal breathing, stimuli could be presented in synchrony with inspiration.[60,64-66] However, one must be aware that responses obtained under these circumstances are contaminated by the so-called contingent negative variation.[67-71] This negativity builds up as a consequence of the expectation of odorous stimulation, which is more likely to happen during inspiration. In addition, there is evidence that the processing of olfactory information differs between inspiratory and expiratory phases.[66]

Considering these extensive prerequisites for an ideal session, it becomes clear why many experiments benefit from a specific adaptation session in which subjects are thoroughly acquainted with these experimental conditions. While these measures clearly optimize recording conditions, and thus the signal-to-noise ratio of the responses, they may be regarded as overkill in situations when subjects are only tested once (e.g., in clinical investigations). Then it may be wise to avoid, for example, velopharyngeal closure and have patients perform normal breathing through the mouth. In this particular situation, it may be that patients are too stressed by the experimental conditions such that the signal-to-noise ratio might be negatively affected. Thus, specific experimental conditions may apply to different experimental conditions.

17.8 RELATION OF OLFACTORY ERP TO STIMULUS CHARACTERISTICS

17.8.1 Stimulus Intensity

There is controversy surrounding how olfactory ERP amplitudes relate to odor concentration. In rats, it has been shown that OERP latencies shorten and amplitudes increase with rising odor concentrations.[72] In humans, some researchers[12,52] used odorants with mixed olfactory-trigeminal properties. Specifically, they used odorants such as linalool or citral that activate intranasal trigeminal afferents in addition to olfactory activation.[19,73] Others have employed odorants with little or no trigeminal activity, but had methodological shortcomings.[43,74,75] These studies have been limited to small sample sizes and the lack of statistical analyses,[75] whereas others have employed too low flow rates to observe all but the late positivity of the OERP.[43] In one study,[74] low concentrations of linalool were presented in ascending sequence, conceivably superimposing adaptation or habituation on the obtained responses.[76-78] Studies by this author in 15 subjects[79] have indicated that olfactory ERP amplitudes (both early and late components of the ERP) discriminate between different concentrations of vanillin (see Table 17.2). Using a much larger sample of 36 subjects,[80] results obtained for vanillin were confirmed in that greater olfactory ERP amplitudes N1 and P2 are obtained in response to higher H_2S stimulus concentrations (another stimulus that, at the concentrations used, appears to exclusively activate the olfactory system[81]). In addition, olfactory ERP latencies were found to shorten in a concentration-related manner. However, in line with previous research (e.g., References 75

TABLE 17.2
Descriptive Statistics of Olfactory ERPs (at position Cz) and Intensity Ratings

	28% Vanillin		56% Vanillin		84% Vanillin	
Parameter	**M**	**SD**	**M**	**SD**	**M**	**SD**
A-P1N1	10.5	5.6	12.4	8.0	15.2	9.9
A-N1P2	15.9	6.4	20.5	7.5	37.1	7.3
A-P1	3.2	4.9	4.0	6.1	2.8	6.2
A-N1	7.3	4.4	–8.5	7.2	–12.4	8.5
A-P2	8.6	4.9	12.1	5.3	24.7	9.7
T-P1	331	63	301	49	271	42
T-N1	400	49	373	39	341	40
T-P2	559	74	531	38	495	48
Ratings	45.6	18.0	82.8	17.0	109.8	22.6

Note: Means (M), standard deviations (SD), n = 15. Latencies T-P1, T-N1, and T-P2 in msec, base-to-peak amplitudes A-P1, A-N1, and A-P2 in μV, peak-to-peak amplitudes A-P1N1 and A-N1P2 also in μV. Intensity ratings in estimation units (EU). Compare with Reference 79.

Data from Reference 80.

and 82), results also suggested that latencies exhibit a stronger relationship to changes in stimulus intensity compared to olfactory ERP amplitudes — which is quite common for other sensory modalities as well — indicating the olfactory processing is more strongly related to discrimination between odor qualities than odor intensities.

17.8.2 Stimulus Duration

Olfactory ERPs relate to stimulus onset. Kobal[12] investigated a group of five volunteers presenting the mixed olfactory/trigeminal stimuli isoamylacetate and eucalyptol at constant concentration but different stimulus durations (100, 300, 500, and 700 msec; ISI 40 sec) (Figure 17.6). While odor intensity ratings increased with increasing stimulus duration, there was no difference between olfactory ERPs. This indicated that the olfactory ERP — as ERPs in other sensory modalities — is predominantly determined by early portions of stimulus onset.

17.8.3 Relation to Air Flow

Olfactory ERPs depend on the flow rate used to transport odorants to the olfactory epithelium. This is expected because the olfactory ERP is dependent on the number of odorous molecules presented during early parts of the stimulus (see above). Thus, both odor concentration and air flow should be determinants of the olfactory ERP. Kobal[12] investigated this issue in five healthy subjects using eucalyptol and linalool (at concentrations of 1287 and 6481 ppm, respectively; ISI 40 sec; stimulus duration

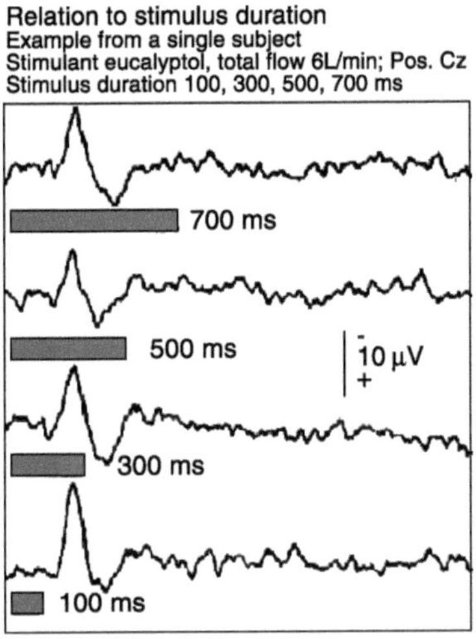

FIGURE 17.6 Olfactory ERP in relation to stimulus duration; example from a single subject; olfactory ERP obtained at recording position Cz to eucalyptol stimuli of different durations, but identical concentration (100, 300, 500, and 700 msec; ISI 40 sec). (Modified from Kobal, G., *Elektrophysiologische Untersuchungen des menschlichen Geruchssinns,* Thieme Verlag, Stuttgart, 1981).

200 msec). Air flow varied between 5 and 277 ml/sec. Both amplitudes and latencies varied as a function of air flow (Figure 17.7).

Considering the above, it becomes clear that the olfactory ERP is critically dependent on stimulus control. It varies as a function of stimulus concentration, stimulus duration, stimulus onset, and air flow which makes it difficult to set up normative data for the interpretation of olfactory ERP.[83] In turn, these characteristics of the olfactory ERP indicate that it is an extremely sensitive measure that can detect subtle changes in the processing of odorous information.

17.8.4 Relation to Odor Quality

Attempts to experimentally determine differences in the shape of chemosensory ERPs have not been successful. Previous work[19] with eight different chemical stimuli indicated that the shape of ERPs is not related to odor quality. However, ERPs to different odorous stimuli exhibit differences in the topographical distribution over the scalp, even when normalized with respect to maximum ERP amplitudes.[56] When additionally considering differences in the topographical distribution of ERP amplitudes and latencies in relation to the stimulated nostril,[57] this complex pattern can be used to investigate stimulus quality. Specifically, differences between latencies and amplitudes of olfactory ERPs to various odorous stimuli (e.g., acetaldehyde, phenyl ethyl alcohol, hydrogen sulfide, eugenol, and vanillin) were found in relation to lateralized stimulation.[57,84,85] As discussed above, this may be a reflection of the activation of different brain areas by different odorants[86-88]; in addition, it may also relate to the hedonic valence/emotional significance of the odorants used.

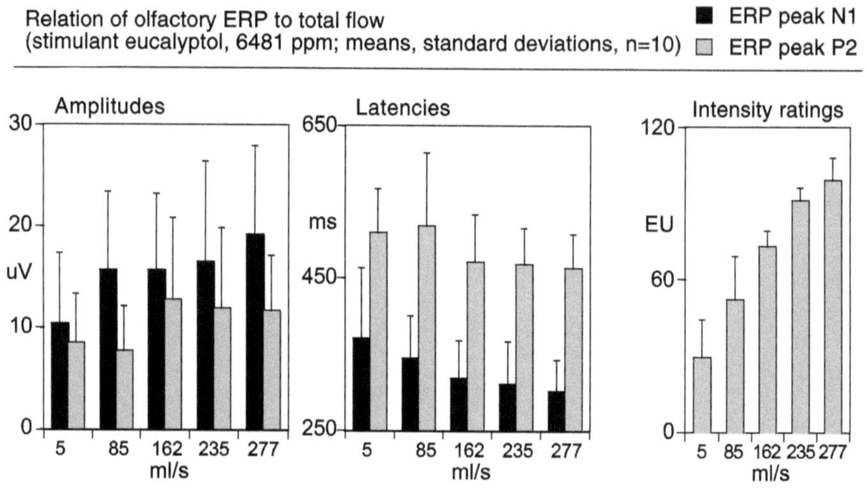

FIGURE 17.7 Olfactory ERP to eucalyptol (200-msec stimulus duration, ISI 60 sec, 6481 ppm) presented at different flow rates (5, 85, 162, 235, 277 ml/sec; means, standard deviations, n = 10). Results are shown for peaks N1 and P2. Correspondingly obtained intensity ratings (magnitude estimations; in estimation units, EU) are shown on the right. (Compare Kobal, G., *Elektrophysiologische Untersuchungen des menschlichen Geruchssinns,* Thieme Verlag, Stuttgart, 1981).

17.8.5 Relation to the Interstimulus Interval

Although the ISI is a significant determinant of olfactory ERP, little research has been done in this area. Kobal[12] used the mixed olfactory/trigeminal stimulant eucalyptol (total flow 94 ml/sec; 12,037 ppm) to investigate this issue. In 18 healthy subjects (age 20 to 41 years) he investigated ISIs of 12, 22, 32, 42, and 52 sec. He found that an increase of the ISI from 12 to 42 sec was accompanied by a marked increase in the amplitudes of the olfactory ERP. A further increase in the ISI had little, if any, effect on the olfactory ERP. Kobal concluded that an ISI of 40 to 50 sec would be ideal for these studies because desensitization was minimal, and, in turn, the time needed to obtain olfactory ERPs was kept as short as possible. Using the odorant isoamylacetate at concentrations that reportedly do not produce trigeminally mediated sensations[89] Morgan et al.[90] confirmed the results provided by Kobal, at least for young subjects. Specifically, they observed no significant differences between olfactory ERPs obtained at ISIs of 45, 60, and 90 sec. However, in terms of olfactory ERPs, a group of eight older males benefited from longer ISIs, female subjects (n = 8) of comparable age exhibited no such benefit. These data provided by Murphy et al.[83] also indicate that olfactory desensitization, as measured with olfactory ERP, may be a highly sensitive indicator of olfactory dysfunction.

While the studies mentioned above investigated the optimal ISI for recording olfactory ERP it clearly is possible to record meaningful data using ISIs of 20 sec[86] or less.[91,92] Krauel et al.[93] varied the ISI to investigate the potential presence of a mismatch negativity[94] in the olfactory ERP. Using an ISI of 15 sec a negative deflection was found that was not detectable at an ISI of 30 sec. The authors interpreted their findings in terms of the presence of a transient storage of olfactory information as reflected by the mismatch negativity.

17.8.6 Relation to Arousal/Vigilance

Olfactory ERPs are related to arousal on different levels. On the one hand, ERPs are dependent on the background signal; on the other hand, cognitive factors relate to the ERP. Considering the background signal, the EEG frequency pattern changes as a function of arousal. In fact, the frequency pattern of the EEG is used as a highly sensitive indicator of vigilance. A decrease of vigilance is typically accompanied by the increased occurrence of lower frequencies. For example, the sedative diazepam produces a dose-related increase of activity in the delta- and theta-range.[95] As one of the prerequisites for the recording of ERP includes a stationary EEG, and, in addition, as the frequency range from 1 to 8 Hz is most important to late near-field ERPs, they may be significantly affected by temporal fluctuations in the lower frequencies of the EEG induced by changes in the subject's vigilance during the recording of olfactory ERPs.

In addition, as olfactory ERPs are responses of the cortex, they are affected by changes in cognition (see Section 17.13). Specifically, when attention is diverted from olfactory stimuli, amplitudes of the olfactory ERP are strongly reduced. Using linalool and eugenol in six subjects, Krauel et al.[54] found that allocation of attention led to a latency decrease of early components and a simultaneous increase of the amplitude of the late positivity of the olfactory ERP (compare Reference 97). Similar findings for ERP to amyl acetate were reported by Geisler and Murphy.[37] In addition, they reported that responses to stimuli with a strong trigeminal component are less subject to attentional influences.

Thus, in this respect, olfactory ERPs are not different from cortical responses to other stimuli.[97] For example, in a study on late near-field ERPs evoked by electrical skin stimulation, Miltner et al.[98] found that ERP amplitudes were larger when subjects focused their attention on the stimulus, and considerably smaller when they ignored it. Thus, stabilization of vigilance during the relatively long recording periods (see Section 17.6) clearly adds to the quality of the outcome of the measurements.

17.8.7 Ultradian Variations of Olfactory Sensitivity

The perception of odors varies in relation to the menstrual cycle.[99,100] Using olfactory ERPs to stimulation with citral, Pause et al.[101] found evidence that the processing of olfactory information varies during the menstrual cycle; they investigated six subjects with cycle lengths ranging from 21 to 39 days. During the ovulatory period (monitored by changes in body temperature), stimuli were perceived as more complex/novel, as indicated by an increased amplitude P3-1 (see also Section 17.13). In addition, the authors reported a shortening of ERP latencies around ovulation and a shortening of peak latencies during the follicular phase. Consequently, the menstrual cycle should be considered when looking for subtle changes in the perception of odors using small sample sizes.

17.8.8 Repetitive Recordings of the Olfactory ERP

Despite the multitude of possible sources of variation, given a proper experimental environment, the test-retest reliability of olfactory ERPs is regarded as high (Figure 17.9). For example, Kobal and Hummel[19] reported very good reproducibility of olfactory ERP in 13 subjects who were tested on three different days, using eight different stimuli (compare Reference 103). Also, when measurements of ERP to H_2S (12 subjects) were repeated before and after application of a non-active (placebo) nasal spray,[80] there was no significant difference between olfactory ERPs obtained during the two consecutive sessions.

17.8.9 Relation to Age

Age-related decrease in chemosensory function (see Reference 104) is indicated by ERP measures. Specifically, the amplitude of the olfactory ERP's late positivity (P2) and the composite amplitudes of the two major peaks (N1P2) decreased with increasing age, while a prolongation of N1 peak latencies was observed. Using amyl acetate Evans et al.[104] reported a decrease in peak-to-peak amplitudes N1P2 in relation to the subject's age. They also reported a significant life-span prolongation of the P2 latency (2.5 msec/year; compare Reference 83), accompanied by a trend of mean N1 latencies to increase with age. In addition, Hawkes and co-workers (personal communication) found both an age-related decrease of N1 and P2 ERP amplitudes and an increase of latencies for the stimulant H_2S ($n = 60$). Because this is confirmed by numerous other studies (e.g., References 82, 105, and 106), it can be assumed that especially earlier ERP components (e.g., N1) reflect loss of olfactory function.

Olfactory ERP data indicate that changes in the processing of olfactory information may appear relatively early in life. Specifically, data from Murphy et al.[59] and Hummel et al.[106] suggest that ERP amplitudes already exhibit a decrease in middle age. In elderly individuals, this may, at least in part, be due to increased adaptation to repetitive olfactory stimuli.[90,107]

In a series of studies, Murphy and co-workers investigated age-related changes in the ERP with a special focus on the late component of the potential, the P3 component.[108] Results indicated that the age-related decrease of the P3 amplitude is related to an overall reduction in neuropsychological performance.[109,110] Thus, the olfactory P3 may serve as an indicator of attentional resources related to olfactory cognitive processing.[58] The question, however, remains as to which specific way the olfactory P3 differs from age-related changes seen in the visual or auditory systems.[111-113]

Murphy et al.[105] investigated chemosensory ERPs to amyl acetate in the elderly (mean age 66 years, $n = 7$) and young adults (mean age 27 years, $n = 7$), respectively. They found differences in amplitudes at central and parietal recording positions, but not at the frontal site Fz. This compares to data provided by Hummel et al.[106] who described a leveling of the topographical differences between ERP amplitudes; that is, differences between groups were largest at the central site Cz, while they were smaller at frontal (Fz) and parietal sites (Pz). The change in this topographical distribution of olfactory ERPs in older subjects may be due to central nervous plasticity. Here, the processing of olfactory information may be reorganized, possibly as a consequence of loss of neurons both in the periphery[114,115] and at central sites.[116,117]

Recently, Murphy et al.[83] investigated a group of 140 subjects in relation to aging effects on the ERP to amyl acetate. With age, they reported a linear prolongation of peak latencies accompanied by a decrease of amplitudes (amplitude N1P2: –1.5 μV/year; amplitude P2: –2.0 μV/year; latency P2 and P3: 2.0 msec/year).

Electrophysiological studies also indicated differences in age-related changes of responses to trigeminal or olfactory stimuli. Specifically, responses to trigeminal and olfactory stimulation may follow a different course across the life span (see Reference 24). However, more research is needed to investigate the behavioral correlate of this change in trigeminal responsiveness.

17.8.10 Relation to Gender

Olfactory ERPs reflect differences in olfactory sensitivity in relation to gender. A study in 35 healthy subjects indicated that ERP amplitudes to both H_2S and vanillin were larger in women when compared to male subjects.[118] In addition, Evans et al.[104] found the peak-to-peak amplitude N1P2 to be larger in women compared to men (compare References 92 and 119). These results are in line with the general finding that, independent of cultural influences,[120] women outperform men on tests of odor identification, detection, and discrimination.[121,122] Thus, electrophysiological investigations appear to confirm these psychophysical findings. However, it cannot be ruled out that the differences in amplitudes might be related to other factors, for example, different states of attention.[5]

17.8.11 Relation to Psychophysical and Neuropsychological Measures

Olfactory ERPs are related to psychophysical measures of olfactory function. A study in 16 healthy subjects indicated[79] that coefficients of correlations between threshold measures and latencies of olfactory ERP to vanillin were highest when the highest concentrations had been used, which may relate to the detectability of ERP peaks at low concentrations. In addition, coefficients of correlations between latencies and thresholds were higher for the early ERP components (P1 and N1) than for the later positivities. Coefficients of correlations were generally higher for latencies than for amplitudes of olfactory ERP. The subject's ability to identify odors was found to correlate with amplitude P2.[123]

Geisler et al.[109] reported significant correlations between tests of neuropyschological performance (Trail Making Test: a test for visual-motor attentional processing[124]; California Verbal Learning Test: a test for short- and long-term memory, free and cued recall, and delayed recognition[125]) and both amplitudes and latencies of the late positivity of the olfactory ERP. The results also indicated that the age-related decrease in olfactory ERP amplitudes and the corresponding increase in latencies were accompanied by reduced neuropsychological performance.

17.8.12 Determination of Thresholds Using Olfactory ERP

While elicited by suprathreshold stimuli, olfactory ERPs may be helpful in the determination of odor thresholds — in analogy to the "cortical evoked response audiometry."[126,127] Methods for the extrapolation of thresholds from dose–response curves have been reviewed by Marin et al.[128] Using eight concentrations of scatole (compare: T&T olfactometer[129]) Wada[130] reported the ability to determine the threshold at which olfactory ERP became detectable. In addition, Wada reported a congruence between psychophysical measures and the results obtained by means of the electrophysiological investigations.

17.9 CORTICAL GENERATORS OF OLFACTORY ERP

Where do olfactory ERPs come from? Using magnatoencephalographic techniques,[131] Kobal and co-workers conducted a series of experiments that addressed this question. As early as 1986,[132] cortical generators of the responses to trigeminal stimulation with CO_2 were localized in the secondary somatosensory cortex. This research also indicated that this brain area was not activated by non-painful stimuli of isoamylacetate.[51] Work by Kettenmann et al.[86,133] yielded reproducible results, indicating that olfactory stimuli (H_2S, vanillin) activated the anterior-central parts of the insula, the parainsular cortex, and the superior temporal sulcus. Specifically, an equivalent current dipole (ECD) equivalent to the P1 peak of the olfactory ERP (268 to 388 msec) was identified in the superior temporal plane and the parainsular cortex, an ECD equivalent to the N1 peak of the olfactory ERP (366 to 474 msec) was found in anterior-central parts of the insula, and another ECD equivalent to the olfactory ERP's P2 peak (472 to 749 msec) was found in the superior temporal sulcus (compare References 89 and 135). Surprisingly, only vanillin activated the left insular cortex, while this was not seen for stimulation with H_2S — possibly reflecting the concept that the right hemisphere is predominantly involved in the processing of olfactory information.[135-137] Alternatively, the results might also relate to a differential role of the hemispheres in the processing of olfactory stimuli with different emotional significance (compare Reference 139). The centro-parietal maximum in the topographical distribution of olfactory ERP amplitudes[56,139] was explained by the ECD's orientation toward parietal regions. These data also confirm the suggestion of connections between primary olfactory areas and the insular cortex.[140,141] Other work[88] has also indicated that eugenol leads to activation of hippocampal areas or the gyrus parahippocampalis.[142]

Using a blast-injection technique, Sakuma et al.[36] analyzed responses to amyl acetate and phenyl ethyl alcohol after the responses to odorless air had been subtracted. Their results indicated that the ECD equivalent to a P1 component of the ERP (at around 300 msec) was found in the inferior parietal lobe around the superior bank of the Sylvian fissure, and in the superior temporal gyrus. The current dipole equivalent to the ERP's N1 component (at around 500 to 600 msec) was found in the superior bank of the Sylvian fissure. Not much difference was seen between ECDs evoked by phenyl ethyl alcohol and amyl acetate. While the results from these studies are incompatible with the research discussed above, the discrepancies may be due to mechanical contamination of the chemosensory stimulation used by Sakuma et al., which cannot be remedied by the subtraction of responses (see Section 17.4).

17.10 APPLICATIONS OF OLFACTORY ERP

17.10.1 Medico-legal Investigations

An important application of olfactory ERPs is their employment in clinical olfactometry to diagnose olfactory deficits. The test procedure standardized by Kobal and Hummel[51] includes the recording of responses to olfactory (e.g., H_2S and vanillin) and trigeminal (e.g., CO_2) stimuli. All three stimulants are applied 15 times, both to the left and the right nostril of the patient. The interstimulus interval is approximately 40 sec. The session, including preparation, lasts for 80 min. The EEG (filtering 0.1 to 30 Hz, electrode impedance <10 kΩ) is recorded from four positions of the international 10/20 system (Fz, Cz, Pz, and Fp2, referenced to A1+A2). EEG records of 2048-msec duration are digitized (sampling frequency 250 Hz) and averaged in six groups according to the three stimulants and the binasal stimulation. EEG records contaminated by eye-blinks (Fp2/A1+A2) or motor artifacts have to be discarded from the average. This procedure (compare Reference 41) has recently been adopted by the working group Olfaction and Gustation of the German ENT Society.[42] Thus far, in all investigated anosmic patients, intranasal trigeminal ERPs could be obtained after stimulation with CO_2, although with significantly smaller amplitudes than in healthy controls.[33,143] In contrast, no olfactory ERPs could be detected after stimulation with the odorants H_2S and vanillin[81] (Figure 17.8). Using identical or slightly different protocols, this technique has been used in numerous labs (e.g., References 119 and 144–150). The results provide significant information in the testing of malingering patients.

17.11 OLFACTORY ERP IN NEURODEGENERATIVE DISORDERS

17.11.1 Parkinson's Disease

Loss of olfactory function in patients with Parkinson's disease (PD) has been reported by several authors.[151-154,222] Using olfactory and trigeminal ERP, Barz et al.[45] investigated PD patients treated with anti-Parkinson drugs (n = 13) and PD patients who received no pharmacological treatment (n = 18). They were compared to age- and sex-matched controls (n = 38). Comparable to previous research, odor identification was impaired in PD patients; it was not influenced by treatment with anti-Parkinson drugs (compare Reference 153). Latencies of olfactory ERP were prolonged in both PD patients taking and not taking anti-Parkinson drugs (compare Reference 55).[45] This effect was more pronounced in PD patients taking anti-Parkinson drugs. In contrast to olfactory ERP, the intranasal chemosensory trigeminal system was neither affected by the neuronal degeneration seen in PD nor by treatment with anti-Parkinson drugs (compare Reference 55). Further, the results indicated that, other than psychophysical measures,[153] electrophysiological measures can be

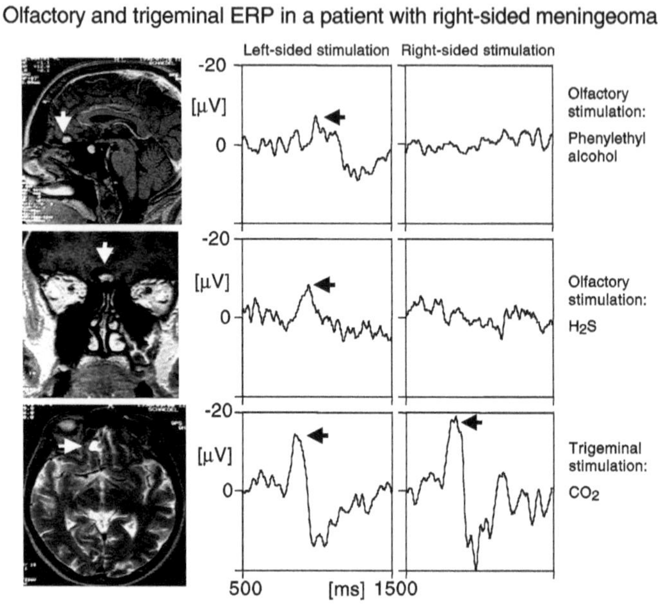

FIGURE 17.8 Olfactory and trigeminal ERPs in a patient (female, 64 years) with right-sided meningeoma. Structural magentic resonance images (left) indicate the intracranial position of the tumor (white arrows). Accordingly, olfactory ERP can only be obtained after left-sided stimulation (stiumulation started 500 msec after beginning the recordings; N1 peaks are indicated with black arrows). However, responses to trigeminal stimualtion with CO_2 can be obtained after left- and right-sided stimulation. In line with psychophysical testing using the "Sniffin' Sticks"[218,219] these data indicate right-sided anosmia and left-sided normosmia (compare to References 70, 218-221.)

used as a measure of the progression of PD,[155] which may relate to the decrease in the number of neurons in the olfactory bulb in relation to disease duration.[156] Similarly, Hawkes and colleagues reported that some patients had normal smell identification test scores but delayed olfactory ERPs,[157] indicating that olfactory ERPs might be a more sensitive measure of olfactory dysfunction than psychophysical tests.

17.11.2 Alzheimer's Disease

Despite the olfactory deficit in early Alzheimer disease (AD),[158-161] thus far there are no formal studies of olfactory ERPs in AD patients. Anecdotally obtained results in our lab indicate the presence of olfactory ERPs in some AD patients, although they tested anosmic when psychophysical techniques were applied. In addition, Hawkes and Shephard[164] reported that they obtained normal results in all four AD patients who could be tested, although all UPSIT scores[165] were abnormal. Thus, it appears that more research is needed in this area, especially as olfactory ERPs appear to be ideally suited for the investigation of patients who may have difficulties in cooperating.

17.11.3 Multiple Sclerosis

Olfactory function has been reported to be impaired in patients with multiple sclerosis (MS).[162,163] Hawkes et al.[164] reported that 25% of the 45 MS patients had delayed olfactory ERPs, while only 15% of MS patients exhibited a decreased ability to identify odors. Similar to the findings reported in PD (see above), this may also indicate that olfactory ERPs are more sensitive to olfactory than psychophysical measures.

17.11.4 Motor Neuron Disease

Hawkes and co-workers[165] investigated olfactory function in 58 patients with motor neuron disease (MND) in comparison to 132 controls. According to psychophysical testing with an odor identification test, olfactory function was slightly decreased in MND patients (compare Reference 166). Olfactory ERPs were found to be absent in 2 of 15 patients, and delayed in 1 of the 15 patients investigated.

17.12 OLFACTORY ERP IN OTHER DISORDERS

A study in temporal lobe epilepsy[136] compared 12 patients with temporal lobe epilepsy (TLE) with a left-sided focus with 10 TLE patients with a right-sided focus. In both groups, prolonged ERP latencies were found after left-sided trigeminal stimulation with CO_2 compared to stimulation of the right nostril. In contrast, a different pattern emerged for olfactory stimuli. After right-sided olfactory stimulation, latencies were prolonged in patients with right-sided epileptical foci. Similarly, when the left nostril was stimulated in patients with a left-sided focus, CSERP latencies were prolonged (Figure 17.9). Thus, it was suggested that the neocortical processing of olfactory, but not trigeminally-mediated, information is affected by functional lesions of the temporal lobe. Moreover, analyses revealed non-overlapping 95% confidence intervals for latency N1 when vanillin was applied to the right nostril. These results indicated that the right temporal lobe may play a different role in the processing of olfactory information compared to the left temporal lobe. Finally, the data indicated that olfactory information is predominantly processed ipsilaterally to the stimulated nostril (compare Reference 170; for review, see Reference 137).

Olfactory ERPs have also been investigated in migraineurs.[167] There, migraineurs exhibited greater ERP amplitudes N1 to trigeminal stimulation, supporting the concept of trigeminal hyperexcitability in migraineurs. In contrast, olfactory ERP amplitudes P1N1 were significantly smaller in migraineurs. A leave-one-out classification procedure on the basis of these two parameters assigned 76% cases correctly. The olfactory ERP amplitude discriminated better between groups than trigeminal ERPs, emphasizing the significance of the olfactory system in migraine. Other uses of olfactory ERPs included the assessment of patients with multiple chemical sensitivities (MCS)/idiopathic environmental intolerances (IEI),[168-170] or Down's syndrome.[171] Further, olfactory ERPs have been used to address chemosensory changes during the course of a common cold[172] and the investigation of drug effects on the olfactory system, for example, local application of decongestive agents in the nasal cavity,[173] and the systemic administration of diazepam.[174]

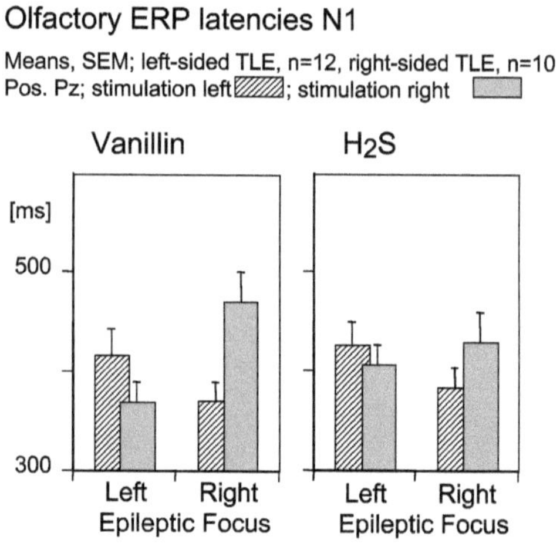

FIGURE 17.9 Olfactory ERP in temporal lobe epilepsy (TLE). Twelve TLE patients with left-sided focus were compared with 10 TLE patients with a right-sided focus. After right-sided olfactory stimulation, latencies were prolonged in patients with right-sided epileptical foci. When the left nostril was stimulated in patients with a left-sided focus, olfactory ERP latencies were prolonged. Compare Hummel, T. et al., *Epilepsia,* 36, 79-85, 1995.

17.13 OLFACTORY ERPS AS A TOOL TO INVESTIGATE THE COGNITIVE PROCESSING OF ODORS

As indicated above, one of the most important areas in the application of olfactory ERPs is the investigation of the cognitive processing of odors. Early investigations of the late positivity of the olfactory ERP were performed by Durand-Lagarde and Kobal.[91] Using the so-called odd-ball paradigm, stimuli were presented at an ISI of 6 to 8 sec such that a frequent stimulus (e.g., vanillin) alternated with a rare stimulus (e.g., H_2S). Subjects were asked to count the occurrence of the rare stimulus. As with other sensory modalities[175] a late positivity occurred within the evoked potential in response to the rare stimulus (compare Reference 51).

Since then, a large number of studies have focused on the late positivity of the olfactory ERP (for review, see Reference 16). Data provided by Pause et al.[53] indicated that this component is modulated by stimulus significance and stimulus probability. Further, results from this study indicated that the P3 component elicited by meaningful stimuli is so large that the P2 component can be completely overlapped. When odorous stimuli are presented at ISIs of 30 to 40 sec, comparable to a single-stimulus paradigm,[176,177] it appears that the olfactory stimulus reaches a significance that will elicit a P3 component,[110,178] which will determine the shape of the olfactory ERP and largely replace the P2. In other words, what is frequently addressed as the P2 peak in the olfactory ERP[50] may largely resemble characteristics of a P300 component.[44] According to hypotheses by Donchin and co-workers,[179] the P3 signifies context updating related to the maintenence of the internal model of the external environment. In this model, the P3 amplitude would reflect the amount of change of the model, while the latency would reflect the duration of stimulus evaluation (compare Reference 110). Although alternative hypotheses have been put forward (e.g., Reference 180), they are all based on the view that P3 is largely a result of the cognitive processing of stimuli.

Major findings in this area of research indicate that:

- The P3 component is larger when subjects thought they had perceived the target odor, i.e., for both hits and false alarms.[53]
- Rare odors elicit larger amplitdues than frequent odors, independent of stimulus quality.[53,181]
- The emotional significance of an odor may contribute to the generation of the P3.[52,74,182]
- In analogy to research in ERPs in other sensory modalities,[61] the P3 component can be divided into a P3-1 and a P3-2 peak, which differ in their topographical distribution (P3-1: fronto-central, P3-2: centro-parietal); P3-1 appears to be more related to the novelty and significance of an odor, whereas P3-2 more resembles features of the classical P3 (e.g., its latency relates to the time required for stimulus evaluation).[16,53]

While the above primarily focused on ERP to olfactory stimuli, the following chapter section briefly describes characteristics and uses of ERPs to intranasal trigeminal stimuli.

17.14 ERPS TO INTRANASAL TRIGEMINAL STIMULI

ERPs to trigeminal stimuli such as CO_2[81,183-186] are highly correlated with pain ratings.[63] One of the sources of these responses has been localized in the somatosensory area SII,[132,142,187] which can be assumed to be a primary projection area for nociceptive afferents.[188] Trigeminal ERPs are thought to predominantly reflect nociceptive information transmitted via A_{delta}-fibers.[193,194]

The ISI has been investigated systematically[195]: between stimuli, it was held constant for each experiment, but varied between experiments (10, 20, 30, 40, 60, and 90 sec). Intensity ratings of pain decreased in relation to the ISI. Specifically, intensity ratings were most strongly reduced at the shortest ISI used. They were largest at an ISI of 90 sec. It was hypothesized that these effects were the result of both habituation and stimulus predictability, which is greater the shorter the interval.

Other studies have focused on the effects of even shorter interstimulus intervals.[189,191] In these experiments, trains of four to six stimuli of constant stimulus duration and concentration were applied. Again, the ISI was constant for each experiment, but varied between experiments (2 to 8 sec). At an ISI of 8 sec, ratings decreased and only stinging sensations were reported. In contrast, at an ISI of 2 sec ratings increased and were accompanied by the buildup of burning pain. This increase in burning sensations after repetitive stimulation at short intervals cannot be explained by an increased firing rate of C-fibers. It has been shown that unmyelinated fibers only have a limited capacity for repetitive firing[192]; at short ISIs, the conduction velocity in human C-fibers slows down and may even fail at intervals shorter than 1 sec.[193] In contrast, responses in dorsal horn neurons increase in response to repetitive stimulation of C-fibers when the ISI becomes smaller than 3 sec.[194,195] Thus, the wind-up of burning painful sensations after repetitive stimulation may be due to central nervous summation of the input produced by stimulation of C-fiber afferents rather than by a peripheral sensitization of nociceptors. That is, summation occurs despite the partial suppression of afferent activity.[191] The overall sensation is therefore critically dependent on temporal summation.

In contrast, no such summation has been reported for stinging painful sensations. Most A_{delta} afferents adapt during sustained painful stimulation[196,197]; this has been observed in response to stimulation with capsaicin, nicotine, or ammonia.[198]. However, desensitization of A_{delta}-fibers seems to be less pronounced when compared to similar processes in C-fibers.[199,200] This suggests that both peripheral adaptation and/or central habituation[201] may eventually lead to the decrease in stinging that is observed after repetitive stimulation.

17.14.1 Assessment of Analgesic Drug Effects

Both opioidergic (e.g., morphine) and non-opioidergic drugs (e.g., aspirin) produce a decrease in the amplitudes of the trigeminal ERP. Changes in ERP latencies were usually observed when opioidergic drugs had been tested (e.g., fentanyl, pentazocine, or morphine[202-204]); however, they were frequently absent when less powerful analgesics were used (e.g., flurbiprofen or ibuprofen[205-207]). When comparing effects of analgesics (with a clinically proven significant effect) on the subjects' intensity ratings and those on the correspondingly obtained ERPs, it appears that trigeminal ERPs are more sensitive than psychophysical measures. This has been shown for aspirin,[202] ibuprofen,[206] tramadol,[208] flurbiprofen,[205] azapropazone,[209] and flupirtine.[210] In all of these studies, however, intensity ratings exhibited a decrease similar to that of trigeminal ERPs. It can be assumed that the missing statistical significance of the psychophysical measures is primarily due to the larger variance of these data. In fact, although the pain relief provided by an aspirin appears to be an everyday experience in many of those suffering from headache, due to the small effect size, a large number of subjects are needed to substantiate the pain relief in a clinical context. For example, more than 150 subjects have been used to demonstrate the effect of non-steroidal antiinflammatory drugs,[211,212] while 20 to 30 subjects are an adequate sample size to show antinociceptive properties of these agents. This fact also indicates that trigeminal ERPs are a measure of high sensitivity; therefore, they can also be used to investigate subtle changes in trigeminal chemoreception, which can be induced by indoor air pollution[169] or the loss of the sense of smell.[33]

Trigeminal ERPs have been tested with both opioidergic and non-opioidergic drugs to address the question of whether they reflect dose-related changes. This has been demonstrated for ibuprofen,[206,207] tramadol,[208,213] flupirtine,[210] and RP60180, a kappa-opioid receptor agonist.[214] It should be noted that in the cases of ibuprofen and tramadol, the drug effects on trigeminal ERPs were reproduced in different studies, a fact that clearly points to the usefulness of this approach in routine investigations of anti-nociceptive effects of analgesics.

With regard to the tracking of analgesic changes over time, comparative studies with sustained- and normal-release tramadol indicated that ERPs discriminated between formulas.[208,213] Specifically, it was shown that the sustained-release formula produced longer-lasting analgesia, which correlated well with the plasma levels of this compound.

Antiinflammatory analgesics produce a decrease in pain induced by dry air. This has been found for ibuprofen,[206] and ketoprofen.[20] In addition, the opioidergic agonist tramadol exerted a significant dose-related effect on pain intensity induced by intranasally applied air.[213] Single-dose studies also indicated that drugs such as dipyrone,[215] and azapropazone[209] lead to a decreased intensity in perceived pain. Overall, these data indicate that inflammatory pain induced by intranasally applied dry air can be used as an indicator of analgesic drug effects.

17.14.2 Specificity of Trigeminal ERPs

It has been demonstrated that the minor tranquilizer diazepam has only marginal effects on the trigeminal ERP to intranasal stimulation with gaseous CO_2; this indicates the nociceptive specificity of the ERP.[95] In this respect, responses to intranasal trigeminal stimulation are strikingly different from responses obtained after electrical tooth-pulp stimulation[95] and cutaneous laser-heat stimulation.[216,217] The reason for this difference may lie in the functional significance of the trigeminal chemosensory system, which acts as a sentinel of the vitally important lower airways; as a consequence, sensations elicited by intranasal trigeminal stimulation may be less susceptible to non-nociceptive influences (compare Reference 34).

A different study compared the pain relieving properties of imipramine (a tricylic antidepressant frequently used in chronic pain), tramadol, and anpirtoline (a serotonergic agent). Anpirtoline and tramadol produced a significant decrease in both ERP amplitude and pain rating. In contrast, after administration of imipramine, no change in ERP amplitudes could be detected, whereas pain ratings decreased significantly. This was accompanied by a significant decrease of arousal, indicating that pain relief produced by acute administration of imipramine was primarily related to its sedative action. Thus, these data also indicated that trigeminal ERPs respond specifically to antinociceptive modulation. This characteristic appears to be of critical importance, for example, regarding future investigations of trigeminal irritation produced by environmental agents.

17.15 PERSPECTIVES

Today, olfactory ERPs are a well-investigated means that allows the investigation of early components in the processing of olfactory information with a special focus on high temporal resolution. In addition to its significance in the investigation of olfactory processing in many clinical situations, this technique is one key to the study of subtle alterations in olfactory perception, odor memory, and odor aversion.

REFERENCES

1. Fleischl von Marxow, E., Mittheilung, betreffend die Physiologie der Hirnrinde, *Zbl. Physiol.,* 4, 537-540, 1890.
2. Berger, H., Über das Elektroenkephalogramm des Menschen, *Arch. Psychiat. Nervenkr.,* 40, 160-179, 1929.
3. Adrian, E.D. and B.H.C. Mathews, The Berger rhythm: potential changes from the occipital lobes in man, *Brain,* 57, 355-384, 1934.
4. Finkenzeller, P., Gemittelte EEG-Potentiale bei olfactorischer Reizung, *Pflügers Arch.,* 292, 76-85, 1965.
5. Näätänen, R., *Attention and Brain Function.* Lawrence Erlbaum Associates, Hillsdale, NJ, 1992.
6. Allison, T. and W.R. Goff, Human cerebral evoked responses to odorous stimuli, *Electroenceph. Clin. Neurophysiol.,* 23, 558-560, 1967.
7. Giesen, M. and D. Mrowinski, Klinische Untersuchungen mit einem Impuls-Olfactometer, *Arch. Hals-Nas-Kehlk. Heilk.,* 196, 377-380, 1970.
8. Herberhold, C., Typical results of computer-olfactometry, *Rhinology,* 14, 109-116, 1976.
9. Cianfrone, G. and L. Subiaco, Olfattometria objettiva computerizzata (a risposte evocate mediate), *Les Cahiers D'O.R.L.,* 13, 963-969, 1978.
10. Smith, D.B., et al., Human odorant evoked responses: effects of trigeminal or olfactory deficit, *Electroenceph. Clin. Neurophysiol.,* 30, 313-317, 1971.
11. Kobal, G. and K.H. Plattig, Methodische Anmerkungen zur Gewinnung olfaktorischer EEG-Antworten des wachen Menschen (objektive Olfaktometrie), *Z. EEG-EMG,* 9, 135-145, 1978.
12. Kobal, G., *Elektrophysiologische Untersuchungen des menschlichen Geruchssinns.* Stuttgart, Thieme Verlag, 1981.
13. Desmedt, J.E., Somatosensory evoked potential, in *Human Event-Related Potentials,* T.W. Picton, Ed., Elsevier, Amsterdam, 1988, 245-261.
14. Chen, A.C.N., C.R. Chapman, and S.W. Harkins, Brain evoked potentials are functional correlates of induced pain in man, *Pain,* 6, 365-374, 1979.
15. Picton, T.W. and S.A. Hillyard, Endogenous event-related potentials, in *EEG-Handbook, Revised Series, Vol. 3,* T.W. Picton, Ed., Elsevier, Amsterdam, 1988, 361-426.
16. Pause, B. and K. Krauel, Chemosensory event-related potentials (CSERP) as a key to the psychology of odors, *Int. J. Psychophysiol.,* 36, 105-122, 2000.
17. Loveless, N.E. and C.M.H. Brunia, Effects of rise-time on late AEP components, in *Psychophysiological Brain Research,* C.M.H. Brunia, A.W.K. Gaillard, and A. Kok, Eds., Tilburg University Press, Tilburg, 1990, 65-68.
18. Prah, J.D., S.B. Sears, and J.C. Walker, Modern approaches to air dilution olfactometry, in *Handbook of Olfaction and Gustation,* R.L. Doty, Ed., Marcel Dekker, New York, 1995, 227-255.
19. Kobal, G. and C. Hummel, Cerebral chemosensory evoked potentials elicited by chemical stimulation of the human olfactory and respiratory nasal mucosa, *Electroenceph. Clin. Neurophysiol.,* 71, 241-250, 1988.
20. Hummel, T., et al., Tonic vs. phasic pain: dose-related effects of ketoprofen, *Eur. J. Clin. Pharmacol.,* 49, 7-14, 1995.
21. Mohammadian, P., et al., Bilateral hyperalgesia to chemical stimulation of the nasal mucosa following unilateral inflammation, *Pain,* 73, 407-412, 1997.
22. Mohammadian, P., et al., Experimentally induced nasal irritation, *Rhinology,* 37, 175-178, 1999.

23. Lötsch, J., et al., Factors affecting pain intensity in a pain model based on tonic intranasal stimulation in humans, *Inflammation Res.,* 47, 446-450, 1998.
24. Hummel, T. and A. Livermore, Intranasal chemosensory function of the trigeminal nerve and aspects of its relation to olfaction, *Int. Arch. Env. Occ. Health,* (in press), 2001.
25. Bauer, L.O. and A.E. Mott, Differential effects of cocaine, alcohol, and nicotine dependence on olfactory evoked potentials, *Drug Alc. Depend.,* 42, 21-26, 1996.
26. Swandulla, D., Einige Aspekte der klinischen Anwendung olfaktorisch evozierter Potentiale (OEP), in *Physiology,* 1986, Erlangen-Nürnberg, Erlangen, Germany.
27. Sakuma, K., K. Nakashima, and K. Takahashi, Olfactory evoked potentials in Parkinson's disease, Alzheimer's disease and anosmic patients, *Psychiatry Clin. Neurosci.,* 50, 35-40, 1996.
28. Harada, H., et al., Topography of the olfactory evoked responses in patients with smell disorder, *Chem. Senses,* 22, 360, 1997.
29. Stone, H. and C.S. Rebert, Observations on trigeminal olfactory interactions, *Brain Res,* 21(1), 138-142, 1970.
30. Bouvet, J.F., J.C. Delaleu, and A. Holley, Does the trigeminal nerve control the activity of the olfactory receptor cells?, *Ann. N.Y. Acad. Sci.,* 510, 187-189, 1987.
31. Livermore, A., T. Hummel, and G. Kobal, Chemosensory event-related potentials in the investigation of interactions between the olfactory and the somatosensory (trigeminal) systems, *Electroencephalogr. Clin. Neurophysiol.,* 83(3), 201-210, 1992.
32. Inokuchi, A., C.P. Kimmelmann, and J.B.J. Snow, Convergence of olfactory and nasotrigeminal inputs on possible trigeminal contributions to olfactory responses in rat thalamus, *Eur. Arch. Otorhinolaryngol.,* 249, 473-477, 1993.
33. Hummel, T., et al., Loss of olfactory function leads to a decrease of trigeminal sensitivity, *Chem. Senses,* 21, 75-79, 1996.
34. Hummel, T. and A. Livermore, Intranasal chemosensory function of the trigeminal nerve and aspects of its relation to olfaction, *Int. Arch. Occ. Env. Health,* (in press), 2001.
35. Herberhold, C., Geruchs- und Geschmacksstörungen, *HNO Praxis heute,* 13, 35-90, 1993.
36. Sakuma, K., et al., Odorant evoked magnetic fields in humans, *Neurosci. Res.,* 27, 115-122, 1997.
37. Geisler, M. and C. Murphy, Event-related brain potentials to attended and ignored olfactory and trigeminal stimuli, *Int. J. Psychophysiol.,* 37, 309-315, 2000.
38. Sato, M., et al., Olfactory evoked potentials: experimental and clinical studies, *J. Neurosurg.,* 85, 1122-1126, 1996.
39. Ishimaru, T., et al., Electrical olfactory evoked potentials in clinical cases, *Chem. Senses,* 22, 77-81, 1997.
40. Silver, W.L., Physiological factors in nasal trigeminal chemoreception, in *Chemical Senses, Vol. 2, Irritation,* B.G. Green, J.R. Mason, and M.R. Kare, Eds., 1991, Marcel Dekker, New York. 21-37.
41. Evans, W.J., et al., Suggestions for collection and reporting of chemosensory (olfactory) event-related potentials, *Chem. Senses,* 18, 751-756, 1993.
42. Hummel, T., et al., Standards für die Ableitung chemosensorisch evozierter Potentiale zur klinischen Diagnostik von Riechstörungen, *HNO,* 48, 481-485, 2000.
43. Prah, J.D. and V.A. Benignus, Olfactory evoked responses to odorous stimuli of different intensities, *Chem. Senses,* 17, 417-425, 1992.
44. Lorig, T.S., et al., Event-related potentials to odor stimuli, *Bull. Psychon. Soc.,* 31, 131-134, 1993.

45. Barz, S., et al., Chemosensory event-related potentials in response to trigeminal and olfactory stimulation in idiopathic Parkinson's disease, *Neurology,* 49(5), 1424-1431, 1997.
46. Getchell, T., F.L. Margolis, and M.L. Getchell, Perireceptor and receptor events in vertebrate olfaction, *Prog. Neurobiol.,* 23, 317-345, 1984.
47. Hummel, T., M. Knecht, and G. Kobal, Peripherally obtained electrophysiological responses to olfactory stimulation in man: electro-olfactograms exhibit a smaller degree of desensitization compared with subjective intensity estimates, *Brain Res.,* 717, 160-164, 1996.
48. Leopold, D.A., et al., Anterior distribution of human olfactory epithelium, *Laryngoscope,* 110(3 Pt 1), 417-421, 2000.
49. Jinks, A. and Laiug, D.G. Temporal processing reveals a mechanism for limiting the capacity of humans to analyze odor mixtures, *Brain Res. Cogn. Brain Res.,* 8, 311-325, 1999.
50. Lorig, T.S., On the similarity of odor and language perception, *Neurosci. Biobehav. Rev.,* 23(3), 391-398, 1999.
51. Kobal, G. and T. Hummel, Olfactory evoked potentials in humans, in *Smell and Taste in Health and Disease,* T.V. Getchell, et al., Eds., Raven Press, New York. 1991, 255-275.
52. Donchin, E., et al., Cognitive psychophysiology and human information processing, in *Psychophysiology: Systems, Processes and Applications,* M.G.H. Coles, E. Donchin, and S.W. Porges, Eds., Guilford Press, New York, 1986.
53. Pause, B.M., et al., The nature of the late positive complex within the olfactory event-related potential, *Psychophysiology,* 33, 168-172, 1996.
54. Krauel, K., et al., Attentional modulation of central odor processing, *Chem. Senses,* 23, 423-432, 1998.
55. Hummel, T., G. Kobal, and T. Mokrusch, Chemosensory evoked potentials in patients with Parkinson's disease, in *New Developments in Event-Related Potentials,* H.J. Heinze, T. Münte, and G.R. Mangun, Eds., Birkhäuser Verlag, Boston, 1993, 275-281.
56. Hummel, T., et al., Chemosensory event-related potentials in man: relation to olfactory and painful sensations elicited by nicotine, *Electroencephalogr. Clin. Neurophysiol.,* 84(2), 192-195, 1992.
57. Kobal, G., T. Hummel, and S. Van Toller, Differences in chemosensory evoked potentials to olfactory and somatosensory chemical stimuli presented to left and right nostrils, *Chem. Senses,* 17, 233-244, 1992.
58. Pause, B., B. Sojka, and R. Ferstl, The latency but not the amplitude of the olfactory event-related potential (OERP) varies with the odor concentration, *Chem. Senses,* 21, 485, 1996.
59. Murphy, C., et al., Age effects on central nervous system activity reflected in the olfactory event-related potential. Evidence for decline in middle age, *Ann. N.Y. Acad. Sci.,* 855, 598-607, 1998.
60. Lorig, T.S., et al., The effects of active and passive stimulation on chemosensory event-related potentials, *Int. J. Psychophysiol.,* 23, 199-205, 1996.
61. Spencer, K.M., J. Dien, and E. Donchin, A componential analysis of the ERP elicited by novel events using a dense electrode array, *Psychophysiology,* 36, 409-414, 1999.
62. Nordin, S., et al., Olfactory event-related potentials in young and elderly adults: evaluation of tracking task vs. eyes open/closed recording, *Chem. Senses,* 24, 459-464, 1999.
63. Kobal, G. and T. Hummel, Brain responses to chemical stimulation of the trigeminal nerve in man, in *Chemical Senses, Vol.2: Irritation,* B.G. Green, J.R. Mason, and M.R. Kare, Eds., Marcel-Dekker, New York, 1989, 123-139.

64. Tonoike, M. and Y. Kurioka, Precise measurements of human olfactory evoked potentials for odorant stimuli synchronized with respirations, *Bull. Electroenceph. Lab.*, 46, 62-73, 1982.
65. Pause, B., et al., Is odor processing related to oral breathing?, *Int. J. Psychophysiol.*, 32, 251-260, 1999.
66. Hummel, T., G. Strehle, and M. Di Benedetto, Responses to olfactory and intranasal trigeminal stimuli: relation to the respiratory cycle, *Chem. Senses,* 25, 605, 2000.
67. Walter, W.G., et al., Contingent negative variation: an electric sign of sensori-motor association and exspectancy in the human brain, *Nature,* 203, 380-384, 1964.
68. Tecce, J.J., Contingent negative variation and psychological processes in man, *Psycholog. Bull.*, 77, 73-108, 1972.
69. Torii, S., et al., Contingent negative variation (CNV) and the psychological effects of odour, in *Perfumery,* S. Van Toller and G.H. Dodd, Eds., Chapman and Hall, London, 1988, 107-121.
70. Lorig, T.S. and M. Roberts, Odor and cognitive alteration of the contingent negative variation, *Chem. Senses,* 15, 537-545, 1990.
71. Auffermann, H., et al., Olfactory evoked potentials and contingent negative variation simultaneously recorded for diagnosis of smell disorders, *Ann. Otol. Rhinol. Laryngol.,* 102, 6-10, 1993.
72. Evans, W.J. and A. Starr, Stimulation parameters and temporal evolution of the olfactory evoked potential in rats, *Chem. Senses,* 17, 61-78, 1992.
73. Doty, R.L., et al., Intranasal trigeminal stimulation from odorous volatiles: psychometric responses from anosmic and normal humans, *Physiol. Behav.,* 20, 175-185, 1978.
74. Pause, B.M., B. Sojka, and R. Ferstl, Central processing of odor concentration is a temporal phenomenon as revealed by chemosensory event-related potentials (CSERP), *Chem. Senses,* 22, 9-26, 1997.
75. Thiele, V. and G. Kobal, Vergleich der objektiven und subjektiven Methoden olfaktometrischer Bestimmungen — Beispiel Schwefelwasserstoff, *Schriftenreihe der Landesanstalt für Immissionsschutz des Landes NW,* 59, 41-47, 1984.
76. Köster, E.P., Adaptation, recovery and specificity of olfactory receptors, *Rev. Laryngol. Otol. Rhinol.,* 86, 880-894, 1965.
77. Köster, E.P. and R.A. de Wijk, Olfactory adaptation, in *The Human Sense of Smell,* D.G. Laing, R.L. Doty, and W. Breipohl, Eds., Springer Verlag, Berlin, 1991, 199-215.
78. Dalton, P. and C.J. Wysocki, The nature and duration of adaptation following long-term exposure to odors, *Percept. Psychophys.,* 58, 781-792, 1996.
79. Tateyama, T., et al., Relation of olfactory event-related potentials to changes in stimulus concentration, *Electroenceph. Clin. Neurophysiol.,* 108, 449-455, 1998.
80. Hummel, T., et al., Effects of the decongestant oxymetazoline on human olfactory and intranasal trigeminal function in acute rhinitis, *Eur. J. Clin. Pharmacol.,* 54, 521-528, 1998.
81. Kobal, G. and T. Hummel, Olfactory and intranasal trigeminal event-related potentials in anosmic patients, *Laryngoscope,* 108, 1033-1035, 1998.
82. Covington, J.W., et al., Normal aging and odor intensity effects on the olfactory event-related potential, *Int. J. Psychophysiol.,* 32(3), 205-214, 1999.
83. Murphy, C., et al., Olfactory event-related potentials and aging: normative data, *Int. J. Psychophysiol.,* 36, 133-145, 2000.
84. Kobal, G., T. Hummel, and E. Pauli, Correlates of hedonic estimates in the olfactory evoked potential, *Chem. Senses,* 14, 718, 1989.

85. Hummel, T. and G. Kobal, Chemosensory event-related potentials: effects of dichotomous stimulation with eugenol and dipyridyl, in *Olfaction and Taste XI,* K. Kurihara, N. Suzuki, and H. Ogawa, Eds., Springer, Berlin, 1994, 659-663.
86. Kettenmann, B., et al., Multiple olfactory activity in the human neocortex identified by magnetic source imaging, *Chem. Senses,* 22, 493-502, 1997.
87. Kettenmann, B., H. Stefan, and G. Kobal, Differences in magnetoencephalographically identified sources of cortical olfactory activity after stimulation with different odorants, *Chem. Senses,* 22, 217-218, 1997.
88. Ayabe-Kanamura, S., et al., Measurement of olfactory evoked magnetic fields by a 64-channel whole-head SQUID system, *Chem. Senses,* 22, 214-215, 1997.
89. Cometto-Muniz, J.E. and W.S. Cain, Nasal pungency, odor, and eye irritation thresholds for homologous acetates, *Pharmacol. Biochem. Behav.,* 39, 983-989, 1991.
90. Morgan, C.D., et al., Olfactory event-related potentials: older males demonstrate the greatest deficits, *Electroencephalogr. Clin. Neurophysiol.,* 104(4), 351-358, 1997.
91. Durand-Lagarde, M. and G. Kobal, P300: a new technique of recording a cognitive component in the olfactory evoked potentials, *Chem. Senses,* 16, 379, 1991.
92. Hummel, T., et al., Do men and women respond differently to repeated olfactory or intranasal trigeminal stimuli?, *Chem. Senses,* 23, 584, 1998.
93. Krauel, K., et al., Interstimulus interval and mismatch negativity in the chemosensory event-related potential. Poster at the *25th Meeting of the German Society for Psychophysiological Methods and Their Applications,* 3-5 June, 1999, Trier, Germany, 1999.
94. Näätänen, R., et al., Attention and mismatch negativity, *Psychophysiology,* 30, 436-450, 1993.
95. Thürauf, N., W. Ditterich, and G. Kobal, Different sensitivity of pain-related chemosensory potentials evoked by stimulation with CO_2, tooth pulp event-related potentials, and acoustic event-related potentials to the tranquilizer diazepam, *Br. J. Clin. Pharmacol.,* 38, 545-555, 1994.
96. Spence, C., et al., Selective attention to the chemosensory modality, *Percept. Psychophys.,* 62, 1265-1271, 2000.
97. Näätänen, R., The role of attention in auditory information processing as revealed by event-related potentials and other brain measures of cognitive function, *Behav. Brain Sci.,* 13, 201-288, 1990.
98. Miltner, W., et al., Somatosensory event-related potentials to painful and no-painful stimuli: effects of attention, *Pain,* 38, 303-312, 1989.
99. Doty, R.L., et al., Endocrine, cardiovascular, and psychological correlates of olfactory sensitivity changes during the human menstrual cycle, *J. Comp. Physiol. Psychol.,* 95, 45-60, 1981.
100. Hummel, T., et al., Changes in olfactory perception during the menstrual cycle, *Experientia,* 47(7), 712-715, 1991.
101. Pause, B., et al., Olfactory information processing during the course of the menstrual cycle, *Biol. Psychol.,* 44, 31-54, 1996.
102. Lago, P., et al., Trigeminal evoked potentials in man: a new oflactory stimulation device, *Funct. Neurol.,* 13, 135-140, 1998.
103. Hummel, T., S. Heilmann, and C. Murphy, Age-related changes of chemosensory functions, in *Olfaction, Taste and Cognition,* C. Rouby, et al., Eds., Cambridge University Press, Cambridge, (in press), 2001.
104. Evans, W.J., L. Cui, and A. Starr, Olfactory event-related potentials in normal human subjects: effects of age and gender, *Electroenceph. Clin. Neurophysiol.,* 95, 293-301, 1995.

105. Murphy, C., et al., Olfactory-evoked potentials: assessment of young and elderly, and comparison to psychophysical threshold, *Chem. Senses,* 19, 47-56, 1994.
106. Hummel, T., et al., Chemosensory event-related potentials change as a function of age, *Electroencephalogr. Clin. Neurophysiol.,* 108, 208-217, 1998.
107. Stevens, J.C., et al., Olfactory adaptation and recovery in old age, *Perception,* 18, 265-276, 1989.
108. Murphy, C., et al., Late (cognitive) components of the olfactory event-related potential are more sensitive to aging than early components, *Chem. Senses,* 21, 648-649, 1996.
109. Geisler, M.W., et al., Neuropsychological performance and cognitive olfactory event-related brain potentials in young and elderly adults, *J. Clin. Exp. Neuropsychol.,* 21(1), 108-126, 1999.
110. Morgan, C.D., et al., Olfactory P3 in young and older adults, *Psychophysiology,* 36(3), 281-287, 1999.
111. Ford, J.M. and A. Pfefferbaum, Age-related changes in event-related potentials, *Adv. Psychophysiol.,* 1, 301-339, 1985.
112. Knight, R.T., Aging decreases auditory event-related potentials to unexpected stimuli in humans, *Neurobiol. Aging,* 8, 109-113, 1987.
113. Tachibana, H., K. Toda, and M. Sugita, Age-related-changes in attended and unattended P3 latency in normal subjects, *Int. J. Neurosci.,* 66, 277-284, 1992.
114. Nakashima, T., C.P. Kimmelman, and J.B. Snow, Jr., Olfactory marker protein in the human olfactory pathway, *Arch. Otolaryngol.,* 111(5), 294-297, 1985.
115. Rawson, N.E., et al., The use of olfactory receptor neurons (ORNs) from biopsies to study changes in aging and neurodegenerative diseases, *Ann. N. Y. Acad. Sci.,* 855, 701-707, 1998.
116. Bhatnagar, K.P., et al., Number of mitral cells and the bulb volume in aging human olfactory bulb: a quantitative morphological study, *Anat. Rec.,* 218, 73-87, 1987.
117. Yousem, D.M., et al., The effect of age on odor-stimulated functional MR imaging, *Am. J. Neuroradiol.,* 20(4), 600-608, 1999.
118. Becker, E., et al., Olfactory event-related potentials in psychosis-prone subjects, *Int. J. Psychology,* 15, 51-58, 1993.
119. Cui, L. and W.J. Evans, Olfactory event-related potentials to amyl acetate in congenital anosmia, *Electroenceph. Clin. Neurophysiol.,* 102, 303-306, 1997.
120. Doty, R.L., et al., A cross-cultural study on sex differences in odor identification ability, *Neuropsychologia,* 23, 667-672, 1985.
121. Koelega, H.S. and E.P. Köster, Some experiments on sex differences in odor perception, *Ann. N.Y. Acad. Sci.,* 237, 234-246, 1974.
122. Cain, W.S., Odor identification by males and females: prediction vs. performance, *Chem. Senses,* 7, 129-142, 1982.
123. Hummel, T., et al., Chemosensory event-related potentials change with age, *Electroencephalogr. Clin. Neurophysiol.,* 108(2), 208-217, 1998.
124. Reitan, R.M., Trail Making Test results for normal and brain-damaged children, *Perc. Mot. Skills,* 33, 575-581, 1971.
125. Delis, D.C., et al., *The California Verbal Learning Test,* Psychological Corporation, New York, 1987.
126. Prasher, D., M. Mula, and L. Luxon, Cortical evoked potential criteria in the objective assessment of auditory threshold: a comparison of noise induced hearing loss with Meniere's disease, *J. Laryngol. Otol.,* 107, 780-786, 1993.
127. Hoth, S., Computer-aided hearing threshold determination from cortical auditory evoked potentials, *Scan. Audiol.,* 22, 165-177, 1993.

128. Marin, A.B., et al., Sensory thresholds: estimation from dose–response curves, *J. Sens. Studies,* 1991, 6, 205-225.
129. Kondo, H., et al., A study of the relationship between the T&T olfactometer and the University of Pennsylvania Smell Identification Test in a Japanese population, *Am. J. Rhinol.,* 12(5), 353-358, 1998.
130. Wada, M., Measurement of olfactory threshold using an evoked potential technique, *Rhinology,* 37(1), 25-28, 1999.
131. Williamson, S.J. and L. Kaufman, Analysis of neuromagnetic signals, in *Handbook of Electroencephalography and Clinical Neurophysiology, Vol. 1: Methods of Brain Electrical and Magnetical Signals,* A.S. Gevins and A.A. Rèmond, Eds., Elsevier, Amsterdam. 1987, 405-448.
132. Huttunen, J., et al., Cortical responses to painful CO_2-stimulation of nasal mucosa: a magnetencephalographic study in man, *Electroenceph. Clin. Neurophysiol.,* 64, 347-349, 1986.
133. Kettenmann, B., et al., Odorants activate the human superior temporal sulcus, *Neurosci. Lett.,* 203, 143-145, 1996.
134. Seo, R., et al., Late components of olfactory event-related magnetic fields, *Abstracts of the 21st Meeting of the Association for Research in Otolaryngology,* February 15-19, 1998, St. Petersburg, FL, U.S.A., 1998, 68.
135. Zatorre, R.J. and M. Jones-Gotman, Right-nostril advantage for discrimination of odors, *Perc. Psychophys.,* 47, 526-531, 1990.
136. Hummel, T., et al., Chemosensory event-related potentials in patients with temporal lobe epilepsy, *Epilepsia,* 36(1), 79-85, 1995.
137. Doty, R.L., et al., Laterality in human nasal chemoreception, in *Cerebral Asymmetries in Sensory and Perceptual Processing,* S. Christman, Ed., North Holland Publishing, Amsterdam, 1997, 497-542.
138. Kobal, G. and T. Hummel, Olfactory evoked potential activity and hedonics, in *Fragrance: the Psychology and Biology of Perfume,* T.S. Van and G.H. Dodd, Eds., Elsevier Applied Science, London, 1992, 175-194.
139. Hummel, T. and G. Kobal, Differences in human evoked potentials related to olfactory or trigeminal chemosensory activation, *Electroenceph. Clin. Neurophysiol.,* 84, 84-89, 1992.
140. Mesulam, M.M. and E.J. Mufson, Insula of the old world monkey. I. Archetectonics in the insulo-orbito-temporal component of the paralimbic brain, *J. Comp. Neurol.,* 212, 1-22, 1982.
141. Carmichael, S.T. and J.L. Price, Sensory and premotor connections of the orbital and medial prefrontal cortex, *J. Comp. Neurol.,* 63, 642-664, 1995.
142. Kettenmann, B., et al., Multichannel magnetoencephalographical recordings: separation of cortical responses to different chemical stimulation in man, *Functional Neuroscience (EEG Suppl.),* 46, 287-290, 1996.
143. Kobal, G., A new method for determination of the olfactory and the trigeminal nerve's dysfunction: olfactory (OEP) and chemical somatosensory (CSEP) evoked potentials, in *Event-Related Potentials in Children,* A. Rothenberger, Ed., Elsevier Biomedical Press, Amsterdam, 1982, 455-461.
144. Matern, G., C. Matthias, and D. Mrowinski, Olfaktorisch evozierte Potentiale (OEP) und Contingent Negative Variation (CNV) bei der Begutachtung von Riechstörungen, *Laryngorhinootologie,* 74(2), 118-121, 1995.
145. Leplow, B., et al., Objective assessment of "olfactory" and "trigeminal" perception of odors in patients with lesions of the olfactory system, *J. Psychophysiol.,* 8, 53-54, 1994.

146. Mata, J., J. Jimenez, and A. Postigo, Potenciales evocados corticales olfatorios, *Acta Otorinolaring. Esp.*, 49, 385-387, 1998.
147. Geisler, M.W., et al., Traumatic brain injury assessed with olfactory event-related brain potentials, *J. Clin. Neurophysiol.*, 16, 77-86, 1999.
148. Hawkes, C.H. and B.C. Shephard, Olfactory evoked responses and identification tests in neurological disease, *Ann. N.Y. Acad. Sci.*, 855, 608-615, 1998.
149. Welge-Lüssen, A., Chemosensorisch evozierte Potentiale, HNO, 5, 453-455, 1999.
150. Welge-Luessen, A., G. Kobal, and M. Wolfensberger, Assessing olfactory function in laryngectomies using the Sniffin' Sticks test battery and chemosensory evoked potentials, *Laryngoscope,* 110, 303-307, 2000.
151. Korten, J. and J. Meulstee, Olfactory disturbances in parkinsonism, *Clin. Neurol. Neurosurg.,* 82, 113-118, 1980.
152. Ward, C.D., W.A. Hess, and D.B. Calne, Olfactory impairment in Parkinson's disease, *Neurology,* 33, 943-946, 1983.
153. Doty, R.L., D. Deems, and S. Steller, Olfactory dysfunction in Parkinson's disease: a general deficit unrelated to neurologic signs, disease stage, or disease duration, *Neurology,* 38, 1237-1244, 1988.
154. Hawkes, C.H., B.C. Shephard, and S.E. Daniel, Is Parkinson's disease a primary olfactory disorder?, *Qjm,* 92(8), 473-480, 1999.
155. Hummel, T., Olfactory evoked potentials as a tool to measure progression of Parkinson's disease, in *Focus on Medicine, Vol. 14: New Developments in the Drug Therapy of Parkinson's Disease,* T. Chase and P. Bedard, Eds., Blackwell Science, Oxford. 1999, 47-53.
156. Pearce, R.K., C.H. Hawkes, and S.E. Daniel, The anterior olfactory nucleus in Parkinson's disease, *Mov. Disord.,* 10(3), 283-287, 1995.
157. Hawkes, C.H., B.C. Shephard, and S.E. Daniel, Olfactory dysfunction in Parkinson's disease, *J. Neurol. Neurosurg. Psychiatry,* 62, 436-446, 1997.
158. Nordin, S., A.U. Monsch, and C. Murphy, Unawareness of smell loss in normal aging and Alzheimer's disease: discrepancy between self-reported and diagnosed smell sensitivity, *J. Gerontol.,* 50, 187-192, 1995.
159. Kesslak, J.P., et al., Olfactory tests as possible probes for detecting and monitoring Alzheimer's disease, *Neurobiol. Aging,* 9, 399-403, 1988.
160. Bacon, A.W., et al., Very early changes in olfactory functioning due to Alzheimer's disease, *Ann. N.Y. Acad. Sci.,* 855, 723-731, 1998.
161. Murphy, C., et al., Olfactory thresholds are associated with degree of dementia in Alzheimer's disease, *Neurobiol. Aging,* 11(4), 465-469, 1990.
162. Doty, R.L., *The Smell Idenfication Test(R) Administration Manual,* 3rd ed., Sensonics, Inc. Haddon Heights, NJ, 1995, 1-57.
163. Doty, R.L., et al., Olfactory dysfunction in multiple sclerosis: relation to longitudinal changes in plaque numbers in central olfactory structures, *Neurology,* 53, 880-882, 1999.
164. Hawkes, C.H., Assessment of olfaction in multiple sclerosis, *Chem. Senses,* 21, 486, 1996.
165. Hawkes, C.H., et al., Olfactory disorder in motor neuron disease, *Exp. Neurol.,* 150, 248-253, 1998.
166. Wenning, G.K., et al., Olfactory function in atypical parkinsonian syndromes, *Acta Neurol. Scand.,* 91(4), 247-250, 1995.
167. Grosser, K., et al., Olfactory and trigeminal and event-related potentials in migraine, *Cephalalgia,* 20, 621-631, 2000.

168. Otto, D.A. and H.K. Hudnell, The use of visual and chemosensory evoked potentials in environmental and occupational health, *Env. Res.,* 62, 159-171, 1993.
169. Dalton, P. and T. Hummel, Chemosensory function and response in idiopathic environmental intolerance, *Occup. Med.,* 15(3), 539-556, 2000.
170. Hummel, T., et al., Intranasal chemoreception in patients with multiple chemical sensitivities: a double-blind investigation, *Reg. Tox. Pharmacol.,* 24, S79-S86, 1996.
171. Wetter, S. and C. Murphy, Individuals with Down's syndrome demonstrate abnormal olfactory event-related potentials, *Clin. Neurophysiol.,* 110(9), 1563-1569, 1999.
172. Hummel, T., et al., Olfactory function in acute rhinitis, *Ann. N.Y. Acad. Sci.,* 855, 616-624, 1998.
173. Hummel, T., et al., Effects of the nasal decongestant oxymetazoline on human olfactory and intranasal trigeminal function in acute rhinitis, *Eur. J. Clin. Pharmacol.,* 54, 521-528, 1998.
174. Hummel, T. and G. Kobal, Chemosensory evoked potentials, in *Chemical Signals in Vertebrates VI,* R.L. Doty and D. Müller-Schwarze, Eds., Plenum, New York, 1992, 565-569.
175. Sutton, S., M. Braren, and J. Zubin, Evoked-potential correlates of stimulus uncertainty, *Science,* 150, 1187-1188, 1965.
176. Polich, J., S.E. Eischen, and G.E. Collins, P300 from a single auditory stimulus, *Elcetroenceph. Clin. Neurophysiol.,* 92, 253-261, 1994.
177. Cass, M. and J. Polich, P300 from a single-stimulus paradigm: auditory intensity and tone frequency effects, *Biol. Psychiatry,* 46, 51-65, 1997.
178. Pritchard, W.S., The psychophysiology of the P300, *Psych. Bull.,* 89, 506-540, 1981.
179. Donchin, E. and M.G.H. Coles, Is the P300 a manifestation of context updating?, *Behav. Brain Sci.,* 11, 357-428, 1988.
180. Verleger, R., Event-related potentials and cognition: a critique of the context updating hypothesis and an alternative interpretation of P3, *Behav. Brain Sci.,* 11, 343-356, 1988.
181. Pause, B.M., et al., Body odor evoked potentials: a new method to study the chemosensory perception of self and non-self in humans, *Genetica,* 104, 285-294, 1999.
182. Krug, R., et al., Selective influence of the menstrual cycle on the perception of stimuli with reproductive significance: an event-related potential study, *Pychophysiology,* 37, 111-122, 2000.
183. Fröhlich, R., Ueber einige Modificationen des Geruchsinnes. *Akad. Wiss. Wien,* math.-nat. CL, 6, 322-328, 1851.
184. Cain, W.S., Olfaction and the common chemical sense: some psychophysical contrasts, *Sensory Processes,* 1, 57-67, 1976.
185. Thürauf, N., et al., The mucosal potential elicited by noxious chemical stimuli: is it a peripheral nociceptive even, *Neurosci. Lett.,* 128, 297-300, 1991.
186. Hummel, T., Assessment of intranasal trigeminal function, *Int. J. Psychophysiol.,* 36, 147-155, 2000.
187. Hari, R., et al., Right-hemisphere preponderance of responses to painful CO_2 stimulation of the human nasal mucosa, *Pain,* 72, 145-151, 1997.
188. Chudler, E.H., W.K. Dong, and Y. Kawakami, Tooth pulp evoked potentials in the monkey: cortical surface and intracortical distribution, *Pain,* 22, 221-223, 1985.
189. Hummel, T., et al., Event-related potentials in response to repetitive painful stimulation, *Electroenceph. Clin. Neurophysiol.,* 92, 426-432, 1994.
190. Hummel, T. and G. Kobal, Chemo-somatosensory (intranasal trigeminal) event-related potentials change in relation to repetitive painful chemical stimulations of the nasal mucosa: effects of the interstimulus interval, *Eur. Arch. Otorhinolaryngol.,* 256, 16-21, 1999.

191. Hummel, T., et al., Peripheral electrophysiological responses decrease in response to repetitive painful stimulation of the human nasal mucosa, *Neurosci. Lett.,* 212, 37-40, 1996.
192. Torebjörk, H.E. and R.G. Hallin, C-fibre units recorded from human sensory nerve fascicles *in situ, Acta Soc. Med. Upsala,* 75, 81-84, 1970.
193. Raymond, S.A., et al., Changes in axonal impulse conduction correlate with sensory modality in primary afferent fibers in the rat, *Brain Res.,* 526, 318-321, 1990.
194. Mendell, L.M., Responses of single dorsal cord cells to peripheral cutaneous unmyelinated fibers, *Nature,* 206, 97-99, 1965.
195. Handwerker, H.O., A. Iggo, and M. Zimmermann, Segmental and supraspinal actions on dorsal horn neurons responding to noxious and non-noxious skin stimuli, *Pain,* 1, 147-165, 1975.
196. Adriaensen, H., et al., Response properties of thin myelinated (A-delta) fibers in human skin nerves, *J. Neurophysiol.,* 49, 111-122, 1983.
197. Sumino, R. and R. Dubner, Response characteristics of specific thermoreceptive afferents innervating monkey facial skin and their relationship to human thermal sensitivity, *Brain Res. Rev.,* 3, 105-122, 1981.
198. Sekizawa, S.I. and H. Tsubone, Nasal receptors responding to noxious chemical irritants, *Resp. Physiol.,* 96, 37-48, 1994.
199. Adriaensen, H., et al., Nociceptor discharges and sensations due to prolonged noxious mechanical stimulation — a paradox, *Hum. Neurobiol.,* 3, 53-58, 1984.
200. Handwerker, H.O., F. Anton, and P.W. Reeh, Discharge patterns of afferent cutaneous nerve fibers from the rat's tail during prolonged noxious mechanical stimulation, *Brain Res.,* 65, 493-504, 1987.
201. Price, D.D., et al., Peripheral suppression of first pain and central summation of second pain evoked by noxious heat pulses, *Pain,* 3, 57-68, 1977.
202. Kobal, G., et al., Effects of pentazocine and acetylsalicylic acid on pain-rating, pain-related evoked potentials and vigilance in relationship to pharmacokinetic parameters, *Agents Actions,* 29 (3/4), 342-359, 1990.
203. Kobal, G., T. Hummel, and M. Hoesl, Pain-related electrical evoked potentials by chemical stimuli: effects of fentanyl, in *Advances in Pain Research and Therapy,* S. Lipton, E. Tunks, and M. Zoppi, Eds., Raven Press, New York, 1990, 95-98.
204. Lötsch, J., et al., Lack of analgesic activity of morphine-6-glucuronide after short-term intravenous administration in healthy volunteers, *Anesthesiology,* 87, 1348-1358, 1997.
205. Lötsch, J., et al., Effects of flurbiprofen enantiomers on pain-related chemo-somatosensory evoked potentials in human subjects, *Br. J. Clin. Pharmacol.,* 40, 339-346, 1995.
206. Kobal, G., et al., Dose-related effects of ibuprofen on pain-related potentials, *Br. J. Clin. Pharmacol.,* 37, 445-452, 1994.
207. Hummel, T., et al., Comparison of the antinociception produced by two oral formulations of ibuprofen: ibuprofen effervescent vs. ibuprofen tablets, *Eur. J. Clin. Pharmacol.,* 52, 107-114, 1997.
208. Hummel, T., et al., Asessment of analgesia in man: tramadol controlled release formula vs. tramadol standard formulation, *Eur. J. Clin. Pharmacol.,* 51, 31-38, 1996.
209. Lötsch, J., et al., Effects of azapropazone on pain-related brain activity in human subjects, *Br. J. Clin. Pharmacol.,* 40, 545-552, 1996.
210. Hummel, T., et al., Dose-related analgesic effects of flupirtine, *Br. J. Clin. Pharmacol.,* 32, 69-77, 1991.
211. Sunshine, A., et al., Comparative study of flurbiprofen, zomepirac sodium, acetaminophen plus codeine, and acetaminophen for the relief of postsurgical dental pain, *Am. J. Med.,* 80, 50-54, 1986.

212. Sunshine, A., et al., Analgesic efficacy of two ibuprofen-codeine combinations for the treatment of postepisiotomy and postoperative pain, *Clin. Pharmacol. Ther.,* 42, 374-380, 1987.
213. Thürauf, N., et al., Dose-dependent time course of the analgesic effect of a sustained release preparation of tramadol on experimental phasic and tonic pain, *Br. J. Clin. Pharmacol.,* 41, 115-123, 1996.
214. Lötsch, J., et al., Analgesic effects of the kappa-opioid agonist RP 60180 in man, *Psychopharmacology,* 20, 224-232, 1997.
215. Kobal, G., et al., Testing the analgesic activity of dipyrone in human subjects using an experimental pain model with tonic and phasic pain stimuli, in New pharmacological and epidemiological data in analgesics research, K. Brune, Ed., Birkhäuser, Basel, 1990, 29-37.
216. Petersen-Felix, S., et al., Psychophysical and electrophysiological responses to experimental pain may be influenced by sedation: comparison of the effectcs of a hypnotic (propofol) and an analgesic (alfentanil), *Br. J. Anaesth.,* 77, 165-171, 1996.
217. Zaslansky, R., et al., Pain-evoked potentials: what do they really measure, *Elcetroenceph. Clin. Neurophysiol.,* 100, 384-391, 1996.
218. Hummel, T., et al., 'Sniffin' sticks': olfactory performance assessed by the combined testing of odor identification, odor discrimination and olfactory threshold, *Chem. Senses,* 22, 39-52, 1997.
219. Kobal, G., et al., Multi-center investigation of 1036 subjects using a standardized method for the assessment of olfactory function combining tests of odor identification, odor discrimination, and olfactory thresholds, *Eur. Arch. Otorhinolaryngol.,* 257, 205-211, 2000.
220. Welge-Luessen, A., et al., Olfactory function in patients with olfactory groove meningioma, *J. Neurol. Neurosurg. Psych.,* 70, 218-221, 2001.
221. Nagel, W.A., Einige Bemerkungen über nasales Schmecken, *Ztschr. f. Psychol.,* 25, 268, 1904.
222. Daum, R., et al., Riechprüfung wit "Suiffin' Sticks" zur Kliuischen Diagnostik des Morlous Parkinson, *Nerveuarzt,* 71, 643-650, 2000.

18 Application of Functional MRI in Olfactory Studies

F. Xu, Charles A. Greer, and Gordon M. Shepherd

CONTENTS

18.1 INTRODUCTION

Magnetic resonance imaging (MRI), introduced in late 1970s, flourished in clinical applications. Knowledge accumulation from many years of basic nuclear magnetic resonance research realized the hope of obtaining information about the biochemistry, perfusion, and metabolism that can be coupled to energy consumption and *functional* activity in the examined tissues (fMRI). First appearing in the early 1990s,[1] fMRI gained almost immediate popularity in both basic research in the life sciences and in the clinical diagnosis of various diseases.

Functional MRI is based on the behavior of protons in a high magnetic field. When placed in a magnetic field, protons act as tiny magnets so that they line up with the external field to stay at the lowest energy level (Figure 18.1). These protons can pick up electromagnetic energy (in the radio frequency range) to jump to a higher energy level. The dynamics of the protons returning to the ground state (relaxation) are affected by their environment, revealing several important properties of the tissue being examined. One of the environmental factors is blood oxygen level.[1] The iron in deoxygenated hemoglobin (rich in venous capillaries and veins)

0-8493-2329-0/02/$0.00+$1.50

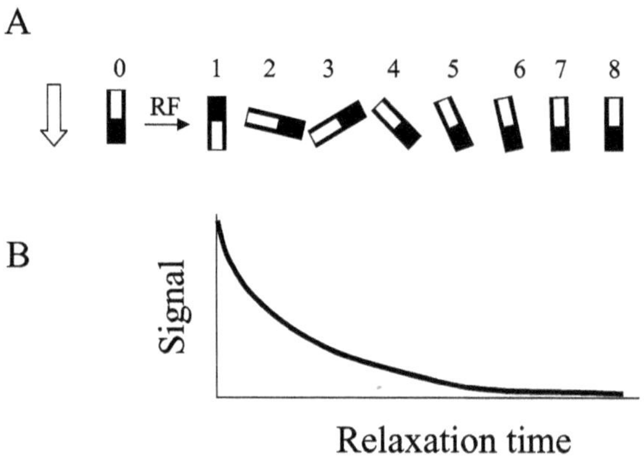

FIGURE 18.1 MRI signal and T_1 relaxation time. All protons in a pixel as a whole act as a compass. At rest, the compass is aligned with the external magnetic field (A0), which is indicated by the arrow. A radio frequency (RF) excites the protons to a higher energy level, causing the compass to rotate 180° to the opposing external magnetic field (A1). Over time, the compass rotates to realign with the external field (A1-8). The rotation rate is greatest at the beginning and decreases exponentially. The signal detected by MRI is plotted in B. The rate constant of the decay is called the T_1 relaxation time. The longer T_1, the faster the compass returns to equilibrium. T_1 is affected by the environment of the protons. The oxygenation and deoxygenation of hemoglobin has a critical effect on T_1.

is paramagnetic, which leads to quicker relaxation and a shorter signal lifetime. However, the oxygen in oxygenated hemoglobin (in arterial capillaries and arteries) eliminates the paramagnetic property of the iron. The recent results that fMRI blood oxygenation level dependent (BOLD) signals are proportional to neuronal activity,[2] neurotransmitter release,[3] and energy consumption[4] provide the basis for fMRI data interpretation in terms of neuronal activities in the brain.[4a]

The application of fMRI in olfactory-related studies, primarily in humans, has increased in recent years. These studies have involved identifying anatomical regions that are activated by odors, examining their responses, and characterizing their roles in olfactory information processing, odor identification, and olfactory memory.

The reported brain regions activated by odors include frontal, orbital, piriform, entorhinal, temporal, insular, cingulate, occipital,[5-7] perisylvian cortices,[8-10] hippocampus,[11] amygdala,[12] and cerebellum.[13] Although some of them are more common (e.g., frontal and orbital cortices), the regions and activities detected in a specific study varied broadly, depending on the properties of the stimuli and the age and sex of the subject.

Asymmetrical activities have been commonly observed in these cortical regions.[7,13,14] The fact that patients with brain lesions in right frontal and orbital cortices exhibit more severe deficits in olfactory discrimination,[15] better discriminative performance with right nostril stimulation, and stronger fMRI activities in the right hemisphere[13] all point to a hemispherical difference in olfaction.

Odor stimulation is commonly associated with emotions. Pleasant and unpleasant odors (clementine vs. isovaleric acid) elicit different activity patterns across the brain.[7] Some of the activated regions overlap (BA32, BA46, BA6, and insula), whereas some are specific (BA8 to unpleasant odor). For the nonspecific regions, the activity levels and/or lateralizations are different. Birbaumer[12] reported that only aversive odors elicit strong response in the amygdala. The amygdala's response to aversive odors is in agreement with clinical evidence that seizures or electrical stimulation of this area often produce unpleasant olfactory hallucinations.[16]

Multiple adaptation mechanisms have been revealed in the olfactory system, from olfactory epithelium to cortices. Using repeated exposures, Yousem[17] and Sobel[6] showed that adaptation occurs in both primary and secondary olfactory cortices. After a strong early transient increase, the signal then decreased within 30 to 40 sec, similar to the time-course in electrophysiological recording.[18]

On standard tests of odor detection and identification, women tend to perform better than men in nearly all age groups.[19] This is reflected in the signal intensity in fMRI.[10] In general, the activity in women is higher than in men. In specific regions (frontal and perisylvian), the number of voxels reaching threshold is up to eight times more in women than in men. The results suggest that fMRI activities correlate with psychological performance.

The results of fMRI studies on patients with Type I hyposmia (able to detect but unable to recognize odors) and Type II hyposmia (able to detect and recognize but with less acuity) demonstrate that Type I patients have weaker activities than normal people in some regions and no significant activities at all in other regions (inferior frontal, orbitofrontal, and cingulate cortex), suggesting a role of these regions in odor recognition.[20] In the same studies, fMRI was used to monitor drug efficacy in Type I hyposmia.

Although these studies contribute to our standing of human brain regions involved in olfactory information processing, odor identification and recognition, investigation in the olfactory bulb (OB) — the first olfactory center — is completely absent. The two major reasons are that the OB in humans is so small that it occupies only several voxels[6,21] and stronger magnets (e.g., >7 Tesla), to enable humans to achieve higher resolution, are not commonly available.

It is generally accepted that odor information is represented in the OB by an activity pattern (other terms used in literature include odor map, odotope map, chemotopic map, etc.) that is distributed across the glomerular layer.[22,23] The concept was suggested a half century ago,[24] and clear evidence was obtained using a 2-deoxyglucose mapping technique.[25] Since then, results from various methods, including 2-deoxyglucose,[26-29] c-*Fos* immunocytochemistry,[30,31] voltage-sensitive dyes,[32-34] intrinsic optical imaging,[35] and molecular biology,[36-38] have supported the hypothesis.

High-resolution fMRI has the capability to visualize the activity throughout the OB with histological correlation for repeated exposures at different odor concentrations and durations; the spatial resolution is sufficient to correlate with the histological lamina of the OB, with different temporal resolution.[23,29] This chapter focuses on the experiments in the investigation of the rat OB, with examples.

18.2 EXPERIMENTAL METHODS FOR fMRI IN THE RAT OB

18.2.1 Animal Preparation

Animals used for fMRI experiments can be awake or anesthetized. In the former situation, dental cement is used to hold the head firmly in place.[40] In manganese-enhanced tracing studies of neuronal projections, because of the long imaging time, the mice are killed after exposures before the imaging experiment.[41] The focus here is on anesthetized rats used by the authors to map the odor-elicited activity patterns in the OB.

Any strain, age, and sex of animals can be used. In our experiments, male Sprague-Dawley rats (200 to 400 g) are anesthetized with 1 to 2% halothane and the skin overlying the OB is removed to expose the skull for precise placement of the RF probe and better fixation to minimize head movement. A tracheal double-cannulation procedure[42] is performed to allow artificial control of respiration and the flow of air through the nasal cavity. A femoral artery is cannulated for monitoring blood pressure, gases, and pH levels. Before the rat is placed in the magnet, the anesthesia is switched to urethane (i.p. 1.5 g/kg initial; 0.1 g/kg/hr supplemental), which is the best of the several agents we have tested.

18.2.2 Odor Delivery

Any of the common odor delivery apparati can be used. It is necessary to eliminate all metal components from the system. The olfactometer we use is made of Teflon and glass to minimize cross-contamination. The dead volume should be as small as possible to decrease the delay time (ours has a delay of 0.24 sec). Mineral oil, which has been used in other olfactory studies[43] and does not produce any significant fMRI signal in control experiments, is used as the solvent to prepare odor solutions. The flow rate of the extra-pure air (grade 0.1, <0.1 ppm hydrocarbon) passing over the odor solution is kept at 5 l/min, generating odorized air flow with various vapor-phase concentrations. The concentration of any odorant can be calculated from the mass decrease over a period of time (e.g., 20 min) under the experimental conditions. If the solution concentration is relatively low (e.g., less than 1%) or the odorant has a low boiling point, the solution should be changed frequently to make sure the odorant concentration does not alter significantly. The odorized air is delivered through the nasal cavities by a ventilator with a fixed volume and rate of 3 ml/stroke and 60 strokes/min, while the lung is ventilated separately to maintain the blood gases ($p_{CO_2} \approx 38$ and $p_{O_2} \approx 100$ mmHg), arterial blood pressure (~100 mmHg), and blood pH value (~7.4) throughout the experiment. Because the amount of odorant molecules delivered depends on multiple factors — the odorant concentration in the vapor phase, exposure duration, partition coefficient of odorant in vapor and mucus, air flow rate, and the anatomical structure of the nasal cavity and the olfactory epithelium — the actual odor concentration at the olfactory epithelium is not known.

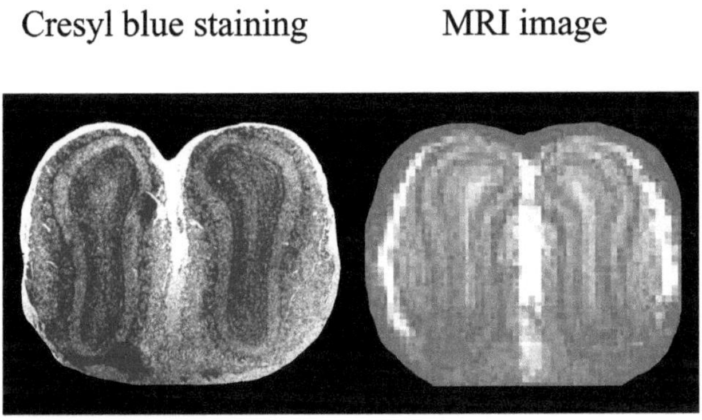

FIGURE 18.2 Correlation laminar structure of the OB revealed by cresyl blue staining and MRI. Twenty T_1-weighted images with 0.25-mm thickness (total of 5.0 mm) cover the entire OB. Then the same rat is fixed and the OB sectioned at 20 mm. Each slice is counted so as to correlate the location of the two images in the OB.

18.2.3 Imaging Experiments

Imaging experiments are performed in a modified horizontal-bore 7T (Bruker, Billerica, MA) spectrometer. For activity pattern mapping, all images are obtained with the fast low-angle single-shot (FLASH) gradient-echo imaging sequence. The rat head is secured in a holder designed to minimize movement. The body temperature is maintained by a water-heated blanket. The rat is placed prone on a cradle, with its nose loosely fit into a odor delivery funnel. A circular radio-frequency transmit-receive surface coil probe (10-mm diameter) is centered on top of the skull above the OB. The magnetic field homogeneity is carefully shimmed so that the half-peak width is less than 20 Hz. This is one of the most important requirements to obtain a desirable signal-to-noise ratio and reproducibility of the collected images.

T_1 (longitudinal relaxation time of tissue water) weighted FLASH anatomical images are obtained (image dimension = 128×128 pixels; in-plane resolution = 110×110 μm; slice thickness = 250 μm; repetition delay = 5.0 sec; echo time = 16 msec; flip angle = 60°) with variable inversion recovery weighting per slice.

Figure 18.2A (left) shows an example of T_1-weighted image, which reveals the laminar structure of the OB. These layers correlate well with the structure shown by cresyl blue staining of the tissue at a similar site (Figure 18.2B, right). Using electrochemically deposited iron as land marker, more accurate correlation has been achieved (Xu et al. unpublished results).

Each fMRI experiment consists of a series of T_2^* (apparent transverse relaxation time of tissue water) weighted FLASH images (image dimension = 64×64 pixels; in-plane resolution = 220×220 μm; slice thickness = 250 μm; repetition delay = 25 to 125 msec; echo time = 16 msec; flip angle = 5 to 15°). Figure 18.3A and B show the T_1- and T_2-weighted images of the same slice, respectively. The temporal resolutions are 8 and 1.5 sec/image for single and 20 slices, respectively. The spatial resolution depends on the field of view, the image dimension, and the slice thickness.

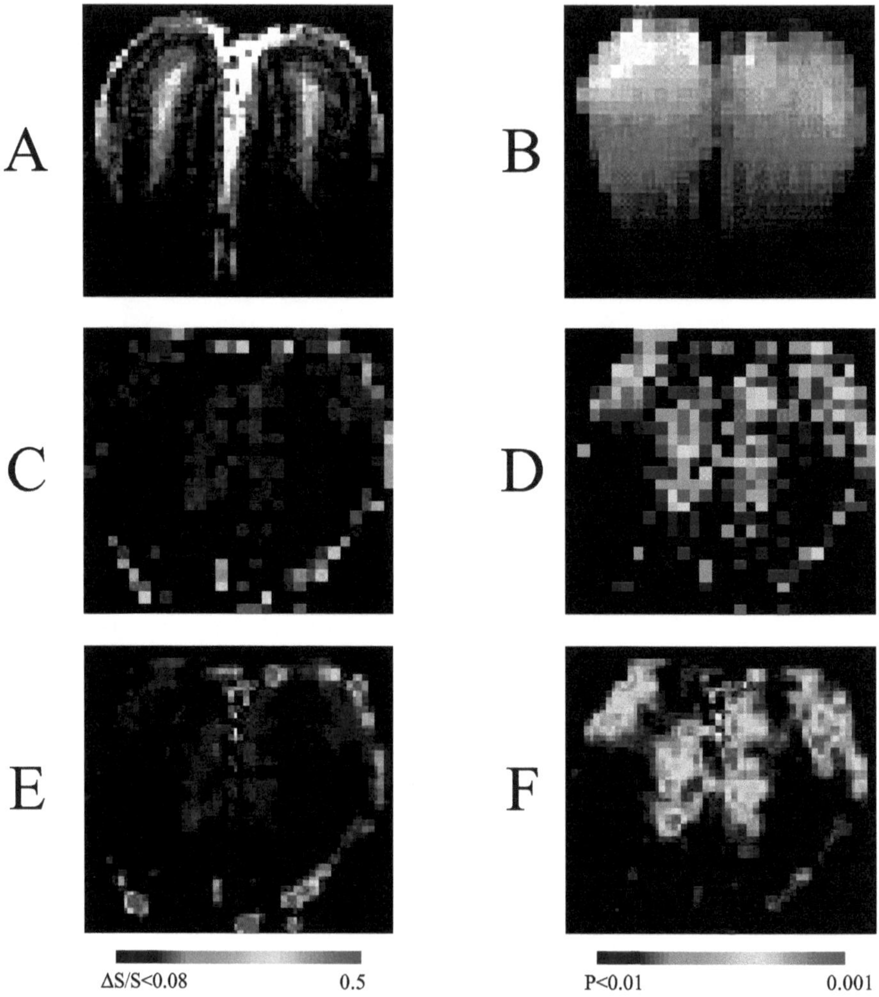

FIGURE 18.3 (see color insert following page 238) Data processing in which all images are from the same slice. A: T_1-weighted image of a slice located about 2/3 of the OB from anterior. B: T_2-weighted image. C: $\Delta S/S$ image; the rat is exposed to isoamylacetate for 2 min. D: *t* Image in the same exposure experiment as in C. E: Odor map produced by overlaying C on A. F: Odor map produced by overlaying D on A.

18.2.4 Data Processing

The data processing step differs broadly, depending on the imaging system and software being used. A detailed description of data processing is beyond the scope of this chapter. Briefly, Fourier transformed data are assessed by a center-of-mass analysis to examine the head movement artifacts. Images are discarded from analysis if the movement is greater than 25% of a pixel (45 μm). For each experiment, the mean image of the control images (S) is subtracted from the stimulated images (ΔS+S) on a pixel-by-pixel basis to create ΔS/S images or Student's *t* images. Figure 18.3C and D show the ΔS/S images and Student's *t* images of the same slice as shown in Figure 18.3A. Although the inner regions of the two images are highly similar, the signal of the peripheral region in the ΔS/S image is stronger than that in the corresponding *t* image because the baseline at the edge of the OB is more bumpy. Therefore, the *t* image is more frequently used in presenting odor-elicited activity patterns.

To demonstrate the odor map, each *t* image (or ΔS/S image) is linearly interpolated to a 128×128 image matrix and overlaid onto the corresponding anatomical image to locate the activity in different bulbar layers. Figure 18.3E (ΔS/S image/T_1 image) and 18.3F (*t* image/T_1 image) show the isoamylacetate-elicited activity pattern of the same slice as in Figure 18.3A. Clearly, the activities are primarily localized within the glomerular layer, but extend into the neighboring olfactory nerve layer and external plexiform layer.

Time-courses are plotted for highly activated foci to reveal the dynamic patterns (Figure 18.4B). For analysis of similarity between odor maps, the BOLD signal in each pixel is divided by the highest value to form a normalized odor map, and then linear Pearson correlation coefficients between paired functional maps are calculated based on spatially binned data. Another statistical method, Spearman's ranking coefficient, can also be used to compare the images.

18.2.5 Stimulation Protocols

Depending on the purposes of the experiments, stimulation protocols, with different stimulation periods or/and interstimulation intervals, are used. The protocols we have used include: (1) a single, short exposure (18 to 64 sec); (2) a train of short exposures (two to five in one run); (3) a long exposure (8 to 28 min); and (4) a combination of (1) and (4). In each experiment, four dummy scans are used to stabilize the system, then 8 to 24 prestimulation control scans are collected to obtain a baseline. The same number of post-stimulation control scans is used to find out whether the baseline is stable. During the interval or resting periods, the rat breathes moisturized air.

18.2.6 Responses to a Train of Short Exposures

Example results from an fMRI study of an odor map are described. If the activity pattern in the OB serves as the basis for odor perception, discrimination, and identification, it should be reproducible or predictable in repeated exposures. A train of short exposures with equal interstimulation intervals is used to reveal this property. Figure 18.4 shows the activity pattern elicited by isoamylacetate in a single slice with four 0.6-min exposures and 3.5-min intervals. Although the activity decreased in the following exposures, the general features of the activity pattern are conserved. Figure 18.4B shows the highly activated focus circled in Figure 18.4A; Spearman's rank correlation analysis indicates the patterns of the focus are highly similar ($r_{1st\ vs.\ 2nd} = 0.72$, $p < 0.001$; $r_{2nd\ vs.\ 3rd} = 0.74$, $p < 0.001$; $r_{3rd\ vs.\ 4th} = 0.71$, $p < 0.001$). Figure 18.4C shows the intensity of the foci

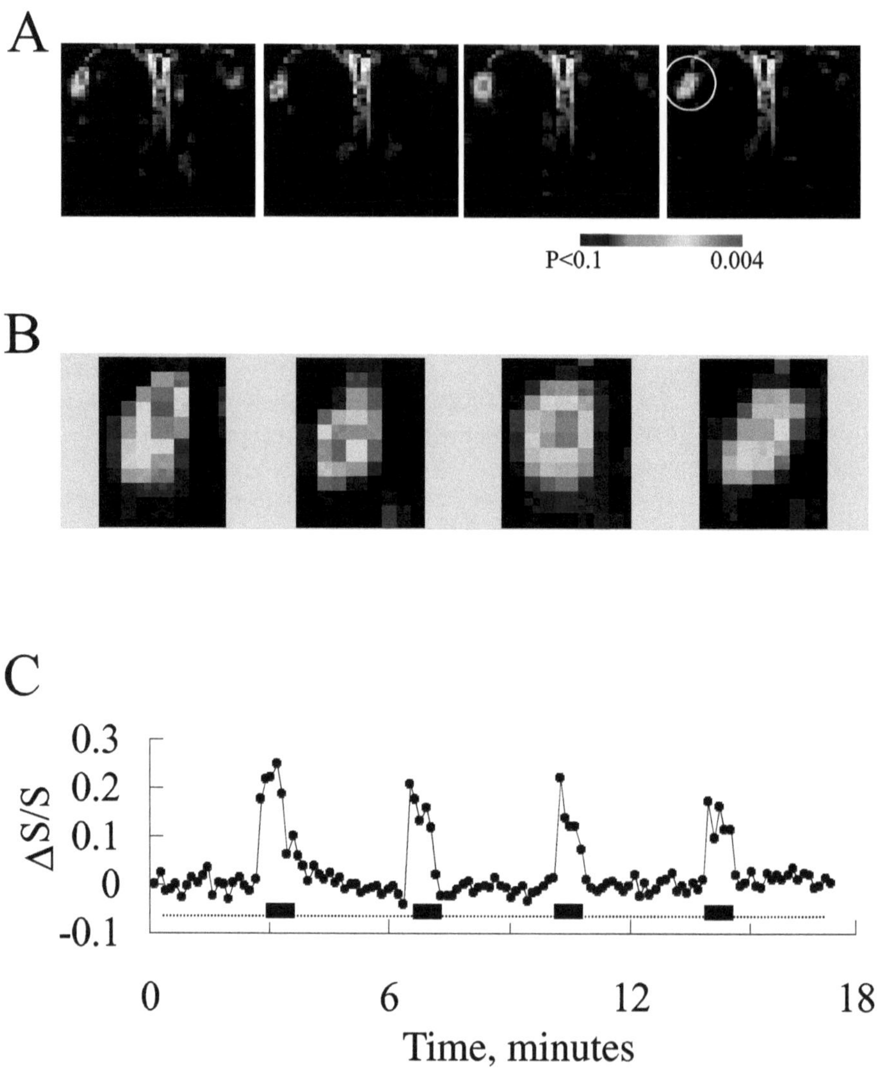

FIGURE 18.4 (see color insert following page 238) Odor maps of one slice generated by four repeated exposures (36 sec) of isoamylacetate. A: The activity patterns. B: The enlarged view of the highly activated focus circled in A. C: The corresponding time-course of B.

in Figure 18.4B. In all rats, the increase and decrease of odor-elicited fMRI signal changes occurred within seconds of stimulation onset and offset.

18.3 fMRI COMPARED WITH OTHER METHODS

An ideal method to study odor maps should provide five-dimensional (5-D) information about the activity patterns, that is, the intensity (1-D) at any position in the bulb (3-D) at any time (1-D) for different experiments in the same animal. Each method used in activity mapping of the OB has its advantages and limitations. Briefly, the 2-DG mapping can monitor the entire bulb with histological correlation in the waking animal, but allows only one experiment per animal with poor temporal resolution (~40 min[29,44]). c-*Fos* mRNA and protein mapping has similar advantages and disadvantages to 2-DG, but with even lower spatial resolution and indirect relation to neuronal activity.[31] Electrophysiological recordings, an invasive method, offers excellent temporal resolution and sensitivity for repeated exposures and/or different odors, but can only examine a few cells obtained from the selected areas and with variable histological correlation in the bulbar layers.[45] Voltage-sensitive dye imaging, also an invasive method, provides good temporal and spatial resolutions for repeated experiments but yields images of only part of the OB.[46,47] The intrinsic optical imaging method is similar in character to dye imaging, with better spatial but lower temporal resolution.[35] High-resolution fMRI fits most of these requirements reasonably well, although there are trade-offs between experimental parameters (spatial and temporal resolution, sensitivity, and amount of the examining tissues; see below). The reproducibility of patterns from repeated exposures shown here and the high topographical similarity of the patterns from an odorant with a 100-fold difference in concentration indicate the reliability of fMRI for mapping.[23] The most important limiting factor for fMRI is the cost and maintenance of the entire system.

18.4 PROPERTIES OF fMRI IN MAPPING ACTIVITY PATTERNS

Different imaging sequences with different properties are available for fMRI. For example, FLASH gradient-echo imaging sequence can be used for high spatial resolution (down to 50 μm) with decent temporal resolution (seconds level), whereas echo planar imaging (EPI) sequence can provide higher temporal resolution (down to 60 msec) at the expense of either spatial resolution or amount of tissue (or signal-to-noise ratio). Therefore, the selection of sequences is based on one's purpose. For example, if one wants to examine the rapid adaptation behavior of a specific focus in the OB already identified, one can choose an EPI sequence for a single slice so as to obtain decent spatial resolution and sensitivity with excellent temporal resolution.

REFERENCES

1. Ogawa, S., Lee, T.M., Nayak, A.S. and Glynn, P., Oxygenation-sensitive contrast in magnetic resonance image of rodent brain at high magnetic fields, *Magn. Reson. Med.*, 14, 68, 1990.
2. Rees, G., Friston, K., and Koch, C., A direct quantitative relationship between the functional properties of human and macaque V5 [see comments], *Nat. Neurosci.*, 3, 716, 2000.

3. Shulman, R.G. and Rothman, D.L., Interpreting functional imaging studies in terms of neurotransmitter cycling, *Proc. Natl. Acad. Sci. U.S.A.*, 95, 11993, 1998.
4. Kida, I., Kennan, R.P., Rothman, D.L., Behar, K.L., and Hyder, F., High-resolution CMR(O_2) mapping in rat cortex: a multiparametric approach to calibration of BOLD image contrast at 7 Tesla, *J. Cereb. Blood Flow Metab.*, 20, 847, 2000.

4a. Logothetis, N.K., Pauls, J., Augath, M., Trinath, T., and Oeltermann, A., Neurophysiological investigation of the basics of the fMRI signal, *Nature,* 412, 150-157, 2001.

5. Koizuka, I., Yano, H., Nagahara, M., Mochizuki, R., Seo, R., Shimada, K., Kubo, T., and Nogawa, T., Functional imaging of the human olfactory cortex by magnetic resonance imaging, *ORL J. Otorhinolaryngol. Relat. Spec.*, 56, 273, 1994.
6. Sobel, N., Prabhakaran, V., Zhao, Z., Desmond, J.E., Glover, G.H., Sullivan, E.V., and Gabrieli, J.D., Time course of odorant-induced activation in the human primary olfactory cortex, *J. Neurophysiol.*, 83, 537, 2000.
7. Fulbright, R.K., Skudlarski, P., Lacadie, C.M., Warrenburg, S., Bowers, A.A., Gore, J.C., and Wexler, B.E., Functional MR imaging of regional brain responses to pleasant and unpleasant odors, *AJNR Am. J. Neuroradiol.*, 19, 1721, 1998.
8. Yousem, D.M., Maldjian, J.A., Hummel, T., Alsop, D.C., Geckle, R.J., Kraut, M.A., and Doty, R.L., The effect of age on odor-stimulated functional MR imaging, *AJNR Am. J. Neuroradiol.*, 20, 600, 1999.
9. Yousem, D.M., Geckle, R.J., Bilker, W.B., Kroger, H., and Doty, R.L., Posttraumatic smell loss: relationship of psychophysical tests and volumes of the olfactory bulbs and tracts and the temporal lobes, *Acad. Radiol.*, 6, 264, 1999.
10. Yousem, D.M., Maldjian, J.A., Siddiqi, F., Hummel, T., Alsop, D.C., Geckle, R.J., Bilker, W.B., and Doty, R.L., Gender effects on odor-stimulated functional magnetic resonance imaging, *Brain Res.*, 818, 480, 1999.
11. Small, D.M., Jones-Gotman, M., Zatorre, R.J., Petrides, M., and Evans, A.C., Flavor processing: more than the sum of its parts, *Neuroreport.*, 8, 3913, 1997.
12. Birbaumer, N., Grodd, W., Diedrich, O., Klose, U., Erb, M., Lotze, M., Schneider, F., Weiss, U., and Flor, H., fMRI reveals amygdala activation to human faces in social phobics, *Neuroreport.*, 9, 1223, 1998.
13. Sobel, N., Prabhakaran, V., Desmond, J.E., Glover, G.H., Goode, R.L., Sullivan, E.V., and Gabrieli, J.D., Sniffing and smelling: separate subsystems in the human olfactory cortex, *Nature*, 392, 282, 1998.
14. Zald, D.H. and Pardo, J.V., Emotion, olfaction, and the human amygdala: amygdala activation during aversive olfactory stimulation, *Proc. Natl. Acad. Sci. U.S.A.*, 94, 4119, 1997.
15. Zatorre, R.J. and Jones-Gotman, M., Human olfactory discrimination after unilateral frontal or temporal lobectomy, *Brain*, 114, 71, 1991.
16. Andy, O.J., Jurko, M.F., and Hughes, J.R., The amygdala in relation to olfaction, *Confin. Neurol.*, 37, 215, 1975.
17. Yousem, D.M., Williams, S.C., Howard, R.O., Andrew, C., Simmons, A., Allin, M., Geckle, R.J., Suskind, D., Bullmore, E.T., Brammer, M.J., and Doty, R.L., Functional MR imaging during odor stimulation: preliminary data, *Radiology*, 204, 833, 1997.
18. Wilson, D.A., Odor specificity of habituation in the rat anterior piriform cortex, *J Neurophysiol.*, 83, 139, 2000.
19. Doty, R.L., Applebaum, S., Zusho, H., and Settle, R.G., Sex differences in odor identification ability: a cross-cultural analysis, *Neuropsychologia*, 23, 667, 1985.
20. Levy, L.M., Henkin, R.I., Lin, C.S., and Finley, A., Rapid imaging of olfaction by functional MRI (fMRI): identification of presence and type of hyposmia, *J. Comput. Assist. Tomogr.*, 23, 767, 1999.

21. Kobal, G. and Kettenmann, B., Olfactory functional imaging and physiology, *Int. J. Psychophysiol.*, 36, 157, 2000.
22. Hildebrand, J.G. and Shepherd, G.M., Mechanisms of olfactory discrimination: converging evidence for common principles across phyla, *Annu. Rev. Neurosci.*, 20, 595, 1997.
23. Xu, F., Greer, C.A., and Shepherd, G.M., Odor maps in the olfactory bulb [in process citation], *J. Comp. Neurol.*, 422, 489, 2000.
24. Adrian, E.D., The electrical activity of the mammalian olfactory bulb, *Electroencephalogr. Clin. Neurophysiol.*, 2, 377, 1950.
25. Sharp, F.R., Kauer, J.S., and Shepherd, G.M., Local sites of activity-related glucose metabolism in rat olfactory bulb during olfactory stimulation, *Brain Res.*, 98, 596, 1975.
26. Stewart, W.B., Kauer, J.S., and Shepherd, G.M., Functional organization of rat olfactory bulb analysed by the 2-deoxyglucose method, *J. Comp. Neurol.*, 185, 715, 1979.
27. Jourdan, F., Duveau, A., Astic, L., and Holley, A., Spatial distribution of [^{14}C]2-deoxyglucose uptake in the olfactory bulbs of rats stimulated with two different odours, *Brain Res.*, 188, 139, 1980.
28. Greer, C.A., Stewart, W.B., Kauer, J.S., and Shepherd, G.M., Topographical and laminar localization of 2-deoxyglucose uptake in rat olfactory bulb induced by electrical stimulation of olfactory nerves, *Brain Res.*, 217, 279, 1981.
29. Johnson, B.A., Woo, C.C., Hingco, E.E., Pham, K.L., and Leon, M., Multidimensional chemotopic responses to n-aliphatic acid odorants in the rat olfactory bulb, *J. Comp. Neurol.*, 409, 529, 1999.
30. Onoda, N., Odor-induced *fos*-like immunoreactivity in the rat olfactory bulb, *Neurosci. Lett.*, 137, 157, 1992.
31. Guthrie, K.M., Anderson, A.J., Leon, M., and Gall, C., Odor-induced increases in c-*fos* mRNA expression reveal an anatomical "unit" for odor processing in olfactory bulb, *Proc. Natl. Acad. Sci. U.S.A.*, 90, 3329, 1993.
32. Trombley, P.Q. and Westbrook, G.L., Voltage-gated currents in identified rat olfactory receptor neurons, *J. Neurosci.*, 11, 435, 1991.
33. Cinelli, A.R. and Kauer, J.S., Salamander olfactory bulb neuronal activity observed by video rate, voltage-sensitive dye imaging. II. Spatial and temporal properties of responses evoked by electric stimulation, *J. Neurophysiol.*, 73, 2033, 1995.
34. Cinelli, A.R. and Kauer, J.S., Voltage-sensitive dyes and functional activity in the olfactory pathway, *Annu. Rev. Neurosci.*, 15, 321, 1992.
35. Rubin, B.D. and Katz, L.C., Optical imaging of odorant representations in the mammalian olfactory bulb, *Neuron*, 23, 499, 1999.
36. Ressler, K.J., Sullivan, S.L., and Buck, L.B., Information coding in the olfactory system: evidence for a stereotyped and highly organized epitope map in the olfactory bulb, *Cell*, 79, 1245, 1994.
37. Vassar, R., Chao, S.K., Sitcheran, R., Nunez, J.M., Vosshall, L.B., and Axel, R., Topographic organization of sensory projections to the olfactory bulb, *Cell*, 79, 981, 1994.
38. Mombaerts, P., Wang, F., Dulac, C., Chao, S.K., Nemes, A., Mendelsohn, M., Edmondson, J., and Axel, R., Visualizing an olfactory sensory map, *Cell*, 87, 675, 1996.
39. Yang, X., Renken, R., Hyder, F., Siddeek, M., Greer, C.A., Shepherd, G.M., and Shulman, R.G., Dynamic mapping at the laminar level of odor-elicited responses in rat olfactory bulb by functional MRI, *Proc. Natl. Acad. Sci. U.S.A.*, 95, 7715, 1998.
40. Khubchandani, M., Mallick, H.N., Kumar, V.M., and Jagannathan, N.R., Preliminary results of simultanous electroencephalography and functional magnetic resonance imaging in rat brain during sleeping-wakefulness, presented at *Proc. Int. Soc. MRM*, Denver, CO, 2000.

41. Pautler, R.G., Silva, A.C., and Koretsky, A.P., *In vivo* neuronal tract tracing using manganese-enhanced magnetic resonance imaging, *Magn. Reson. Med.*, 40, 740, 1998.
42. Ravel, N. and Pager, J., Respiratory patterning of the rat olfactory bulb unit activity: nasal vs. tracheal breathing, *Neurosci. Lett.*, 115, 213, 1990.
43. Mori, K., Relation of chemical structure to specificity of response in olfactory glomeruli, *Curr. Opin. Neurobiol.*, 5, 467, 1995.
44. Johnson, B.A., Woo, C.C., and Leon, M., Spatial coding of odorant features in the glomerular layer of the rat olfactory bulb, *J. Comp. Neurol.*, 393, 457, 1998.
45. Imamura, K., Mataga, N., and Mori, K., Coding of odor molecules by mitral/tufted cells in rabbit olfactory bulb. I. Aliphatic compounds, *J. Neurophysiol.*, 68, 1986, 1992.
46. Friedrich, R.W. and Korsching, S.I., Combinatorial and chemotopic odorant coding in the zebrafish olfactory bulb visualized by optical imaging, *Neuron*, 18, 737, 1997.
47. Friedrich, R.W. and Korsching, S.I., Chemotopic, combinatorial, and noncombinatorial odorant representations in the olfactory bulb revealed using a voltage-sensitive axon tracer, *J. Neurosci.*, 18, 9977, 1998.

19 Functional Imaging of Olfactory Activation in the Human Brain

Birgit Kettenmann, Thomas Hummel, and Gerd Kobal

CONTENTS

19.1 INTRODUCTION

The cerebral representation of odor perception in man is poorly understood, especially when compared to other sensory modalities such as hearing, vision, and touch. In part, this results from the lack of appropriate instruments for specific stimulation of chemosensors, making investigations of these systems extremely difficult. Primary olfactory structures, namely the olfactory bulbs and tracts and the pyriform cortex that have been studied in animal experiments, have also been investigated extensively in humans.[1] However, little is known of other cortical regions that might be involved in the processing of olfactory information. Findings in patients with brain lesions indicate that the frontal and temporal lobes are of critical importance in the perception of odors.[2-6]

Recent progress in the field of neuronal imaging has provided the opportunity to study the functional topography of the olfactory system in greater detail. Results obtained by positron emission tomography (PET) showed for the first time that, in

0-8493-2329-0/02/$0.00+$1.50

addition to a bilateral increase in cerebral blood flow (CBF) in the pyriform cortex, the insular cortex and the right orbitofrontal cortex were also activated by odorous stimulation.[7] Also, using functional magnetic resonance imaging (fMRI), a significant bilateral increase in CBF in the pyriform cortex and the orbitofrontal cortex[8] was demonstrated. Although the number of PET and fMRI studies of the olfactory system still remains rather small when compared to studies of other sensory modalities, there are a growing number of investigations that provide significant insight into the neural correlates and epiphenomena of olfactory processes in man.

A major disadvantage of PET and fMRI is their relatively low temporal resolution, especially when compared to electrophysiological techniques. Using olfactory event-related potentials (OERPs),[9-13] (for a review, see Chapter 17 in this book), it has become possible to non-invasively record the activity of cortical neurons in fractions of a second. A similar time resolution is achieved by magnetic source imaging (MSI), which is used to identify brain areas activated after olfactory stimulation.[14,15]

As the temporal resolution of the recording techniques increases there is an increasing demand for precise time control of reproducible olfactory stimuli. The lack of specific stimulators that provide control over stimuli comparable to auditory or visual instruments has been a major problem that slowed down progress in chemosensory research.

In this chapter, we discuss the strengths and weaknesses of PET, fMRI and MSI and describe some of the findings that provided new insights into the olfactory system, or, that give rise to new questions concerning functional properties and relationships of brain areas that are activated after an odorous substance interacts with chemosensors in the nasal mucosa.

19.2 METHODS

19.2.1 Positron Emission Tomography (PET)

PET is a functional, *in vivo* neuroimaging technique. It allows one to obtain quantitative maps of a number of physiological parameters, for example, blood flow, protein synthesis, glucose metabolic rate, receptor density, and affinity. Whether PET is a non-invasive technique is still a question of debate. Despite this, it is extensively used in neuroscience research in both healthy volunteers and patients.

PET is an autoradiographic technique. Molecules of biological interest are labeled with cyclotron-produced, short-lived, positron-emitting nuclei before they are injected intravenously into a human subject. Positron emitters have an excess of protons compared to the number of neutrons. Consequently, the nucleus undergoes a nuclear transition whereby the proton is converted to a neutron. As a result, a positron is emitted that is a positively charged particle (the anti-particle of the electron). After the positron has lost its energy by collisions, it combines with a free electron. Then the electron-positron pair undergoes a phenomenon called annihilation and produces a pair of anti-parallel photons of 511 keV energy each. These photons are detected by means of a positron tomograph.

The positron tomograph consists of sensors surrounding the subject's head; it is a circular array of gamma-ray detectors that dynamically records coincident anti-parallel photon pairs of 511 keV. After data correction, image processing, and statistical analyses, the three-dimensional distribution of the absolute concentration of the positron emitter in the brain is provided.[16] Depending on the factor or process of interest, a number of different tracers are used.[17] For example, hemodynamic tracers are used to map cerebral blood flow (e.g., ^{15}O); metabolic tracers are used to map the metabolism of glucose or amino acids (e.g., ^{18}F-deoxyglucose); and neuropharmacological tracers are used to investigate the dynamics of neurotransmitters/drugs (e.g., ^{18}F-DOPA).[18,19] The most frequently used radioisotopes are ^{11}C, ^{13}N, ^{15}O, ^{18}F, and ^{68}Ge.[20] Because of their short half-life (e.g., ^{15}O has a half-life of 2.2 min), the handling of many positron emitters is difficult. Thus, production sites must be as close as possible to the site of application.

Many PET tracers are produced by dedicated radiochemistry labs. In terms of safety, in the European Union, the total dose delivered to a subject participating in PET research is strictly regulated. It must not exceed the average natural dosage received by the general population over 1 year (5 mSv). Additionally, before using it in humans, for each PET tracer, approval must be obtained.[20]

In the mid-1980s, PET emerged as the functional imaging method of choice. Later, in the mid-1990s, functional MRI (fMRI) was established (see below). Due to its superior spatial/temporal resolution and the large availability of commercial magnetic resonance scanners, fMRI became increasingly important for the mapping of the functional anatomy of the human brain. Further, major disadvantages of PET include its costs, the requirement of a nearby cyclotron, its relatively slow temporal resolution (in the range of 2 to several minutes), and the subjects' exposition to certain doses of radioactivity.[20] However, when compared to fMRI, PET also has advantages such as:

- It has the potential to provide absolute values of physiological parameters.
- It is suitable for investigating *in vivo* neurotransmission,[21] which is only partly possible with fMRI.
- It allows one to simultaneously image neural information from the entire brain, especially from the ventral temporal and orbitofrontal cortices — which are of special interest in olfaction.

In addition, during PET measurements, it is easily possible to simultaneously record EEG, which is more difficult during fMRI measurements due to the strong magnetic fields.

19.2.2 Functional Magnetic Resonance Imaging (fMRI)

While magnetic resonance imaging (MRI) has become a standard in structural brain imaging, the emergence of functional magnetic resonance imaging (fMRI) methods for localizing brain activity has significantly expanded the scientific and clinical significance of MRI. This more recent technique produces images of activated brain regions by detecting the indirect effects of neural activity on local blood volume, flow, and oxygen saturation. Thus, it is a promising new tool to understand the relationship between brain structure, function, and pathology. The non-invasive nature of this technology, and its potentially simple integration into neuroscience, make it an appealing technique. The appearance of MR images reflects a complex relationship between the MR sequence design and the detailed physical, chemical, and physiological nature of the tissue. This is particularly true for MR images representing functional activation. Details characterizing this relationship for this relatively new technique are still under study.

MRI makes use of the magnetic properties of some atomic nuclei, especially hydrogen nuclei from water molecules. When placed in a strong magnetic field (such as the one in a MR scanner), the "natural" magnetic fields of hydrogen nuclei orient themselves along the strong magnetic field of the scanner (see Chapter 18). This orientation can be perturbed by exciting those nuclei with a burst of electromagnetic energy (e.g., radio-frequency pulses). As these excited nuclei realign themselves with the scanner's strong magnetic field, they emit a radio-frequency signal that can be detected by a receiver coil. Because the nuclei emit different signals depending on their surroundings, by using MRI it is possible to distinguish, for example, between gray and white matter.

To perform MRI, an individual lies on a horizontal surface with the head in the center of the scanner. The MR scanner visualizes the brain in consecutive sections (i.e., a series of two-dimensional slices) through the application of three mutually orthogonal local magnetigradients. The combination of radio-frequency and magnetic gradient pulses form an MRI sequence. Typical sequences are gradient-echo and spin-echo sequences, depending on the way the MRI signal (echo) is generated (i.e., from gradient or radio-frequency pulses). With echo-planar imaging (EPI), images can be obtained within hundreds of milliseconds (for a comprehensive review of EPI, see Reference 22).

Several methods have been used to sensitize MRI to blood flow, and specifically to changes in local blood flow in relation to brain activity. These methods are the basis of fMRI. Among them are techniques using contrast agents,[23] the inflow approach,[24] the EPISTAR technique,[25] magnetic resonance spectroscopy,[26] and blood oxygen level dependent (BOLD) contrast.[27]

The most common approach to fMRI of brain activation has been the one using the BOLD contrast. This technique is based on the fact that although oxyhemoglobin is diamagnetic (i.e., has no effect on the MR signal), deoxyhemoglobin, which is paramagnetic as a result of its four unpaired electrons,[28] becomes slightly magnetized in the magnetic field of the MR scanner. As deoxyhemoglobin is confined to red blood cells, it acts like a natural endogenous paramagnetic contrast agent, which is present in the blood stream at high concentrations and which is modulated by variations in oxygen supply (blood flow) and oxygen utilization (tissue metabolism). As in the case with a paramagnetic contrast agent, deoxyhemoglobin causes a difference in magnetic susceptibility between blood and surrounding tissue, which results in a small signal drop.[29] Conversely, a decrease in the blood deoxyhemoglobin concentration (because of an increase in blood flow and blood oxygenation) leads to a small, but measurable signal increase.

The BOLD effect is better seen at high field strength (such as 4 Tesla) and with gradient echo-based sequences.[30] The increases in blood flow and in tissue oxygen consumption during neuronal activation have opposite effects on the deoxyhemoglobin content of capillary and venous blood: tissue oxygen consumption increases very little whereas blood flow increases up to 50%, so that the net effect is a marked decrease in deoxyhemoglobin concentration.[31] Thus, the main concern raised by some investigators is that much of the BOLD contrast signal has no direct relationship to the activated neural tissue, but may predominantly reflect signals from large draining veins.[32-33a] The pertinent question is, however, how localized the hemodynamic response is, or, in other words, which blood vessels dominate the BOLD signal. The answer depends in part on the MRI technique used. The conventional gradient-echo sequences were found to be more sensitive to deoxygenated blood in large veins and to inflow effects in arteries.[34,35] However, with proper tuning (e.g., choice of field strength, pulse sequence, acquisition timing, and image-encoding scheme), one can greatly reduce inflow effects in large blood vessels and increase the contribution of capillaries.[36,37]

A typical functional neuroimaging experiment consists of acquiring a series of brain images while the subject is performing a task. Using EPI, a set of multiple slices covering the entire brain can be collected in a few seconds. These sets of images are acquired over several minutes while the paradigm switches several times between rest and activation. As changes in fMRI signal intensity during activation are small — typically 1 to 10% — statistical treatment of the data is essential to distinguish the pattern of signal intensity changes evoked by the stimulus/task condition from the resting state.

The significance of brain activation has been assessed using parametric and nonparametric statistics, as well as through characterization of the information content. The problem is that the signal distribution is generally unknown, and successive measurements in time are not independent, but are correlated to each other due to the (slow) hemodynamic nature of the response.[38] Consequently, attribution of p-values to the measurements remains a difficult task.

All data processing strategies, however, make use of the time-dependent nature of the activation paradigms,[38,39] that is, the fact that the onset and offset of different conditions occur within a controlled time sequence. One can thus use correlation measurements of the MR signal time-course with an idealized response derived from the activation paradigm.[38] Further improvement might be to consider the size of the activated regions (in terms of clusters and pixels) to improve statistical power. This approach, however, may require validation because stimulus and task parameters, as well as the specific area of brain examined, may strongly influence the time-course and size of the evoked signal changes.[40]

19.2.3 Magnetic Source Imaging (MSI)

The general goal of magnetic source imaging (MSI) is to localize generators of magnetic fields measured at the surface of the scalp.[41] These generators may be a number of cerebral neurons (electrically) active at the same time. Hence, other than PET and fMRI, MSI allows one to directly assess the activity of the neurons involved in the processing of sensory information. Assuming that the human head is a spherical volume conductor[42] and electrical neuronal activity has the property of a current dipole possessing an electric field and a magnetic field, it becomes possible to calculate the location the orientation and the strength of the current sources creating the magnetic field measured at the surface of the sphere.[43,44] In this model, not knowing the number of active dipoles, the magnetic field can be generated by one dipole or several different ones (the inverse problem, first discussed by von Helmholtz in 1853[45]). Unfortunately, this is the situation in the active brain and therefore a unique solution cannot be obtained. Additionally, a radial dipole does not produce a magnetic field that can be measured with the usual order of the magnetometers. Such dipoles are silent and cannot be localized.

Hence, on the basis of a measured magnetic field, a dipole position is estimated and followed by a nonlinear fit strategy such as a Marquardt[46] or Powell[47] algorithm. Using these procedures, the location, strength, and orientation of the theoretically determined dipole is iteratively changed such that the magnetic field produced by this calculated dipole corresponds to the dipole measured by the magnetometer. The statistical procedure leading to the so-called equivalent current dipole is a least-squares search. An extensive description of the data analysis applied can be found in Scholz and Oppel.[48] As with most recording techniques, the accuracy strongly depends on the signal-to-noise ratio of the measured data — which decreases with increasing depth of the dipole within the brain. Also, localization accuracy is better for a dipole centered below the measurement grid than below the outer part of the measurement grid. With regard to the situation in the human brain, the size of this error can only be estimated by simulations.[49-52]

By linking the magnetically defined equivalent current dipoles to the anatomical data provided by MRI,[53] it is possible to visualize the location of activated areas in the subject's brain and to check them for both anatomical and physiological plausibility. Aside from these basic issues related to the physics of magnetic fields, the reliability of the estimations is influenced by external error sources, such as magnetic noise caused by electronic devices or artifacts caused by movements, heart activity, etc.[42,54] (for a review see Reference 55).

19.3 GENERAL ISSUES IN IMAGING

At present, most investigators using imaging techniques focus on activation in brain areas that are known from experimental animals (e.g., rodents and primates) to process olfactory information. Among them is the primary olfactory cortex, the orbitofrontal cortex, the amygdala, the entorhinal cortex, and the hippocampus. Activities in other brain areas, more or less closely related to olfaction, have also been reported. It is interesting to note that the cerebellum, which receives sensory inputs from other modalities, seems to get input from the olfactory system as well.[56,57] Some techniques such as MSI are more suited to investigate transient activities of the temporal lobe and the hippocampus.

It is quite clear that we are far from understanding the functional meaning of these activities and their relationships. Depending on the time resolution of the imaging technique, we do not even know the temporal sequence of activations or whether the activations occur sequentially. From our anatomical understanding, we assume that the olfactory activity starts in the bulb and from there on is mediated to the primary olfactory cortex. It is quite uncertain what happens from there. There might be parallel processing in different brain areas as well as cross-connections and loops. Such sequential processing will be blurred, depending on the time window of the recording technique. This window can be very long, such as in PET (a minute or more), in the medium range such as in fMRI (a second and more) or short as in MSI (in the range of milliseconds). Although MSI has an extremely good time resolution, the technique can only "see" activity that does not last longer than 1 or 2 sec. Longer-lasting processes are not recordable because MSI usually needs synchronized neuronal activity in order to separate it from background noise through averaging procedures.

19.4 FINDINGS USING fMRI

This chapter section reviews some of the results reported to date.

Koizuka et al.[8] studied the effects of five odorants in five subjects. They applied the same stimulation technique that Zatorre and co-workers used in their PET study.[7] Subjects were asked to inhale odors from a cotton swab. This was compared to inhalation from odor-free cotton swabs. The authors reported activation of the pyriform cortex, orbitofrontal cortex, and medial frontal lobe. Because it appears that the authors did not analyze their data statistically, little is known about the validity of these findings. In addition, the authors failed to provide anatomical details about the activitated areas. In another study[58] in 17 subjects, responses to three olfactory stimuli were measured using fMRI of three coronal brain sections. The authors reported orbitofrontal cortex, entorhinal cortex, and "other" frontal cortex activity. Activity was lower in women than in men but no difference in brain regions activated by pleasant compared to unpleasant odors could be made. There was also no lateralization of the responses. Yousem and colleagues[57] investigated five male subjects using pure odorants and odorants with mixed properties, i.e., odors known to activate both the olfactory and the trigeminal nerve. Olfactory stimuli mildly activated the right orbitofrontal cortex and the cerebellum. Odorants that also stimulated the trigeminal nerve produced additional cingulate, temporal, cerebellar, and occipital activation. It is noteworthy that only one of these early studies claimed to demonstrate activity in the pyriform cortex. From these first experiments, it became clear that recording pyriform activity seems to be a major difficulty in this developing field of science.

Shortly after these early experiments, a study aimed to investigate the coding of odor-induced emotions.[59] Hence, the purpose of the study was to identify regional activation of the human cerebral cortex in response to pleasant (positive hedonic value) and unpleasant (negative hedonic value) odors. Thirteen subjects underwent fMRI of frontal and anterior temporal brain regions separated in 11 regions of interest. Clementine and isovaleric acid were used for olfactory stimulation. Both odors activated the insula, the anterior aspect of precentral gyrus and the posterior aspect of medial border of superior frontal gyrus, the lateral aspects of middle frontal gyrus, and the anterior paracingulate cortex. In addition, clementine activated the superomedial aspect of superior frontal gyrus and middle third of middle frontal. The authors hypothesized that the frontal activation was due to secondary or higher-order olfactory sites, respectively, which link olfactory networks with complex processes such as cognition, perception, and motor functions. The lack of activation of the orbitofrontal cortex or the medial temporal lobes was explained as a possible result of magnetic susceptibility artifacts in these areas. This pointed at fMRI-inherent difficulties in recording from the frontal parts of the brain that are close to air-filled bones. Possible approaches to this problem include the use of oblique sections instead of coronal or sagittal ones.[60]

The perception of odors is tied to the intranasal flow of air. Sobel and colleagues[60] were able to provide evidence that pyriform areas are active even in the absence of odor when subjects were actively sniffing. Specifically, fMRI measurements were performed while subjects were sniffing odorless air at different rates. Sniffing in the absence of odor induced significant signal changes in the pyriform area, as well as in the entorhinal and parahippocampal cortices. Further, sniffing rate consistently determined the activity in the pyriform area. A series of control experiments indicated that this activity was not attributable to artifacts or movements. Because the pyriform activity was abolished by topical anesthesia of the nasal passages, it was shown that this activation was closely related to the somatosensory input provided by the sensation of air flow. The authors concluded that the pyriform cortex is driven by sensory input related to sniffing. This effect is in agreement with electrophysiological findings in several species.[61,62] The authors speculated that sniffing serves an exploratory function for the olfactory system, and that the activation in pyriform areas during sniffing may comprise a form of perceptual priming to facilitate odor processing.

Sobel et al.[60] also reported that using an fMRI paradigm in which they presented subjects with decanoic acid and vanillin stimulation with these odor results in activation of (predominantly right-sided) anterior and lateral orbitofrontal areas, although additional signal increase was also detected in pyriform areas of the peri-insular region, superior temporal, and various ports of the limbic system. The same group[56] could also verify earlier findings of Yousem and colleagues[57] in demonstrating cerebellar olfactory activities. Both vanillin and propionic acid induced concentration-dependent activation, primarily in the posterior lateral hemispheres. The action of sniffing non-odorized air induced significant activation in the anterior cerebellum, primarily in the central lobe. The authors hypothesized that the cerebellum maintains a feedback mechanism that regulates sniff volume in relation to odor concentration.

Even an undetected air-borne chemical[63] was shown to induce activition in the brain — although not in the pyrifom cortex. In this experiment, the authors investigated the brain structures involved in chemical sensing that occurs below a threshold of conscious detection. A low and a high concentration of an air-borne estrogen derivative oestra-1,3,5(10),16-tetraen-3-yl acetate (which is supposed to be a human pheromone) was presented to eight subjects. After the experiment, all subjects reported that they could not detect any odor. However, forced choice detection performed during the presentations revealed above-chance detection of the high concentration, but no better than chance detection of the low concentration compound. Both concentrations induced significant brain activation, primarily in the anterior medial thalamus and inferior frontal gyrus. Activation in the inferior frontal gyrus during the high concentration was significantly greater in the right than in the left hemisphere. A trend toward greater thalamic activation was observed for the high concentration than for the low concentration compound.

Aging is a major determinant of olfactory function.[64] Using more sophisticated olfactometry than in previous studies, Yousem and colleagues addressed this question in terms of age-related fMRI responses to odorants.[65] Two different age groups of five right-handed subjects each were investigated. One group consisted of subjects with a mean age of 73 years, the other group of subjects with a mean age of 24 years. Given similar olfactory task paradigms (phenyl ethyl alcohol alternating with hydrogen sulfide), the older subjects showed the largest number of activated voxels in the right perisylvian region, followed by the right superior frontal region. Single voxels of activation were noted in the right inferior frontal and left superior frontal zones. There was a right-sided preponderance of activation across all these regions, even on individual maps. The younger subject group showed greater volumes of activation at all sites. The right perisylvian, right superior frontal, right inferior frontal, and left superior frontal regions were also activated to the highest degree in the younger subject group. Additionally, activation was found in the left superior frontal, left perisylvian, and both cingulate regions. The authors concluded that age might be a factor with regard to the results of odor-stimulated fMRI studies, a factor that needs to be controlled in future studies. In this study, so far not acknowledged brain areas were added to the set of zones involved in olfactory information processing, i.e., the cingulate and perisylvian areas. This finding is in accordance with findings of other laboratories using MSI techniques. The cingulate cortex was also identified in a study by Kettenmann et al.[66] In this study, imaging was performed in seven subjects. Two odorants (vanillin and H_2S) were delivered to both nostrils by a specialized olfactometer in five bursts within an "ON" period of 5 sec. This was followed by a 15-sec "OFF" period when non-odorous air was delivered. In addition to the activation of primary olfactory, orbitofrontal, and temporal areas, the authors identified significant signals arising from the cingulate cortex in all subjects for both odorants. Activated zones were not homogenous across subjects. Anterior as well as posterior parts of the cingulum were involved. The authors assumed that, due to the small number of subjects, no systematic differences between odorants were detectable. Interestingly, the cingulate gyrus has also been reported to be of critical importance in the processing of painful sensations.[67] Thus, one might speculate that emotions induced by either odors or pain relate to the similarity in the pattern of brain activation seen after odorous or nociceptive stimulation.

In addition to aging, gender has a major impact on olfactory acuity.[68] These effects were investiagetd by the same group of investigators.[69] Volumes of activated brain areas in male and female subjects were compared. Eight right-handed male and eight right-handed female subjects took part in the experiments. They were stimulated with eugenol, phenyl ethyl alcohol, or phenyl ethyl alcohol alternating with hydrogen sulfide. In total, there were ten paradigms employing the phenyl ethyl alcohol/hydrogen sulfide combination, three paradigms employing just phenyl ethyl alcohol, and two paradigms employing eugenol. For specific regions of the brain (frontal and perisylvian regions), activation maps in women showed up to eight times more activated voxels than those in men. The left and right inferior frontal regions exhibited increased activation in women.

These basic investigations were followed by studies focusing on psychological aspects of odor-induced brain activitation. Levy and colleagues[70] used fMRI to measure brain activation in response to the imagination of odors. The study was performed by measuring multislice FLASH MRI in three coronal sections. The smell of amylacetate and menthone was compared to the subjects' imagination of banana and peppermint. The same paradigm was used in hyposmic patients before and after theophylline treatment, which returned smell function to normal, or close to normal, respectively. These results led the authors to conclude that (1) odors can be imagined and (2) similar brain regions are activated by odors regardless of whether they are imagined or actually perceived. However, imagination of odors appeared to elicit quantitatively less brain activation than actual smells of the corresponding odors. Absolute brain activation in men by odor imagination is greater than in women for some odors, but on a relative basis, the ratio for odor imagination to actual smelling is twice as great in women than in men. Odor imagination, once the odor has been experienced, is present, recallable, and capable of inducing a relatively constant degree of brain activation even in the absence of the ability to recognize an actual corresponding odor. The cortical areas showing activation in the three MRI sections were labeled orbitofrontal cortex, other frontal cortex, cingulate gyrus, pyriform and entorhinal cortex, hippocampus, and amygdala. Kettenmann et al.[71] also compared brain activation in response to odor stimuli with the response to the imagination of the same odor by fMRI. In this study, imaging was performed in 13 healthy volunteers. The odorant eugenol was presented by a specialized olfactometer to the left nostrils in four 800-msec bursts within an "ON" period of 48 sec. This was followed by a 42-sec "OFF" period when non-odorous air was delivered. During a second "ON" period, the subjects were asked to imagine/recall the previously smelled odorant. This second "ON" period lasted for 48 sec, the same as the first one. This procedure was then repeated three times. Group analysis of all subjects revealed surprising similarity in cortical activation during stimulation and imagination/recall with/of eugenol. The areas activated in this experiment were mainly secondary/tertiary cortical areas related to association, i.e., the supramarginal gyrus, medial frontal gyrus, inferior frontal gyrus, the transition of the orbitofrontal gyrus and the insula, and the cingulate gyrus[72] (Figure 19.1). The authors concluded that this was due to the type of analysis used in this study. Intensity of activation during the two conditions was similar except for that of the medial frontal gyrus and the cingulate gyrus, which were more activated during the imagination/recall task (Figure 19.2).

Another important area in the perception of odors is the highly integrative interaction between olfaction and other sensory modalities. Francis and colleagues[73] investigated the processing of somatosensory stimuli that feel pleasant (soft velvet touch to the hand) and compared them with wood, presumably an affectively neutral touch. To separate the affective aspects of the tactile stimuli from modality-specific aspects, two positively affective stimuli in other sensory modalities (taste and smell) were also investigated. It was shown that a pleasant touch to the hand produced a stronger activation of the orbitifrontal cortex than a touch considered affectively neutral. In contrast, the affectively neutral, but more intense touch produced more activation in the primary somatosensory cortex than the pleasant stimulus. The authors concluded that parts of the orbitofrontal cortex are concerned with the representation of positive affective stimuli. The comparison of these data with the results obtained from gustatory and olfactory stimulation showed that the orbitofrontal areas activated

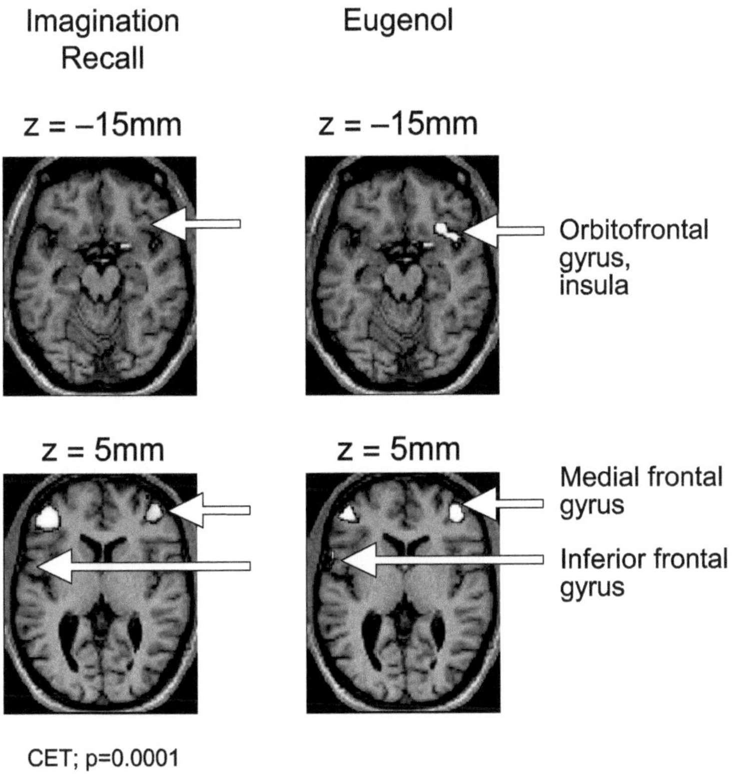

FIGURE 19.1 (see color insert following page 238) FMRI comparison between activations induced by the perception of eugenol vs. imagination of the smell of eugenol. Group analysis of 13 subjects using SPM'99. Activity differences between perception vs. non-perception and imagination vs. non-imagination. Two selected horizontal slices showing activities in the frontal and insular cortex. Shaded arrows indicate missing activities either in the perception or the imagination condition. The inferior and orbitofrontal activities are missing in the imagination condition. Not shown in this graph is that the study revealed activity in the posterior part of the medial frontal cortex only present in the imagination condition.

during pleasant touch is different from the orbitofrontal area activated by taste and smell. This finding was extended in another study by the same group in cooperation with Kobal and colleagues in that satiety-related olfactory activation of the human orbitofrontal cortex was investigated.[74] The paper describes the response of the orbitofrontal cortex to odors when food was eaten to satiety. Activation of regions in the human orbitofrontal cortex was related to sensory-specific satiety. In particular, activation of the orbitofrontal cortex decreased to the odor of a food eaten to satiety (banana), but not to another odor that had not been eaten at all (vanilla). The particular aspect of the sensory stimuli that had the strongest effect in this experiment was the pleasantness of the taste or smell of the food. Eating food to satiety did not alter the intensity of the corresponding odor. The authors concluded that when food is eaten to satiety, its reward value and its pleasantness decreases. This experiment identified orbitofrontal areas that specifically signal olfactory information, but only

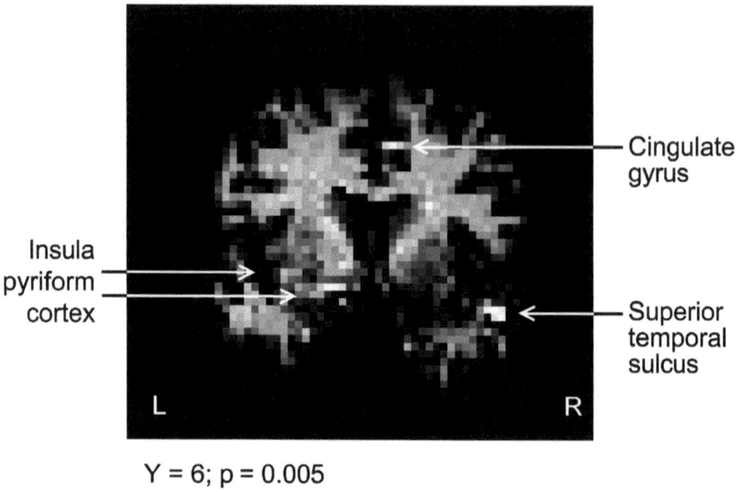

FIGURE 19.2 (see color insert following page 238) fMRI example of a single subject. Activities induced by the perception of the unpleasant hydrogen sulfide: pyriform cortex, insula, superior temporal sulcus, and also the cingulate cortex are significantly activated. The cingulate activity is more diffusely distributed in the posterior/anterior direction interindividually.

under certain circumstances — namely, when the individual is hungry. This finding is in line with results from electrophysiological studies in monkeys.[75] It would be highly interesting to study patients with eating disturbances using this experimental paradigm.

Recently, a clinical study was published by Birbaumer and colleagues.[76] fMRI was used to determine the activation of the amygdala and the thalamus in seven social phobics and five healthy controls exposed to slides of both neutral faces and aversive odor stimuli. The amygdala was selectively activated in the social phobics during presentation of the faces. In contrast, thalamic activations showed no significant differences between groups. From this data one can conclude that the amygdala is active in human phobics when they are exposed to potentially fear-relevant stimuli.

19.5 FINDINGS USING PET

Zatorre et al.[7] were among the first to explore the basic functional anatomy of odor processing using PET in healthy volunteers. Their goal was to identify the human primary and secondary olfactory cortices. The experimental paradigm contrasted two conditions: (1) a control condition in which subjects sniffed in the presence of an odorless cotton swab; and (2) an active condition in which subjects inhaled while being presented with a series of cotton swabs, one every 7 sec, containing one of eight different odorants. The odorants were selected from a larger set to ensure that the odors were different in terms of quality and pleasantness, difficult to name but not completely unfamiliar, and similar in perceived intensity. Subjects were informed that the baseline condition would not contain odors; and during active odor presentation conditions, subjects were also familiarized with the odor stimuli prior to scanning. The experimenters observed highly significant CBF (cerebral blood flow) increases bilaterally in a region at the junction of the temporal and frontal lobes, corresponding to the pyriform cortex, and unilaterally in the right orbitofrontal cortex.

In addition, a small area of activity was observed within the left orbitofrontal area, more lateral than the one on the right, but this area of activation failed to reach statistical significance. The authors interpreted these findings in the context of behavioral lesion experiments in patients with focal cortical excisions that indicated a possible functional asymmetry favoring the right hemisphere in some aspects of odor processing.[3,4,77-80] Hence, they proposed that bilateral activation of primary regions represents basic sensory processes, whereas the asymmetric activity in the orbitofrontal cortex is related to more complex analyses of stimulus properties that preferentially recruit right-hemispheric mechanisms.

Odors are not perceived through the olfactory nerve alone; typically, they lead to activation of other sensory inputs. Thus, interactions between taste and smell are of crucial importance in the functioning of chemical senses. Hence, Small and colleagues[81] wanted to obtain insight into the phenomenon of flavor that constitutes the marriage of taste and smell. They attempted to disentangle the neural correlates of these two aspects of flavor. Odors were presented in an "olfactory" PET scan, tastes were presented in a "gustatory" scan, and, in two other scans, the odors were presented simultaneously with the taste to create a sensation of flavor. In one of the two flavor conditions, the odors and tastes were congruent or matched (e.g., a bitter taste with a coffee odor), while in the other they were incongruent, or mismatched (a sour taste with a coffee odor). When the matched condition was subtracted from the mismatched condition, increases in rCBF were observed bilaterally in the basal forebrain (probably in the nucleus basalis Meynert) and, notably, unilaterally in the left amygdala. Given that the mismatched condition was reported by most subjects to be an unpleasant sensory experience, the left amygdala activation was considered to reflect a lateralized role of the amygdala in an aversive affective situation. This is consistent with the finding by Zald and Pardo[82] of left amygdala activation in response to an aversive odor. Surprisingly, subtraction of the unimodal conditions (smell alone or taste alone) from the bimodal matched condition did not yield significant decreases bilaterally in the anterior insula/frontal operculum, in the right caudolateral orbitofrontal cortex, and in a so-called secondary olfactory cortex. These results showed that there was greater activity during unimodal taste or smell processing than in the bimodal flavor processing. When interpreting this pattern of CBF decreases, the authors postulated that the primary olfactory cortex has a role in discriminating odor as food vs. non-food, in which case the pyriform cortex would be activated equally in the odor or flavor condition. In contrast, taste stimuli are necessarily associated with ingestion because a stimulus must be in the mouth in order to be perceived, with the result being that food vs. non-food discriminations are not required by the gustatory system. Although individual taste and smell stimuli were identical in both conditions, there were differences in the activations seen in subtraction of the unimodal from the bimodal (matched) condition. The authors concluded that the differential processing was not due to analysis of individual stimuli, but rather to the emerging flavor.

In the report by Zald and Pardo,[82] the authors focused attention on emotional aspects of olfaction and the human amygdala. They examined CBF responses associated with the inhalation of a highly unpleasant mixture of sulfides. The most salient finding was that highly unpleasant odors elicited bilateral activity in the amygdala in addition to activation of the left orbitofrontal region. A response in the right orbitofrontal region was also detectable, but only at a low level of statistical significance. Presentation of less aversive but still unpleasant odors also increased CBF in the left orbitofrontal area, but did not result in activation of the amygdala. Importantly, CBF in the left amygdala and in the left orbitofrontal area were found to correlate with the subjects' subjective ratings of perceived unpleasantness, but not with perceived intensity, thus lending support to the interpretation that these regions were specifically involved in processing the emotional response. Again, these findings suggest that there may be hemispheric asymmetries associated with the hedonic valence of the stimulus. The data from this paper were re-analyzed and the covariance among amygdala and orbitofrontal regions in both hemispheres was calculated in order to estimate functional connectivity.[83] Among the relevant findings was that the left orbitofrontal area and the left amygdala responded to aversive odorants relative to odor detection in a functionally similar manner. Furthermore, the functional coupling that existed between the left and right orbitofrontal regions during rest and during odor detection decreased substantially during exposure to aversive odors. This may reflect relatively independent processing of these stimuli by the two homologous areas.

A detailed investigation of odor memory was conducted by contrasting encoding with recognition conditions.[84] Twelve subjects were investigated who had undergone training sessions with six odorants four days prior to PET scanning. During scanning, subjects were tested under four conditions: odor encoding, long-term odor recognition, short-term odor recognition, and a no-odor sensorimotor control task. Results of the two recognition conditions revealed consistent activity within the pyriform and right orbitofrontal cortex, but no such activity was observed during encoding. The lack of CBF change in olfactory cortices during encoding is puzzling because subjects demonstrated good recognition in later testing, indicating that they indeed had encoded the stimuli. Hence, the results suggest either that the activity was present but only at a very weak level (not detectable by PET) or that encoding results in a subtle pattern not revealed by the imaging technique used. It is also notable that this condition did not elicit left frontal-lobe activity as predicted on the basis of findings in encoding studies in other sensory modalities.[85]

Another cognitive aspect of odor perception is semantic processing.[86] To study this aspect, a series of odors was presented to subjects under three conditions and the subjects were asked to determine if the presented odor was familiar, edible, or whether an odor was presented at all (detection task). The major finding was that familiarity judgments were associated with activity in the right orbitofrontal area, as well as regions of the left frontal lobe and the cingulate gyrus. Judgments of edibility, as contrasted with the other two conditions, activated the left inferior frontal cortex as well as several areas of the visual cortex. The latter finding was unexpected, but suggests that subjects may have had recourse to a visual processing strategy in order to decide whether or not an odor corresponds to a food item. It is of interest that activation in the right orbitofrontal regions was observed during the familiarity judgment, although all of the conditions incorporated odor presentation. The results of this study suggest that the right orbitofrontal cortex may be specifically involved in the processing of stimulus features that are necessary for judgment of familiarity. This is different from other imaging studies, which consistently indicate that this region is active even in simpler odor presentation conditions that presumably also would include the detection task used by Royet et al.[86] The finding of right orbitofrontal activity suggests that this processing occurs to a greater degree when judgments elicit activity in left inferior frontal cortical regions (Brodmann area 47), not identified in earlier studies. The authors interpreted these findings as a reflection of semantic processing. While it is likely that this is required by the tasks employed, both familiarity and edibility correlate strongly with naming, so that activity within the left area 47 is also likely to reflect naming, which may have occurred implicitly or explicitly.

Another study[87] also addressed this semantic aspect of olfactory information processing by investigating the functional anatomy of olfactory and visual naming and matching in humans. One baseline control task without olfactory or visual stimulation, one control task with simple olfactory and visual stimulation without cognition, one set of olfactory and visual naming tasks, and one set of olfactory and visual matching tasks were administered to eight normal subjects. In the olfactory naming task, odors from familiar items associated with some verbal label were to be named. Hence, it required long-term olfactory memory retrieval for stimulus recognition. The olfactory matching task involved differentiating a recently encoded unfamiliar odor from a sequentially presented group of unfamiliar odors. This required short-term olfactory memory retrieval for stimulus differentiation. This simple olfactory and visual stimulation resulted in activation of the left orbitofrontal region, the right pyriform cortex, and the bilateral occipital cortex. During olfactory naming, activation was detected in the left cuneus, right anterior cingulate gyrus, left insula, and bilaterally in the cerebellum. It appears that the effort to identify the origin of an odor involved semantic analysis and some degree of mental imagery. During olfactory matching, activation was observed in the left cuneus and bilaterally in the cerebellum. This task was performed to identify brain areas activated during differentiation of one unlabeled odor from the others. In cross-task analysis, the region found to be specific for olfactory naming was the left cuneus. The results showed recruitment of the visual cortex in olfactory naming and olfactory matching tasks, most likely related to the imagery component of these tasks. The cerebellar role in cognitive tasks has been recognized, but because this was the first PET study suggesting that the human cerebellum may have a role in cognitive olfactory processing as well, the authors could not provide further details of the mechanisms that might have been involved. Another study[88] aimed to disentangle the processing of odor discrimination, odor perception, and odor recognition in humans. Five different olfactory tasks were applied: monorhinal smelling of odorless air and single odors, discrimination of odor intensity, discrimination of odor quality, and odor recognition memory. Single-odor stimulation activated amygdala-pyriform, orbitofrontal, insular and cingulate cortices, and right thalamus. Intensity and quality tasks engaged the left insula and right cerebellum. The quality task also involved other areas, including the right caudate and subiculum. The odor recognition memory did not activate the insula, but instead the pyriform cortex. With the exception of the caudate and subiculum, it shared the remaining activations with the quality task and engaged, in addition, the temporal and parietal cortices. The authors concluded that olfactory functions are organized in a parallel and hierarchical manner. No difference between right or left nostril stimulation was seen.

19.6 FINDINGS USING MSI

Employing a whole-head neuromagnetometer, Kettenmann et al.[14] found bilateral activation in the superior temporal sulcus at approximately 700 msec using vanillin, phenylethyl alcohol, and hydrogen sulfide. These three substances were chosen because they are specific odorants, i.e., they do not activate the trigeminal nerve and therefore appear to be more appropriate for studying the sense of smell than those used by Zatorre and Koizuka.[89-93] This study confirmed the results obtained with the whole-head magnetometer by employing a planar 37-channel sensor array,[94] and additionally identified the neuronal generators underlying the earlier components of the event-related potentials in the time interval between 200 and 700 msec following olfactory stimulation. Ten healthy volunteers participated in the experiments (five males and five females; 20 to 40 years of age, mean age 28 years). They were carefully trained in avoiding eye-blinking and other movement artifacts. Respiratory air flow in the nose that might have quantitatively influenced the stimuli was avoided by the subjects performing a special breathing technique (velopharyngeal closure[95]).

For olfactory stimulation, an apparatus was employed that delivered odorants without altering the mechanical or thermal conditions at the mucosa.[13,95] This monomodal olfactory stimulation was achieved by mixing pulses of the odorants in a constantly flowing air stream (total flow rate 140 ml/sec) with controlled temperature and humidity (36.5°C, 80% relative humidity). Stimuli were applied non-synchronously to breathing. This technique avoids any synchronous noise related to motor activity or the somatosensory sensation of air flow in the nose. On the other hand, priming of the olfactory areas by the sniffing procedure itself is eliminated. This might reduce the signal-to-noise ratio, namely in the pyriform cortex as shown in fMRI studies.[60] Odor concentration was defined once in the beginning of the experiment by gas chromatography for vanillin. In the case of H_2S, predefined concentrations in pressure gas cylinders (22 ppm) were used. The olfactometer was specially designed for this experiment. Twelve separately temperature-controlled Teflon tubings were fed through a grid with 3 × 3-cm holes incorporated in the wall of the shielded room. This did not alter the quality of shielding (measured by frequency distribution). All parts of the olfactometer consisted of nonmagnetic materials such as Teflon, silicon, Plexiglas, glass, gas, and water. Even at a distance of 5 m between the outlet of the instrument and the solenoid valves, the olfactometer can provide steep increases of odor pulses, which were verified by measuring CO_2 pulses with a thermistor (for details, see Reference 95). Subjects were comfortably seated in a magnetically shielded, ventilated chamber that was monitored by means of a video system. White noise (approximately 60 db SPL) applied to both ears was used to mask the switching clicks of the stimulator. The sound was transferred via air conduction through two hollow plastic tubings that were used as a connection between an acoustic stimulator outside the shielded room and the subject's ears. Each subject participated in six experimental sessions on each of four experimental days (three experimental sessions for one nostril). In one experimental day, either one of the two substances — hydrogen sulfide (0.8 ppm) or vanillin (2.1 ppm) — was tested with the sensor array positioned over one hemisphere. Sites of sensor location (left/right hemisphere MEG recording) and application of both odorants were randomized across experimental days. Three sessions were performed for each nostril. During each of the three sessions, 30 stimuli of the same substance were applied by way of Teflon tubing (6-cm length, 2-mm inner diameter). To avoid habituation, an interstimulus interval (ISI) of 40 sec and a stimulus duration of 200 msec were used so that there was no residual odorant left in the tubings. Additionally, the three sessions performed for one nostril were separated by breaks of 10 min. Subsequently, data were averaged across these sessions if the head position of the subject had not changed (controlled by laser motion detector).

Olfactory event-related potentials (ERP) were obtained from electrodes positioned at the vertex (Cz/A1). They were used as temporal markers during the analysis of the magnetic responses. The olfactory event-related magnetic fields (ERMF) were obtained by means of a biomagnetic system (Siemens Krenikon™) with a sensor array of 37 first-order gradiometers (baseline 7 cm) in a circular arrangement of 19 cm diameter.[96] It was possible to obtain isocontour plots from the region of interest without repositioning. Electroencephalographical (EEG) and MEG data were recorded with a sampling rate of 500 Hz, a bandpass filter 0.1–70 Hz, and filtered offline with a 30-Hz lowpass digital filter. The source analyses were based on signals averaged from artifact-free records ($30 < n < 90$).

For data evaluation, we used a map correlation ≥ 0.92, defined as:

$$\frac{\sum_{i=1}^{N}(x_i - \bar{x})(y_i - \bar{y})}{\sqrt{\sum_{i=1}^{N}(x_i - \bar{x})^2 \sum_{i=1}^{N}(y_i - \bar{y})^2}}$$

We used an interval length with a minimum of 0.006 sec. The upper and lower limits for the strength of the reconstructed source was defined as >0.007 and <0.03 mA*mm (deduced from electrophysiology). Furthermore, an upper limit for the rate of change of the source strength was defined (factor <10). The acceptable localization error was set to 20 mm as a maximum (defined as the area in which the equivalent current dipole has its possible location).

A magnetic field was called consistent if it lasted over a minimum time period of 10 msec. As a result, the field patterns were individually and interindividually reproducible, which means that equivalent current dipoles (ECDs) could be localized at nearly the same latency in about the same anatomical structure.

The time of best fit (TOBF) for one ECD estimated gives the time point were the isofield map of this special ECD shows its best map correlation, meaning that at this TOBF, the measured map of the ECD has the best correlation with the theoretically calculated map and it has the smallest localization error and the best signal-to-noise ratio. However, this does not mean that the neuronal source is only active at this time point. It can already start being active several milliseconds before and after this TOBF; but depending on the actual situation, it can be masked by additional neuronal activity from other sources.

The signal-to-noise ratio (SNR) of the data determines the accuracy in biomagnetic source reconstruction. To increase the SNR, we used the averaging procedure.[97] An additionally recorded vertical electro-oculogram was used to detect signals coinciding with eye-blinks or movements. Records contaminated with EOG deflections >200 mV were rejected. The noise level of the pre-trigger period of the averaged session was estimated and used as a baseline. The noise level of all 64 measurements was in the range of 0.00003 pT^2 and 0.000488 pT^2, with a mean value of 0.000222 pT^2 and a standard deviation of 0.000128 pT^2. The signals of the ECDs were in the range of eight to ten times higher than the calculated noise. No average across sessions was performed.

To provide accurate images of the brain's anatomy, magnetic resonance imaging (MRI) was employed. For the alignment of the functional MEG information with this tomomographical information, a head coordinate system, defined by the nasion and the two preauricular points, was created for each subject. Contrast-giving water pellets (0.3 mm diameter) were attached to the subject's skin during the MRI and subsequently their locations were projected into the coordinate system of the MEG data. Individually shaped realistic conductor models, which might provide better localization accuracy, were not supported by the Krenikon™ software. To deduce the positioning of the subject's head with respect to the sensor array after head positioning, the three landmarks were digitized with a 3-D digitizer (Isotrak, Polhemus Navigation Sciences, Colchester, VT) and aligned with the coordinate system received from the MRI. This method of alignment presupposes that the subject is not moving during the measurement since dislocations cannot be detected (controlled by a laser motion detector). To minimize the possibility of moving, the head was stabilized with a vacuum pillow.

Control measurements were carried out to exclude the possible contamination of the measured signals by tactile and/or auditory artifacts. Humidified blanks delivered to the nostrils did not result in activity in the MEG nor in the EEG measurements.

The signal quality of the evoked responses was sufficient for source localization in eight out of ten subjects. Consistent magnetic fields were identified in both hemispheres following the stimulation of each nostril. In 60% of the measurements, reproducible dipolar field patterns were obtained 226 to 380 msec after stimulus onset, preceding or following the first major positive electric deflection P1 of the event-related potential (ERP). This equivalent current dipole was labeled ECD I. In 44% of the measurements, a reproducible dipolar distribution was obtained 306 to 486 msec after stimulus onset, corresponding to the ascending or descending slope of the N1 component of the ERP, which was named ECD II. In the left hemisphere, this dipole was not identifiable in any of the subjects after stimulation with hydrogen sulfide. It was identifiable only in the right hemisphere in 36% of the measurements.

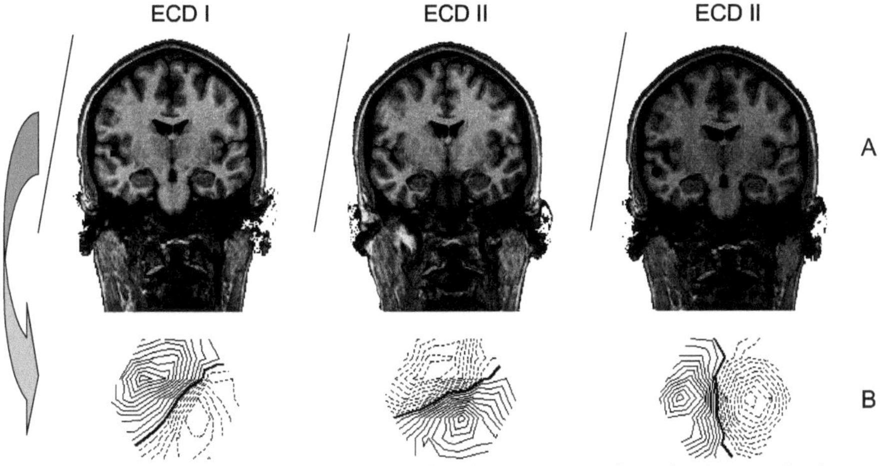

FIGURE 19.3 (see color insert following page 238) MSI activities after vanillin stimulation. Example from a single subject. A: MRI graphs including the localizations (size of spots indicates strength of dipole) of the three major equivalent current dipoles (ECD I, ECD II, ECD III) in the superior temporal plane the insular cortex, and the superior temporal sulcus. The activities are separated by approximately 100 msec, starting with ECD I at 200 msec after stimulus onset. B: Isomagnetic field line maps (dashed and solid lines indicate the polarity of the magnetic field) recorded with a 36-channel sensor system from the surface of the left hemisphere.

The most stable dipolar field pattern appeared 518 to 730 msec after stimulus onset (in 66% of the measurements; ECD III) corresponding to the P2 component of the ERP (Figure 19.3).

In five subjects and 14% of all measurements, all three ECDs could be identified during one session. In all other cases in which ECD I was identified, it was localized in the area between the superior temporal plane and the parainsular cortex. In seven of eight subjects, a generator was localized in the left hemisphere after stimulation of the left nostril with vanillin. This was statistically significant (chi-square test: 4.5; df: 1; $p = 0.034$). ECD II was localized in the anterior-central parts of the insula. In seven of eight subjects, this generator was localized in the right hemisphere after stimulation of the right nostril with vanillin. This was also statistically significant (chi-square test: 4.5; df: 1; $p = 0.0339$). ECD III was obtained in the superior temporal sulcus. In seven of eight subjects, a statistical significance was obtained for this generator localized in the left hemisphere after stimulation of the left nostril with hydrogen sulfide (chi-square test: 4.5; df: 1; $p = 0.0339$). Intraindividually, spatial differences of localization sites for one type of ECD were less than 20 mm. Individually, the angle of orientation varied between 10° and 30° in all three dimensions. The dipole strength varied between 0.009 and 0.03 mA*mm.

Hence, localization results of the olfactory ERMFs demonstrated that odorants specifically activate neocortical areas; i.e., during the period of time when the ERP is obtained, areas between the superior temporal plane and the parainsular cortex, anterior-central parts of the insula, and the superior temporal sulcus were active. These electro(magneto)physiological data confirm the hypothesis that there is a direct connection between primary olfactory areas and the insular cortex.[98] Physiological and axonal tracer experiments in animals have consistently demonstrated inputs from the primary olfactory cortical areas to the ventral agranular insular area.[99] The results of this study describing activity in the superior temporal sulcus (ECD III) also confirmed findings of a previously published report using a whole-head neuromagnetometer.[14] These data indicated, for the first time by different measurements, the activation of the superior temporal sulcus after olfactory stimulation. This finding is supported by the fact that there are axonal connections between the inferior orbitofrontal areas that receive input from specific olfactory structures, i.e., the primary olfactory cortex[100] and the superior temporal sulcus. With one exception, both the dipole locations and the dipole orientations were reproducible for the responses after stimulation with both odorants. Hence, both odorants appear to activate similar cortical areas. The data also showed bilateral neocortical activation. This must be considered when discussing the hypothesis that olfactory input is primarily processed ipsilaterally to the stimulated nostril[101-103] (for a review, see Reference 80). Also, in animal studies, it was demonstrated that there are bilateral connections between the anterior olfactory nucleus, the anterior part of the pyriform cortex, and neocortical areas. The fibers cross in the anterior commissure and are distributed equally in both hemispheres.[1]

However, there was one exception in the bilaterality of our results. We were not able to identify the ECD II in the left hemisphere after stimulation with hydrogen sulfide. This might indicate the coding of qualitative differences between hydrogen sulfide and vanillin. Of course, this hypothesis must be extensively tested using a larger scale of odorants. In an earlier study, it had been demonstrated that amplitudes and latencies of OERPs elicited by vanillin and hydrogen sulfide were correlated with their hedonic properties.[104] Responses to hydrogen sulfide had significantly shorter latencies and smaller amplitudes after stimulation of the left nostril compared to responses after stimulation of the right nostril. In contrast, latencies were longer when the left nostril was stimulated with vanillin, and the amplitudes of the N1/P2 components were significantly larger when compared to the responses after stimulation of the other nostril. This finding was reproduced in another experiment in which subjective estimates of the odorants showed that after stimulation of the left nostril, the more pleasant the perception of the odor, the larger the amplitudes and the longer the latencies.[105] In light of the results of the present study, this could mean that different cortical areas were activated after stimulation with odorants of different hedonic properties. There is also some evidence for lateralization of hedonic information from clinical studies. Pleasant emotions appear to be predominantly processed by the left hemisphere, whereas unpleasant emotions appear to be predominantly processed by the right hemisphere.[106,107] The missing ECD II in the left insular cortex following hydrogen sulfide stimulation might justify speculation that the observed differences in the responses to vanillin and hydrogen sulfide have their origin in the asymmetrical activation of the insular region. Of course, this does not yet explain why the differences in the OEP could only be observed when left and right nostril stimulation was compared.

Another indication for lateralization was found by Hummel et al.[103] investigating patients with left- or right-sided temporal lobe epilepsy. When olfactory stimuli were presented ipsilaterally to the epileptic focus, peak latencies of the OERP were prolonged in comparison to OERPs obtained after contralateral stimulation. Additionally, the topographical distribution of the P2 component changed in patients with right-sided foci. In these patients, the amplitude of P2 was maximal at Cz, whereas in previous studies in normal subjects, maximum P2 amplitudes were always obtained at Pz.[108,109]

The significance of olfaction in humans can be stressed by the fact that olfactory dysfunctions can produce severe impairments in the quality of life. Infections (e.g., postviral anosmia), degeneration (e.g., hyposmia in Alzheimer's and Parkinson's disease[110,111]) or a tumor affecting the olfactory bulb[112] belong to those diseases not yet sufficiently curable. In addition, the sense of smell plays a role in complex and controversial disorders such as idiopathic environmental intolerances (IEI).[113] It might be interesting to hypothesize that the underlying pathology results in changes of cortical functions. Possible dislocations and plastic changes might easily be studied using the technique of MSI.

In sum, for the first time, brain areas have been identified that generate some of the components of olfactory bioresponses (olfactory ERPs and ERFs) in humans. Deduced from our results we have strong evidence to believe that apart from the known primary olfactory areas, these neocortical areas are of specific olfactory origin.

Only a few other groups have studied the olfactory system with MSI. Tonoike and colleagues[114] used a whole-cortex 122-channel biomagnetometer. Amyl acetate gas (approx. 1%) was administered for 300 msec into either the right or left nostril in synchronization with respiration using a mask and an optical fiber sensor. Clear olfactory event-related magnetic fields were asymmetrically obtained on both sides of the forehead in all six subjects. The generators of olfactory magnetic fields were estimated at two regions located fairly asymmetrically near the bilateral frontal deep areas. The goodness-of-fit was better for the two-dipole model than the one-dipole model in all experiments. In almost all subjects, the latency and intensity of ipsilateral olfactory magnetoencephalography (MEG) responses were shorter and larger, respectively, than those of the contralateral responses.

Pulses of odorant air containing amyl acetate or phenethyl alcohol, and odorless air were administered to 14 subjects through a nasal tube in a study that confirmed the findings of Kettenmann and colleagues.[15] A clear and consistent olfactory ERF component was identified in all subjects; a second component was detected in seven subjects; but no consistent component was identified in response to the odorless air. The peak latencies were approximately 320 and 630 msec, respectively. Their equivalent current dipoles (ECDs) were estimated in the regions around the Sylvian fissure symmetrically in both hemispheres.

19.7 CONCLUSION

Biomagnetic fields directly reflect electrophysiological events. Other neuroimaging techniques (PET, FMRI) reflect either changes in blood flow or changes in metabolism that are epiphenomena of neuronal activity. Hence, the two latter techniques only offer indirect insight into the functioning of brain circuits. Neuromagnetic recordings offer several advantages over alternative noninvasive imaging modalities. Magnetoencephalography (MEG), which MSI is based on, as well as electroencephalography (EEG), provide real-time, direct assessment of brain physiology because it more or less directly measures signaling among neurons. This signaling consists of electrochemical events that take place at the synapses, in the axon and the dendrites of neurons. While neurotransmitter release and uptake at the synapses do not directly involve electrical activity, most other events involve movement or flow of electrically charged particles/ions resulting in a physical, potentially measurable quantity — namely, electrical current. Therefore, apart from its unparallelled temporal resolution, MSI provides the significant complement to structural imaging modalities such as computer tomography (CT) and magnetic resonance imaging (MRI), and to hemodynamic and metabolic techniques such as fMRI, and PET.

MSI affords certain advantages over ESI (electric source imaging, which is based on electroencephalogram recordings), which has not been discussed in detail in this chapter. The most salient of advantages is that neuromagnetic signals penetrate the skull and scalp without significant distortion. The spatial resolution of MEG is higher than that of the EEG, and mathematical analyses of the spatial pattern of the neuromagnetic field can offer accurate spatial localization of the neurons responsible for generating neuromagnetic signals. In short, of the available noninvasive neuroimaging techniques, MEG appears to provide an excellent balance of temporal and spatial sensitivity to brain physiology. During the past two decades, most MEG investigations have focused on basic research in sensory and cognitive neuroscience. However, with the recent development of large-array biomagnetometer systems capable of simultaneous recording of the magnetic signal over large portions of the head, clinical applications are moving to the forefront.

fMRI measures changes in the blood oxygen level dependent (BOLD) signal, which allows the assessment of the functional status of brain regions. The BOLD signal reflects changes in the relative proportion of oxyhemoglobin and deoxyhemoglobin that occurs when a brain region increases its blood flow during increased neural activity.[27] fMRI possesses several advantages over PET:

- It is highly accessible and less costly than PET imaging.
- Because there is no radioactive exposure, subjects can be scanned repetitively.
- Its spatial resolution typically exceeds that of PET.
- For acquisition of the entire brain volume, the temporal resolution can be kept as low as 4 sec. Typically, the fMRI signal rises above noise baseline 1 to 2 sec after neural activity has begun, and reaches a plateau at about 8 sec. The higher temporal resolution becomes a tremendous aid in studies measuring responses to single trial and in studies directly probing the time course of activation. Such studies have already begun to provide information that would not have been available with PET. If cells in certain areas show transient or rapid adapting responses, the time course of many fMRI and PET studies would blur together periods with increased signal and periods with no increased signal. Such temporal blurring would dilute any signal increases.

The major disadvantage of fMRI studies is that the ventral, frontal, and temporal brain areas are often compromised by artifacts. Specifically, underlying bone and air produce field inhomogeneities that can result in signal dropouts/distortions. This may render evaluation of these regions impossible — which appears to be a clear disadvantage in measurements based on PET. Such artifacts arise due to the different magnetic susceptibilities of brain tissue and the underlying bone and air.[115] Depending on the imaging parameters, the field inhomogeneity in these regions can produce signal dropouts, geometric distortions, aliasing, blurring, and smearing.[116,117] These problems are so severe in the orbitofrontal cortex that this area is unaccessible to many fMRI studies.[117] Pulsatile and gross movement artifacts can also make it difficult to image this region. In fact, pulsatile and other subject movements can cause signal changes that exceed those produced by actual activation.[118] Pulsatile artifacts pose particular problems for olfactory studies due to their link to respiration. The failure of some fMRI studies to observe activity in regions such as the orbitofrontal cortex likely reflects these problems.

However, that is not to say that these areas cannot be imaged by fMRI. A number of techniques appear to master these problems. For example, local shimming and post-processing can reduce some of the artifacts associated with field inhomogeneities.[119] Alignment of scans parallel to the orbital plane and utilization of thin slices[120] can also reduce these problems by limiting the degree to which slices include areas with dramatically different field properties. Other artifacts, such as those associated with pulsatile movement, can be alleviated using special image acquisition parameters, such as spiral pulse sequences.[118,121] Thus, while application of fMRI techniques to the study of olfaction presents a number of unique challenges, these problems can be overcome with careful attention to imaging methodology.

In conclusion, at this point in time, we are facing a new epoch of functional investigations of the brain. Those who first started to use these techniques have the feeling that their eyes have been uncovered and that — although not yet clear — one has gained new insight into brain function that Berger[122] in the 1920s only would have dreamt of.

REFERENCES

1. Price, J.L., Olfactory system, in *The Human Nervous System*, Paxinos, G., Ed., Academic Press, San Diego, 1990, 979-998.
2. Eichenbaum, H., Morton, T.H., Potter, H., and Corkin, S., Selective olfactory deficits in case H.M, *Brain*, 106, 459-472, 1983.
3. Eskenazi, B., Cain, W.S., Novelly, R.A., and Mattson, R., Odor perception in temporal lobe epilepsy patients with and without temporal lobectomy, *Neuropsychologia*, 24, 553-562, 1986.
4. Jones-Gotman, M. and Zatorre, R.J., Olfactory identification deficits in patients with focal cerebral excision, *Neuropsychologia*, 26, 387-400, 1988.
5. Jones-Gotman, M. and Zatorre, R.J., Contribution of the right temporal lobe to odor memory, *Epilepsia*, 29, 661, 1988.
6. Jones-Gotman, M. and Zatorre, R.J., Odor recognition memory in humans: role of right temporal and orbitofrontal regions, *Brain Cogn.*, 22, 182-98, 1993.

7. Zatorre, R.J., Jones-Gotman, M., Evans, A.C., and Meyer, E., Functional localization and lateralization of human olfactory cortex, *Nature*, 360, 339-340, 1992.
8. Koizuka, I., Yano, H., Nagahara, M., Mochizuki, R., Seo, R., Shimada, K., Kubo, T., and Nogawa, T., Functional imaging of the human olfactory cortex by magnetic resonance imaging, *ORL J. Otorhinolaryngol. Relat. Spec.*, 56, 273-275, 1994.
9. Finkenzeller, P., Gemittelte EEG-Potentiale bei olfactorischer Reizung, *Pflügers Arch.*, 292, 76-85, 1965.
10. Allison, T. and Goff, W.R., Human cerebral evoked responses to odorous stimuli, *Electroenceph. Clin. Neurophysiol.*, 23, 558-560, 1967.
11. Kobal, G. and Plattig, K.H., Methodische Anmerkungen zur Gewinnung olfaktorischer EEG-Antworten des wachen Menschen (objektive Olfaktometrie), *Z. EEG-EMG*, 9, 135-145, 1978.
12. Kobal, G., *Elektrophysiologische Untersuchungen des menschlichen Geruchssinns*, Thieme Verlag, Stuttgart, 1981.
13. Kobal, G. and Hummel, C., Cerebral chemosensory evoked potentials elicited by chemical stimulation of the human olfactory and respiratory nasal mucosa, *Electroenceph. Clin. Neurophysiol.*, 71, 241-250, 1988.
14. Kettenmann, B., Jousmäki, V., Portin, K., Salmelin, R., Kobal, G., and Hari, R., Odorants activate the human superior temporal sulcus, *Neurosci. Lett.*, 203, 143-145, 1996.
15. Sakuma, K., R, K., Kaneoke, Y., Hoshiyama, M., Koyama, S., Nagata, O., Takeshima, Y., Ito, Y., and Nakashima, K., Odorant evoked magnetic fields in humans, *Neurosci. Res.*, 27, 115-122, 1997.
16. Bendriem, J.C., Frackowiak, R.S.J., Herholz, K., Jones, T., Lammertsma, A.A., Mazoyer, B., and Wienhard, K., Use of PET methods for measurement of cerebral energy metabolism and hemodynamics in cerebrovascular disease, *J. Cereb. Blood Flow Metab.*, 9, 723-742, 1989.
17. Stöcklin, G. and Pike, V.W., *Radiopharmaceuticals for Positron Emission Tomography. Methodological Aspects*. Kluwer Academic, Dordrecht, 1993.
18. Phelbs, M.E., Huang, S.C., Hoffamn, E.J., Selin, C., Sokoloff, L., and Kuhl, D., Tomographic measurement of local cerebral glucose metabolic rate with (F-18)2-fluoro-2-deoxy-D-glucose: validation of the method, *Ann. Neurol.*, 6, 371-388, 1979.
19. Phelbs, M.E., Mazziottaq, J.C., and Schelbert, H., *Positron Emission Tomography and Autoradiography. Principles and Application for the Brain and Heart*, Raven Press, New York, 1986.
20. Crivello, F. and Mazoyer, B., Positron emission tomography of the human brain, in *Modern Techniques in Neuroscience Research*, Windhorst, U. and Johansson, H., Eds., Springer, Berlin, 27-50, 1999.
21. Frost, J.J. and Wagner Jr., H.N., *Neuroreceptors, Neurotransmitters and Enzymes*, Raven Press, New York, 1990.
22. Stehling, M.K., Turner, R., and Mansfield, P., Echo-planar imaging: magnetic resonance imaging in a fraction of a second, *Science*, 254, 43-50, 1991.
23. Belliveau, J.W., Kennedy, D.N., McKinstry, R.C., Buchbinder, B.R., Weisskoff, R.M., Cohen, M.S., Vevea, J.M., Brady, T.J., and Rosen, B.R., Functional mapping of the human visual cortex by magnetic resonance imaging, *Science*, 254, 716-719, 1991.
24. Williams, D.S., Detre, J.A., Leigh, J.S., and Koretsky, A.P., Magnetic resonance imaging of perfusion using spin inversion of arterial water, *Proc. Natl. Acad. Sci. U.S.A.,* 89, 212-216, 1992.
25. Edelman, R.R., Siewert, B., Darby, D.G., Thangaraj, V., Nobre, A.C., Mesulam, M.M., and Warach, S., Qualitative mapping of cerebral blood flow and functional localization with echo-planar MR imaging and signal targeting with alternating radio frequency, *Radiology*, 192, 513-520, 1994.

26. Prichard, J., Rothman, D., Novotny, E., Petroff, O., Kuwabara, T., Avison, M., Howseman, A., Hanstock, C., and Shulman, R., Lactate rise detected by ^{1}H NMR in human visual cortex during physiologic stimulation, *Proc. Natl. Acad. Sci. U.S.A.,* 88, 5829-5831, 1991.
27. Ogawa, S., Lee, T.M., Kay, A.R., and Tank, D.W., Brain magnetic resonance imaging with contrast dependent on blood oxygenation, *Proc. Natl. Acad. Sci. U.S.A.*, 87, 9868-9872, 1990.
28. Pauling, L. and Coryell, C.D., The magnetic properties and structure of hemoglobin, oxyhemoglobin and carbon monoxyhemoglobin, *Proc. Natl. Acad. Sci. U.S.A.,* 22, 210-216, 1936.
29. Thulborn, K.R., Waterton, J.C., Matthews, P.M., and Radda, G.K., Oxygenation dependence of the transverse relaxation time of water protons in whole blood at high field, *Biochim. Biophys. Acta*, 71, 265-270, 1982.
30. Turner, R., Jezzard, P., Wen, H., Kwong, K.K., Le Bihan, D., Zeffiro, T., and Balaban, R.S., Functional mapping of the human visual cortex at 4 and 1.5 Tesla using deoxygenation contrast EPI, *Magn. Reson. Med.*, 29, 277-279, 1993.
31. Fox, P.T. and Raichle, M.E., Focal physiological uncoupling of cerebral blood flow and oxidative metabolism during somatosensory stimulation in human subjects, *Proc. Natl. Acad. Sci. U.S.A.,* 83, 1140-1144, 1986.
32. Lai, S., Hopkins, A.L., Haacke, E.M., Li, D., Wasserman, B.A., Buckley, P., Friedman, L., Meltzer, H., Hedera, P., and Friedland, R., Identification of vascular structures as a major source of signal contrast in high resolution 2D and 3D functional activation imaging of the motor cortex at 1.5T: preliminary results, *Magn. Reson. Med.*, 30, 387-392, 1993.
33. Segebarth, C., Belle, V., Delon, C., Massarelli, R., Decety, J., Le Bas, J.F., Decorps, M., and Benabid, A.L., Functional MRI of the human brain: predominance of signals from extracerebral veins, *Neuroreport*, 5, 813-816, 1994.

33a. Logothetis, N.K., Pauls, J., Augath, M, Trinath, T., and Oeltermann, A., Neurophysiological investigation of the basics of the fMRI signal, *Nature,* 412, 150-156, 2001.

34. Kennan, R.P., Zhong, J., and Gore, J.C., Intravascular susceptibility contrast mechanisms in tissues, *Magn. Reson. Med.*, 31, 9-21, 1994.
35. Weisskoff, R.M., Zuo, C.S., Boxerman, J.L., and Rosen, B.R., Microscopic susceptibility variation and transverse relaxation: theory and experiment, *Magn. Reson. Med.*, 31, 601-610, 1994.
36. Menon, R.S., Ogawa, S., Tank, D.W., and Ugurbil, K., Tesla gradient recalled echo characteristics of photic stimulation-induced signal changes in the human primary visual cortex, *Magn. Reson. Med.*, 30, 380-386, 1993.
37. Kim, S.G., Ashe, J., Georgopoulos, A.P., Merkle, H., Ellermann, J.M., Menon, R.S., Ogawa, S., and Ugurbil, K., Functional imaging of human motor cortex at high magnetic field, *J. Neurophysiol.*, 69, 297-302, 1993.
38. Friston, K.J., Holmes, A.P., Poline, J.B., Grasby, P.J., Williams, S.C., Frackowiak, R.S., and Turner, R., Analysis of fMRI time-series revisited, *Neuroimage*, 2, 45-53, 1995.
39. Bandettini, P.A., Jesmanowicz, A., Wong, E.C., and Hyde, J.S., Processing strategies for time-course data sets in functional MRI of the human brain, *Magn. Reson. Med.*, 30, 161-173, 1993.
40. Bandettini, P.A., Wong, E.C., Hinks, R.S., Tikofsky, R.S., and Hyde, J.S., Time course EPI of human brain function during task activation, *Magn. Reson. Med.*, 25, 390-397, 1992.
41. Cohen, D., Magnetoencephalography: detection of the brain's electrical activity with a superconducting magnetometer, *Science*, 175, 664-666, 1972.

42. Abraham Fuchs, K., Schneider, S., and Reichenberger, H., MCG inverse solution: influence of coil size, grid size, number of coils, and SNR, *IEEE Trans. Biomed. Eng.*, 35, 573-576, 1988.
43. Williamson, S.J. and Kaufman, L., Analysis of neuromagnetic signals, in *Handbook of Electroencephalography and Clinical Neurophysiology, Vol. 1. Methods of Brain Electrical and Magnetical Signals*, Gevins, A.S. and Rèmond, A.A., Eds., Elsevier, Amsterdam, 1987, 405-448.
44. Romani, G.L., Williamson, S.J., and Kaufman, L., Biomagnetic instrumentation, *Rev. Sci. Instrum.*, 53, 1815-1845, 1982.
45. Helmholtz, C.V., Über einige Gesetze der Verteilung elektrischer Ströme in körperlichen Leitern, mit Anwendung auf die thierisch elektrischen Versuche, *Ann. Phys. Chem.*, 89, 353-377, 1853.
46. Marquardt, D.W., An algorithm for least-squares estimation of nonlinear parameters, *J. Soc. Ind. Appl. Math.*, 11, 431-441, 1963.
47. Powell, M.J.D., An efficient method for finding the minimum of a function of several variables without calculating derivatives, *The Computer J.*, 7, 155-162, 1964.
48. Scholz, B. and Oppelt, A., Probability based dipole localization and individual localization error calculation in biomagnetism, *Proc. Ann. IEEE Med. Biol. Soc.*, 14, 1766-1767, 1992.
49. Hari, R., Joutsiniemi, S.L., and Sarvas, J., Spatial resolution of neuromagnetic records: theoretical calculations in a spherical model, *Electroencephalogr. Clin. Neurophysiol.*, 71, 64-72, 1988.
50. Janday, B.S. and O'Connell, J., Locating current sources within skulls using magnetometry, *Radiography*, 53, 287-291, 1987.
51. Meijs, J.W., Peters, M.J., Boom, H.B., and Lopes da Silva, F.H., Relative influence of model assumptions and measurement procedures in the analysis of the MEG, *Med. Biol. Eng. Comput.*, 26, 136-142, 1988.
52. Barth, D.S., Sutherling, W., Broffman, J., and Beatty, J., Magnetic localization of a dipolar current source implanted in a sphere and a human cranium, *Electroencephalogr. Clin. Neurophysiol.*, 63, 260-273, 1986.
53. Stefan, H., Schneider, S., Abraham Fuchs, K., Bauer, J., Feistel, H., Pawlik, G., Neubauer, U., Rohrlein, G., and Huk, W.J., Magnetic source localization in focal epilepsy. Multichannel magnetoencephalography correlated with magnetic resonance brain imaging, *Brain*, 113, 1347-1359, 1990.
54. Hansen, J.S., Ko, H.W., Fisher, R.S., and Litt, B., Practical limits on the biomagnetic inverse process determined from *in vitro* measurements in spherical conducting volumes, *Phys. Med. Biol.*, 33, 105-111, 1988.
55. Hämäläinen, M., Hari, R., Ilmoniemi, R.J., Knutila, J., and Lounasmaa, O.V., Magnetoencephalography — theory, instrumentation, and applications to noninvasive studies of the working human brain, *Rev. Mod. Phys.*, 65, 413-498, 1993.
56. Sobel, N., Prabhakaran, V., Hartley, C.A., Desmond, J.E., Zhao, Z., Glover, G.H., Gabrieli, J.D., and Sullivan, E.V., Odorant-induced and sniff-induced activation in the cerebellum of the human, *J. Neurosci.*, 18, 8990-9001, 1998.
57. Yousem, D.M., Williams, S.C., Howard, R.O., Andrew, C., Simmons, A., Allin, M., Geckle, R.J., Suskind, D., Bullmore, E.T., Brammer, M.J., and Doty, R.L., Functional MR imaging during odor stimulation: preliminary data, *Radiology*, 204, 833-838, 1997.
58. Levy, L.M., Henkin, R.I., Hutter, A., Lin, C.S., Martins, D., and Schellinger, D., Functional MRI of human olfaction, *J. Comput. Assist. Tomogr.*, 21, 849-856, 1997.

59. Fulbright, R.K., Skudlarski, P., Lacadie, C.M., Warrenburg, S., Bowers, A.A., Gore, J.C., and Wexler, B.E., Functional MR imaging of regional brain responses to pleasant and unpleasant odors, *AJNR Am. J. Neuroradiol.*, 19, 1721-1726, 1998.
60. Sobel, N., Prabhakaran, V., Desmond, J.E., Glover, G.H., Goode, R.L., Sullivan, E.V., and Gabrieli, J.D., Sniffing and smelling: separate subsystems in the human olfactory cortex, *Nature*, 392, 282-286, 1998.
61. Bressler, S.L. and Freeman, W.J., Frequency analysis of olfactory system EEG in cat, rabbit, and rat, *Electroencephalogr. Clin. Neurophysiol.*, 50, 19-24, 1980.
62. Ueki, S. and Domino, E.F., Some evidence for a mechanical receptor in olfactory function, *J. Neurophysiol.*, 24, 12-25, 1961.
63. Sobel, N., Prabhakaran, V., Hartley, C.A., Desmond, J.E., Glover, G.H., Sullivan, E.V., and Gabrieli, J.D., Blind smell: brain activation induced by an undetected air-borne chemical, *Brain*, 122, 209-217, 1999.
64. Murphy, C., Razani, J., and Davidson, T.M., Aging and the chemical senses, in *Taste and Smell Disorders*, Seiden, A.N., Ed., Thieme, New York, 1997, 172-193.
65. Yousem, D.M., Maldjian, J.A., Hummel, T., Alsop, D.C., Geckle, R.J., Kraut, M.A., and Doty, R.L., The effect of age on odor-stimulated functional MR imaging, *Am. J. Neuroradiol.*, 20, 600-608, 1999.
66. Kettenmann, B., Francis, S.T., Bowtell, R.W., McGlone, F., Renner, B., Ahne, G., Rolls, E., and Kobal, G., Olfactory activity in the human cingulate cortex identified by FMRI, *Chem. Senses*, 24, 539, 1999.
67. Peyron, R., Laurent, B., and Garcia-Larrea, L., Functional imaging of brain responses to pain. A review and meta-analysis, *Neurophysiol. Clin.*, 30, 263-288, 2000.
68. Doty, R.L., Gender and endocrine-related influences upon olfactory sensitivity, in *Clinical Measurement of Taste and Smell*, Meiselman, H. and Rivlin, R.S., Eds., MacMillan, New York, 1986, 377-413.
69. Yousem, D.M., Maldjian, J.A., Siddiqi, F., Hummel, T., Alsop, D.C., Geckle, R.J., Bilker, W.B., and Doty, R.L., Gender effects on odor-stimulated functional magnetic resonance imaging, *Brain Res.*, 818, 480-487, 1999.
70. Levy, L.M., Henkin, R.I., Lin, C.S., Hutter, A., and Schellinger, D., Odor memory induces brain activation as measured by functional MRI, *J. Comput. Assist. Tomogr.*, 23, 487-498, 1999.
71. Kettenmann, B., Wiesmann, M., Heuberger, E., Yousry, I., Nolte, A., Ilmberger, J., Yousry, T., and Kobal, G., Comparison of brain activity induced by stimulation and imagination, *Chem. Senses*, 25, 623, 2000.
72. Wiesmann, M., Kettenmann, B., Yousry, I., Heuberger, E., Nolte, A., Ilmberger, J., Yousry, T., and Kobal, G., Functional magnetic resonance imaging of human olfaction, *Neuroimaging Clin. N. Am.*, (in press) 2001.
73. Francis, S., Rolls, E.T., Bowtell, R., McGlone, F., O'Doherty, J., Browning, A., Clare, S., and Smith, E., The representation of pleasant touch in the brain and its relationship with taste and olfactory areas, *Neuroreport*, 10, 453-459, 1999.
74. O'Doherty, J., Rolls, E.T., Francis, S., Bowtell, R., McGlone, F., Kobal, G., Renner, B., and Ahne, G., Sensory-specific satiety-related olfactory activation of the human orbitofrontal cortex, *Neuroreport*, 11, 893-897, 2000.
75. Rolls, E.T., Critchley, H.D., and Treves, A., Representation of olfactory information in the primate orbitofrontal cortex, *J. Neurophysiol.*, 75, 1982-1996, 1996.
76. Birbaumer, N., Grodd, W., Diedrich, O., Klose, U., Erb, M., Lotze, M., Schneider, F., Weiss, U., and Flor, H., fMRI reveals amygdala activation to human faces in social phobics, *Neuroreport*, 9, 1223-1226, 1998.

77. Abraham, A. and Mathai, K.V., The effect of right temporal lobe lesions on matching of smells, *Neuropsychologia*, 21, 277-281, 1985.
78. Rausch, R., Serafetinides, E.A., and Crandall, P.H., Olfactory memory in patients with anterior temporal lobectomy, *Cortex*, 13, 445-452, 1977.
79. Martinez, B.A., Cain, W.S., De Wijk, R.A., Spencer, D.D., Novelly, R.A., and Sass, K.J., Olfactory functioning before and after temporal lobe resection for intractable seizures, *Neuropsychology*, 7, 3551-3560, 1993.
80. Doty, R.L., Bromley, S.M., Moberg, P.J., and Hummel, T., Laterality in human nasal chemoreception, in *Cerebral Asymmetries in Sensory and Perceptual Processing*, Christman, S., Ed., North Holland Publishing, Amsterdam, 1997, 497-542.
81. Small, D.M., Jones-Gotman, M., Zatorre, R.J., Petrides, M., and Evans, A.C., Flavor processing: more than the sum of its parts, *Neuroreport*, 8, 3913-3917, 1997.
82. Zald, D.H. and Pardo, J.V., Emotion, olfaction, and the human amygdala: amygdala activation during aversive olfactory stimulation, *Proc. Natl. Acad. Sci. U.S.A.*, 15, 4119-4124, 1997.
83. Zald, D.H. and Pardo, J.V., Functional neuroimaging of the olfactory system in humans, *Int. J. Psychophysiol.*, 36, 165-181, 2000.
84. Dade, L.A., Jones-Gotman, M., Zatorre, R.J., and Evans, A.C., Human brain function during odor encoding and recognition. A PET activation study, *Ann. N.Y. Acad. Sci.*, 855, 572-574, 1998.
85. Tulving, E., Markowitsch, H.J., Kapur, S., Habib, R., and Houle, S., Novelty encoding networks in the human brain: positron emission tomography data, *Neuroreport*, 5, 2525-2528, 1994.
86. Royet, J.P., Koenig, O., Gregoire, M.C., Cinotti, L., Lavenne, F., Le Bars, D., Costes, N., Vigouroux, M., Farget, V., Sicard, G., Holley, A., Mauguiere, F., Comar, D., and Froment, J.C., Functional anatomy of perceptual and semantic processing for odors, *J. Cogn. Neurosci.*, 11, 94-109, 1999.
87. Qureshy, A., Kawashima, R., Imran, M.B., Sugiura, M., Goto, R., Okada, K., Inoue, K., Itoh, M., Schormann, T., Zilles, K., and Fukuda, H., Functional mapping of human brain in olfactory processing: a PET study, *J. of Neurophysiol.*, 84, 1656-1666, 2000.
88. Savic, I., Gulyas, B., Larsson, M., and Roland, P., Olfactory functions are mediated by parallel and hierarchical processing, *Neuron*, 26, 735-745, 2000.
89. Tucker, D., Olfactory, vomeronasal and trigeminal receptor responses to odorants, in *Olfaction and Taste*, Zotterman, Y., Ed., Pergamon Press, Oxford, 1963, 45-69.
90. Silver, W.L., Mason, J.R., Adams, M.A., and Smeraski, C.A., Nasal trigeminal chemoreception: responses to n-aliphatic alcohols, *Brain Res.*, 37, 221-229, 1986.
91. Kobal, G. and Hummel, T., Olfactory evoked potentials in humans, in *Smell and Taste in Health and Disease*, Getchell, T.V., Doty, R.L., Bartoshuk, L.M., and Snow, J.B.J., Eds., Raven Press, New York, 1991, 255-275.
92. Hummel, T., Livermore, A., Hummel, C., and Kobal, G., Chemosensory event-related potentials in man: relation to olfactory and painful sensations elicited by nicotine, *Electroencephalogr. Clin. Neurophysiol.*, 84, 192-195, 1992.
93. Doty, R.L., Brugger, W.P.E., Jurs, P.C., Orndorff, M.A., Snyder, P.J., and Lowry, L.D., Intranasal trigeminal stimulation from odorous volatiles: psychometric responses from anosmic and normal humans, *Physiol. Behav.*, 20, 175-185, 1978.
94. Kettenmann, B., Hummel, C., Stefan, H., and Kobal, G., Magnetoencephalographical recordings: separation of cortical responses to different chemical stimulation in man, *Funct. Neurosci. (EEG Suppl.)*, 46, 287-290, 1996.
95. Kobal, G., Pain-related electrical potentials of the human nasal mucosa elicited by chemical stimulation, *Pain*, 22, 151-163, 1985.

96. Hoenig, H.E., Daalamans, G.M., Bär, L., Bömmel, F., Paulus, A., Uhl, D., Weisse, H.J., Schneider, S.H.S., Reichenberger, H., and Abraham-Fuchs, K., Multichannel DC SQUID sensor array for biomagnetic applications, *IEEE Trans. Magn.*, 27, 2777-2785,
97. Abraham Fuchs, K., Harer, W., Schneider, S., and Stefan, H., Pattern recognition in biomagnetic signals by spatio-temporal correlation and application to the localisation of propagating neuronal activity, *Med. Biol. Eng. Comput.*, 28, 398-406, 1990.
98. Mesulam, M.M. and Mufson, E.J., Insula of the old world monkey. III. Efferent cortical output and comments on function, *J. Comp. Neurol.*, 212, 38-52, 1982.
99. Carmichael, S.T. and Price, J.L., Architectonic subdivision of the orbital and medial prefrontal cortex in the macaque monkey, *J. Comp. Neurol.*, 346, 366-402, 1994.
100. Carmichael, S.T. and Price, J.L., Limbic connections of the orbital and medial prefrontal cortex in macaque monkeys, *J. Comp. Neurol.*, 363, 615-641, 1995.
101. Gordon, H.W. and Sperry, R.W., Lateralization of olfactory perception in the surgically separated hemispheres of man, *Neuropsychologia*, 7, 111-120, 1969.
102. Youngentoub, S.L., Kurtz, D.B., Leopold, D.A., Mozell, M.M., and Hornung, D.E., Olfactory sensitivity: is there laterality?, *Chem. Senses,* 7, 1-9, 1982.
103. Hummel, T., Pauli, E., Stefan, H., Kettenmann, B., Schüler, P., and Kobal, G., Chemosensory event-related potentials in patients with temporal lobe epilepsy, *Epilepsia*, 36, 79-85, 1995.
104. Kobal, G., Hummel, T., and Van Toller, S., Differences in chemosensory evoked potentials to olfactory and somatosensory chemical stimuli presented to left and right nostrils, *Chem. Senses,* 17, 233-244, 1992.
105. Kobal, G., Hummel, T., and Pauli, E., Correlates of hedonic estimates in the olfactory evoked potential, *Chem. Senses,* 14, 718, 1989.
106. Davidson, J.D., Affect, cognition, and hemispheric specialization, in *Emotions, Cognition, and Behavior*, Izard, C.E., Kagan, J., and Zajonc, R.B., Eds., Cambridge University Press, Cambridge, 1984, 320-365.
107. Dimond, S.J., Farrington, L., and Johnson, P., Differing emotional response from right and left hemispheres, *Nature*, 261, 690-692, 1976.
108. Hummel, T. and Kobal, G., Differences in human evoked potentials related to olfactory or trigeminal chemosensory activation, *Electroencephalogr. Clin. Neurophysiol.*, 84, 84-89, 1992.
109. Livermore, A., Hummel, T., and Kobal, G., Chemosensory event-related potentials in the investigation of interactions between the olfactory and the somatosensory (trigeminal) systems, *Electroencephalogr. Clin. Neurophysiol.*, 83, 201-210, 1992.
110. Murphy, C., Loss of olfactory function in dementing disease, *Physiol. Behav.*, 66, 177-182, 1999.
111. Ansari, K.A. and Johnson, A., Olfactory function in patients with Parkinson's disease, *J. Chron. Dis.*, 28, 493-497, 1975.
112. Welge-Luessen, A., Temmel, A., Quint, C., Moll, B., Wolf, S., and Hummel, T., Olfactory function in patients with olfactory groove meningioma, *J. Neurol. Neurosurg. Psych.*, 70, 218-221, 2001.
113. Hummel, T., Roscher, S., Jaumann, M.P., and Kobal, G., Intranasal chemoreception in patients with multiple chemical sensitivities: a double-blind investigation, *Reg. Tox. Pharmacol.*, 24, S79-S86, 1996.
114. Tonoike, M., Yamaguchi, M., Kaetsu, I., Kida, H., Seo, R., and Koizuka, I., Ipsilateral dominance of human olfactory activated centers estimated from event-related magnetic fields measured by 122-channel whole-head neuromagnetometer using odorant stimuli synchronized with respirations, *Ann. N.Y. Acad. Sci.*, 855, 579-590, 1998.

115. Frahm, J., Merboldt, K.D., and Hanicke, W., Direct FLASH MR imaging of magnetic field inhomogeneities by gradient compensation, *Magn. Reson. Med.*, 6, 474-480, 1988.
116. Farzaneh, F., Riederer, S.J., and Pelc, N.J., Analysis of T2 limitations and off-resonance effects on spatial resolution and artifacts in echo-planar imaging, *Magn. Reson. Med.*, 14, 123-139, 1990.
117. Ojemann, J.G., Akbudak, E., Snyder, A.Z., McKinstry, R.C., Raichle, M.E., and Conturo, T.E., Anatomic localization and quantitative analysis of gradient refocused echo-planar fMRI susceptibility artifacts, *Neuroimage*, 6, 156-167, 1997.
118. Glover, G.H. and Lee, A.T., Motion artifacts in fMRI: comparison of 2DFT with PR and spiral scan methods, *Magn. Reson. Med.*, 33, 624-635, 1995.
119. Morrell, G. and Spielman, D., Dynamic shimming for multi-slice magnetic resonance imaging, *Magn. Reson. Med.*, 38, 477-483, 1997.
120. Young, I.R., Cox, I.J., Bryant, D.J., and Bydder, G.M., The benefits of increasing spatial resolution as a means of reducing artifacts due to field inhomogeneities, *Magn. Reson. Imaging*, 6, 585-590, 1988.
121. Noll, D.C., Cohen, J.D., Meyer, C.H., and Schneider, W., Spiral K-space MR imaging of cortical activation, *J. Magn. Reson. Imaging*, 5, 49-56, 1995.
122. Berger, H., Über das Elektroenkephalogramm des Menschen, *Arch. Psychiat. Nervenkr.*, 40, 160-179, 1929.

Index

A

D

E

F

G

J

N

O

P

Q

R

S

T

U

V

W

X

Y

Z